Design Technologies for Wheeled and Tracked Vehicles (MMBC 2019)

IOP Conference Series: Materials Science and Engineering
Volume 820

Moscow, Russia
1 - 2 October 2019

ISBN: 978-1-7138-3161-7
ISSN: 1757-8981

Printed from e-media with permission by:

Curran Associates, Inc.
57 Morehouse Lane
Red Hook, NY 12571

CURRAN ASSOCIATES INC. proceedings.com

Some format issues inherent in the e-media version may also appear in this print version.

Institute of Physics
Dirac House, Temple Back
Bristol BS1 6BE UK

Phone: 44 1 17 929 7481
Fax: 44 1 17 920 0979

techtracking@iop.org

Additional copies of this publication are available from:

Curran Associates, Inc.
57 Morehouse Lane
Red Hook, NY 12571 USA
Phone: 845-758-0400
Fax: 845-758-2633
Email: curran@proceedings.com
Web: www.proceedings.com

TABLE OF CONTENTS

IOP Conf. Series: Materials Science and Engineering **820** (2020) 011001 doi:10.1088/1757-899X/820/1/011001

Design Technologies for Wheeled and Tracked Vehicles

The conference "Design Technologies for Wheeled and Tracked Vehicles" is held annually at Bauman Moscow State Technical University. The conference is organized by the Department of "Wheeled vehicles" and the Department of "multi-purpose tracked vehicles and mobile robots". Every year, representatives of various scientific, educational and industrial organizations in Russia and other countries come to us with reports and speeches. The main topics of the reports are issues related to the design of wheeled and tracked vehicles. We are glad to see scientists from all countries and organizations in the field of education, science and industry at our conference.

Sincerely, A. Protopopov

Peer review statement

All papers published in this volume of *IOP Conference Series: Materials Science and Engineering* have been peer reviewed through processes administered by the proceedings Editors. Reviews were conducted by expert referees to the professional and scientific standards expected of a proceedings journal published by IOP Publishing.

Analysis of the diagrams and main parameters of the existing structures of the articulated tracked vehicles

V Belyakov[1], V Makarov[1], V Klubnichkin[2] and E Klubnichkin[2,3]

[1]Nizhny Novgorod State Technical University n.a. R.E. Alekseev, Minin str., 24, 603950, Nizhny Novgorod, Russia
[2]Bauman Moscow State Technical University

[3]E-mail: klubnichkin@mgul.ac.ru

Abstract. The articles reviews structures of the existing all-terrain articulated tracked vehicles (ATV). It includes the existing classification of the ATVs by the type of pivot coupling mechanism (PCM). It analyses the vehicles produced by the principal manufacturers of ATV available on the Russian market. It offers statistical dependencies for the main relationships of the basic power, dimensional characteristics as well as ground pressure depending on the ATV gross weight. It presents a block diagram of the method of rational selection of the vehicle type, maneuverability control system, provision of the total required power, capacity, maximum travel speed.

Introduction

The functional tasks performed by the transport-technological and special vehicles in certain branches of real economy as well as operational-tactical missions of the special and law-enforcement agencies in most cases are linked with operation on temporary roads or off-road.

The emerging energy-packed multi-operational vehicles allow for mechanization of most works in the forestry, agricultural, oil-gas and mining industries as well as other branches of the national economy and security. Employment of articulated transport-technological systems is one of the promising lines in solving many tasks evolving during operation of the transport and technological complexes under off-road conditions and especially during winter and muddy seasons as well as other specific conditions of their functioning.

An articulated tracked vehicle (ATV) is a special case of an articulated transport system (ATS) [1]. The latter concept is very wide and includes a range of transport and technological vehicles as well as transport-technological complexes featuring at least two hinged sections with one or more degree of freedom.

Today there are no many technical and specialized publications which describe ATV. First of all they are the publications produced by the manufacturers of this type of vehicles [2,3]. There is a number of thesis researches dedicated to ATV [4–6]. Some aspects concerning selection of parameters of the running gear of those vehicles are presented in the works [7, 8].

ATV classification

The number and location of the coupling hinge points [9] (Fig. 1) are taken as a basis for the ATV classification.

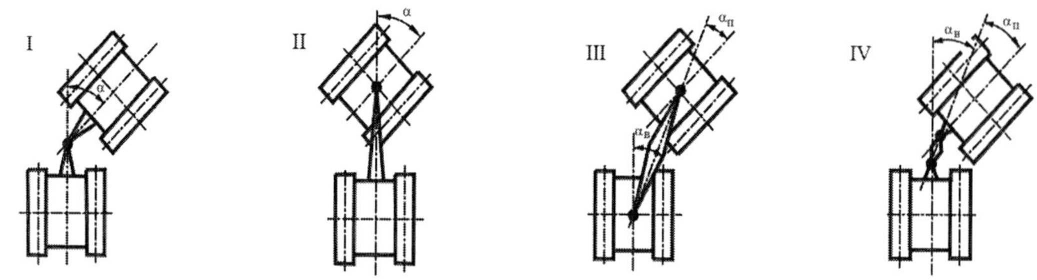

Figure 1. ATV classification [9]: I — one-hinged symmetrical (towed) ATV, II — one-hinged saddle-type ATV, III — two-hinged (wagon-type) ATV, IV — two-hinged ATV with intermediate member [9]

Today there are few diagrams of ATV presented in Fig. 2 [2]. Each of those options has its advantages and disadvantages. This article reviews approaches to selection of rational parameters of ATV at an early designing stage.

a

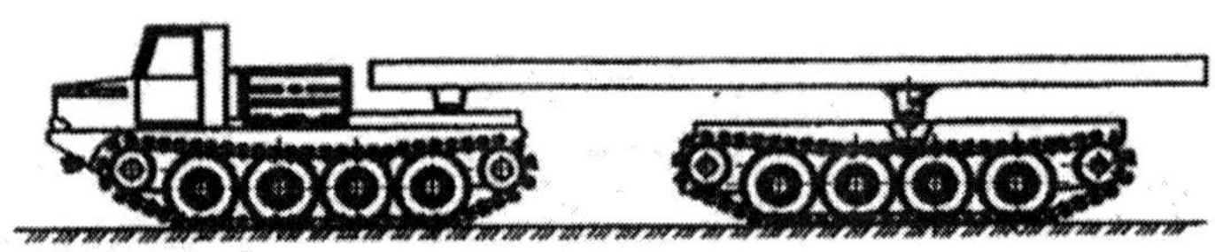

b

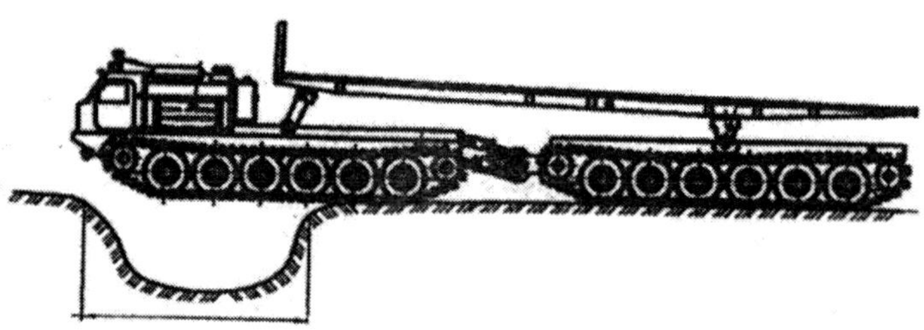

c

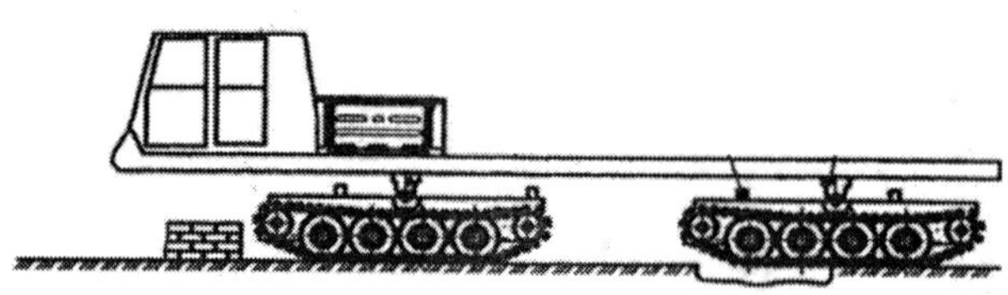

d

Figure 2. Diagrams of known ATVs: a — one-hinged towed ATV, b — one-hinged saddle-type ATV, c — towed (platform) ATV, d — bolster-type two-hinged (wagon-type) ATV [2]

Reviews approaches to selection of rational parameters of ATV at an early designing stage
Curvilinear motion of this class vehicles is performed with use of: a pivot-coupling mechanism (PCM) or a pivot-saddle mechanism (PSM). The first type of a pivot mechanism shall provide for jackknifing of elements in three planes: horizontal, longitudinal vertical (modes: shock-absorbers; forced jackknifing; locked state), lateral vertical. The PCM design is much dependent on the designer's preferences, dimensions, ATV requirements to profile passability, design limitations.

When designing articulated transport means, one should be guided by the practice of designing of the existing vehicles. Today the main manufacturers of this type of vehicles are as follows (Fig. 3): JSC MK Vityaz, Zavolzhksky Plant of Tracked Tractors, CJSC Transport, LLC All-Terrain Vehicle Plant, LLC TransMash, LLC EZSM Continent, Foremost Industries LP (Canada), AB Haegglunds (Sweden).

Figure 3. Main manufacturers of ATV

Analysis of parameter processing of the main characteristics of ATV
From the analysis of parameter processing of the main characteristics of ATV we have obtained dependencies and made conclusions about relationship between the capacity and the engine power and travel speed.

Fig. 4 shows that the points are grouped characteristically. 3 groups can be distinguished: 1- one-hinged ATV, 2 — two-hinged ATV, 3 — special purpose ATV (perform wide range of technological and special operations). Basing on those dependencies we can select vehicle's parameters [10].

In the course of the research we have also obtained statistical dependencies to find speed parameters of the ATV presented in Fig. 5.

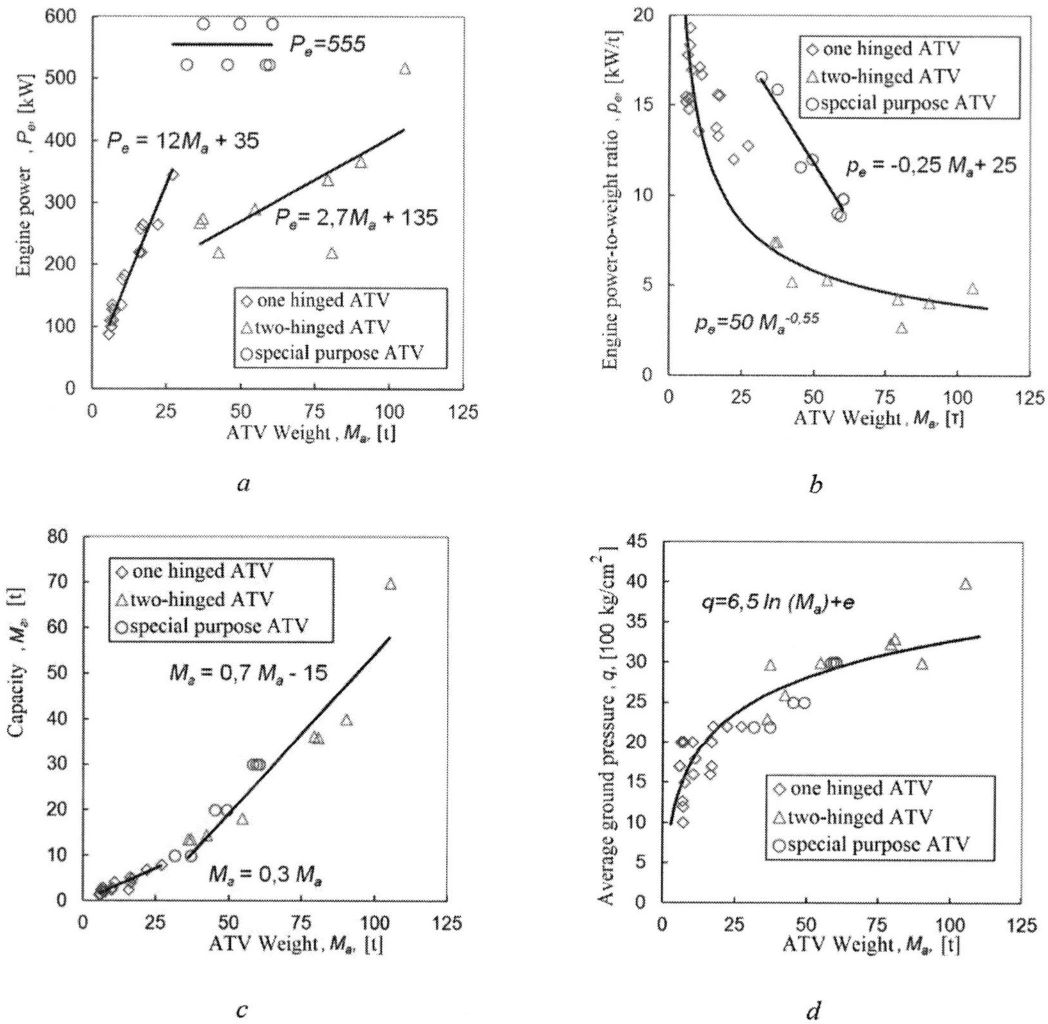

Figure 4. ATV parameters dependence on gross weight: a — engine power, b — engine power-to-weight ratio, c — capacity, d — average ground pressure

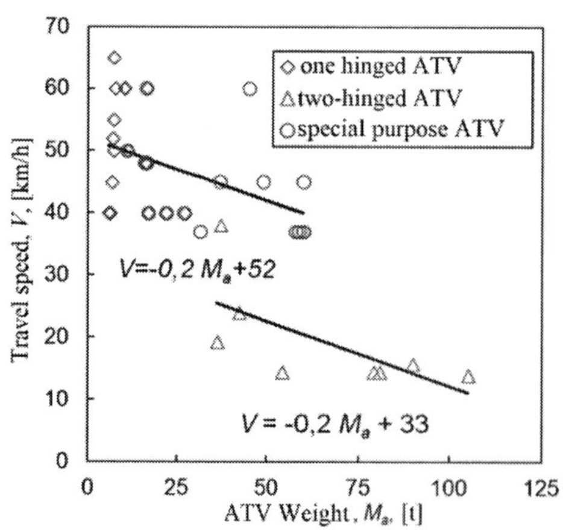

Figure 5. Travel speed dependence on ATV weight

Proposed block diagram of the method of rational selection ATV pivot-coupling mechanism
Figure 6 shows a block diagram of the method of rational selection of parameters of the ATV pivot-coupling mechanism. Lengths of parts of PCM, location of a hinged point of turn relative to PCM center, location of a vertical jackknifing hinge, etc. (in accordance with Fig. 6) are found as a result of task solving consisting in finding rational parameters which provide for required vehicle mobility factors [11, 12], including: ATV passability, maneuverability, smooth ride and efficiency. With that it is proposed to use special mathematical models given in the works [1-10, 13-21] which describe variable operation conditions of the transport-technological vehicles, including ATV. The obtained parameters serve as initial data for finding parameters of the hydro cylinders of turn (depending on required drawbar pull at turns) and hydro cylinder of vertical jackknifing and estimating hydraulic equipment, PCM frame layout.

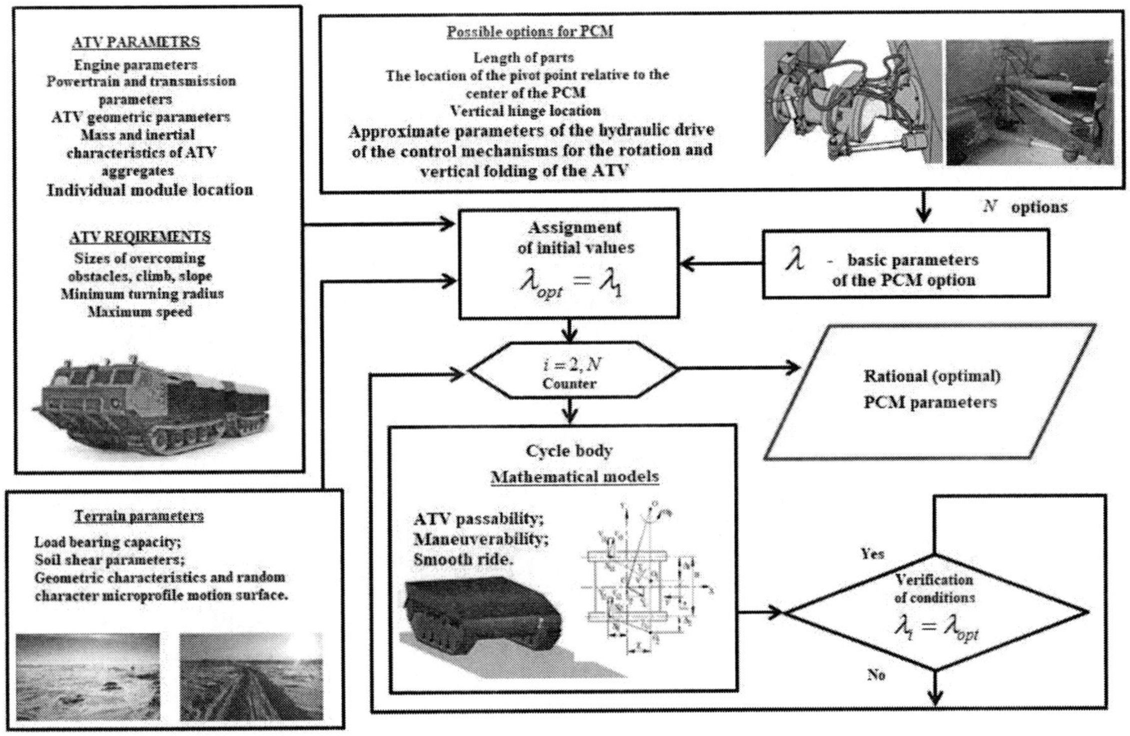

Fig. 6. Block diagram for selecting ATV parameters

Conclusion
Thus, the general ATS design solutions determine selection of a type of ATV, PCM and PSM systems, diagrams of power distribution by modules, provision of required power, capacity, maximum travel speed that dictates mobility, output and efficiency of ATS.

References
[1] Bazhenov E.E. Theory of a Vehicle and a Tractor/E.E. Bazhenov. — Yekaterinburg: UGTU-UPI, 2000. — 125 p.
[2] Oskolkov K.V. First Domestic Articulated Tracked Vehicles — Two-Link Carriers Vityaz. Ufa: Slovo Publishing House, 2005. — 280 p.
[3] Veselov N.B. All-road transport-technological vehicles. Designs. Designing and Calculation.
[4] Shmakov A.Yu. Forecasting Characteristics of Curvilinear Motion of Articulated Vehicles.Candidate's Thesis (Technical Sciences) — M.: MGTU, 2000.

[5] Sarach E.B. Developing Scientific Methods of Creation of a Complex Cushion System of Highly Mobile Two-Link Tracked Vehicles: Abstract of a Doctoral Thesis (Technical Sciences)
[6] Bazhenov E.E. Developing Scientific Methods of Forecasting of Operational Properties of Articulated Ground Transport-Technological Vehicles. Doctoral Thesis (Technical Sciences) — N.Novrogod, NGTU named after R.E. Akekseev. 2010.
[7] Kulyashov A.P. Ecological Performance of Undercarriage of Transport-Technological Vehicles. / A.P. Kulyashov, E.V. Kolotilin. — M.: Mashinostroenie, 1993. — 203 p.
[8] All-Terrain Transport-Technological Vehicles. Foundations of Motion Theory / V.V. Belyakov [and others]; Under general editorship of V.V. Belyakov and A.P. Kulyashov. N. Novgorod, 2004.
[9] Farobin Ya.E. Theory of Turn of Transport Vehicles/Ya.E. Farobin. — M.: Mashinostroenie, 1970. — 276 p.
[10] Makarov V.S. Developing Scientific-Based Technical Solutions of Creating Coastal Area Monitoring Mobile Systems. Doctoral Thesis (Technical Sciences): 05.05.03. MGTU named after N.E. Bauman, 2017. 321 p.
[11] Belyakov V.V. Concept of Mobility of Ground Transport-Technological Vehicles / Belyakov V.V., Belyaev A.M., Bushueva M.E., Vakhidov U.Sh., Goncharov K.O., Zezyulin D.B, Kolotilin V.E., Leliovskiy K.Ya., Makarov V.S., Papunin A.V., Tumasov A.V., Fedorenko A.V. Works of NGTU named after R.E. Akekseev. 2013. No. 3 (100). P. 145–174
[12] Balyakov V.V., Zezyulin D.V., Kolotilin V.E., Makarov V.S. Mobility of Ground Transport-Technological Vehicles. Works of NGTU named after R.E. Akekseev. 2013. No. 4. P. 72–77
[13] Bekker M.G. Introduction to Terrain-Vehicle Systems. — M.: Mashinostroenie, 1973.
[14] Zabavnikov N.A. Foundations of the Theory of Transport Tracked Vehicles. — M.: Mashinostroenie, 1967.
[15] Belyakov V.V. Interaction of Elastic Undercarriage of Special-Purpose Transport Vehicles with Snow Cover. Doctoral Thesis (Technical Sciences): 05.05.03. NGTU, 1999. 485 p.
[16] Alyabiev A F, Kotov A A, Klubnichkin V E, Klubnichkin E E, Bykovskiy M A 2019 Experimental determination of soil shear force by track model depending on track surface to soil friction coefficient IOP Conf. Series: Journal of Physics: Conf. Series 1177 012033
[17] Klubnichkin E E, Klubnichkin V E, Kotiev G O 2018 Theoretical research of soil packing by timber harvester running gear IOP Conference Series: Materials Science and Engineering 386 012025
[18] Bekker M G 1969 Introduction to Terrain-Vehicle Systems (Ann Arbor: The University of Michigan Press)
[19] Kotiev G, Sarach E, Beketov S 2018 Methods for road microprofile statistical data transformation MATEC Web of Conferences 224 02096
[20] Wong J Y 2009 Terramechanics and Off-Road Vehicle Engineering 2 488
[21] Klubnichkin V E, Klubnichkin E E, Kotiev G O, Beketov S A, Makarov V S 2018 Interaction between elements of the track ground contacting area with the soil at curvilinear motion of the timber harvesting machine IOP Conference Series: Materials Science and Engineering 386 012016

Prediction of fatigue life of suspension parts of the semi-trailer in the early stages of design

D Vdovin[1,2], Y Levenkov[1] and I Chichekin[1]

[1]Bauman Moscow State Technical University

[2]E-mail: vdovin@bmstu.ru

Abstract. Cargo delivery by trucks is a main method of transportation of goods to the cities and other places. Many goods are transported by articulated lorries. Thus, improving the design methods of thearticulated lorries is relevant. Trailer suspension is the object under study.Afatigue life prediction of the trailer suspension components is a urgent task at the early stage of design concept development. The formation principles of the MBD model is essential in obtainingbest results.In thispapertwo types of models are considered: containing only the rigid bodies; containing the flexible bodies.Researchobjective:the investigation of the effect of the load distribution that was obtained using different MBD models on the durability of suspension parts. A comparison of the results obtainedusing the different suspension modelswas performed. The research has shown that loads are reducedusing the flexible bodies in the MBD model. Using flexible bodies permits to specify the results of the durability analysis.

1. Introduction

Cargo delivery by trucks is a main method of transportation of goods to the cities and other places. Many goods are transported by articulated lorries. Thus, improving the design methods of articulated lorries is relevant[1-5].

It's knownthatmassreduction of the unspring parts significantly improves the driving performancesuch as theride comfort, average speed, fuel efficiency [6-15].

The durability of the suspension parts refers to the fatigue initiation life definedas the number of cyclesto a specific crack length of the component under cyclic loads. OftentheS-Ncurveofmetalisdescribedby a power function (Basquin's equation) which means a highly nonlinear dependence of the number of cycles on stress. The refinement of dynamic models by usingthe flexible parts (suspension elements and a bearing system) can significantly affect the load distribution in the system, which may affect the results of the fatigue life [16-23].

Researchobjective: estimation of the influence of the flexible bodies on loadswhich wereobtained using the dynamic model of the semi-trailers on the durability of the suspension parts.

In this papera dynamic model of the articulated lorries is developedwhich takes into account theflexible suspension parts, nonlinear stiffness of the suspension mounts, the flexible axils,frame andtank.The loadcalculationis performed using the MBD program. The MBD model was performed in AltairHyperWorks software(Motion solve). The movement on the differentrigid uneven roadswas simulated.

Finite-element (FE) modelsand the stress-strain analysis are performed in AltairHyperWorks software. FE modelswerecreatedusing Hyper mesh, Forthe definition of stress Opti struct wasused. Fatigue life analysis was performed in CodeDesignLife software.

2. Mathematical model of the articulated lorry

Dynamic model of the articulated lorry is shown in . This model was performed in the MBD program. Components of the mathematicalmodelof the vehicle are shown in Figure 2. Mathematical model of the articulated lorry includes: models of the truck, tank, suspension and wheels.

In the model, the following assumptions are made:

– parts ofthe truck arerigid;
– articulatedlorryismoving at a constant speed (60kph);
– forgrossvehicleweightcalculations are performed;
– joints haven't internal friction;
– the liquid occupies the entire volume of the tank;
– centerofgravityofthetraileris constant;
–trailer suspension is equalizing;
– suspension characteristic isnonlinear;
– deflection of the tire is taken into account. Loads are applied to the wheel axle.

Roadirregularitiesare defined by its spectral density Dq = 5.33 cm^2.RMSheight of the irregularities areq C = 2.3 cm, coefficientsα = 0.15 1/mandβ = 0. Theroad profileis taking into account the tire enveloping properties [24].

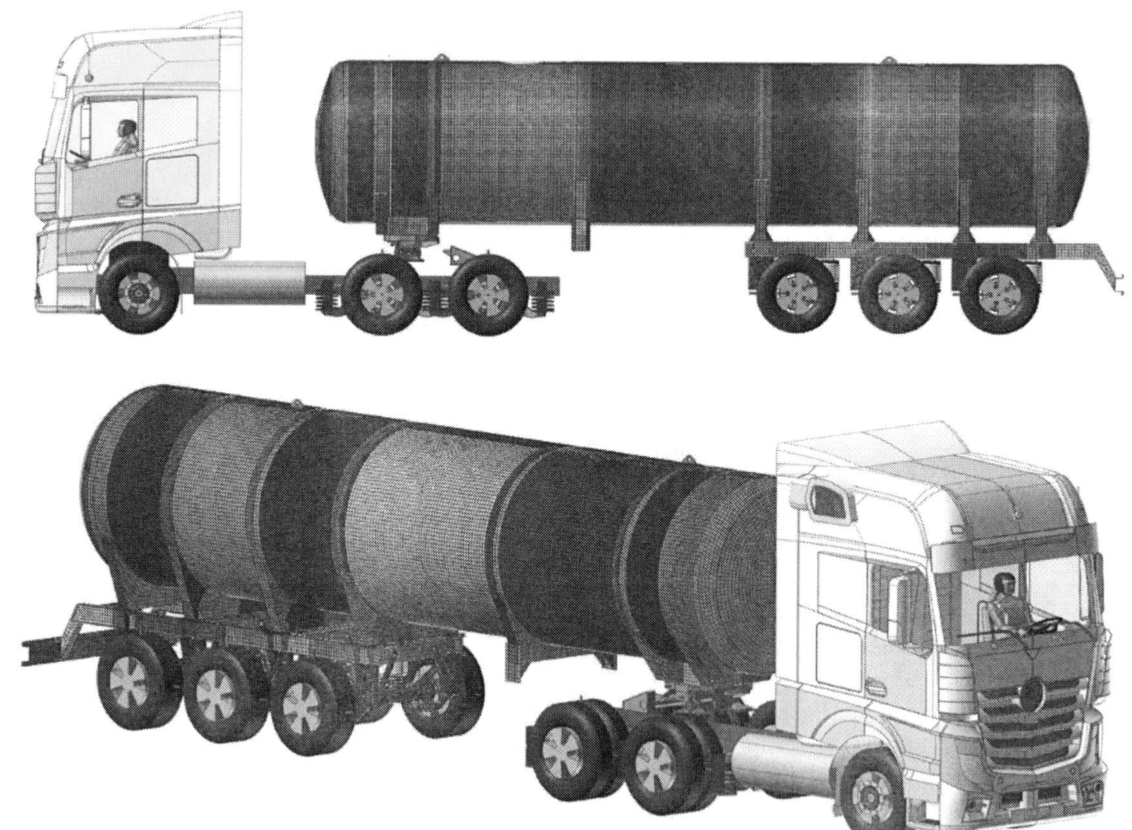

Figure 1. MDB model of the articulated lorry with flexible bodies

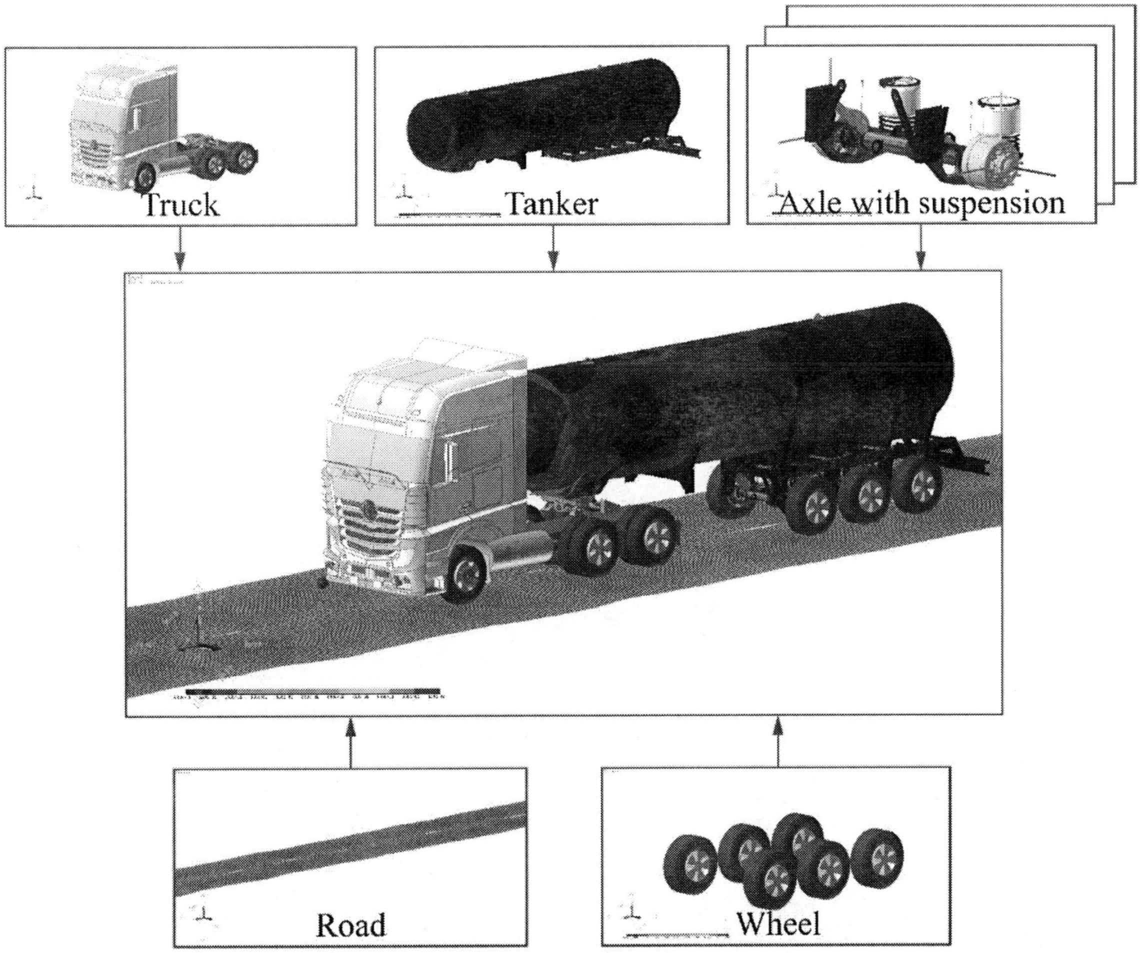

Figure 2. General elements of the mathematical model of the articulated lorry

Vehicle specifications:

Gross vehicle weight	54400 kg

Gross vehicle weight distribution:

Front suspension of the truck	6400 kg
Rear suspension of the truck	21000 kg
Suspension of the semi-trailer	27000 kg
Cargo weight	38000 kg
Tire of the truck	KAMA 315/80 R22.5 NR-201
Tire of the semi-trailer	KAMA 385/65 R22.5 NT-201 160K

Semi-trailersuspension:

Type	Dependent suspension
Springtype	Air spring is 2927V
Damper	Shock absorber isWabco 4386

Theequalizingbeamsuspensionwas created. Thesuspensionmodelwas implemented as external DLL-file. The file implements an algorithm for calculating the forcethat occurs on each air spring. Themagnitudeofaforcedepends on the volume variation of all air springs of the semi-trailer. The geometric dimensions of the 2927V air spring are used.Additional volumes (receivers) and control mechanisms are not taken into account.The model transfers equal forces to each air spring. Similarly, the rear suspension of the truck was modeled. The front suspension of a truck is spring.

The mathematic model of each suspension of the semi-trailer includes the shock absorbers 1, air springs 2, axil 3 and arms 4.The armsarefixedlyconnectedtotheaxis. Also, the each suspension arms are connected to the frame bracket 6 via a mount 7. Tension springs with variable stiffness were used tolimitthewheelrebound. If critical displacement is not achieved then the stiffness of the spring is about zero. If critical displacement is achieved then the stiffness is high.Thehubbearingsaremodeledviatherevolutejoint.

To identify the effect of the compliance ofthe suspension parts on load distribution in the joints theflexiblebodiesare used. Flexible bodiesare defined by Craig-Bampton method, based on specially developed finiteelement models of the suspension parts.

Figure 3. The equalizing beam suspension of the semi-trailer

To investigate the motion simulation along an uneven road with a distance of about 10 km was performed.Comparisons of known mathematical models of the interaction of wheels with an uneven rigid surface were made [25-28]. To simulate the interaction of the wheels with the uneven road MF-Tire model was chosen, which is described in detail in the paper [29].

IOP Conf. Series: Materials Science and Engineering **820** (2020) 012002 doi:10.1088/1757-899X/820/1/012002

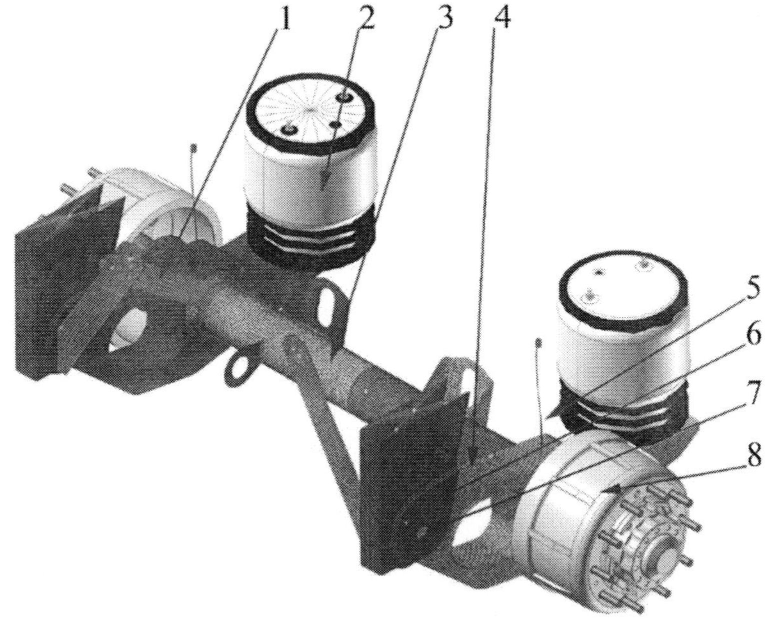

Figure 4. Components of the mathematical model of the suspension
1 – shock absorber; 2 – air spring; 3 – axil; 4 – arm; 5 – cable (limiter);
6 – bracket; 7 – mount; 8 – brake;

3. Estimation of the results

As a result of modeling the movement of the articulated lorry on an uneven rigid road, joint loadswere obtained.As an example, vertical force spectrum in the hub bearing of the left front wheel for the flexible and rigid bodies is shown in . Results processing was performed. Diagrams of magnitude of loads from their frequency of appearancewere obtained. For clarity purposes, results are presented in graph form (Figure 6).

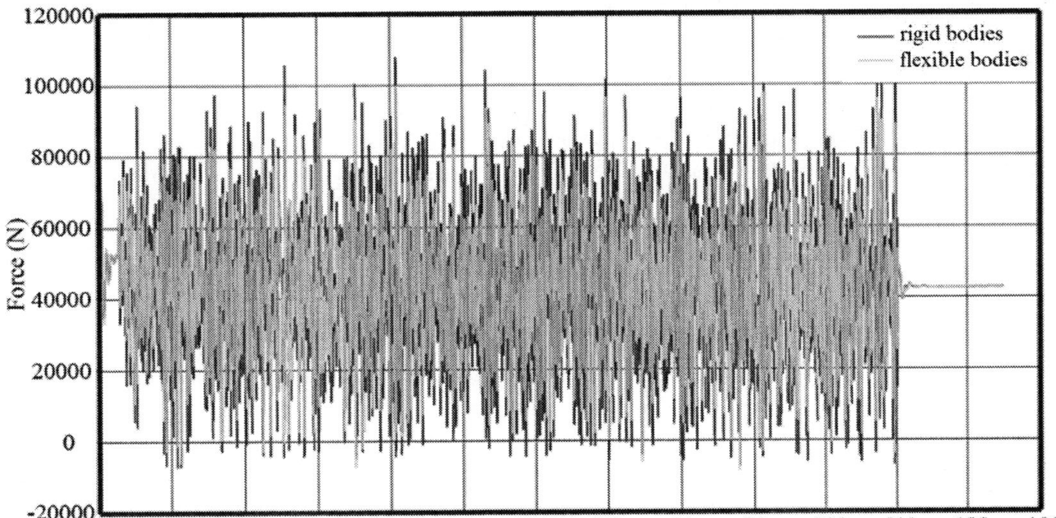

Figure5. Vertical force in the bearing of the front left wheel hub of the semi-trailer trolley for models with the rigid and flexiblebodies

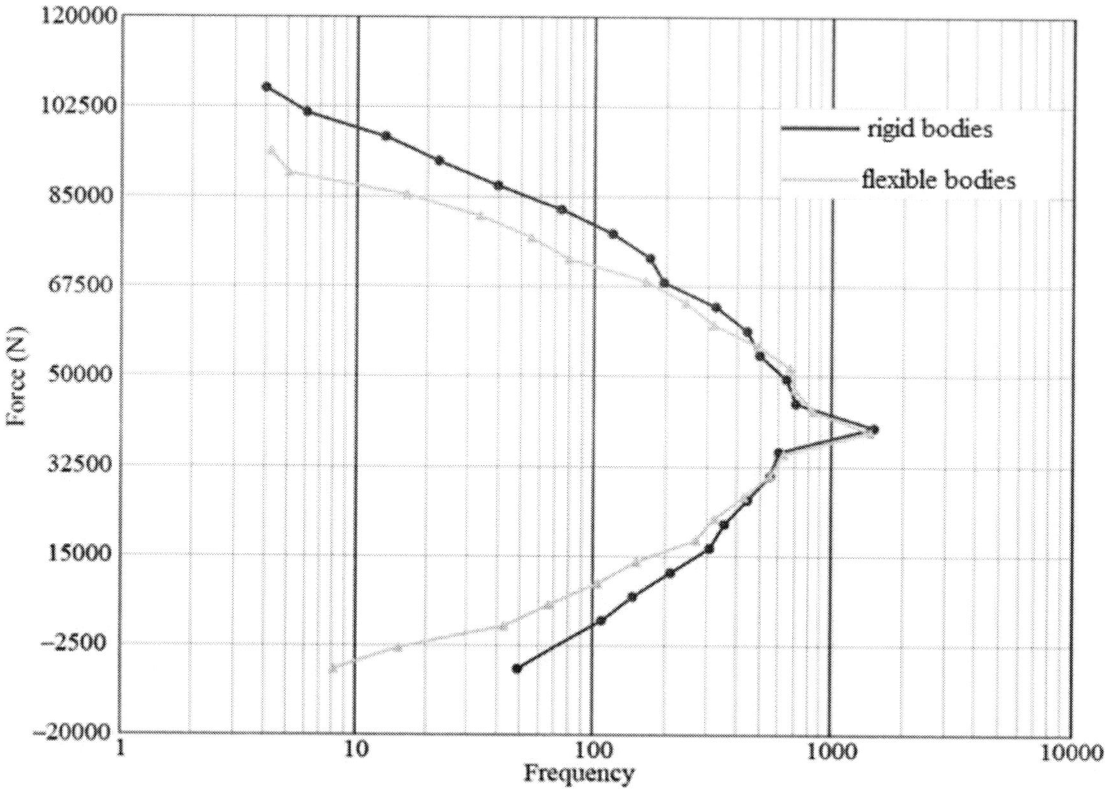

Figure6. Comparison of vertical force in the bearing of the front left wheel hub

Based on the results of the research, the following conclusions were made:

1. The RMS vertical reaction force of the hub bearingdecreases by 5,5% when using flexiblebodies. Similar for other joints.
2. Maximum loads decrease by 10,9%.
3. Frequency of appearance of large loads is less.

Asanexample, for a load whose magnitude is 90kN, frequency of appearance is 10 times less using flexible bodies.

4. FE model of suspension

FE model is shown in . Suspension arms which are made of metal sheets were modeled via shell elements (QUAD4). Arms levees weremodeledviasolidelements (HEX8).

Loads application nodes is shown in : 1* - load node of the air spring; 2* - load node of the brake; 3* - load node of the frame bracket;4* - load node of the shockabsorber;5* - load node of the hub bearing; 6* - load node of the midpoint axil.

The loads acting on the bearing from the hub are transferred to the outer ring through RBE2. Contact interaction of the bearing and the hub was taken into account. The pretension load of the bearing was modeled via negative clearance between nut and the end face of the bearing. Thepretensionloadis 90kN. FE model of the bearing is shown in Figure **9**.

Outer and inner bearing ringswere modeled via solid elements. Contact interaction of rings with rollers wasmodeled by RBE2 elements, its dependent nodes located along the contact line. RBE2 elements of the outer and inner bearing rings is connectedthrough independent nodes by rod elements ROD.

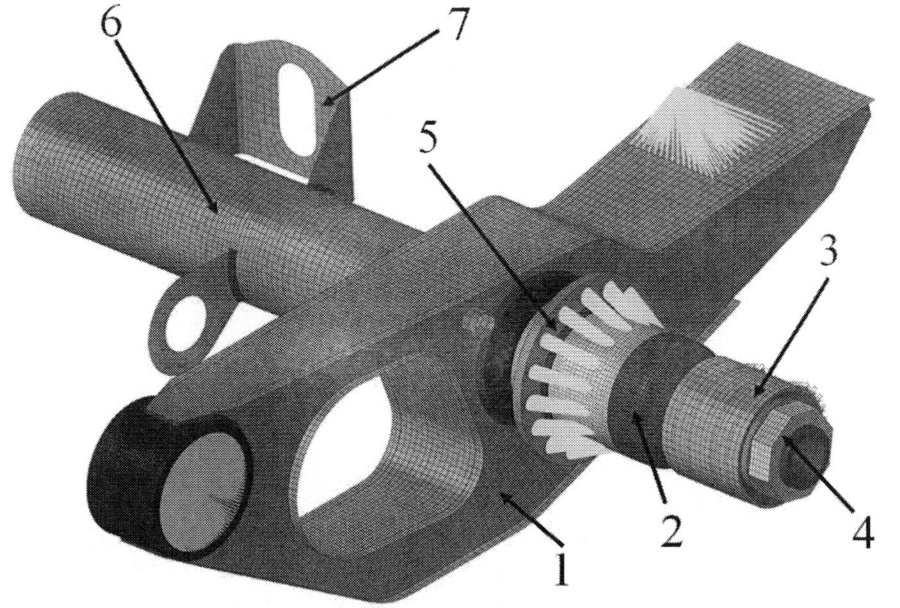

Figure 7. FE model of the suspense on: 1 – arm; 2 – pin; 3 – bearing; 4 – nut; 5 – brake flange; 6 – axil; 7 – bracket

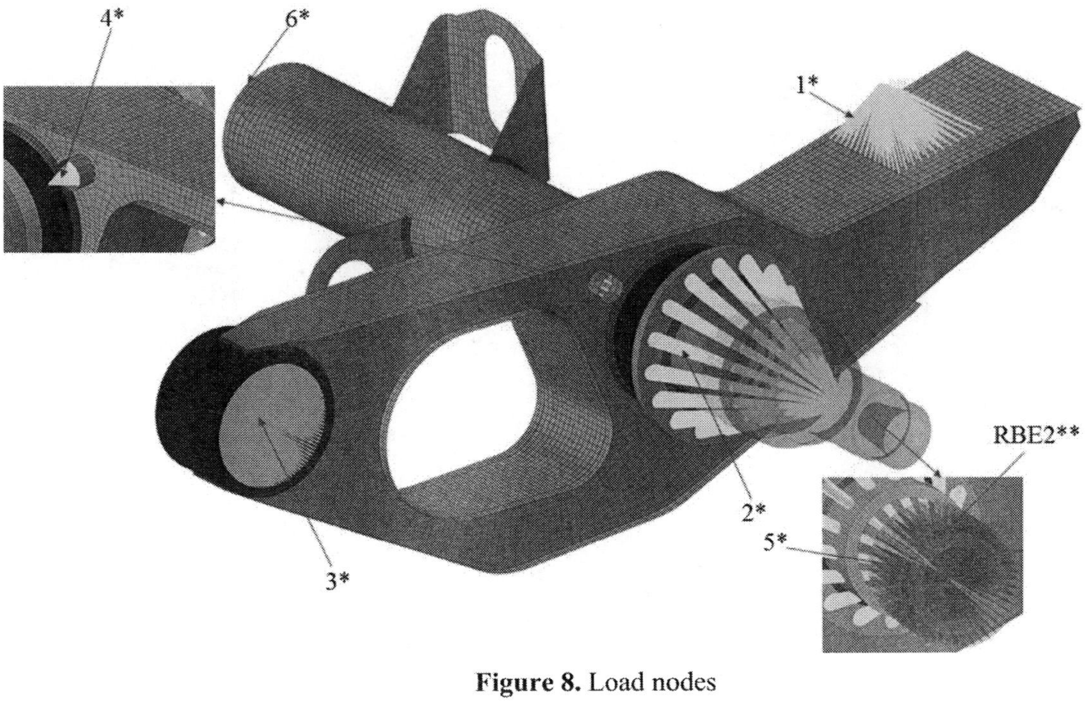

Figure 8. Load nodes

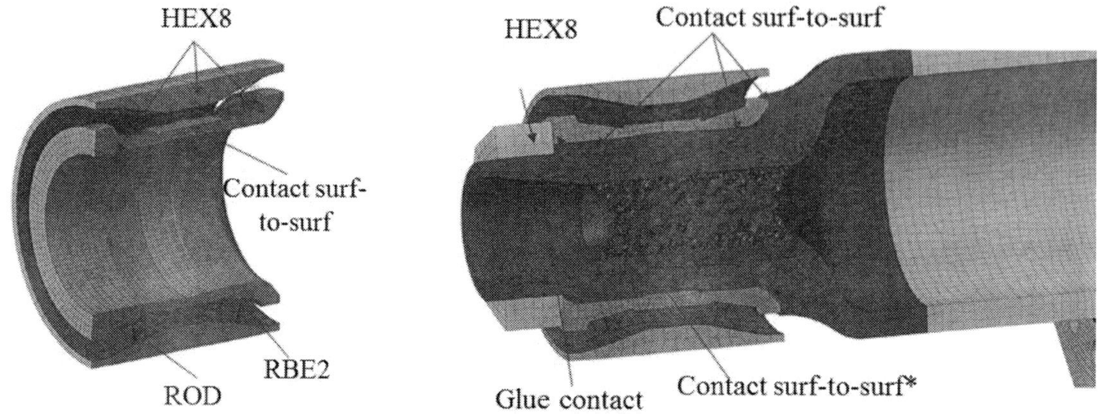

Figure 9. FE model of bearing

5. Fatigue life analysis

Consider the effect of the loads that were obtained usingrigid bodies or flexible bodiesfor the MBD model. Fatigue life analysisof suspension parts was performed.In this paper, estimationofdurabilityoftheseam-welded jointsisnotconsidered.For fatigue life analysis stress-based approach was used.This is sufficiently for a quantitative and qualitative comparison of the results.Mean stress effect was taken into account according to the Goodman's model.

Calculation results are presented in Figures Figure (rigid bodies) andFiguresFigure (flexible bodies).The calculation results showthat the loads that are obtained using different models (rigid or flexible bodies) significantly affect the calculation results of the fatigue life prediction. In this example, the number of cycles the minimum number of cycles differs almost 3 times (10239 cycles for rigid bodies and 32698 cycles for flexible bodies).

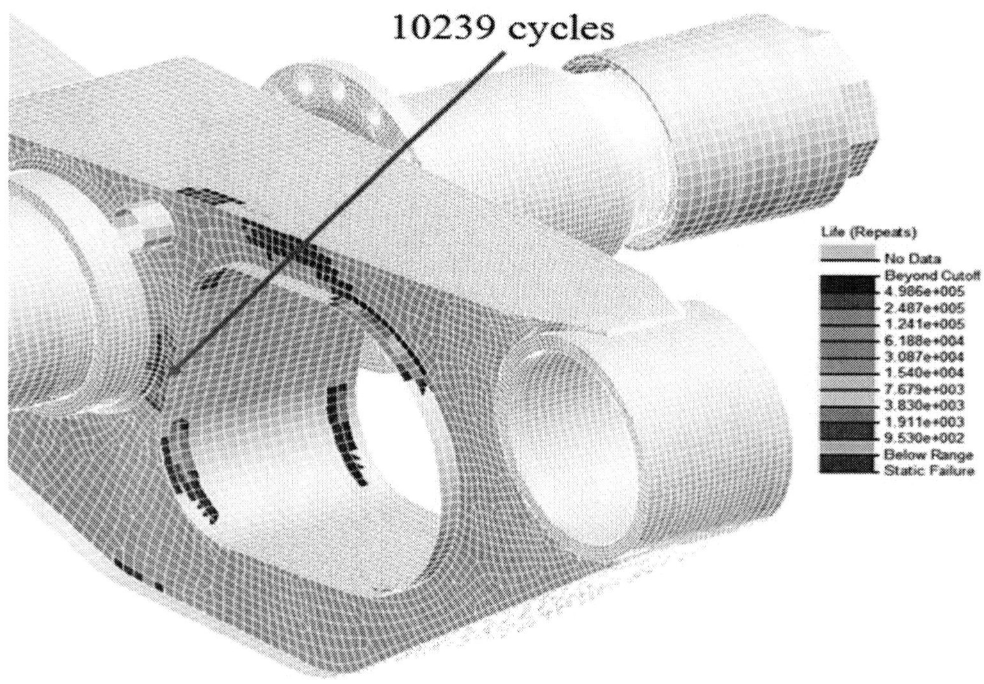

Figure 10. – Calculation results for rigid bodies loads

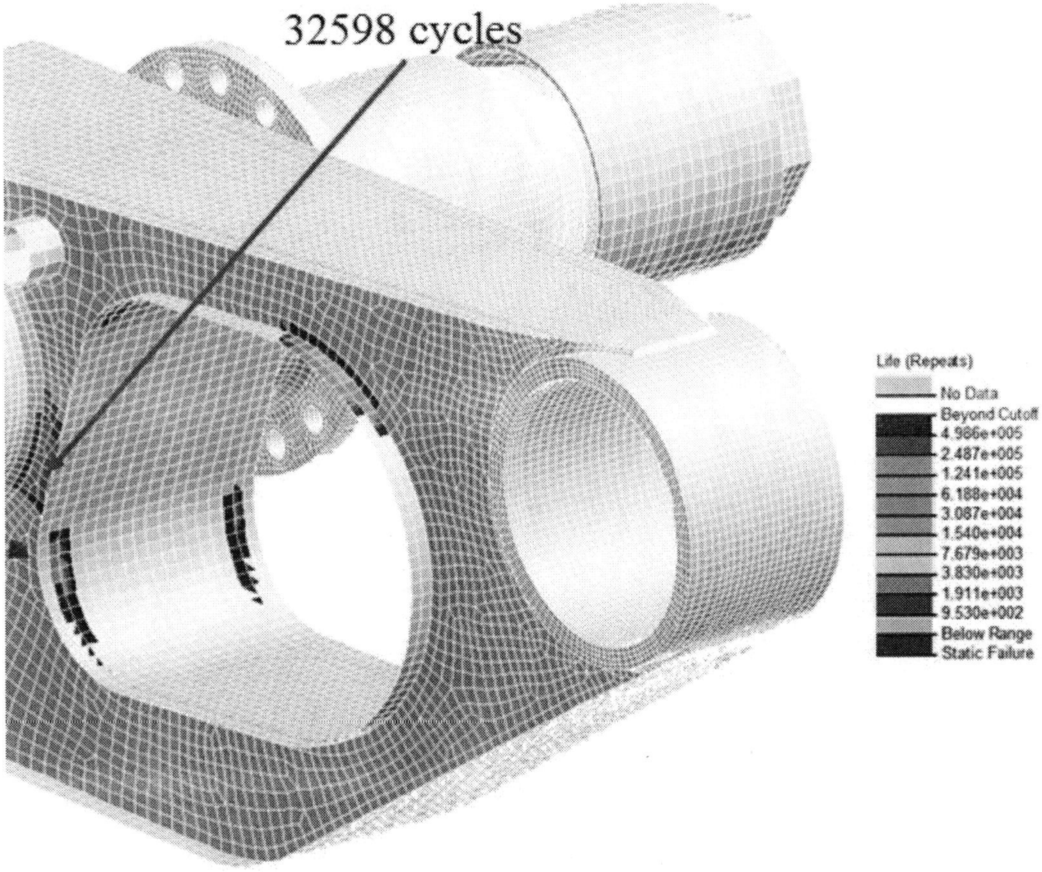

Figure 11. – Calculation results for flexible bodies loads

6. Conclusions

1. The RMS vertical reaction force of the hub bearingdecreases by 5,5% when using flexiblebodies. Similar for other joints.
2. Maximum loads decrease by 10,9%.
3. Frequency of appearance of large loads is less.
4. The number of cycles the minimum number of cycles differs almost 3 times: 10239 cycles for rigid bodies and 32698 cycles for flexible bodies.

References

[1] Hashemi E, Pirani M, Khajepour A, Kasaiezadeh A2016. *Vehicle System Dynamics***54** 1736-61
[2] Wei Y, Liu Y, Li X, Oertel C2016.*Vehicle System Dynamics***54** 463-73
[3] Taheri M, Ahmadian M2016.*Vehicle System Dynamics***54** 653-66
[4] Rodriguez J, Freeman PT, Wagner J, Pidgeon P, Alexander K2016.*Int. J. of Automotive Technol.***17** 71-81
[5] Xia X, Xiong L, Sun K, Yu ZP2016.*Int. J. of Automotive Technol.***17**991-1002
[6] Diakov AS, Kotiev GO 2018. *MATEC Web of Conf.***224**02096
[7] Evseev K B, Kartashov A B, Dashtiev I Z and Pozdeev A V 2018. *MATEC Web of Conf.***224**02039
[8] Kotiev GO, Padalkin BV, Kartashov AB, Diakov AS 2017.*ARPN J. of Eng. and Appl. Sci.***12** 1064-71.
[9] Sarach EB, Kotiev GO, Beketov SA 2018. *MATEC Web of Conf.***224** 04009

[10] KlubnichkinVE,Diakov AS, Klubnichkin EE, Zakharov AY, Vakhidov U Sh, Suchenina AS and Basmanov IV 2019. J. of Phys.: Conf. Series**1177** 012048
[11] Volskaya NS, ZhileykinMMandZakharov AY 2018.*IOP Conf. Series: Materials Science and Engineering***315** 012028
[12] Klubnichkin EE, Klubnichkin VE, Kotiev GO 2018. *IOP Conf. Series: Materials Science and Engineering***386** 012025
[13] Zhileykin M M, Kotiev G O, Nagatsev M V 2018. *IOP Conf. Series: Materials Science and Engineering***315** 012031
[14] Kotiev GO, Butarovich DO, Kositsyn BB 2018. *IOP Conf. Series: Materials Science and Engineering***315** 012014
[15] SkotnikovGI, JileykinMMandKomissarovA.I 2018.*IOP Conf. Series: Materials Science and Engineering***315** 012027
[16] Ejsmont J, Taryma S, Ronowski G, Swieczko-Zurek B2016. *Int. J. of Automotive Technol.* **17**237
[17] Vdovin D, Chichekin I 2016. *Procedia Engineering***150** 1276-79.
[18] Vdovin D S, Chichekin I V,Levenkov Y Y 2018.*Trudy NAMI***1** 36
[19] Vdovin, DS, Chichekin IV, Levenkov YY, Shabolin, ML 2019. *IOP Conf. Series: Materials Science and Engineering***534(1)** 012024
[20] VdovinD, ChichekinI and RyakhovskyO 2019. *IOP Conf. Series: Materials Science and Engineering***589(1)**012025
[21] VdovinD, LevenkovYandChichekinV 2019. *IOP Conf. Series: Materials Science and Engineering***589(1)**012026
[22] VolskayaNS, ChichekinIV2019. *IOP Conf. Series: Materials Science and Engineering***534(1)** 012022
[23] LeeY.-L, BarkeyM E, KangH.-T 2011 *Metal fatigue analysis handbook. Practical problem-solving techniques for computer-aided engineering* (Oxford, Elsevier) p 581
[24] Pacejka, HB 2006.*Tyre and Vehicle Dynamics. Second Edition*(Oxford: Butterworth-Heinemann)
[25] GorelovVA, Komissarov AI 2016.*Procedia Engineering***150** 1322-28.
[26] Gorelov VA, Komissarov AI, Miroshnichenko AV2015.*Procedia Engineering***129** 300-7
[27] Levenkov YY, Vol'skaya NS, Rusanov OA 2019.*IOP Conf. Series: Materials Science and Engineering***534(1)**012023
[28] Keller A V, Gorelov V A, Anchukov V V 2015. *Procedia Engineering***129** 280-7.
[29] Pacejka, HB, BesselinkIJM 1997.*Supplement to Vehicle System Dynamics***27** 234

Design Technologies for Wheeled and Tracked Vehicles (MMBC) 2019 IOP Publishing
IOP Conf. Series: Materials Science and Engineering **820** (2020) 012003 doi:10.1088/1757-899X/820/1/012003

The algorithm of diagnosing the development of a skid when driving a two-axle vehicle

A Antonyan[1], **M Zhileykin**[1] **and A Eranosyan**[1,2]

[1]Bauman Moscow State Technical University

[2]E-mail: artem_bmstu@mail.ru

Abstract. Vehicle controllability and stability are the most important operational properties and components of active traffic safety, improving these qualities around the world are given great importance. One of the major challenges when designing algorithms for active safetyvehicle systems is obtaining reliable information about the quantitative vehicle parametersvalues, allowing to judge how these parameters match the specified driver, to predict theemergencysituations incident moment and diagnose these situations (e.g., front or rear axles kidding, the tipping risk, etc.). One of the most common parameters used in practice, characterizing themultiaxial wheeled vehiclesmotion conditions, is thevectordeviation angle of the vehicle center of gravity (CoG) actual velocity from the vector of its theoretical (kinematic) speed. However, determining the actual velocity vector direction of the CoGis associated with great computational difficulties due to the need to build complex predictive Kalmanfilters. A method for determining the two-axle wheeled vehiclesmotion parameters to ensure the dynamic stabilization system operation, allowing reliable and economical diagnosis of skid front or rear axles of a two-axle vehicle.

1. Introduction

Vehicle controllability and stability are the most important operational properties and active traffic safety components, increasing these qualities around the world are given great importance [1]. As everyone knew, the main purpose of vehicle active safety systems is to prevent emergency situations.

The use of dynamic stabilization systems (DSS) allows in various critical situations to maintain control over the vehicle or, in other words, to maintain the vehicle stability and controllability.

One of the major problems in the of algorithmsdevelopment of work DSS is to obtain reliable information about the quantitative values of the ofwheeled vehicle (WV) motionparameters, which allows to judge how these parameters match the specified driver, to predict the occurrencetime of emergency situations and diagnose these situations (e.g., front or rear axles kidding, the tipping risk, etc.).

One of the most common parameters used in practice, characterizing thewheeled vehicles movement conditions, is the angle of deviation of the actual speed vector V_F vehicle CoG from vector V_T theoretical (kinematic) velocity $\Theta_T - \Theta_F$ (fig. 1) [2–10]. Andangle

value Θ_T simply determine from the kinematic ratios, knowing the geometric chassisparameters and the steered wheel srotation law, then to determine Θ_F it is necessary to determine the current vehicle speed.

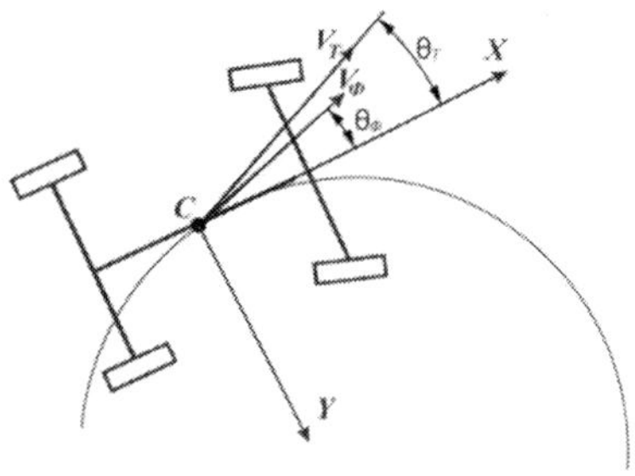

Fig. 1. The deviationangle of the speed vector from the vehicle longitudinal axis: O – instant center of rotation; C – vehicle CoG.

The advantages of this method are the simplicity of diagnosing abnormal situations [11–16]. However, determining the actual velocity vector direction of the CoG is associated with great computational difficulties [17–20] due to the need to build complex observer models or predictive Kalman filters.

It is obvious that a computational procedure to determine the motion parameters WV and compute control actions should be economical, i.e. time spent in computation, and should be less throttle.

The main task of the information system DSS is the vehicle conditiondiagnosis, i.e. sensors need to determine which of the three possible States belongs to the current set of received data:

1. the situation is regular, no adjustments are required;
2. growth of the front axle skid process, adjustment required;
3. growth of the back axis skid process, adjustment is required.

The purpose of this work is to develop a method for determining the motionparameters of two-axle wheeled vehicles to ensure thedynamic stabilization system operation.

2. Problem statement

To achieve this goal it is necessary to solve the following tasks.

1. Propose a meter, i.e. parameter characterizing thevehicle property to maintain the course and trajectory stability.
2. To justify the choice of thestability preservation indicator, i.e. the number characterizing the measurer value, its quantitative value.
3. To develop a diagnostic feature, i.e. a rule that allows unambiguously for all of motionmodes of a wheeled vehicle to uniquely determine to which of the three possible States mentioned above belongs the current set of data obtained.
4. Develop a cost-effective computational procedure that allows real-time rate to estimate thevehicle state and decide on the need to introduce corrective actions.
5. Determine the necessary set of parameters required for the diagnostic system and to be measured during the vehiclemovement.
6. To conduct a study of the proposed algorithm performance by simulation methods.

3. Choice justification of the measurer and the indicator of course andtwo-axle vehicle trajectory stability

Consider the growthprocess of awheel vehicle. If the vehicle is moving at a speed of V_a, growth skid front (fig. 2a) or rear (fig. 2b) axis at a speed of $V_з$, that is as a result ofgeometric speeds addition $V_a$and$V_s$this axis moves in the resulting velocity direction V_R. Because the second axis is still moving at a speed of V_a, this causes the car to turn around the center $O$and the appearance of centrifugal force $P_C$and inertial moment M.Thus, it can be argued that when driving without skidding linear velocity vehicle axes centers will be approximately the same in absolute value. In case of skidding, the modulus of theaxiscenter linear velocity vector that enters the skidding will always be greater than the same index for the axis that is not skidding. This reasoning can be extended to any two points, for example, the of the front and right axiswheelscenters.

To determine the ratio between linear speeds, consider the «ideal» rotation scheme of the two-axle vehicle (Fig. 3) relative to the instantaneous center$O$.Take the pole as the vehicle CoG(point$C$). Then the vector of linear velocity of the front axis center (point A) V_A will be equal to the sum of vectors: portable (linear velocity of point) V_C and the relative velocity of point A relative to pole C V_{AC} (fig. 4a).

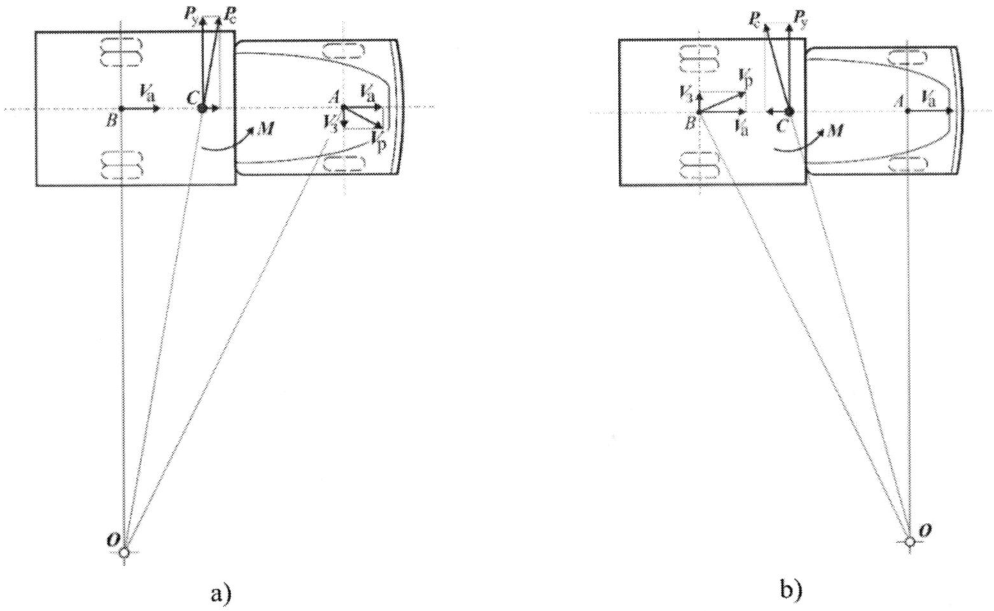

Fig. 2. Skid of front (a) and rear (b) axles of the vehicle: *C* – vehicle CoG; *O* – instant center of rotation; *A*, *B* – the middle of the respectively front and rear axis WV; $\boldsymbol{V}_a$ – the movement WV velocity vector; $\boldsymbol{V}_s$ – skid velocity vector; $\boldsymbol{V}_R$ – resultant velocity vector; $\boldsymbol{P}_C$– centrifugal force vector; *M* – inertial moment; $\boldsymbol{P}_y$– projection of centrifugal force on y axis

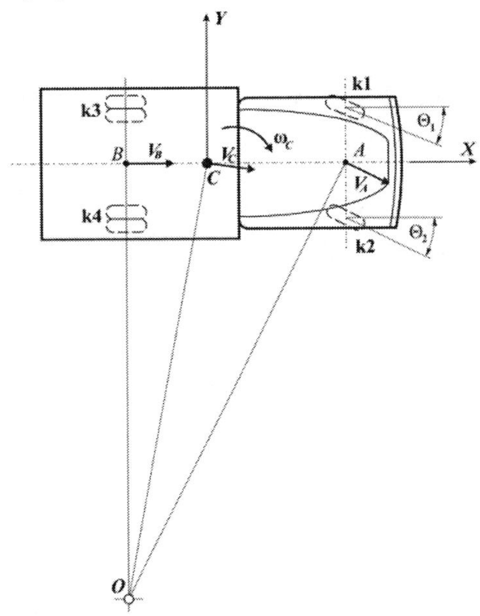

Fig. 3. Scheme of two-axle vehicle rotation: w1...w4 – numbers of wheels; *C* – vehicle CoG; *O* – instant center of rotation; *A*, *B* – the middle of the respectively front and rear axis WV; $\boldsymbol{V}_A$, $\boldsymbol{V}_B$, $\boldsymbol{V}_C$– velocity vectors of *A*, *B*, *C*points; ω_C – vehicle yaw rate; Θ_1, Θ_2 – steering wheel angles of the vehicle front axle

Fig. 4. Speed plans for the centers of the front (a) and rear (b) vehicle wheels

$$V_A = V_C + V_{AC}. \quad (1)$$

If we consider the "perfect" turn, when the vehicle wheels move without withdrawal and without sliding, you can write

$$|V_A| = \omega_A r_1;$$

$$|V_{AC}| = \omega_C |AC|, \quad (2)$$

where $\omega_A = (\omega_1 + \omega_2)/2$ – average angular speed of front axle wheels; r_1 – dynamic wheel radius.

Dependencies similar to (1) and (2) can be written for the center of the rear axis – point B (Fig. 4b)

$$V_B = V_C + V_{BC}; \quad (3)$$

$$|V_B| = \omega_B r_2;$$

$$|V_{BC}| = \omega_C |BC|, \quad (4)$$

whereω_B=$(\omega_3 + \omega_4)/2$ – average angular speed of rear axle wheels; r_2 – dynamic wheel radius.

The distances AC and BC (Fig. 3) for a particular vehicle are known and do not change in the drivingprocess.

4. Development of a diagnostic feature to detect the onset of skidding of thevehicle front or rear axles

The main idea of the proposed method is to estimate the linear velocity vectorV_C two methods: first, by using expressions (1) and (2) – obtained value$|V_{C1}|$, and then using expressions (3) and (4)– obtained value$|V_{C2}|$.It is logical to assume that for a vehicle moving without skidding, the condition is fulfilled

$$|V_{C1}| \approx |V_{C2}|. \quad (5)$$

Consider now the case

$$|V_{C1}| > |V_{C2}|. \quad (6)$$

Taking into account that the evaluation vectorsV_{AC} andV_{BC} no doubt (because yaw rate ω_C can be easily measured, a distance ACandBCunchanged).Means,

$$|V_A| > \omega_A r_1 \text{ and } (or) |V_B| < \omega_B r_2. \quad (7)$$

At the same time, given (6), it is logical to assume that

$$|V_A| > |V_B|. \quad (8)$$

On the basis of conclusions made in section 3, have a skid for the front axle. This case

$$|V_{C1}| < |V_{C2}|. \quad (9)$$

and reasoning similarly, we come to the conclusion that

$$|V_A| < |V_B|. (10)$$

This means that there was a rear axleskid.

Thus, the diagnostic signs of skidding of the front or rear axles of the car can be formulated as follows. If

$$|\delta_V| = ||V_{C1}| - |V_{C2}|| \leq \Delta V, (11)$$

Then there is no skid, the control system reaction is not required. Dead zone ΔVit is introduced, firstly, to compensate for the errors of calculations associated with the failure to account for the removal of tires, etc., and, secondly, from the need to avoid self-oscillations that may occur during the operation of the dynamic stabilization system.

If

$$\delta_V = |V_{C1}| - |V_{C2}| > \Delta V, (12)$$

that is diagnosed skidding of the front axle. If

$$\delta_V = |V_{C1}| - |V_{C2}| < -\Delta V, (13)$$

it is diagnosed by a skid of the rear axle.

5. Development of a computational procedure that implements the vehicle motion statediagnosis

Consider writing expressions (1)–(4) in projections On the X and Y axes associated with the center of mass of the car. For the center of the front axle

$$|V_{C1X}| = \omega_A r_1 cos\Theta_{1cp},$$

$$|V_{C1Y}| + \omega_C |AC| = \omega_A r_1 sin\Theta_{1cp}, (14)$$

where $\Theta = (\Theta_1 + \Theta_2)/2$ – the average angle of rotation of the front steering wheels.

For the center of the rear axle

$$|V_{C2X}| = \omega_B r_2,$$

$$|V_{C2Y}| - \omega_C |BC| = 0. (15)$$

Then the velocity modules $|V_{C1}|$ and $|V_{C2}|$

$$|V_{C1}| = \sqrt{|V_{C1X}|^2 + |V_{C1Y}|^2},$$

$$|V_{C2}| = \sqrt{|V_{C2X}|^2 + |V_{C2Y}|^2}, (16)$$

The resulting expressions (14)–(16) are quite simple and do not require cumbersome computational procedures, which suggests that the proposed methods of ware implementation will be quite economical.

6. Justification of the necessary set of parameters required for the diagnostic system and to be measured in the drivingprocess

On the expressions basis (1)–(14), constituting the computational procedure of the diagnosing process the occurrence of skidding of the front or rear axle of the car, it is possible to make aphysical signals list to be measured during the movement.

1. Angular speeds of rotation of wheels of the car.
2. Angular vehicle yaw rate of the relative to the vertical axis passing through its CoG.
3. The steering wheel angle.

It is also necessary to know the dynamic wheels radius in order to perform calculations in motion, but it is very difficult to measure these values. However, since the dynamic radius can slightly differ from static, with a reasonable accuracydegree when performing calculations, you can use the static radius, ignoring their dependence on possible changes in the vehicle CoG in motion, the redistribution of loads between the axles, etc.

Constants required for system operation:

1. distancesACandBC (fig. 4);
2. the static radius of the wheels.

7. Results

1. A method for determining the motion parameters of two-axle wheeled vehicles to ensure the operation of the dynamic stabilization system, allowing reliable and economical diagnosis of skid front or rear axles of a two-axle vehicle.

References

[1] National Highway Traffic Safety Administration. Motor Vehicle Traffic Crash Injury and Fatality Estimates, Technical Report. NCSA (National Center for Statistics and Analysis) Advanced Research and Analysis, USA, 2003.

[2] A.Y. Ungoren, H. Peng, and H.E. Tseng, “A study on lateral speed estimation methods, ”Int. J. Veh. Auton. Syst., vol.2, no.1/2, pp.126–144, 2004.

[3] Karogal I., Ayalew B. Independent Torque Distribution Strategies for Vehicle Stability Control, in World Congress of the Society of Automotive Engineers. 2009, SAE, Inc.: Detroit, MI.

[4] M. Hiemer, A. von Vietinghoff, U. Kiencke, and T. Matsunaga, “Determination of the vehicle body slip angle with non-linear observer strategies,” in Proc. SAE World Congress, Detroit, MI, 2005, paper no. 2005-01-0400.

[5] M. C. Best, T.J. Gordon, and P.J. Dixon, “An extended adaptive Kalman filter for real-time state estimation of vehicle handling dynamics,” Vehicle Syst. Dyn., vol.34, no.1, pp. 57–75, 2000.

[6] W. Klier, A. Reim, and D. Stapel, “Robust estimation of vehicle sideslip angle–an approach w/o vehicle and tire models,” in Proc. SAEWorldCongress,Detroit,MI,2008,paper no.2008-01-0582.

[7] C. Sentouh, Y. Sebsadji, S. Mammar, and S. Glaser, “Road bank angle and faults estimation using unknown input proportional-integral observer,” in Proc. Eur. Contr. Conf., Kos, Greece, 2007, pp. 5131–5138.

[8] L. Imsland, H.F. Grip, T.A. Johansen, T.I. Fossen, J.C. Kalkkuhl, and A.Suissa, “Nonlinear observer for vehicle velocity with friction and road bank angle adaptation–validation and comparison with an extended Kalman filter,” in Proc. SAE World Congress, Detroit,MI,2007,paper no.2007-01-0808.

[9] L. Imsland, T.A. Johansen, H.F. Grip, and T.I. Fossen, “On nonlinear unknown input observers–applied to lateral vehicle velocity estimation on banked roads, ”Int. J. Contr., vol. 80, no.11, pp.1741–1750, 2007.

[10] J. Jung, S. Taehyun, J. Gertsch. Vehicle full-state estimation and prediction system using state observers. IEEE Transactions onVehicularTechnology,vol.58,no.8,pp.4078–4087,2009.

[11] D. Odenthal, T. Bunte, J. Ackermann. Nonlinear steering and braking control for vehicle rollover avoidance. InProceedingsofEuropeanControlConference,Karlsruhe,Germany,1999.

[12] Belousov, B.N., Merkulov, I.V., Fedotov, I.V. Synthesis of dynamic control system for active mountings of multiaxis ATM (2004) Avtomobil'nayaPromyshlennost, (4), pp. 15-17.

[13] Belousov, B., Demik, V., Kozlova, A., Ksenevich, T., Naumov, S., Medvedev, E., Lyushnin, S.,Kuzminkov, K.The schematic diagrams of actuators of an mechatronic wheel steering system(2018) FISITA World Automotive Congress 2018.

[14] Belousov, B.N., Boldorev, A.G. All-wheel steering with autonomous drive(2005) Avtomobil'nayaPromyshlennost, (11), pp. 16-17.

[15] Belousov, B.N., Boldorev, A.G.System for automatic control of steering for wheel automobile(2005) Avtomobil'nayaPromyshlennost, (9), pp. 15-18.

[16] Ergin, A.A., Kolomejtseva, M.B., Kotiev, G.O. Antiblocking control system of the brake drive of automobile wheel(2004) PriboryiSistemyUpravleniya, (9), pp. 11-13.

[17] Ryu J., Gerdes J.C. Estimation of vehicle roll and road bank angle // American Control Conference, 2004. Proceedings of the 2004 (Volume:3) IEEE Boston, MA, USA. June 30 2004 – July 2, 2004. – P.P. 2110 – 2115.

[18] M. Oudghiri, M. Chadli, A. ElHajjaji. Robust observer-based fault-tolerant control for vehicle lateral dynamics. InternationalJournalofVehicleDesign,vol.48,no.3–4,pp.173–189,2008.

[19] S. Solmaz, M. Corless, R. Shorten. A methodology for the design of robust rollover prevention controllers for auto-motive vehicles: Part2–Activesteering. In Proceedings of the American Control Conference, IEEE, New York, USA, pp.1606–1611,2007.

[20] J. Yoon, W. Cho, B.G. Koo, K.Yi. Unified chassis control for rollover prevention and lateral stability. IEEETransactionsonVehicularTechnology,vol.58,no.2,pp.596–609,2009.

Experimental evaluation of the effect of cryogenic treatment of brake discs of the vehicle on the performance characteristics

V Dygalo[1,4], I Zhukov[1], A Diakov[2] and K Chutkov[3]

[1] Volgograd State Technical University, Lenin avenue, 28, Volgograd, Russia, 400005
[2] Bauman Moscow State Technical University
[3] Combined Arms Academy of the Armed Forces of the Russian Federation

[4] E-mail: dygalo@vstu.ru

Abstract. The article presents the conclusions and prerequisites of the experimental evaluation of friction pairs in the aspect of fading. The most part of kinetic energy of the vehicle with ABS is extinguished at the expense of friction work in the brake mechanism. Thus, the overheating of the brake mechanism, namely its pair of friction is peculiar to vehicles not equipped with ABS (for example, ATV) and leads to the phenomenon of critical fading, accompanied by a sharp decrease in braking torque. The reduction of the influence of this phenomenon is a very difficult task in terms of both taking into account the cost of the braking mechanism and its minimal complication. The authors suggest the use of cryogenic processing of brake disks to reduce the impact of thermal loading on the emergence of fading.

Introduction

The widespread use of automated braking systems in vehicles, in addition to the obvious benefits in terms of active safety, also causes a number of problems due to changes in the workflow [1]–[7]. This increases the amount of kinetic energy that needs to be extinguished during braking.

As against the traditional method of the braking and skidding, the main part of the kinetic energy of a car with ABS is extinguished by friction in the braking mechanism during ABS operation, which inevitably leads to an increase in their thermal load, especially when used by manufacturers of vehicle braking systems, traditional elements of basic models.

The problem of increased heating of friction pairs of brakes is also relevant for such specific vehicles as ATVs, even in spite of the absence of ABS [8], [9]. And this is due to the movement of these vehicles on the terrain with complex terrain, which requires frequent use of brakes. The situation is aggravated by moisture penetration on friction vapors, which in case of their high temperature can lead to accelerated failure (for example, warping of brake discs), despite the fact that the brake discs are made of stainless steel instead of sliding cast iron. The statistics shows that the friction pair life of the ATV is several times lower than that of a car with comparable mileage. Another reason for this is the penetration of sand and soil particles between the friction pairs and in the ventilation holes of the disc, which affects both the wear and tear of the friction pairs and their temperature conditions.

It is known from the literature that the overheating of brake friction pairs contributes to the emergence of critical fading, accompanied by a sharp decrease (up to 50%) in the friction coefficient of brake linings, as well as increased wear of counter bodies, with the formation of macroscopes [10].

The carried out analysis of the influence of the principal change of the working process of braking of the car wheel with ABS on the distribution of work required to extinguish the kinetic energy of the vehicle as an example is shown in Fig. 1. It is designated: change of kinetic energy of the vehicle with ABS at braking from initial speed of 60 km/h on dry asphalt concrete in the course of braking (Wk), the work spent on a friction in contact of the all-terrain tire (ATR) with road and on hysteresis losses in the tire (AFF), and also absorbed in the brake mechanism of aggregate trailer mass (ATM).

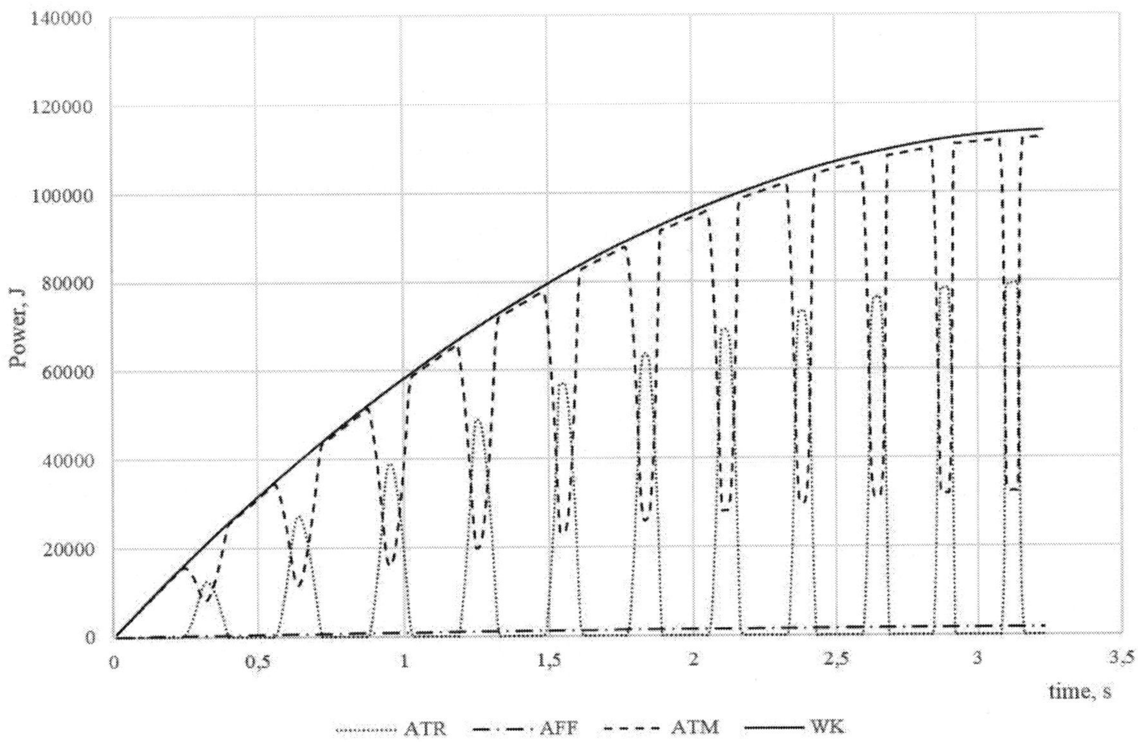

Figure 1. Distribution of the work required to extinguish the kinetic energy of an ABS vehicle during braking

The graph shows that when braking a car with ABS, in contrast to braking and skidding, most of the kinetic energy is damped by friction in the friction couples of the brake mechanism.

It is necessary to make changes in the design to solve the problems associated with fading, after the designer with the help of calculation methods has revealed a problem connected and increased heat loaded braking system. The calculation methods of evaluation were considered by the authors in [11]–[14].One of the ways may be to change the surface area of the friction pairs of the brake mechanism, which may entail a change in the neighboring components (hub, wheel rim), which leads to a rise in the cost of the finished structure. The next approach to solving the problem of increased thermal loading of brakes can be the use of friction pairs of materials whose properties allow to avoid these problems. But materials with higher characteristics have a higher cost. For example, brake discs made of composite materials instead of antifriction grey cast iron used in production vehicles. The authors suggest cryogenic processing of brake discs made of grey sliding cast iron as a cheaper way to improve the material properties.

Test Methods

The article [15] presents the test results after cryogenic processing of cutting tools for turning. There was a decrease in thermal EMF in comparison with untreated tools. From this it is possible to draw a conclusion that the change of properties of a material of a cutter after cryogenic processing influences thermal processes in a cutting zone. In order to be sure of the positive effect on the properties of the brake mechanism of the cryogenic treatment of the brake disk, it is necessary to test the samples by comparison and to draw conclusions on the measured parameters.

There are various normative documents for friction pair testing, for example, RD 50-662-88. Where a set of methods for assessing friction compatibility of lubricants and (or) structural materials used in friction materials is installed.

Figure 2. Installation of test samples in the friction machine SMC-2. Symbols: 1 — movable sample (disk); 2 — friction machine rotating shaft; 3 — grip; 4 — stationary sample (block).

The tests were carried out using the following main test equipment:

1. SMC-2 friction machine;
2. Inductive friction torque sensor;
3. Laptop for capturing the video signal from a thermal imaging camera;
4. SAT HotFind-L thermal imaging camera.
5. Electronic analytical scales

Materials used for testing

As a rotating sample were used discs with a diameter of 36 mm and a thickness of 6 mm obtained from the original brake disk of the Lada Granta by waterjet and turning. This made it possible to obtain a roughness and accuracy of the test surface comparable to the working surface of the original brake disk.

Results and discussion

One of the factors that confirms the adequacy of the tests is that they are friction pair materials. The chemical analysis of these brake discs (Table 1) has been carried out, which showed full compliance of the chemical composition with the requirements of the regulatory documentation for this grade of cast iron, all parameters were in the middle of the permissible ranges. ASF-4 cast iron grade, which means grey sliding cast iron with lamellar graphite.

Table 1. Results of chemical analysis of Lada Granta's brake discs

Parameters	Significance	Requirements of GOST 1585-85
Material	Antifriction cast iron ASF-4	
Carbon	3,46%	3-3,5%
Manganese	0,79%	0,4-0,8%
Silicon	2,06%	1,4-2,2%
Chrome	0,06%	is not regulated
Phosphorus	0,05	no more than 0.3
Sulphur	0,05	0,12-0,2

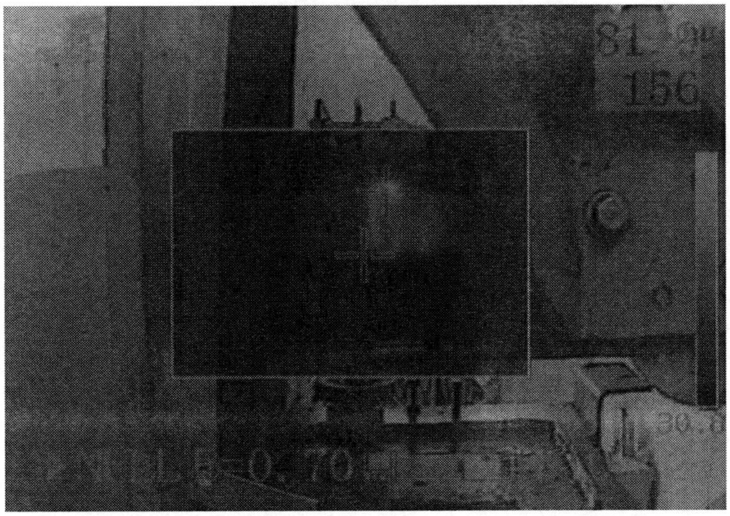

Figure 3. A still from a thermal imaging camera. Symbols: 1 — Temperature at the center of the viewfinder, 4 — Maximum temperature in the captured area.

The movable sample is a disc that rotates at a constant speed of 1000 rpm (16.6 s^{-1}). It is pressed against a fixed specimen fixed in the carriage, which is a rectangular segment of the brake shoe. In the experiment, two manufacturers of brake pads were used slightly lower than average (company Riginal) and the upper price segment with confirmed authentication (company Ferrodo).

According to RD 50-662-88 "Methods of experimental assessment of friction compatibility of friction materials of rubbing interfaces" it is said that if the wear of a moving sample is less than 0.02 mm, the duration of the tests is doubled. Thus, this thesis should be taken into account with the accuracy of measuring the weight of the moving sample, because the wear of the stationary sample (pad) is greater.

The heat load of friction pairs during the test was assessed using a SAT HotFind-L thermal imaging camera.

The temperature of the surface of a moving sample is of great interest. The coefficient of correction for emissivity was chosen to be 0.7, which, according to the documentation for the thermal imaging camera, corresponds to the treated grey cast iron.

According to the test results, it was possible to show the influence of modes, braking time, properties of materials on fading. It was noted that the brake pad material has a significant influence on the torque with the same contact force and comparable areas, even though the surface temperature of the brake disk is too high.

The influence of pad material on the surface temperature of the brake disc is less than that of the cryogenic treatment of the disc. In a vehicle with cryogenically treated brake discs that work in

IOP Conf. Series: Materials Science and Engineering **820** (2020) 012004 doi:10.1088/1757-899X/820/1/012004

conjunction with quality brake pads (Ferodo), fading may not occur for a sufficiently long braking time or may reduce the risk of braking, as shown in these tests. An increase in braking torque with the same brake pad pressure can be achieved by using quality brake pads.

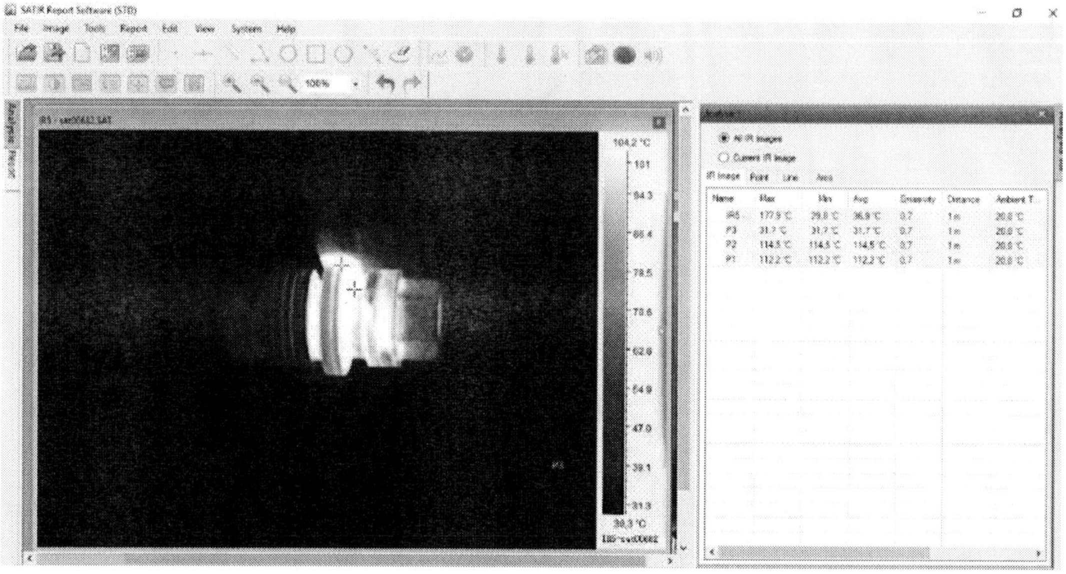

Figure 4. Thermogram processing in SatIrReport

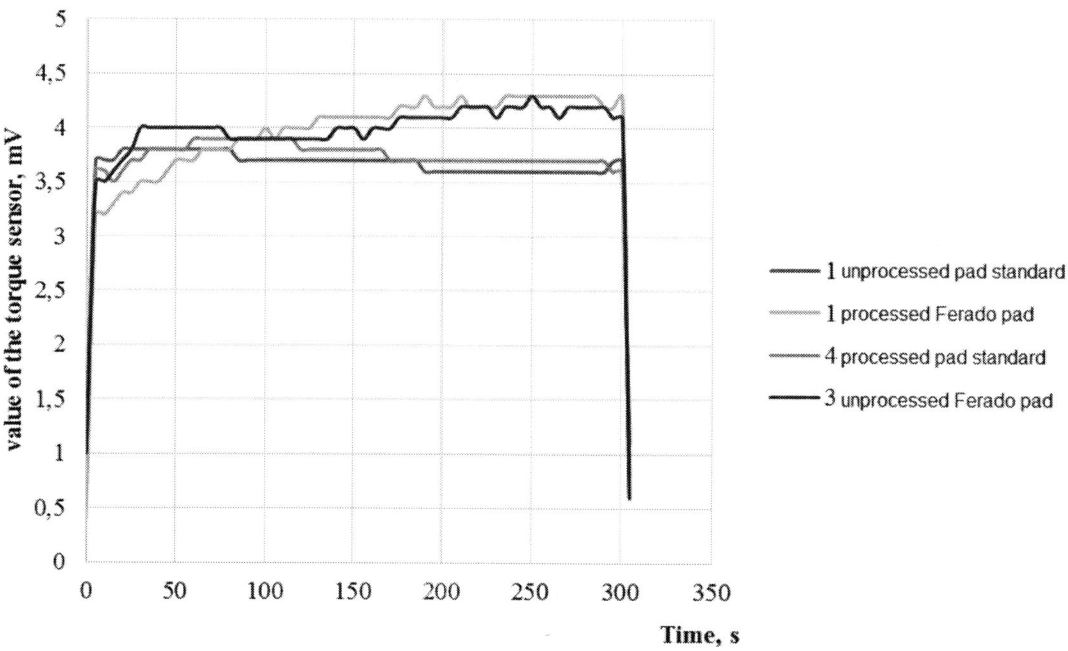

Figure 5. Test torque sensor values for 5 minutes for a combination of different moving and stationary samples.

One of the most important features of the brake system is its reliability. A particular case of faultless braking is the stability of the braking system, e.g. the brake torque. For this purpose, we will use the parameter of stability of the friction coefficient of the pad disk in accordance with PD 50-662-88.

They calculate the coefficient of friction and the stability of the coefficient of friction by the formula

$$a = \sum_{i=1}^{n} M_i \,/\, \sum_{i=1}^{n} M_{max} \tag{1}$$

where n — number of re-tests.

After calculating the coefficient of friction stability, we obtain the friction coefficient for different combinations of test samples of the disc and pad:

– for standard disc and standard shoe 0.968,
– for cryogenically processed disc and standard block 0.974,
– for standard disc and Ferodo pad 0.959,
– for cryogenically processed disc and Ferodo 0.969 pads.

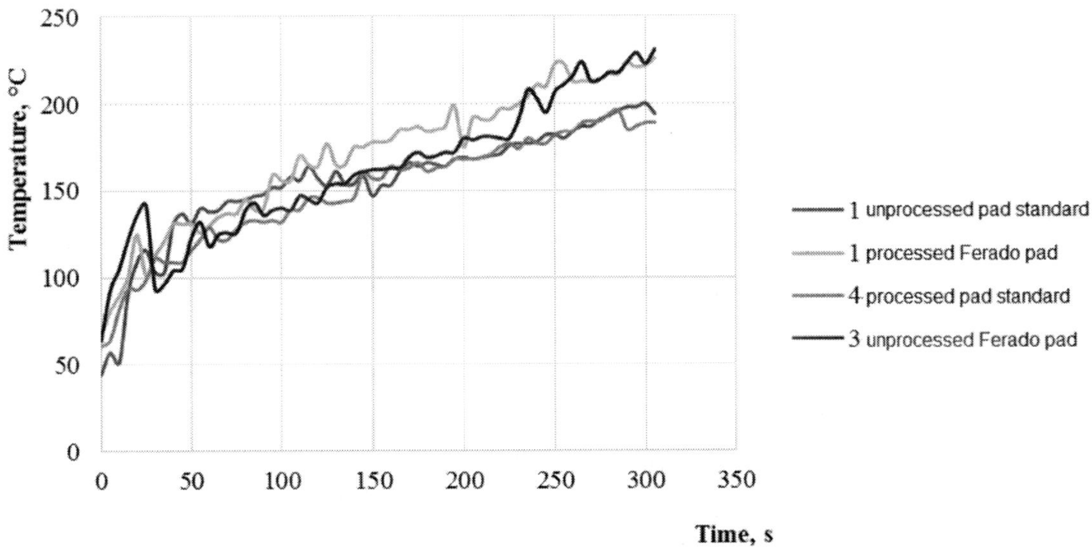

Figure 6. Maximum temperatures of moving samples within 5 minutes for a combination of different moving and stationary samples.

The results of the calculation of a show that the moving cryogenically processed sample has a more stable coefficient of friction and therefore a more stable braking torque. The decrease in the coefficient of friction pairs with Ferodo pads can be explained by the higher values of the realized braking torque at the same clamping force of the stationary sample to the moving one (see Fig. 6) and the high surface temperatures of the moving sample (see Fig. 7).

That is, the braking torque range increases proportionally with the increase in its values. But just as with a standard Ferodo pad, so too, does a cryogenically processed sample have a higher coefficient of friction stability than a pair of Ferodo pads and a standard disc.

Conclusions

The brake lining wear depends on the temperature according to the literature, although the results of the tests have shown the ambiguity of this statement. For the untreated moving sample, which had a higher temperature, the wear of the pads of the standard and Ferodo was significantly lower than for the treated sample, which had a lower surface temperature.

Table 2. Results of wear measurement of moving samples in combination with different pads.

№	Sample type	Disk weight before testing	Disk weight after testing	Disc wear and tear	The weight of the pads before the tests	The weight of the pads after the tests	Wear pads
1	standard disc, standard pad	33,417	33,409	0,008	278,506	277,973	0,533
2	processed disk, standard pad	33,370	33,366	0,004	277,917	277,765	0,152
3	standard disc, standard pad	33,373	33,363	0,01	283,945	283,777	0,168
4	processed disk, standard pad	33,192	33,183	0,009	283,752	283,613	0,139
5	processed disk, Ferodo pad	33,368	33,359	0,009	286,154	286,120	0,034
6	standard disc, Ferodo pad	33,334	33,316	0,018	286,064	286,035	0,029
7	standard disc, standard pad	33,401	33,397	0,004	278,072	278,019	0,053
8	processed disk, standard pad	33,288	33,286	0,002	277,973	277,910	0,063

References

[1] Novikov V.V., Pozdeev A.V., Diakov A.S. Research and testing complex for analysis of vehicle suspension units // Procedia Engineering. 2015. Vol. 129. P. 465–470. DOI: 10.1016/j.proeng.2015.12.153

[2] Kotiev G.O., Padalkin B.V., Kartashov A.B., Dyakov A.S. Designs and development of Russian scientific schools in the field of cross-country ground vehicles building // ARPN Journal of Engineering and Applied Sciences. 2017. T. 12. № 4. C. 1064–1071

[3] Study of active vehicle safety properties by simulation modeling method / A.V. Tumasov, A.M. Groshev, S.Y. Kostin, M.I. Saunin, Yu.P. Trusov, V.G. Dygalo // Journal of automotive engineers. — 2011. — № 2. — p. 34–37.

[4] Kotiev G.O., Diakov A.S. Advanced development and testing of off-road vehicle // DEStech Transactions on COMPUTER SCIENCE and ENGINEERING. 2017 2nd International Conference on Computer, Mechatronics and Electronic Engineering (CMEE 2017) pp. 464–467. ISBN: 978-1-60595-532-2. DOI 10.12783/dtcse/cmee2017/20021.

[5] Diakov A.S., Kotiev G.O. Establishment of production of special wheel and track technology for extreme natural-climate conditions of the Arctic // MATEC Web of Conferences Volume 224, 30 October 2018, 02096 DOI: 10.1051/matecconf/201822402096

[6] D.A. Chumakov, K.V. Chernyshov, V.V. Novikov, A.S. Diakov and A.S. Suchenina. Mathematical model of motor vehicle air suspension with a combined damping system. IOP Conf. Series: Journal of Physics: Conf. Series 1177 (2019) 012049. DOI:10.1088/1742-6596/1177/1/012049.

[7] Diakov A.S., Novikov V.V., Pozdeev A.V. The main directions of the development of snowmobiles in the Russian Federation// MATEC Web of Conferences Volume 224, 30 October 2018, Issue 02080 DOI: 10.1051/matecconf/201822402080

[8] Pozdeev A V, Chumakov D A, Novikov V V, Golyatkin I A, Chernyshov K V, Gavrilov A E, Leonard A V 2019 Thermographic bench tests of rubber-cord pneumatic spring *IOP*

Conference Series: Materials Science and Engineering (ToPME, Moscow, IMASH RAS) **489** 9 p

[9] Novikov V V, Chumakov D A, Golyatkin I A 2019 Experimental Modular Unit for Testing Nodes of Air Suspension with Combined Amortization *Proceedings of the 4th International Conference on Industrial Engineering (ICIE 2018)* (Moscow, Russia) ed A A Radionov (Lecture Notes in Mechanical Engineering (LNME) Springer) pp 385–392

[10] Revin A.A. Methodology for monitoring the technical condition of the ABS braking system in operation / A.A. Revin, I.S. Zhukov, V.S. Shkarupelov // Izvestiya VolgTU, series: ground transport systems. — 2012. — № 2 (89). VOL. 5 p. 90–93.

[11] Dygalo, V.G. Estimation of the thermal loading of the friction pairs of the automated braking system of a car / V.G. Dygalo, I.S. Zhukov // Progress of vehicles and systems — 2018: materials of the international scientific-practical conf. (Volgograd, October 9–11, 2018) / Editorial: I. A. Kalyaev, F. L. Chernousko, V. M. Prikhodko [and others]; VolgGTU, RFBR, "FNPC "Titan-Barrikady". — Volgograd, 2018. — p. 103–105.

[12] V.E. Klubnichkin, A.S. Dyakov, E.E. Klubnichkin, A.Yu. Zakharov, U.Sh. Vakhidov, A.S. Suchenina and I.V. Basmanov. Experimental evaluation of speed and brake properties of domestic and foreign made utility terrain vehicles. IOP Conf. Series: Journal of Physics: Conf. Series 1177 (2019) 012048. DOI:10.1088/1742-6596/1177/1/012048.

[13] V.E. Klubnichkin, A.S. Dyakov, E.E. Klubnichkin, A.Yu. Zakharov, U. Sh. Vakhidov, A.S. Suchenina and I.V. Basmanov. Experimental evaluation of stability and controllability of domestic and foreign made utility terrain vehicles. IOP Conf. Series: Journal of Physics: Conf. Series 1177 (2019) 012045. DOI:10.1088/1742-6596/1177/1/012045.

[14] Shumsky K P 1974 Vacuum apparatus and instruments for chemical engineering *(Moscow, Mechanical engineering)* 576 p

[15] Solonenko, V.G. ThermoEDS and friction coefficient as a result of the cryogenic processing of the turning tools (in Russian) / V.G. Solonenko, E.A. Krivonos // ISTC "Problems of research and design of machines". Penza: PDZ, 2005. p. 89–91.

IOP Conf. Series: Materials Science and Engineering **820** (2020) 012005 doi:10.1088/1757-899X/820/1/012005

Stationary simulation method of road conditions in the problem of interaction prediction of the wheel of the vehicle support surface

N Volskaja[1,2], **M Zhileykin**[1] **and A Zakharov**[1]

[1] Bauman Moscow State Technical University

[2]E-mail: volskaja52@mail.ru

Abstract. A method for estimating the deformation properties of the support surface of the operation proposed area of a four-wheel drive wheeled vehicle is presented. These properties influence the vehicle efficiency. The method is based on the use of physical characteristics of soils specified in the probabilistic form. The estimation of the vehicle movement efficiency is carried out on the average probabilistic speed on the reference route. Features of vehicle rectilinear and curvilinear movement are considered.

Introduction

There are methods for calculating the wheeled vehicles patency (WV) with different wheel formulas [1–17]. One of the main traction indicators is the track depth. At the same time it is important to be able to calculate its numerical value depending on the passes number when the wheels move "track to track" [2]. The more specific are the expected road-ground conditions (RGC) of the movement surface, the more accurate the calculated forecast of this indicator. Thus it is possible asking the RGC by varying the propeller parameters, the calculations to find the preferred combination of effective design solutions for the evolving pilotless vehicle with high cross (PVWHC). An effectiveness indicator of the developed design PVWHC is a higher value of the probabilistic speed on the calculated reference route.

In this regard, it is relevant to solve the problem of patency assessment PVWHC on the ground surfaces of the intended operation region in General, taking into account the heterogeneity of the soil, surface roughness and the curvilinear trajectories possibility.

The purpose of the study: increased permeability of PVWHC for heterogeneous, uneven dirt surfaces and changing the deformation properties trajectory-based prediction of the intended area of operation RGC.

Problem statement

1. Develop a method for evaluating the deformation properties of the intended area of operation soil surface, providing a patency rates prediction (primarily the depth of the rut after the wheel passage) at the design stage PVWHC.

2. Development of a methodology for assessing the effectiveness of KM traffic for a specific operation area.

The tasks decision should be based on the unified approach to the soil physical-mechanical condition assessment of any type, taking into account seasonal operation PVWHC.

Stationary modelling RGC held in conjunction with the separately solved problems of pneumatic wheel interaction modeling with deformable ground (straight and curved types of motion).

Method of mechanical properties evaluation of the proposed area soil surface of operation PVWHC

Evaluation method of the deformation properties of the intended area soil surface of operation PVWHC based on the use of soils physical characteristics are presented in probabilistic form; as information sources are used long-term observations at meteorological stations, topographic and soil maps, results of field experiments, the technical literature [3]–[6].

Well-known vehicle movement mathematical models on soft soils differ from each other by different characteristics of soil deformation properties. The main requirement for them - independence from the wheel characteristics. In this study, the soil characteristics are selected: wetness — W, volumetric density of soil — ρ, modulus of deformation — E, internal friction angle — φ_0, internal cohesion in the soil — c_0, soft layer thickness — H_Γ. They fully correspond to the soil mechanics characteristics [7–12].

Soils are heterogeneous primarily in wetness and density. Long-term observations on soil moisture and density are conducted at weather stations. It was found that the humidity varies by month according to the normal law with a fluctuation relative to the mathematical expectation not exceeding 6 %. Therefore, for different country regions it is possible to predict the physical and soils mechanical characteristics in probabilistic and statistical form. Soil type distribution boundaries can be determined from soil maps. The weather station location is also known according to weather reports.

Depending on the type and physical condition of soils, their graphical calculation models give a different response under vertical loading and shear (fig. 1, a: 1 — porous soil, 2 — dense soil; 3 — soil with a solid base; fig. 1, b: 1 — soil with low internal adhesion; 2 — soil with high internal coupling properties.

To describe the stress-strain state in the contact patch "of the tire-soil" can be selected calculation model proposed by J.S. Ageikin [13]:

$$q_\beta = \left[\frac{(H_\Gamma - z)}{b(1+1{,}75\varphi_0)(k_{\beta_1}\cdot b\cdot\rho\cdot X_1 + k_{\beta_2}\cdot c_0\cdot X_2 + k_{\beta_3}\cdot\rho\cdot X_3\cdot z)\cdot\cos\beta} + \frac{a\cdot b}{E\cdot z} arctg \frac{(H_\Gamma - z)}{a\cdot b\cdot\cos\beta} \right]^{-1};$$

$$\tau = \left[q_\beta \cdot tg\,\varphi_0 + c_0\left(1 - \frac{S_{t0}}{S_t}\right)\right]\left[1 - \exp\left(-\frac{S_{t0}}{k_\tau}\right)\right], \text{ where } k_{\beta 1} = \frac{\pi - 4\cdot\beta\cdot tg\varphi_0}{\pi + 4\cdot\beta\cdot tg\varphi_0},$$

$$k_{\beta 2} = k_{\beta 3} = \frac{3\cdot\pi - 2\cdot\beta}{3\cdot\pi + 2\cdot\beta},\ X_1 = \frac{1 - X^4}{X^6};\ X_2 = \frac{1{,}3[X^2 + 1]}{X^4};\ X_3 = \frac{1}{X^6}, \text{ where } X = tg(\pi/4 - 0{,}5\varphi_0),$$

where q_β — the pressure in the contact patch; τ — shear stress; z — track depth; a — coefficient characterizing the stresses attenuation in the ground; b — the tire contact width with the ground; k_τ — the soil tangential elasticity coefficient; S_t — the grouser step; S_{t0} — grouser shift lug; β — angle between load vector and normal to ground surface. These dependences are confirmed by experiments conducted on sand or sandy loam.

The equations include the soil mechanical parameters φ_0, c_0, E, H_Γ. Statistics on these parameters are limited. At the same time, there is great potential for the statistical data accumulation on the soils physical parameters for specific regions. In order to use these opportunities, relationship studies between the physical and soils mechanical parameters were carried out. As a result, the following equations are obtained:

$$E = K_{E_1}\cdot\frac{W_T\cdot\rho_C^2}{(W - 0{,}05)} - K_{E_2}(W\cdot W_T)^2;\ \varphi_0 = K_{\varphi 1}\sqrt[4]{\frac{K_{\varphi 4}}{W_T}} - K_{\varphi 2}\cdot\left(1 - \frac{\rho_C}{\rho_T}\right) - K_{\varphi_3}\cdot W\cdot\sqrt{\frac{W_T}{K_{\varphi_4}}};$$

$$c_0 = K_{C_1} \cdot \frac{W_T}{W} \rho_C - K_{C_2} \cdot W \cdot W_T,$$

where $K_{E\text{-}\varphi}$ — coefficients; W_T — the wetness of yield point; ρ_c and ρ_T — the density of a skeleton of soil solid particles.

Therefore, it is possible to predict the numerical value of mechanical parameters of connected and disconnected soils depending on changes in soil moisture (seasonal forecasting) [18]–[21].

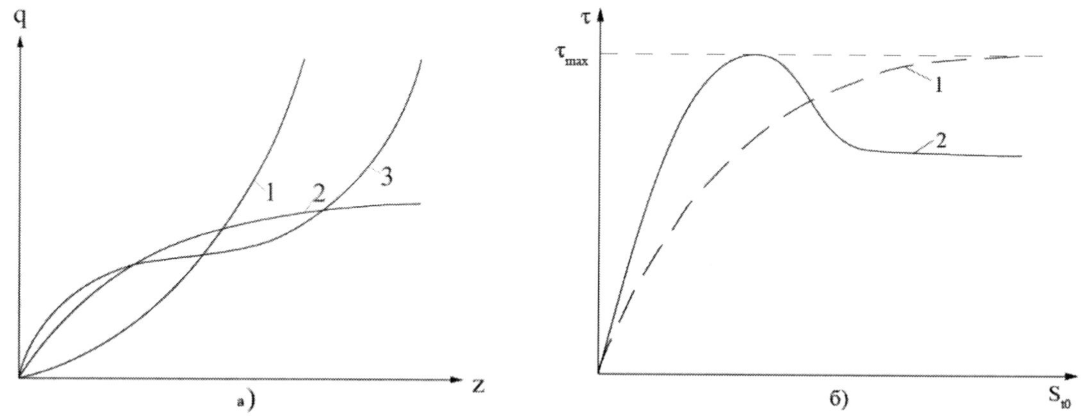

Figure 1. Graphical models of soil types

A refined method for estimating the action time of the load and Cycling loading soil

Consideration of the load action time influence on the soil deformation properties is possible due to the introduction into the equation $q_\beta = f(z)$ the coefficient of dynamic $k_Д$ depending on φ_0: $q_Д = q_\beta \cdot k_Д$, $k_Д = \dfrac{1}{1 + \dfrac{V}{l \cdot \varphi_0 \cdot 2n}}$, $t = \dfrac{l \cdot n}{V}$, where: l — the contact length; n — number of passes per track (or number of axle); V — speed of PVWHC; t — the duration of the load.

When driving even a single KM is characterized by cyclic soil loading. When loading the soil, there is a change in its density and humidity, i.e. its mechanical properties. The loading cyclicity of can be identified with the time of single load action, but with a large effect, which can be reflected by the corresponding coefficient of transition from the time of its action to the cycles number.

Example of loading cyclicity influence (number of passes) and load action time t to the track depth z shown in fig. 2, a, b. Track depth is determined by dynamic pressure $q_д$, fig. 2, a on the set V, l, φ_0, n and $q_\beta = f(z)$. The values are then determined φ_0, c_0 and E for connection equations.

The method of database presentation on the physical and mechanical characteristics of each soil type distributed in the intended area territory of vehicle operation is as follows:

1. According to soil maps, the main soils types are determined (it can be one or two soils). Soils are classified into: clay, loam, sandy loam, sand, snow.

2. According to the weather stations materials the wetness is statistically represented $F(W)$, $\rho(W)$ for a specific month (for example, the time of the thaw is April).

3. Graphs are represented by the above dependencies $F(W)$, ρ, E, φ_0, c_0. A graphical view of the statistical database on soil type silt is shown in figure 3.

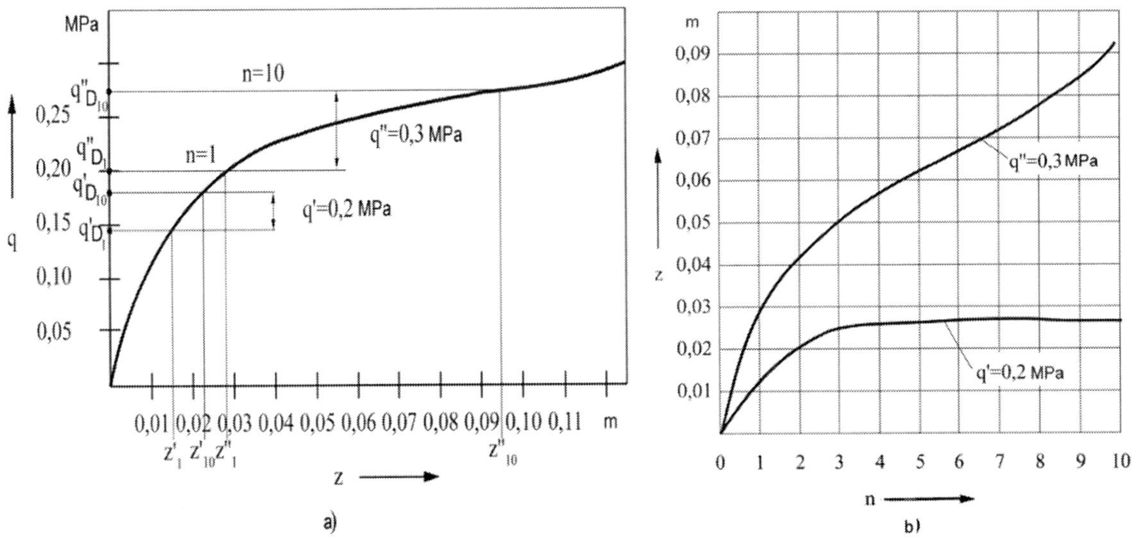

Figure 2. An example of the cyclical loading effect

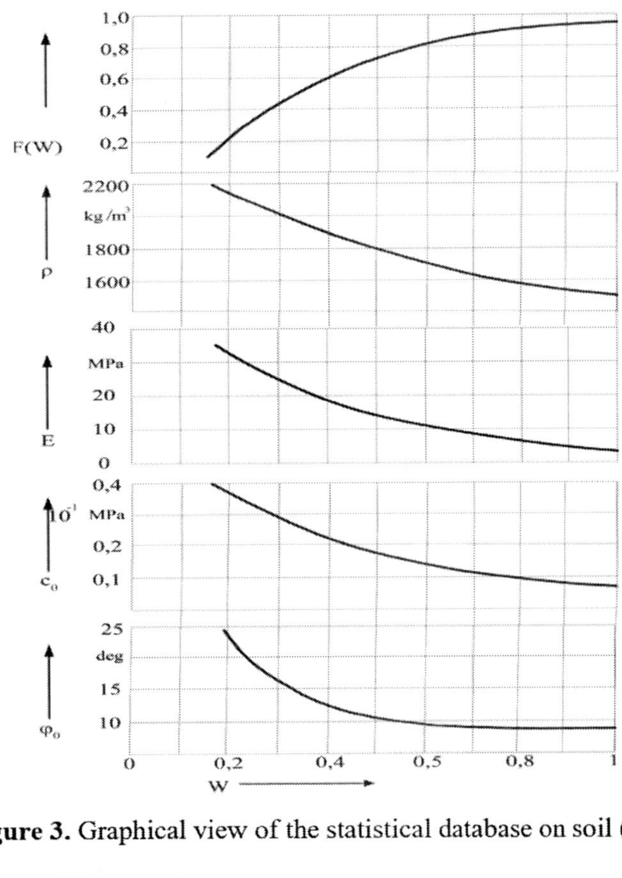

Figure 3. Graphical view of the statistical database on soil (Silt)

Mathematical model of RGC operation area

It is proposed to assess the heterogeneity of the RGC operating area using a virtual reference route. The mathematical model of the route consists of several sections. Soil maps determine the boundaries and areas of main types of soils distribution of the operation intended area (one or more types of soils). Taking the hypothesis of the equal probability of movement PVWHC at any terrain point, it is possible to determine the density of the distribution of soil types: $P(j)=\frac{S_j}{S_{общ}}$, where S_j — distribution area j-th soil; $S_{общ}$ — region area in question. As a analysis result of soil maps to assess the prevalence of soil type, a decision is made to introduce it into the characteristics of the sites.

Reference route — a set of sections, each of which is characterized by a "weighting factor" — the relative length P_i (selected by analyzing soil, topographic and road maps). The prevailing soil type (or several soils) is assigned to the parcel numbers. Each site is characterized by probabilistic characteristics of soil parameters, slopes, degree of evenness. On one of the sites in the probabilistic form is given curvilinear motion. The average probabilistic speed on the reference route is determined by:

$$V_{cp}=\sum\frac{1}{\frac{P_i}{V_{вcpi}}},$$

where P_i — relative length of the section in the reference route; $V_{вci}$ — probabilistic average speed on i-th site.

Mathematical model of PVWHC interaction wheel mover with soil in rectilinear motion

For determining z — track depth and h — tire deformation is used, the above equation models from J.S. Ageikin and the calculated scheme of interaction between wheel and soil (fig.4).

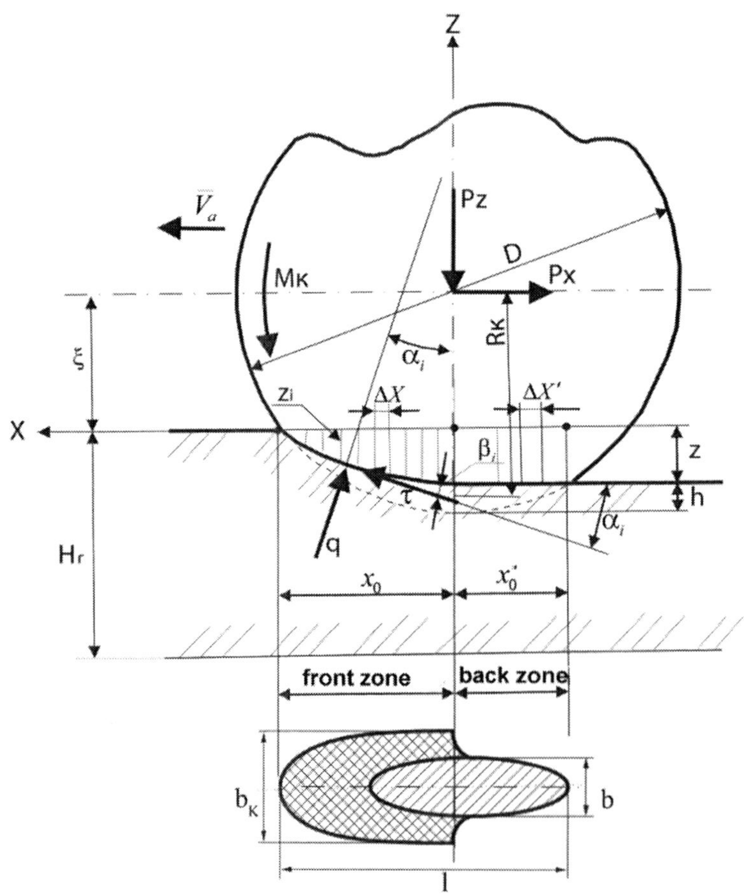

Figure 4. Design scheme of wheel-ground interaction

Determining rut depth and deformation of the tire for the front axle PVWHC for subsequent axes, with the condition of a constant gauge, determine the parameters of the contact problem of a wheel-soil, but the physical and mechanical characteristics of the soil are recalculated according to the number of the axis (or passage) in the dependencies $\rho_n = f(n;q_\partial;\rho_m)$; $E,\varphi_0,c_0 = f(\rho_n;W_m;W)$.

After determining h and z for each vehicle wheel are defined: tire rolling resistance coefficient, $f_{Ш}$; the rolling resistance coefficient of soil, f_Γ; total rolling resistance coefficient f; free relative thrust force on the hook ψ_T:

$$f_{Ш} = 1{,}75 \cdot P_X \cdot \psi_1 \cdot h^2 \left(B^2 + 1{,}5H^2\right) \frac{B - 0{,}3h}{P_Z \cdot H \cdot B^2};\quad f_\Gamma = q \cdot z \cdot 0{,}3\left(\frac{b + b_k}{P_Z}\right);$$

$$f = f_{Ш} + f_\Gamma;\quad \psi_T = \left[k_H \cdot \mu_1 + (1 - k_H)\left(tg\,\varphi_0 + \frac{c_0}{q}\right)\right] - f_\Gamma,$$

where B, H — width and height of tire profile; E_1 — the tangential modulus of soil deformation; k_H — coefficient of saturation of the tread pattern; μ_1 — coefficient of friction of rubber on the ground; ψ_1 — coefficient of hysteresis losses in the rubber-cord bus sheath.

The calculations results show that if we solve the problem of patency estimation (for example, when determining the engine optimal parameters in the given RGC), then in the probabilistic formulation a solution significantly different from the final result in the averaged formulation will be obtained.

Assessing patency of PVWHC when moving along a curved path on a deformable surface

For the case of curvilinear motion developed: 1) the mathematical model of rolling elastic wheels on rigid and deformable grounds; 2) mathematical model of PVWHC motion on deformable soil.

In assessing the pneumatic wheel kinematic and power parameters of the curvilinear motion takes into account its more complex loading and change in the nature of the movement in comparison with the rectilinear.

The wheel power balance is recorded as follows:

$$M_k\omega_k = P_x\omega_k r_k + P_z f_0 r_k^c \omega_k + P_x\omega_k(r_k^c - r_k) + K_y\delta^2\omega_k r_k + M_{kj}\omega_k,$$

where $M_k\omega_k$ — power supplied to the drive wheels from the transmission; $P_x\omega_k r_k$ — power spent on performing useful work; $P_z f_0 r_k^c \omega_k$ — rolling resistance losses due to radial tire deformation in free mode; $P_x\omega_k(r_k^c - r_k)$ — rolling resistance losses from tangential deformations and sliding in the contact spot in the longitudinal direction; $K_y\delta^2\omega_k r_k$ — losses due to lateral withdrawal; $M_{kj}\omega_k$ — losses in acceleration.

In the evaluation of pneumatic wheel kinematic and force parameters of curvilinear motion on deformable soil deformation is considered in three directions as the wheels and the ground. In contrast to the pneumatic wheel deformability, the soil deformability in the normal, tangential and lateral directions is considered in the relationship (fig.5).

Method of calculating the average probabilistic speed of the PVWHC on a route that includes typical sections of paved roads, dirt roads and terrain

As an example of a reference design route, consider the road-ground conditions, which are characterized by the main soil (background) — loam. RGC of wheeled vehicle movement are represented by seven characteristic sites (fig. 6). The sections length and the RGC of each section are given in probabilistic form. Methods for determining the maximum possible speed of movement in the areas are different and depend on the deformation properties of the support surface and its roughness.

IOP Conf. Series: Materials Science and Engineering **820** (2020) 012005 doi:10.1088/1757-899X/820/1/012005

The cumulative distribution function of velocities F(V), RMS values of the dynamic suspension h_{Dc} and vertical vibration acceleration $\ddot{Z}_C$ depending on the vehicle speed in question on each of the sections of the route are presented on fig. 6.

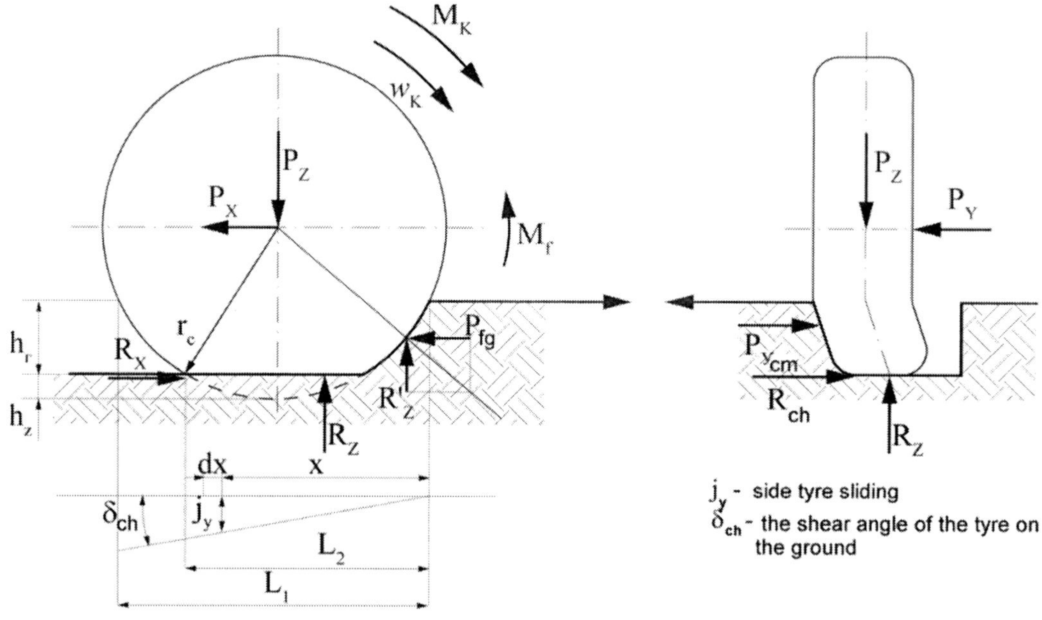

Figure 5. The design scheme of the tire interaction with the ground in curvilinear motion

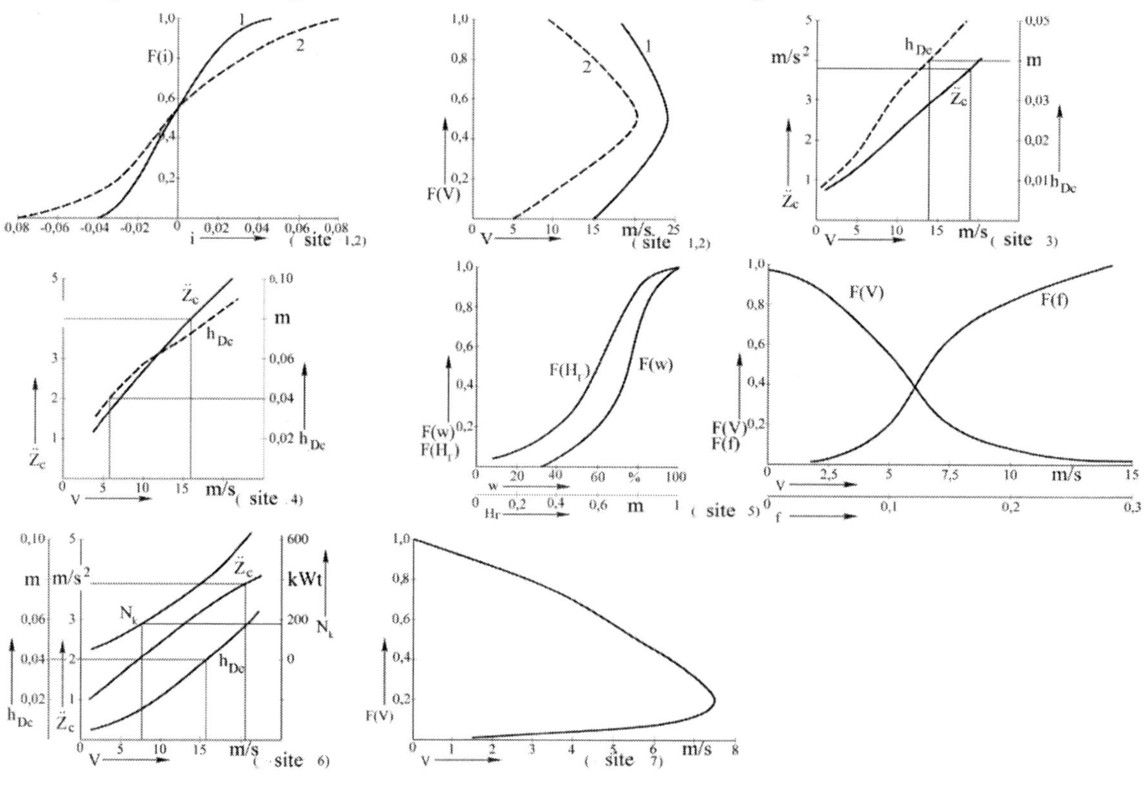

Figure 6. Example of forming the structure of the reference route

In section 1, the speed is determined from the vehicle power balance equation. On the other six sites according to special developed methods. The average probabilistic speed on the route is determined taking into account the possible variability of the nature of the trajectory and types of RGC.

Probabilistic characteristics of seven sections of the reference route: F(i), F(V), F(W), F($H_Г$), F(f) — integral functions of slope distribution, velocity, humidity, soft layer thickness, rolling resistance coefficient, respectively. The table shows an example of the reference route structure formation.

Table

Section number reference route's, N	Surface plot characteristics	The relative length, $P(N)$
1	Paved road, smooth	0,3
2	Dirt road, flat	0,05
3	The road is paved, uneven	0,3
4	Dirt road, uneven	0,2
5	Unpaved surface smooth (clay loam)	0,1
6	Ground surface, uneven	0,025
7	Curved sections of ground surfaces, smooth	0,025

The results reliability was confirmed by assessing the adequacy of mathematical models of the elastic wheel interaction with the deformable soil and the calculation results of the moving PVWHC permeability to the results of the field experiment. Experimental studies were carried out using wheeled vehicle of high permeability on loamy and sandy soils in the floodplain of the Oka river, on muddy and sandy soils of the ZIL landfill in Feodosia.

Conclusion

1. The basic principles of mathematical models formation that allow the estimated by evaluating the deformation properties of a soil surface area exploitation for developing wheeled vehicle. The method allows in stationary simulation by calculation to predict the expected numerical values of permeability. Calculations can be carried out for both unmanned and manned wheeled vehicle.

2. It is offered to estimate efficiency of PVWHC in the area of intended operation by means of the settlement reference route. An appropriate methodology has been developed. Route sections with a certain relative length are divided depending on the type of road-ground surface, the degree of roughness, ruggedness, soil parameters. Each site is represented by statistical characteristics. The criterion of efficiency of BT SVP is a probabilistic average speed at each site and probability speed of traffic on the route.

References

[1] Bekker M.G. Introduction to terrain-vehicle systems. — Ann Arbor: University of Michigan Press, 1969. — 520 p

[2] Volskaya, M.M. Zhileykin and A.Y. Zakharov. Mathematical model of rolling an elastic wheel over deformable support base. IASF-2017. IOP Conf. Series: Materials Science and Engineering 315 (2018) 012028 doi: 10.1088/1757-899X/315/1/012028.

[3] Rybansky, M. The impact of terrain on cross-country mobility geographic factors and their characteristics / M. Rybansky [et. al] // 18th International Conference of the International Society for TerrainVehicle Systems, ISTVS 2014. — Seoul, 2014.

[4] Shaheb M.R., Grift T., Godwin R., Dickin E., White D., Misiewicz P. Effect of tire inflation pressure on soil properties and yield in a corn - soybean rotation for three tillage systems in the Midwestern United States. — 2018. doi: 10.13031/aim.201801834.

[5] Shoop S., Kestler K., Haehnel R. Finite element modeling of tires on snow // Tire Science and Technology. 2006. Vol. 34, no. 1. P. 2–37.

[6] Botero J., Gobbi M., Mastinu G. A new mathematical model of the traction force in pneumatic tire snow chain systems // Associazione Italiana Per L'Analisi Delle Sollecitazioni / Dipartimento di Meccanica, Politecnico di Milano. 2005. 10p. www.aiasonline.org/AIAS2005/Articoli/art084.pdf.

[7] Belousov, B., Ksenevich, T.I., Vantsevich, V., Komissarov, D. 8×8 platform for studing terrain mobility and traction performance of unmanned articulated ground vehicles with steered wheels (2013) SAE Technical Papers, 9.

[8] Belousov, B.N., Shelomkov, S.A., Ksenevich, T.I., Kupreyanov, A.A. Experimental verification of a mathematical model of the wheel-supporting surface interaction during nonstationary rolling motion (2009) Journal of Machinery Manufacture and Reliability, 38 (5), pp. 501–505.

[9] Keller, A.V., Gorelov, V.A., Vdovin, D.S., Taranenko, P.A., Anchukov, V.V. Mathematical model of all-terrain truck (2015) Proceedings of the ECCOMAS Thematic Conference on Multibody Dynamics 2015, Multibody Dynamics 2015, pp. 1285–1296.

[10] Benes L., Hermanek P., Novak P. Tensile resistance of wheeled combine harvester. Engineering for Rural Development // MM Science Journal, Volume. — 2018, pp. 2481–2483; doi:10.17973/mmsj.2018_10_201848.

[11] Špokas L., Adamčuk V., Bulgakov V., Nozdrovický L. The experimental research of combine harvesters. Res. Agr. Eng., 62: (2016): p. 106–112.

[12] Kaneko, T. and Kageyama, I. 2003 A study on the braking stability of articulated heavy vehicles, JSAE Review, Vol. 24, pp.157–164

[13] Ageikin J.S. The flow of cars. — M.: Mashinostroenie, 1981. — 232 pp.N.S.

[14] Gorelov, V.A., Komissarov, A.I. Mathematical model of the straight-line rolling tire - Rigid terrain irregularities interaction (2016) Procedia Engineering, 150, pp. 1322–1328. DOI: 10.1016/j.proeng.2016.07.309

[15] Gorelov, V.A., Komissarov, A.I., Miroshnichenko, A.V. 8×8 wheeled vehicle modeling in a multibody dynamics simulation software (2015) Procedia Engineering, 129, pp. 300–307. DOI: 10.1016/j.proeng.2015.12.066.

[16] Keller, A.V., Gorelov, V.A., Anchukov, V.V. Modeling truck driveline dynamic loads at differential locking unit engagement (2015) Procedia Engineering, 129, pp. 280–287. DOI: 10.1016/j.proeng.2015.12.063

[17] Kotiev, G.O., Padalkin, B.V., Kartashov, A.B., Diakov, A.S. Designs and development of Russian scientific schools in the field of cross-country ground vehicles building (2017) ARPN Journal of Engineering and Applied Sciences, 12 (4), pp. 1064–1071.

[18] Kupreyanov, A.A., Morozov, M.V., Belousov, B.N., Ksenevich, T.I., Vantsevich, V.V. Experimental research of tire elastomer-surface tribological properties(2014) Proceedings of the ASME Design Engineering Technical Conference, 3.

[19] Klubnichkin, V.E., Klubnichkin, E.E., Kotiev, G.O., Beketov, S.A., Makarov, V.S. Interaction between elements of the track ground contacting area with the soil at curvilinear motion of the timber harvesting machine (2018) IOP Conference Series: Materials Science and Engineering, 386 (1), article № 012016.

[20] Wong, J.Y. Theory of Ground Vehicles / J.Y. Wong. — New York: Wiley IEEE, 2001. — 560 p.

[21] Sarach, E., Kotiev, G., Beketov, S. Methods for road microprofile statistical data transformation (2018) MATEC Web of Conferences, 224, article № 04009.

Selection of the combat vehicle individual wheel drive operation modes under required mobility criteria

G Kotiev[1], B Padalkin[1], A Miroshnichenko[2], A Stadukhin[1] and B Kositsyn[1,3]

[1]Bauman Moscow State Technical University
[2]OKB "Tekhnika", 6 Novaya Zarya st., 115191, Moscow, Russia

[3]E-mail: kositsyn_b@bmstu.ru

Abstract. With the development of the electric drive systems and vehicle on-board power sources the development of the combat vehicles with individually driven wheels is becoming a promising area. In the presented article, the authors illustrate an approach to selection of the requirements for power plants of such vehicles by an example of an 8x8 combat vehicle with full weight of 25 t. The approach is based on the use of the real-time driving simulator allowing the operator to drive the vehicle along typical routes and gather statistical data on the operation modes of the electric motors in the conditions close to the real operation conditions. The virtual driving for this purpose is performed along the routes which are generated from the known statistical data on the excitation inputs during the vehicle motion. The collected data can be used for the selection of the electric motor mechanical characteristics providing given mobility criteria of the vehicle and for the statement of the requirements for new electric motor specifications which would provide its most effective use on the vehicles being developed.

1. Introduction

The need to increase the mobility of the combat vehicles inevitably directs engineering teams towards experimenting with electric transmissions.

It is known that the operation of an electric motor is characterized by modes differing in duration, torque, rotor speed, currents, voltage, etc. In well-known applications (urban transport, conveyor drives, elevators, traction drives of the quarry machinery, etc.) the modes are standardized or converted to standardized ones. In the case of the traction drive of combat vehicles, the selection of the long and short-term modes of operation at the design stage in statistically specified driving conditions to ensure the required level of mobility is not possible without the use of mathematical modeling methods.

To solve this problem, a real time driving simulator was created. It using a mathematical model of the dynamics of flat curvilinear motion of a wheeled vehicle [1-11] which allows to simulate motion in "real time" on a computer controlled by a human (driver-operator).

2. Driving Simulator

In the mathematical model [5], the authors consider the wheeled vehicle as a solid body moving in the horizontal plane on a planar undeformed ground surface. The motion of the wheeled vehicle is the result of the translational motion of its center of mass and the rotational motion around the center of mass (figure 1). The principle of virtual displacements is used in the model to account for the redistribution of normal reactions from the action of the air resistance force, rolling resistance moments, acceleration of the center of mass and gravity. The wheels interact with the ground by means of an unilateral constraint.

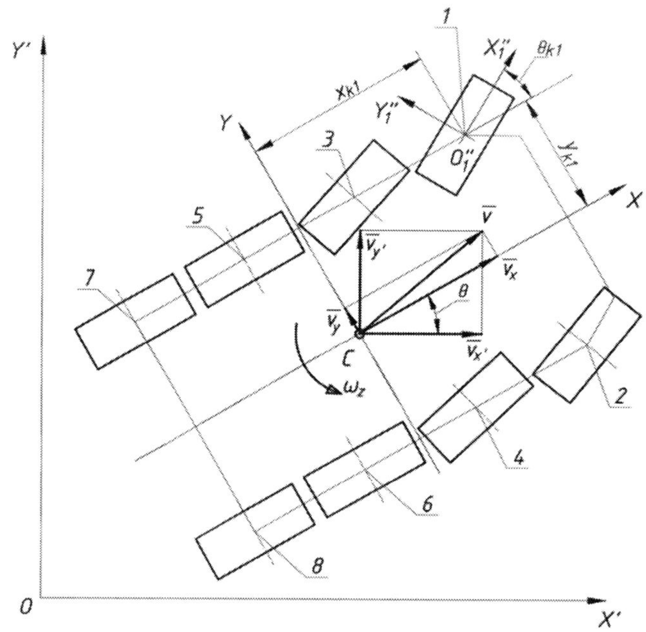

Figure 1. Planar motion of the wheeled vehicle.

In accordance with the presented diagram of the vehicle motion (figure 1), the system of equations describing the planar motion of the wheeled vehicle has form (1). This system allows calculation of the current acceleration of the center of mass from the values of the forces and moments acting on the vehicle.

$$\begin{cases} a_x = \dfrac{dv_x}{dt} - \omega_z v_y = \dfrac{1}{m}\left(\sum\limits_{i=1}^{n} R_{xi} - P_w\right) \\ a_y = \dfrac{dv_y}{dt} + \omega_z v_x = \dfrac{1}{m}\left(\sum\limits_{i=1}^{n} R_{yi}\right) \\ J_z \dfrac{d\omega_z}{dt} = \sum\limits_{i=1}^{n} M\left(R_{yi}\right) - \sum\limits_{i=1}^{n} M\left(R_{xi}\right) - \sum\limits_{i=1}^{n} M_{cpi} \quad \text{in}(\theta) + v_y \cos(\theta), \\ v_{x'} = \dfrac{dx'}{dt} = v_x \cos(\theta) - v_y \sin(\theta) \\ v_{y'} = \dfrac{dy'}{dt} = v_x \, \text{s} \\ \omega_z = \dfrac{d\theta}{dt} \end{cases} \tag{1}$$

where m is the vehicle mass; J_z is the vehicle mass moment of inertia relative to the z-axis passing through the center of mass; a_x, a_y are the projections of the vehicle center of mass acceleration onto the x — and y-axes; v_x, v_y are the projections of the vehicle center of mass velocity onto the x — and y-axes; $v_{x'}$, $v_{y'}$ are the projections of the acceleration of the vehicle center of mass onto the x' — and y'-axes; $\frac{dv_x}{dt}, \frac{dv_y}{dt}$ — projections of the relative derivative of the velocity vector of the vehicle center of mass onto the x — and y-axes; ω_z is the projection of the angular velocity vector of rotation of the vehicle onto the vertical z-axis; θ is the rotation angle, in the fixed coordinate system; x', y' are the coordinates of the vehicle mass center in the coordinate system $x'-y'$; R_{xi}, R_{yi} are the longitudinal and lateral components of the reaction acting at the contact patch of the i-th wheel; P_w is the projection of the air resistance force onto the x -axis; M_{cpi} is the moment of resistance to rotation of the contact patch of the i-th wheel around a vertical axis passing through the patch center; n is the number of the vehicle wheels.

The force of interaction between the wheels and the ground in the base plane is determined by an approach based on the concept of "friction ellipse", according to which the force of interaction with the support surface is directed opposite to the sliding velocity [12, 13, 14].

The connection of the steering angle with the steering angles of the driven wheels is carried out "according to Ackerman" [15], taking into account the gear ratio of the steering mechanism.

The following system of equations describes the connection of the drive motor torque of each wheel with the value of the control parameter and the rotor speed:

$$M_d\left(\omega_{ed}, h\right) = \begin{cases} M_{ed}^{max} \cdot h \text{ , если } \omega_{ed} < N_{ed}^{max} / M_{ed}^{max} \text{ и } \omega_{ed} \geq 0 \\ \dfrac{N_{ed}^{max} \cdot h}{\omega_{ed}} \text{, если } \omega_{ed} \geq N_{ed}^{max} / M_{ed}^{max} \text{ и } \omega_{ed} < \omega_{ed}^{max} \\ 0 \text{ , если } \omega_{ed} \geq \omega_{ed}^{max} \end{cases} \tag{2}$$

where M_{ed}^{max} is the maximum traction/braking torque generated by the motor; N_{ed}^{max} is the maximum traction/braking power of the motor; ω_{ed}^{max} is the limit rotation frequency of the rotor; ω_{ed} is the current rotation frequency of the rotor; h is the the value of the control parameter ($h \in [-1,1]$).

Figures 2 and 3 show the driving simulator layout and its graphical interface.

In order to get statistical data on the drive motor operation in conditions of long amd short duration a lot of numerical experiments with the use of the driving simulator must be carried out. These experiments consist of simulation runs of the wheeled vehicle model for stochastically generated roads in typical operation conditions performed by an operator or by a group of operators.

The roads are generated by means of the mathematical statistics methods with the use of the known stochastic characteristics (correlation functions) of the model inputs affecting the dynamical behavior of the wheeled vehicle. Based on the stochastic data on the road curvature k_r, maximum road friction μ_{smax}, coefficient ψ of the total motion resistance these parameters are distributed along the road by the non-canonical decomposition [16, 17, 18, 19]. Figure 4 shows an example of the driving route generated from the realization of the random function of the road curvature $k_r(s)$.

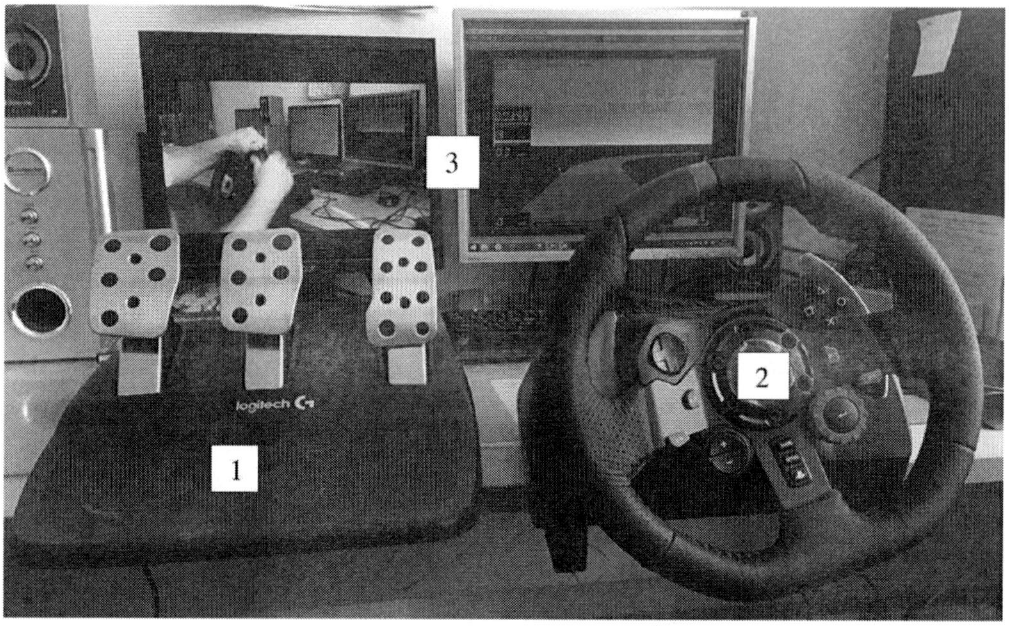

Figure 2. Driving simulator layout:
1,2 — controls; 3 — graphic information output station

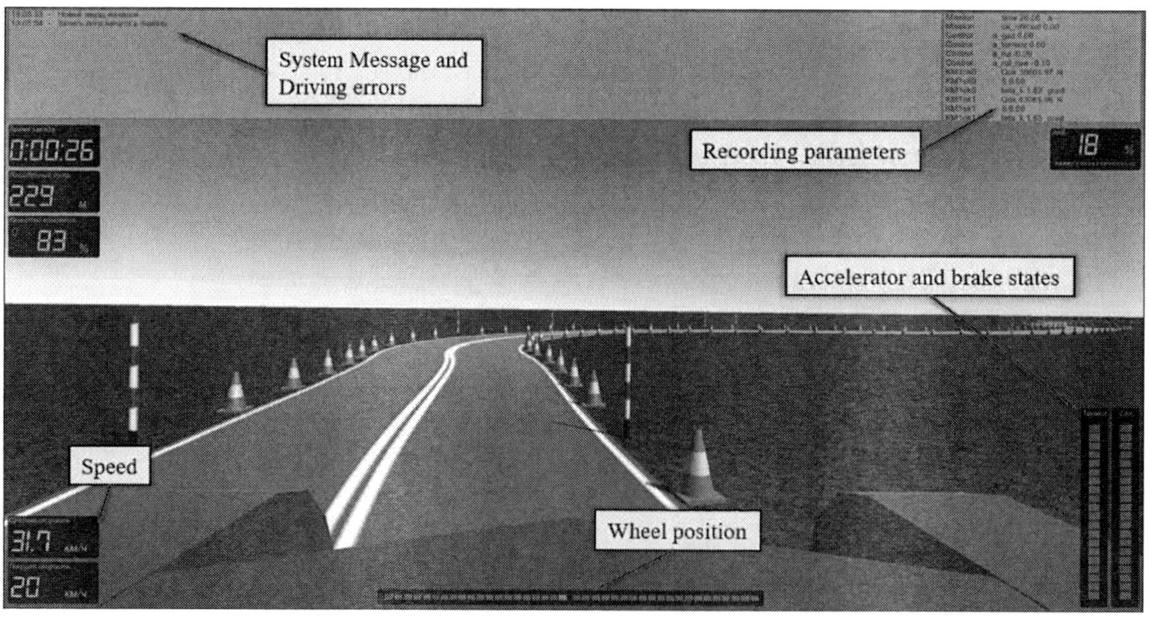

Figure 3. Graphical output of the driving simulator

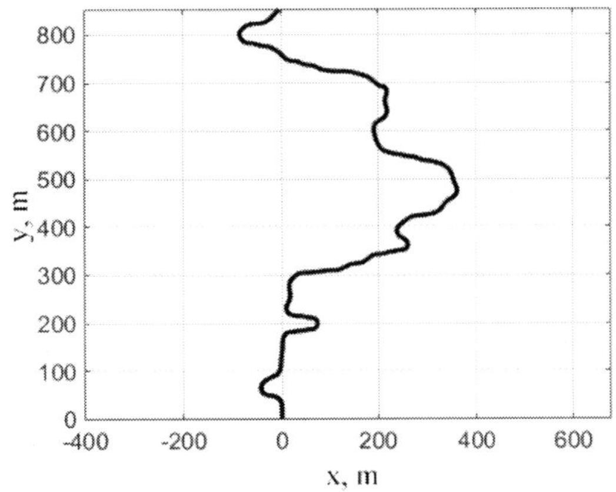

Figure 4. An example of the generated driving route.

3. Selection of the required mechanical characteristics of the electric motors

The driving simulator provides statistical data on the operation modes of the electric motors. This data can be used for selection of the mechanical characteristics of the electric motors of the combat vehicles with individual wheel drive.

A wheeled vehicle with gross mass of 25 t (figure 5) is used as an example. Technical specification of the vehicle is shown in table 1.

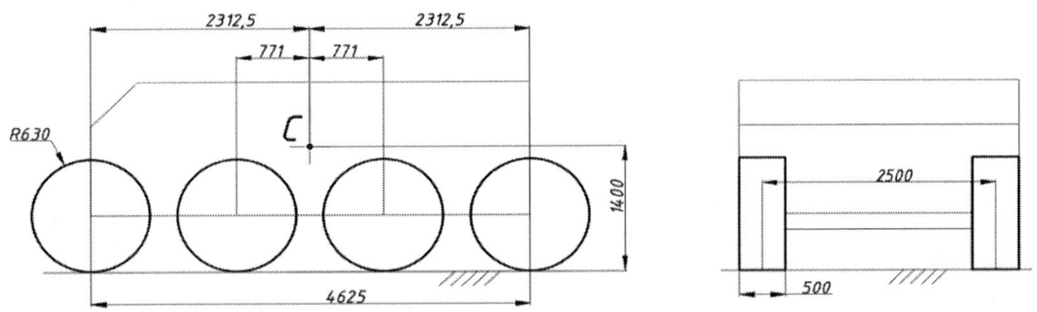

Figure 5. Vehicle dimensions.

Table 1. Technical specification of the 8x8 vehicle.

Parameter	Value
Gross mass, kg	25000
Wheelbase, mm	4625
Track width, mm	2500
Height of the center of mass, mm	1400
Inertia moment about the cenral vertical axis, kg m^2	57580
Loaded tire radius, mm	630
Steering system configuration	12-00
Wheel width, mm	500
Powertrain type	Individual electrically driven wheels
Total number of electric motors	8
Maximum speed, km/h	120
Air drag coefficient	0.7
Frontal projection area, m^2	5.3
Height of the center of pressure, mm	1700

IOP Conf. Series: Materials Science and Engineering **820** (2020) 012006 doi:10.1088/1757-899X/820/1/012006

The speed — torque characteristic of the electric drive providing the pre-defined mobilty is selected from the results of the simulation of the vehicle operation on typical roads.

For this purpose the maximum power of the electric drive is changed iteratively (for instance by the bisection method) until the power providing the required avarege speed of the vehicle is found.

As the initial approximation for simulation, the drive motor reference torque — speed curve (figure 6) providing the pre-defined tractive effort of the vehicle was analytically found, and the gear ratios of the transmission in driving and technological gear ranges were selected 11.88 and 18.81 respectively. The vehicle was assumed to be equiped with a two-range transmission. The purpose of the driving gear range is to provide the maximum speed and the purpose of the technological gear range is to provide the maximum tractive effort on the wheels [20].

Figure 7 shows the tractive effort to weight ratio — speed curve of the 8x8 vehicle (for the case of the drive motor with the torque — speed curve shown in figure 6). The mechanical efficiency the transmission was assumed to be 0.95.

Next, in order to calculate the maximum power and torque — speed curve of the drive motor providing the pre-defined mobility criterion during the vehicle operation the driving simulator is used.

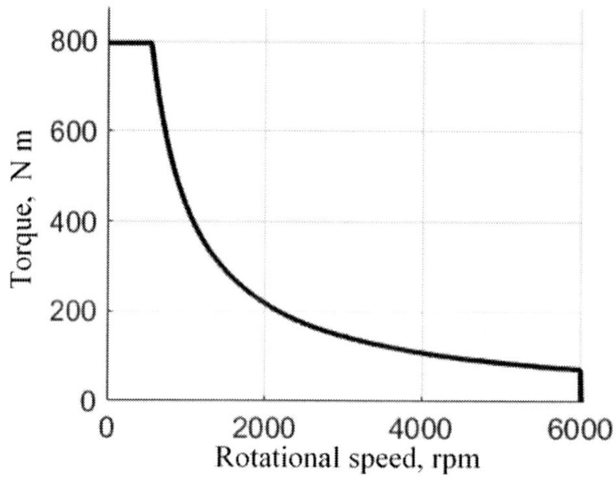

Figure 6. Drive motor theoretical torque — speed curve providing tractive effort of the 8x8 vehicle.

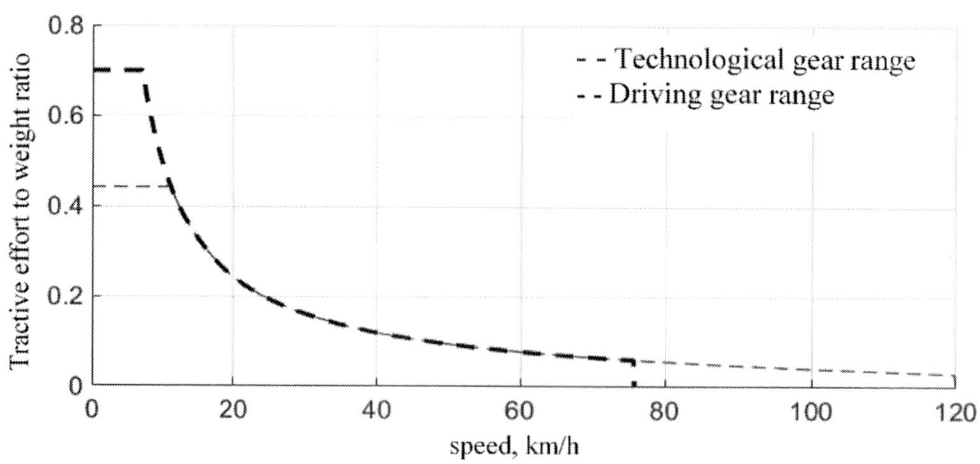

Figure 7. Tractive effort to weight ratio — speed curve of the 8x8 vehicle.

At the first stage, the electric motor maximum power variation range is selected based on the reference torque — speed curve of the motor: from the minimal value (initial power N_{emax}^{rqd}) providing given maximum speed to the maximum attainable value N_e^{MAX} defined by the limitations on the maximum torque of the motor and maximum rotational speed of its rotor (figure 8).

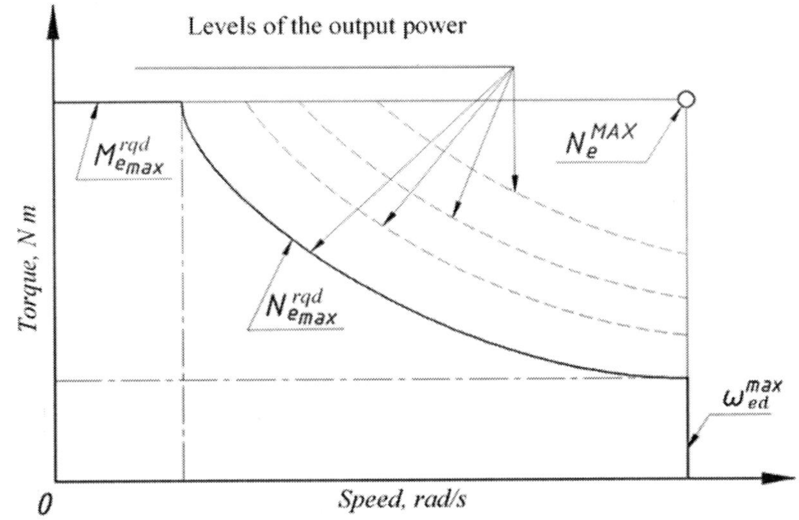

Figure 8. Electric motor maximum power variation range .

In the resulting range, several power levels are selected, for each of which the virtual tests are carried out in different road conditions.

To illustrate this, let us consider the results of the simulation runs for the 8x8 vehicle obtained on the macadam road. The dependence of the average speed of the vehicle on this road on the utilized portion of the maximum power of the electric drive is shown in figure 9.

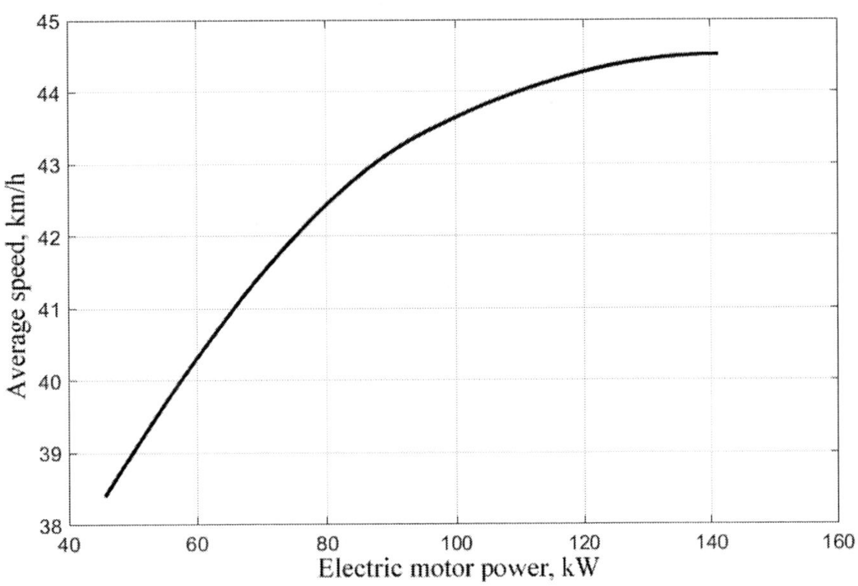

Figure 9. The dependence of the average speed of the vehicle on the utilized portion of the maximum power of the electric drive on the macadam road

The distribution of the electric drive operation modes, as well as the distribution functions of power, torque and angular velocity of the rotor of the electric motor in the traction mode, for one of the performed simulation runs are shown in figure 10, where:

a) distribution of the electric drive operation modes (the modes with relative duration lower than 5 % of the simulation time are discarded);

b) electric drive traction power distribution function;

c) distribution function of the electric drive torque;

d) distribution function of the angular velocity of the rotor.

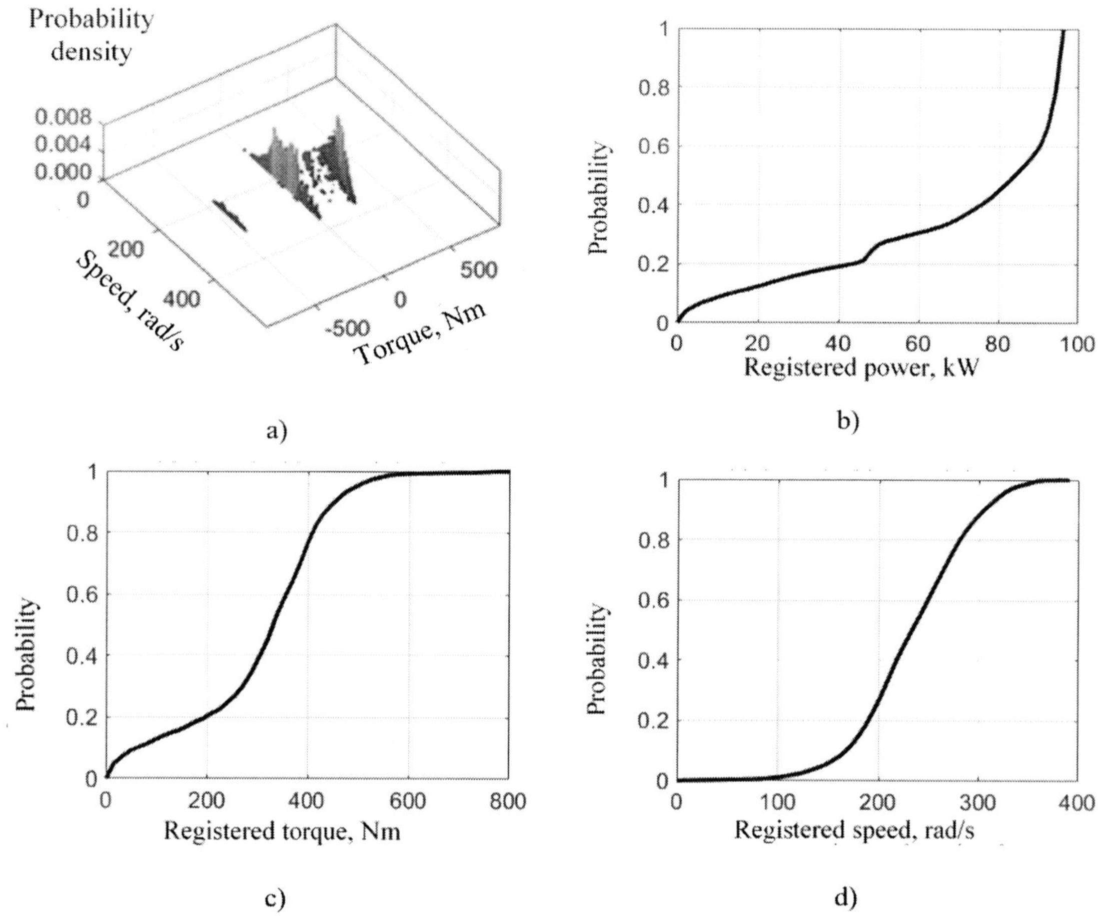

Figure 10. Simulation results for the 8x8 vehicle (electric motor rated power 95.6 kW; average speed 43.5 km/h).

These results show that with the growth of the maximum power of the electric drive, the probability that the vehicle would use the external speed — torque curve of the electric motor during its motion decreases. When the probability of using the external speed — torque curve of the electric motor is reduced to ~50-60%, the increase in the average speed becomes insignificant. Thus, when the 8x8 vehicle drives on a macadam road there is no reason to make the electric drive motor power higher than 96 kW (power-weight ratio is 30.6 kW/t).

The distribution functions of torque and rotor angular velocity for the selected maximum power level determine the torque — speed characteristic of the electrical motor and the distribution of the electric drive operation modes in the traction mode define the region of the motor high efficiency (figure 11).

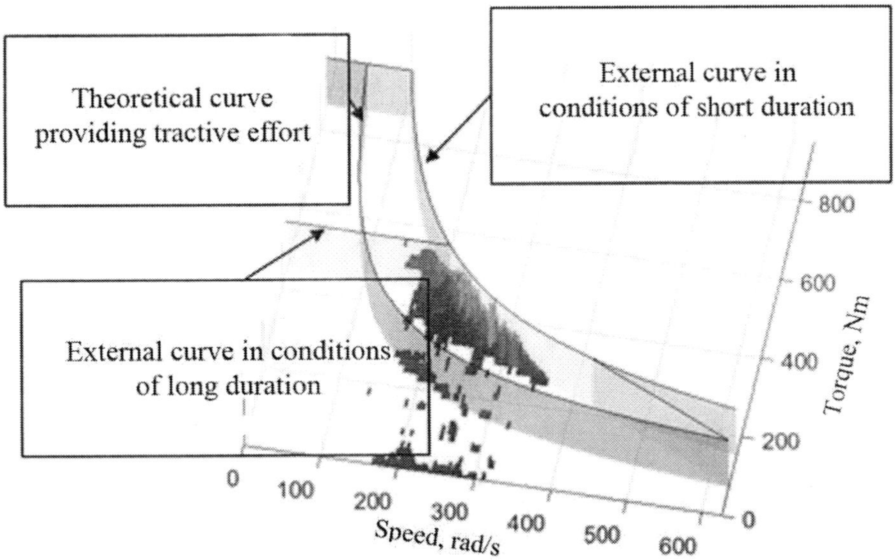

a)

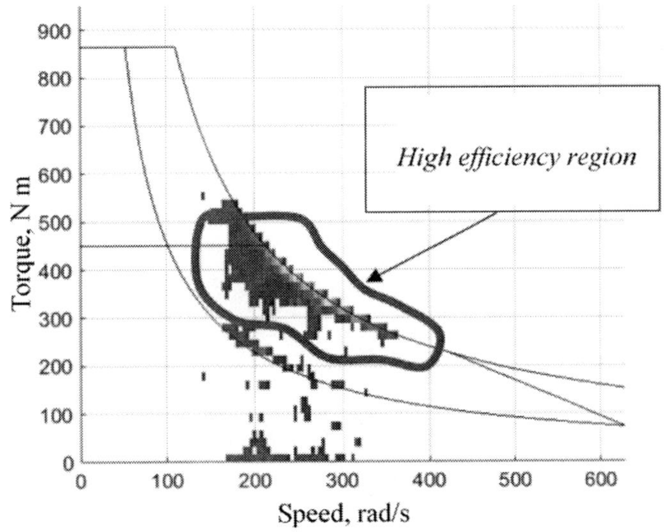

b)

Figure 11. a) required torque — speed curve of the drive motor; b) high efficiency region.

The max torque in operating conditions of long duration is lower than the torque having the probability of 90% (in our example this torque equals 448 Nm). The maximum power in both cases is limited by the calculated value of 95.6 kW. The maximum continous output of the drive motor during operation with the maximum rotor speed ω_{ed}^{max} corresponds to the calculated power N_{emax}^{rqd} = 45.6 kW providing the maximum speed of the the straight-line motion of the vehicle. The rotor speed after which the motor output power can be lowered corresponds to the maximum rotor speed having probability of 90% (in our example 304 rad/s).

The dependence of the average speed of the vehicle on different roads on the utilized portion of the maximum power of the electric drive is shown in figure 12.

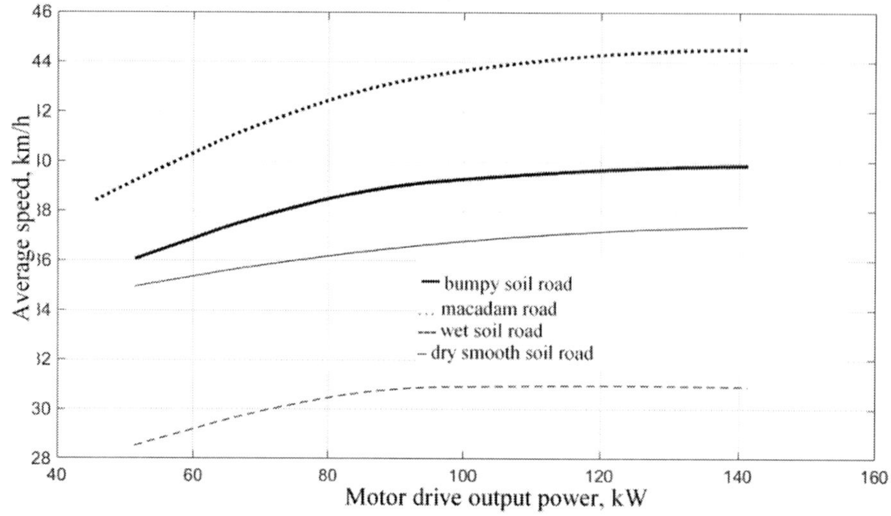

Figure 12. Dependancy of the average speed of the 8x8 vehicle on the utilized portion of the maximum output power of its drive motors for different types of terrain

This data show that when the 8x8 vehicle moves on these types of terrain there is no reason selecting the output power of its drive motors higher than 95.6 kW.

The simulation provided the required torque — speed curves of the drive motor for each type of the terrain. For convenience, the curves are schematically shown in figure 13 and in table 2.

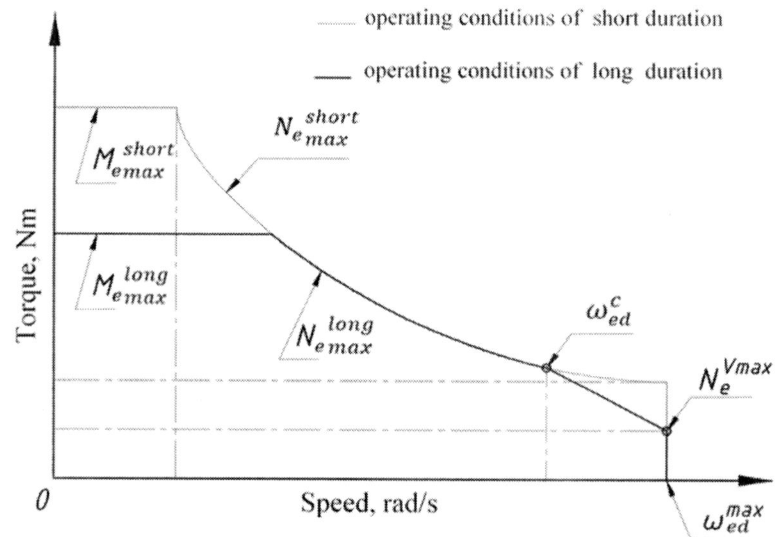

Figure 13. Schematic torque — speed curves of the drive motor for operating conditions of long and short duration

The figure uses the following nomenclature:

N_{emax}^{short}, N_{emax}^{long} is maximum output power of the drive motor in condidtions of short and long duration respectively;

M_{emax}^{short}, M_{emax}^{long} is maximum output torque of the drive motor in condidtions of short and long duration respectively; ω_{ed}^{max} is the maximum rotor speed of the drive motor;

ω_{ed}^{c} is the drive motor rotor speed after which the maximum output power of the motor can be lowered. N_e^{Vmax} is the motor power providing the maximum velocity on a rigid road.

Table 2. Parameters of the torque — speed curves of the 8x8 vehicle motors on different roads

Terrain index	$N_{emax}^{short} = N_{emax}^{long}$, kW	M_{emax}^{short}, Nm	M_{emax}^{long}, Nm	ω_{ed}^{max}, rad/s	ω_{ed}^{c}, rad/s	N_e^{Vmax}, kW
1	95.6	796.1	448	628	304	45.6
2	95.6		484		259	
3	95.6		484		285	
4	95.6		547		230	

1 — macadam road
2 — dry smooth soil road
3 — bumpy soil road
4 — wet soil road

Thus the resulting torque — speed curve in conditions of long and short duration for the 8x8 vehicle is the sum of the torque — speed curves for every road taken with weighting factor depending on the probability of operation on the road.

The resulting torque — speed curve of the drive motor is shown in figure 14 and in table 3 (power to weight ratio 30.6 kW/t).

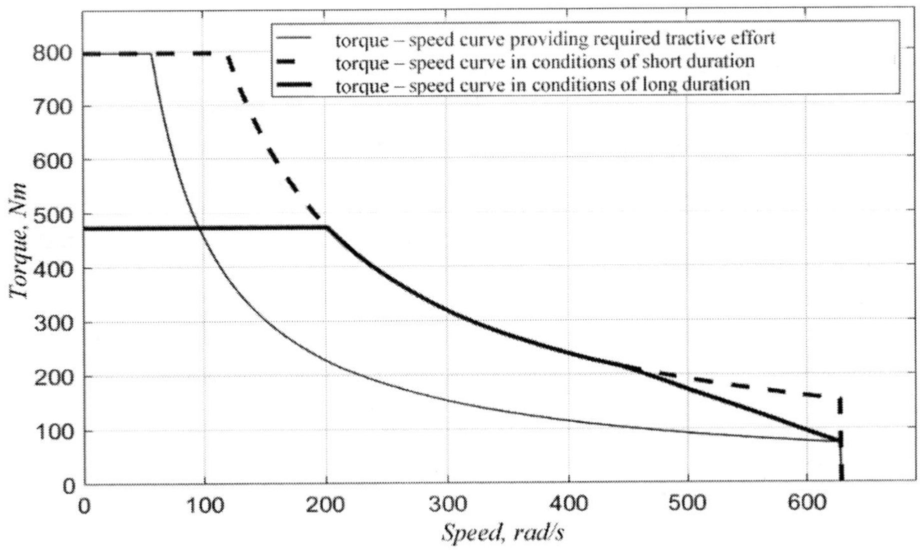

Figure 14. Resulting torque — speed of the 8x8 vehicle drive motor.

Table 3. Parameters of the resulting torque — speed of the 8x8 vehicle drive motor in conditions of long and short duration

$N_{ed\,max}^{short\Sigma} = N_{ed\,max}^{long\Sigma}$, kW	$M_{ed\,max}^{short\Sigma}$, Nm	$M_{ed\,max}^{long\Sigma}$, Nm	$\omega_{ed}^{max\Sigma}$, rad/s	$\omega_{ed}^{c\Sigma}$, rad/s	N_{ed}^{Vmax}, kW
95.6	796.1	473.8	628	278.5	45.6

4. Conclusions

The simulation has shown that the average speed of the vehicle saturates at a certain level of the maximum output power of the drive motor and then remains constant or rises insignificantly. This result allows selecting maximum reasonable output power of the drive motor for a vehicle designed to operate on a pre-defined road or a group of roads.

During the simulation, the 8x8 vehicle operated in the running-down mode during considerable time and almost never braked using its drive motor in the generator mode or with its mechanical brakes. This can be explained by the fact that the maximum speed was limited by the vehicle stability condition. Further rise of the average speed can be provided by more frequent braking, which implies a higher qualification of the driver, or by using a dynamic stability control system to prevent skidding, anti-lock braking system and traction control system.

The stochastic data obtained during the simulation was used for selection of the torque — speed curve of the electric motor operating in conditions of long and short duration. In conditions of long duration the maximum required output power of the electric motor can be selected lower in the region of the torques and rotor speeds the probability of which is not less than 90 %. Maximum output power of the electric motor can be selected such that the probability of the motor to be operatating on the external torque — speed curve when the vehicle moves on a pre-defined road or on a group of roads is 50–60 %.

The analysis has also shown that the electric motor operates in the region of average speeds at the maximum output power, which requires the motor to have a high efficiency in this region of the torque — speed curve.

References

[1] Kotiev G O, Gorelov V A, Miroshnichenko A V A control algorithm for individually steered wheels of transport vehicles *Herald of the BMSTU Engineering*, 2011, 5, 39–58.

[2] Kotiev G O, Butarovich D O, Kositsyn B B 2018 Energy efficient motion control of the electric bus on route *IOP Conference Series: Materials Science and Engineering*, 315(1) DOI: 10.1088/1757-899X/315/1/012014.

[3] Zhileykin M.M., Kotiev G.O., Nagatsev M.V. Comparative analysis of the operation efficiency of the continuous and relay control systems of a multi-axle wheeled vehicle suspension // IOP Conference Series: Materials Science and Engineering. 2018. V. 315, N.1. DOI: 10.1088/1757-899X/315/1/012030.

[4] Zhileykin M.M., Kotiev G.O., Nagatsev M.V. Synthesis of the adaptive continuous system for the multi-axle wheeled vehicle body oscillation damping // IOP Conference Series: Materials Science and Engineering. 2018. V. 315, N.1. DOI: 10.1088/1757-899X/315/1/012031.

[5] Gorelov V.A., Komissarov, A.I, Miroshnichenko, A.V. 8×8 wheeled vehicle modeling in a multibody dynamics simulation software // Procedia Engineering. 2015. V.129. P. 300–307. DOI: 10.1016/j.proeng.2015.12.066.

[6] Keller A.V., Gorelov V.A., Anchukov, V.V. Modeling truck driveline dynamic loads at differential locking unit engagement // Procedia Engineering. 2015. V.129. P. 280–287. DOI: 10.1016/j.proeng.2015.12.063.

[7] Skotnikov, G.I., Jileykin, M.M., Komissarov, A.I. Increasing the stability of the articulated lorry at braking by locking the fifth wheel coupling (2018) IOP Conference Series: Materials Science and Engineering, 315 (1), article № 012027.

[8] Chumakov, D.A., Chernyshov, K.V., Novikov, V.V., Diakov, A.S., Suchenina, A.S. Mathematical model of motor vehicle air suspension with a combined damping system (2019) Journal of Physics: Conference Series, 1177 (1), article № 012049.\

[9] Keller, A.V., Gorelov, V.A., Vdovin, D.S., Taranenko, P.A., Anchukov, V.V. Mathematical model of all-terrain truck (2015) Proceedings of the ECCOMAS Thematic Conference on Multibody Dynamics 2015, Multibody Dynamics 2015, pp. 1285–1296.

[10] Kotiev G. O., Miroshnichenko A. V., Stadukhin A. A. and Kositsyn B. B. Estimating operation modes for the individual wheel electric drive of the all-wheel drive vehicle with the use of the driving simulator (2019) IOP Conf. Series: Materials Science and Engineering 534, article № 012204 DOI:10.1088/1757-899X/534/1/012004
[11] Buzunov N. V., Kotiev G. O. and Padalkin B. V. Wheel vehicle dynamics real-time simulation for on-board stand-alone moving control system realization (2019) IOP Conf. Series: Materials Science and Engineering 534, article № 012021 DOI:10.1088/1757-899X/534/1/012021
[12] Ellis J.R. 1969 Vehicle dynamics.
[13] Yu.L. Rozhdestvenskiy, K.Yu. Mashkov, Generation of the reaction forces in a flexible wheel rolling on the rigid road *BMSTU proceedings*, 390 (1982) 56–64.
[14] Dik A B, 1988 Raschet statsionarnykh i nestatsionarnykh kharakteristik tormozyashchego kolesa pri dvizhenii s uvodom, Ph.D. thesis. Omsk.
[15] Raimpel I Vehicle chassis Steering system 1987 *Mashinostroenie.*
[16] Kotiev G O, Sarach E B 2010 Complex suspension systems of high-mobility articulated tracked vehicles. *BMSTU.*
[17] Chernetskiy V I 1968 Accuracy analysis of nonlinear control systems *Mashinostroenie*
[18] Rassheplyaev Yu S, Fandienko V N 1981 Synthesis of random process models for the study of automatic control systems.
[19] Shalygin A S, Palagin Yu I 1986 Applied methods of statistical modeling *Mashinostroenie.*
[20] Kotiev G O, Miroshnichenko A V, Stadukhin A A 2017 Selection of the speed ranges of multi-purpose tracked and wheeled electric vehicles *NAMI proceedings*, 3 (270) 51–55.

IOP Conf. Series: Materials Science and Engineering **820** (2020) 012007 doi:10.1088/1757-899X/820/1/012007

Stabilization of the loading characteristics in the durability analysis of thesuspension components of tracked vehicles

D Dubin[1,3], I Gadolina[2] and T Khusnetdinov[1]

[1]Bauman Moscow State Technical University
[2]Mechanical Engineering Research Institute of the Russian Academy of Sciences, Department of Structural Materials Science, 4 MalyKharitonyevskyPereulok, Moscow 101990, Russian Federation.

[3]E-mail: ddubin@bmstu.ru

Abstract. For a reasonable and reliable assessment of durability at the design or operation stage of the tracked vehicles, it is required information about the loading process, which is usually obtained by registering the load by various methods, such as strain measurement. For a reliable assessment of loading and, consequently, durability, the question arises about the representative length of the loading process recording. The paper proposes a method for determining the necessary and sufficient realization length using various methods of processing of the recorded random loading realization. The efficiency of the proposed method is demonstrated by the example of the evaluation of the random loading processes in torsion shafts of the suspension systems of a high-speed tracked vehicle. The justification of the necessary and sufficient length of the record of the random loading process is given.

1. Introduction

The loads acting on a suspension system of tracked vehicles are random and depend on the movement conditions, as well as road conditions [1]–[4]. To assess the durability of suspension system elements, it is necessary to have information about the operational loading level, which can be obtained both experimentally[5]–[7]and using methods and means of numerical modeling of working processes [8]–[11].

Information about the loading process is usually presented in the form of distribution of stress amplitudes with the corresponding number of repetitions during operation. The distribution of stress amplitudes is usually obtained by processing the loading processes recorded during service with performing of some of cycle counting methods or by considering random processes of the structural elements by using the spectral density [12].

While planning an experiment for the collection and processing of information about the loading, a number of methodological issues arise that have not yet received the final scientific recommendation, in particular, about the necessary and sufficient length of the registered loading process. It should be noted that the analysis of random processes for the purpose of subsequent evaluation of durability has its own specifics due to the fact that information about the amplitude spans and the number of repetitions of spans is the most important for the fatigue evaluation [13]. They are the characteristics of the random process that have the greatest impact on the accumulation of fatigue damage, both at the stage of origin and at the stage of development of fatigue cracks [14]. In work [15] the example of stabilization of the value of mathematical expectation (MO) with an increase in length of realization of the loading process is considered. It is recommended to consider the realization length at which the value of the MO of the loading process is stabilized and does not go beyond the previously accepted error sufficient for a reasonable assessment of realization durability.

The choice of the required length of realization depends on the considered structural elements and their operating conditions. The standard recommends the use of realizations with at least 1000 extremums length. This recommendation needs to be clarified. First, the number of selected extremums strongly depends on the number of partition classes of the random process by ordinate, i.e. is determined by the accuracy of registration of the random loading process. According to GOST the process is previously divided into classes, the width of the class Δ is by the formula:

$$\Delta = \frac{(\sigma_{max} - \sigma_{min})}{M} \quad (1)$$

where σ_{max} and σ_{min} — are the maximum and minimum of the random process in this realization; M — is the classes number, which is chosen in the interval.

Secondly, the experimental data are known [16], when the number of extremes in the realization can be reduced by 10 times with proper selection of the filtration band of the process. In this case, the value of durability, expressed in the number of loading blocks, changed only by 10%.

In the analysis of loading in General, it is necessary to consider a generalized process consisting of partial random processes recorded under different operating options, which will be further used to assess the durability in service [17]. The purpose of this work is to select the required length of realization to assess the durability of the chassis elements of high-speed tracked vehicles. To obtain a stable estimate of the loading it is necessary that at each of the partial loading modes its characteristics should be stable. At the same time, the load characteristics of each mode are evaluated. To justify the selection of the length the realization of random processes it was considered how the changes in the value of the calculated longevity $N_{□□}$with the increasing length of the realization varies. Measure responsible for the value of $N_{□□}$might be a function of accumulated cycles $\Omega = n(\sigma_a)$, or block loading, where n is the number of cycles of repetition of the stress amplitude σa. The function Ω is defined by processing a random process using one-parameter, or two-parameter schematization method [18]. The characteristic of the loading unit Ω is also its length l_b, the value of which, depending on the type of problem, can also be expressed in the numbers of loading cycles, hours, kilometers, etc.

If the realization by which the durability of the stationary process is evaluated is short, the evaluation of the Ω of the resulting load spectrum is unstable, which leads to significant deviations in the durability evaluation. The longer the realization, the less deviation. The evaluation of durability expressed in the number of blocks of loading calculated for a hypothetical infinite realization denoted as $N_\Sigma(\infty)$. We denote the evaluation of durability for the realization of the limited length L as $N_\Sigma(L)$. When increasing the length of the realization of a stationary random process, a limit exists:

$$\lim_{L\to\infty} \frac{N_\Sigma(L)}{N_\infty} = 1 \quad (2)$$

where $N\infty$ is a hypothetical infinite realization; $N_\Sigma(L)$ — realization of the limited length.

Since there is no physical possibility to measure the stresses on infinite realization, it should be limited by the available length of realization, providing the required accuracy ε of the assessment of N_Σ. The error of the evaluation of the durability of ε depending on the length of the realization is set in each case, taking into account the requirements for the accuracy of the evaluation of the durability. In this regard, you can determine the necessary and sufficient length of the realization L_{NS}. If there is a limit (2) for some random loading process, the necessary and sufficient realization length $L = L_{NS}$ can be determined from the condition:

IOP Conf. Series: Materials Science and Engineering **820** (2020) 012007 doi:10.1088/1757-899X/820/1/012007

$$\left|\frac{N_\Sigma^*\left(L_{NS}\right)}{N_\infty}-1\right|<\varepsilon\ (3)$$

where ε ∞ is a accuracy ε of the assessment of N_Σ.

2. Research Methodology

For the durability estimation of the torsion shaft of the suspension system for tracked vehicles, the stress recording was conducted (torsion in the high-cycle domain) while driving tracked vehicles on forest dirt roads (Fig. 1), which corresponds to the actual operating conditions.

Figure 1. The test tracks

Figure 2. Measuring equipment mounted on the hull and suspensions of the tracked vehicle

Determination of kinematic and loading parameters was carried out with the use of specialized measuring equipment installed on the machine hull (Fig. 2), by recording the angles of twist torsion shafts.

The system of reception and processing of data is implemented on the basis of a portable computer. As the sensors of the angle of twist of the torsion shafts resistors - potentiometers R-24N1-B10K firms HUEI SONG with a nominal impedance of 10 kω were used. To convert the analogue signal into digital code, an analogue-to-digital Converter (ADC) E14-140M, was used. The sensors were connected to the ADC via three channels using a differential connection scheme, which ensures optimal suppression of interference from the external environment induced on the connecting wires (Fig. 3).

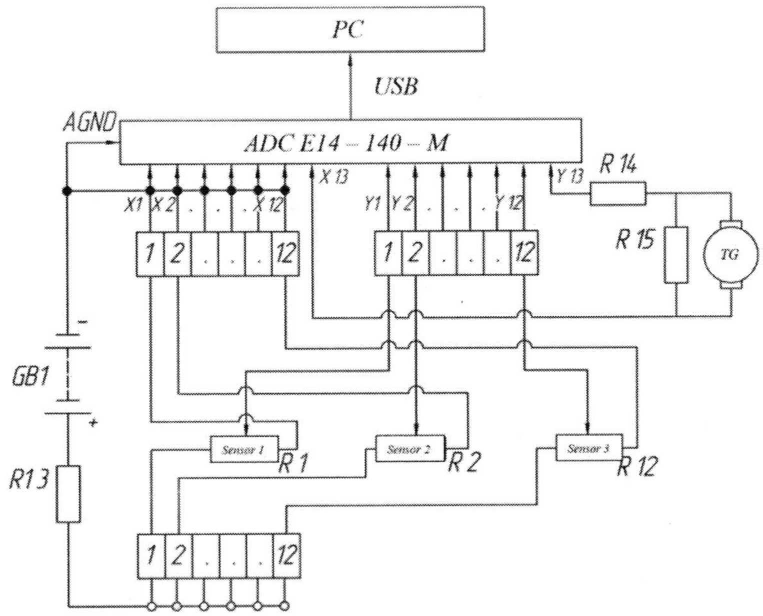

Figure 3. Electrical diagram of the measuring system

The total mileage of the vehicle on the test tracks of latching the input signal was of the order of 50 km. In the course of the experiment was formed by 18 independent samples. Recording of the input signal was carried out at a frequency of 20 Hz, which is sufficient to record low-speed suspensions at high speeds.

A set of programs in the R language was developed to process the obtained random processes [19]. In accordance with the software provides the following operationspreceding cycle-counting:

- Quantization - splitting the process by levels horizontally;
- Discretization- replacing a continuous process with a sequence of random ordinates;
- Selection of extremes;
- Calculation of the coefficient of irregularity χ.

Since the loading processes are random, the spectral analysis was often used for their study. As a rule, in terms of the spectral density of a random process, the researchers use the coefficient of irregularity of the random process χ [12, 20], which serves as a measure of the complexity of the random process and is defined as the ratio of the number of intersections of the average level to the number of extremes:

$$\chi = N_0 / N_e \quad (4)$$

where N_0— is the number of intersections of the middle level process, N_e— is the number of extremums.

IOP Conf. Series: Materials Science and Engineering **820** (2020) 012007 doi:10.1088/1757-899X/820/1/012007

After the preliminary stage, the random sequence is undergone by cycle-counting procedure.The results of the method of extrema, the range method scope and the rain-flow method[18]were compared. As it was shown earlier [21], the longevities, estimated by the method of extrema and the range method, can serve as top (for range method) and lower (for the extrema method) the evaluation of durability. The most preferable method is the rain-flow method. The fatigue curve equation in the following form was used to estimate the durability:

$$\sigma_a^m N = \sigma_{-1}^m N_G \quad (5)$$

where σ_a — is the amplitude of stress, MPa; N —is the limit number of cycles to failure for σ_a; σ_{-1} — is the endurance limit; m —is the fatigue exponent; $N_G = 2.106$.

In this study, the fatigue curve was recorded in relative terms:

$$N = N_G \left(1 / n_p\right)^m \quad (6)$$

where $n_p = \sigma_a / \sigma_{-1}$.

To justify the required realization length, it is necessary to assess the influence of stability of information on loading on the error of durabilityestimation. To assess the durability, it is necessary to know not only the fatigue resistance characteristics of the considered structural element (5) and their scattering, which are obtained experimentally or by calculation, but also the hypothesis of fatigue damage accumulation. It is recommended to use a simplified corrected hypothesis of fatigue damage summation, obtained as a result of the analysis of a large number of experimental data on the evaluation of durability at different distributions of stress amplitudes:

$$\lambda_b \sum_{i=1}^{k} \frac{n_i}{N_i} = 0.25 \quad (7)$$

where λ_b — is the estimated number of blocks to failure; n_i — is the number of cycles of repetition of the amplitude of the stress σ_{ai} in a block loading, N_i —is the limit number of cycles on the fatigue curve obtained with regular loading, corresponding to the amplitude σ_{ai} determined from the equation of the curve of fatigue.

The total number of cycles before failure:

$$N_\Sigma = \lambda_b L \quad (8)$$

Using formula (7) after substitution in (8) we obtain:

$$N_\Sigma = \frac{0.25 L}{\sum_{i=1}^{k} \frac{n_i}{N_i}} \quad (9)$$

In the formula (9) the numerator and denominator increase with increasing length of the realization$L$. The Denominator is increasing, since the longer the realization, the more cycles recorded n_i. With increasing L, the value of the calculated resource N_Σ (L) is stabilizing. Length of realization, in which the value of N_Σ is stabilizing will be necessary and sufficient for a given random process realization of the loading $L_{N\Sigma}$to compute durability with the acceptable accuracy.

3. Research results

As a result of the experimental study, time diagrams of changes in the angles of torsion shafts twist when moving a tracked vehicle over rough terrain are obtained. The conversion of the initial kinematic parameters to the values of shear stresses was carried out according to the following dependence

$$\tau = \frac{((\beta_{st} - \beta_0) + \gamma) \cdot G \cdot d_m}{2L_m} - \tau_{rs} \quad (10)$$

where γ — is the angle of twist of the torsion shaft; β_{st} — is the static angular position of the road arm;G — is the modulus of elasticity of the second kind of material of the torsion shaft; β_0 — isthe angular position of the road arm at zero twist position of the torsion; L_m —is length of torsion shaft; d_m —is the diameter of the torsion shaft; τ_{rs} — is the level of the residual stresses in the torsion shaft.

The value of the residual stresses in the surface layers of the torsion shaft approximately can be determined by the formula:

$$\tau_{rs} = \tau_y \frac{1 - \left(\frac{r_1}{R_T}\right)^3}{3}, \quad (11)$$

where τ_y — is the yield strength; r_1 — is the radius of the elastic layer of the material of the torsion shaft; R_T — is the radius of the torsion shaft.

The most stable mechanical characteristics of the material of the torsion shaft after the operation are obtained at the following ratio, characterizing the depth of plastic deformation:

$$\frac{r_1}{r} \approx 0{,}5 - 0{,}6. \quad (12)$$

The obtained implementations of the random loading process with the indication of the main statistical characteristics are shown in figure 4.

It should be noted that the value χ is very conditional and is strongly dependent on the number of quantization levels of the random process M, as can be seen in Fig.5, where the dependencies χ (M) are shown for the resulting implementations. The coefficient of irregularity of processes determined by the formula (4) on average for the considered processes was estimated as χ= 0.55. As the number of classes M increases, the value χ increases. It is shown that with a decrease in the number of partition classes, the value of χ increases by almost by 1.5 times compared to a more detailed division.

Fig. 6 shows the relationship estimated on the basis of the method of extrema and the method of ranges calculatedlife$N$ to N_d calculated based on the method of rain flow with variable fatigue exponent m. We see that the difference in calculated longevities is quite valuable, especially with increasing of m. This discrepancy is especially important for the processes with complex structure.

In Fig. 7 the dependences of the calculated life obtained for the realizations presented in Fig. 4 are shown. The fatigue curve was recorded in relative form (6). For realizations№2, №3 and №4 the condition (3) is performed at a necessary and sufficient length, that isL_{NS} = 1.5 min. To realization№1 stabilization is not observed even when the record length reach 2.5 min.

IOP Conf. Series: Materials Science and Engineering **820** (2020) 012007 doi:10.1088/1757-899X/820/1/012007

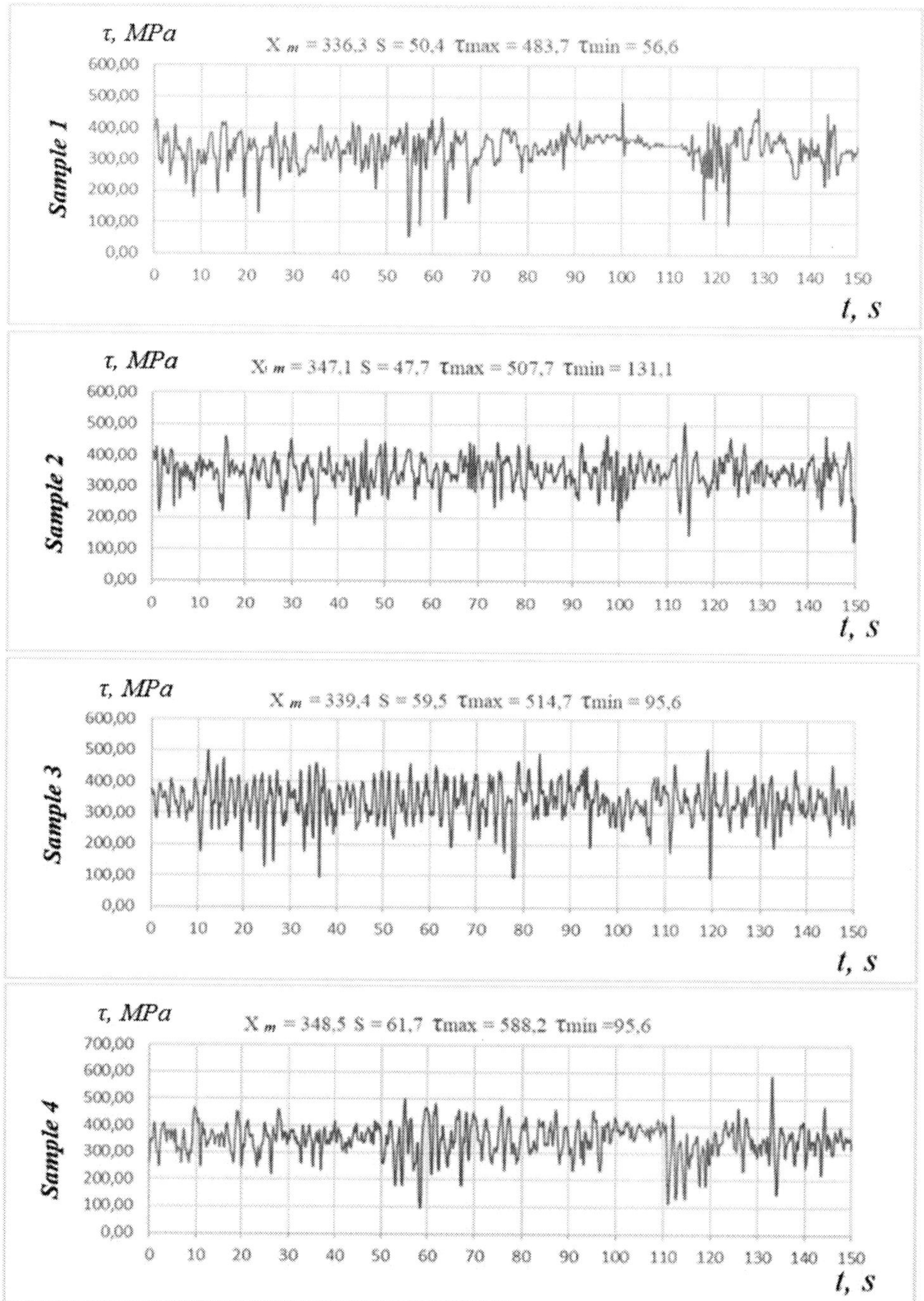

Figure 4. Implementation of random processes of shear stress changes in the torsion shaft of a tracked vehicle

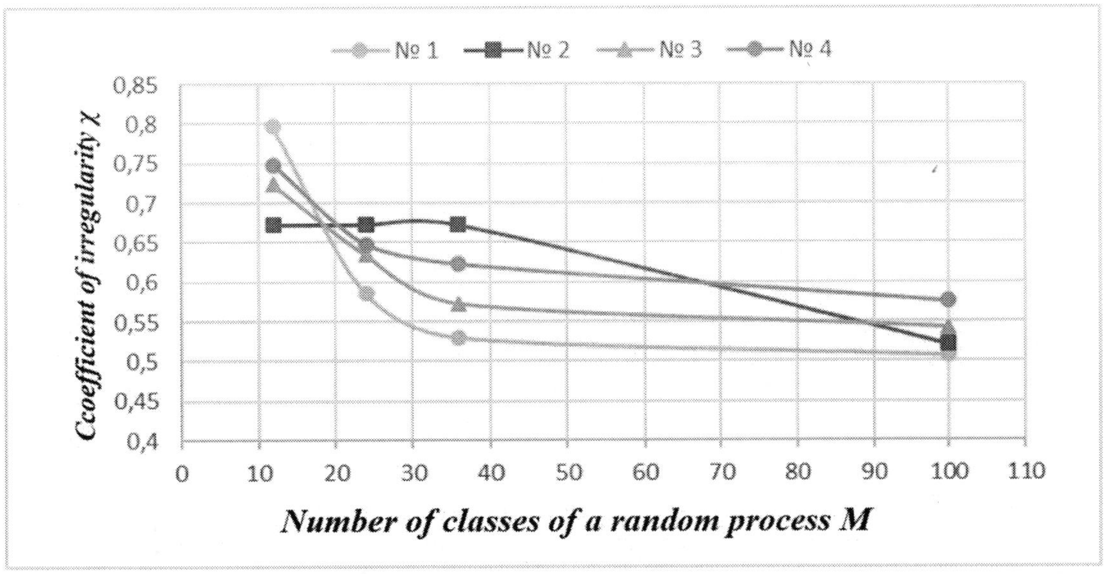

Figure 5. The dependence of the calculated coefficient of irregularity χ on the number of classes of a random process M.

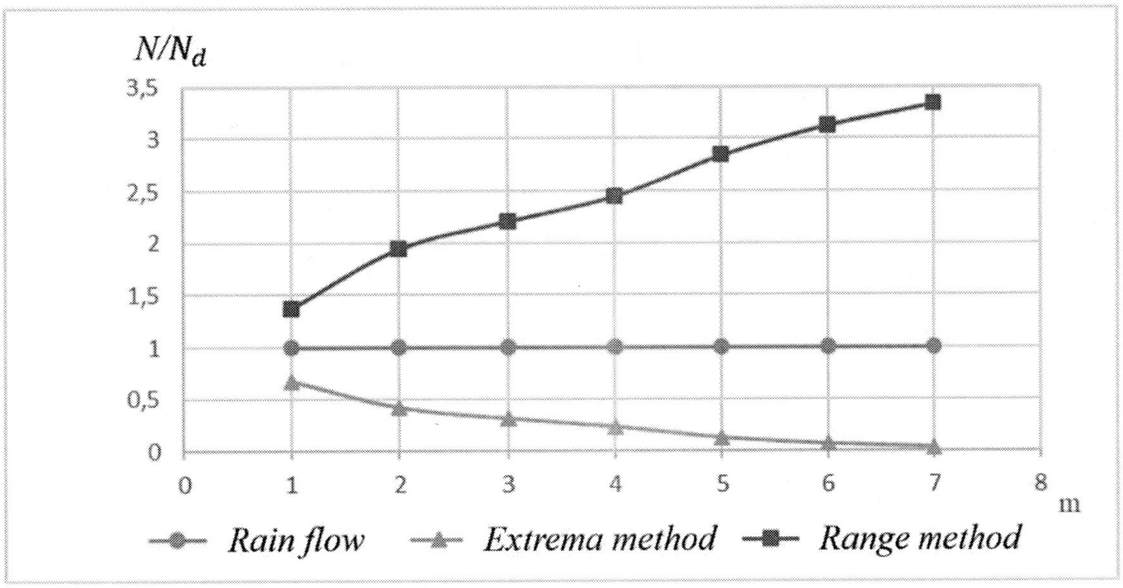

Figure 6. Comparison of relationships of life calculated by the method of extrema and the range method to one, calculated by the method of rain flow

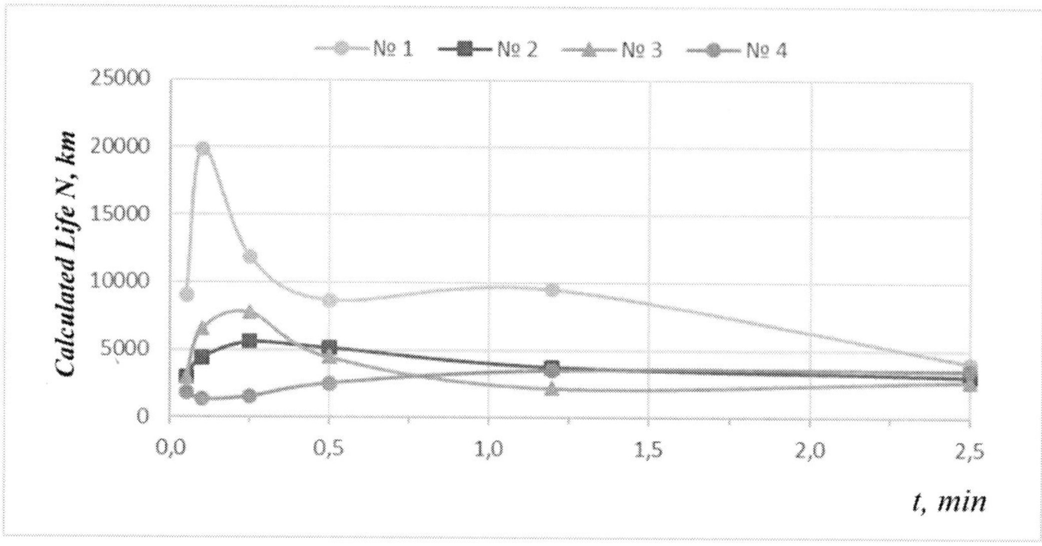

Figure 7. Dependence of calculated durability on realization length, m=6.

4. Conclusion

As a result, the technique on the determination of necessary and sufficient length of realization of a random process is offered. Using the developed technique, the problem of the determining the necessary and sufficient length of the realization of a random loading process is solved by the example of assessing the durability of the torsion shaft of an individual suspension system. The proposed method can be used in planning the investigation of the loading of high-speed tracked vehicles suspension system elements.

References

[1] Naumov V N,Mashkov K Y and Byakov K E 2019 Automatic determination of soil parameters by robotic vehicles Journ. of Phys.: Conf. Ser. 1177 Is.1

[2] Naumov V N,Mashkov K Y andByakov K E 2019 Autonomous tracked vehicles effectiveness estimation IOP Conf. Ser.: Mat. Sc. and Eng. 534 N.1

[3] Klubnichkin V E,Klubnichkin E E and Naumov V N 2018 Improvement of bogie tracks for forestry wheeled vehicles IOP Conf. Ser.: Mat. Sc. and Eng. 386, Is. 1

[4] Kar M K 1987 Prediction of track forces in skid-steering of military tracked vehicles Journ. of Terramech. 24 Is. 1 pp 75–84

[5] Alyabiev A F,Kotov A A,Klubnichkin V E,Klubnichkin E E and Bykovskiy M A 2019 Experimental determination of soil shear force by track model depending on track surface to soil friction coefficient Journ. of Phys.: Conf. Ser. 1177 Is.1

[6] Makarov V S,Markovnina A I,Papunin A V,Belyaev A M,Goncharov K O,Klubnichkin V E,Klubnichkin E E and Belyakov V V 2019 Study of uneven surfaces distribution on forestry roads Journ. of Phys.: Conf. Ser. 1177 Is.1

[7] Longoria R G,Brushaber R and Simms A 2019 An in-wheel sensor for monitoring tire-terrain interaction: Development and laboratory testing Journ. of Terramech. 82 pp 43–52

[8] Vdovin D, Chichekin I, &Ryakhovsky O 2019 Quad bike frame dynamic load evaluation using full vehicle simulation model. *In IOP Conference Series: Materials Science and Engineering*

[9] Vdovin D S, Chichekin I V, Levenkov Y Y, &Shabolin M L 2019 Automation of wheeled vehicles load bearing frames finite-element models loading procedure by using inertia relief method and vehicle multi-body dynamics model. *In IOP Conference Series: Materials Science and Engineering*

[10] Bodin A 1999 Development of a tracked vehicle to study the influence of vehicle parameters on tractive performance in soft terrain Journ. of Terramech. 36 Is. 3 pp 167–181

[11] Song Z,Hutangkabodee S,Zweiri Y H,Seneviratne L D and Althoefer K 2004 Identification of soil parameters for unmanned ground vehicles track-terrain interaction dynamics Proc. of the SICE Ann. Conf. FAI-9-3 pp 1651–1656

[12] MatiazMrsnik, Janko Slavic and MihaBoltezar. Frequency-domain method for vibration-fatigue-life estimation –Application to real data//International Journal of Fatigue. 2013. v.47. p.8–17. DOI: 10.1016/j.ijfatigue.201207.005

[13] Gadolina I.V, Zaynetdinov R. The estimation of the sufficient random loading realization length in the problem of machine parts longevity//The 9th IEEE International Conference on Dependable Systems, Services and Technologies, DESSERT'2018 24-27 May, 2018, Kyiv, Ukraine. pp 166–169.

[14] LebedinskiiS.G. Design Modeling of Propagation of the Fatigue Cracks in the Steel of Molded Parts of the Railway Structures. Journal of Machinery Manufac-ture and Reliability, 2018, Vol. 47, No. 1, pp. 62–66. ISSN 1052-6188, © Aller-ton Press.

[15] 6. BerezinI.Ya., RichterE.E., AbyzovA.A., KhripunovD.V.. Statistical mechanics and reliability of machines / Textbook for the course project. Chelyabinsk. Editionof SUSU. 2011.

[16] Fuchs H.O., D.V.Nelson, M.A.Burke, and T.L. Toomay. Shortcuts in Cumulative Damage Analysis. In book: Fatigue under Complex Loading. SAE, 1977, 207 p.

[17] Petrova I.M., Gadolina I.V. Generation of summarized loading spectra under varied loading modes of exploitation//XVI-th International Scientific and Technical Conference. Readings of memory of V.R. Kubachek. The process equipment for mining and oil and gas industry. Proceedings of the XVI international scientific and technical conference in the framework of the Ural mining decade. Edited by Yu.A. Lagunova. 2018.S. 318–321

[18] ASTM E 1049-85 (Reapproved 2005). Standard practices for cycle counting in fatigue analysis. ASTM International.

[19] R Core Team (2014). R: A language and environment for statistical computing. R Foundation for Statistical Computing, Vienna, Austria. URL http://www.R-project.org

[20] Erpalov, A.V Fatigue-based Classification of Loading Processes / A.V. Erpalov, L.A. Shefer //Procedia Engineering 2nd International Conference on Industrial Engineering (ICIE-2016) Volume 150, 2016. — 2016. — Vol. 150. — P.144–149.

[21] Gadolina I.V., Petrova I.M. Analysis of the influence of processing parameters of a random loading process on the accuracy of resource estimation // Problems of mechanical engineering and machine reliability. 1996. No. 3. P.62–67.

IOP Conf. Series: Materials Science and Engineering **820** (2020) 012008 doi:10.1088/1757-899X/820/1/012008

Method of torque distribution between the axles and the wheels of the rear axle to improve the manageability of two-axle all-wheel drive vehicles

M Zhileykin [1] and A Eranosyan[1,2]

[1]Bauman Moscow State Technical University

[2]E-mail: artem_bmstu@mail.ru

Abstract. Currently, manufacturers of modern cars are constantly increasing the level of control over the parameters of the movement of wheeled vehicles, achieving the maximum level of stability and control. The use of various all-wheel drive systems is becoming increasingly relevant and expedient. The distribution of traction force on all wheels allows you to use the entire weight of the car as a coupling, which has a positive effect on the dynamic qualities of the car, terrain, and allows you to implement control algorithms that improve the handling and directional stability of the car. The aim of the work is to substantiate the methods of controlling the curvilinear motion of two-axle vehicles with various transmission patterns that provide for improving their controllability. A method for improving the controllability of four-wheel drive two-axle vehicles with the transmission of torque to the front axle and the distribution of torque between the wheels of the rear axle is proposed. The effectiveness and efficiency of the proposed method are proved by the methods of simulation of motion of two-axle 4x4 vehicles with a connected front axle and redistribution of torque between the wheels of the rear axle.

Introduction

Investigating trends of development of modern automotive industry, it is possible to see that producers constantly increase the level of control over parameters of the movement of wheel vehicles, trying to obtain the maximum level of stability and road ability of cars.

Now actively are engaged in questions of creation of systems of active safety of the cars providing increase in stability and road ability abroad. Probes are conducted in two directions. First, the system of dynamic stabilization (SDS) which mode of functioning is based on change of the torques brought to driving wheels are created. And expedient use of various systems of all wheel drive is more and more relevant. Distribution of traction effort to all wheels allows to use all weight of the car as coupling that positively affects dynamic qualities of the car, flotation ability, and allows to realize the control algorithms improving road ability and road-holding ability of the car. Works are devoted to development of these methods, for example, [1–6].

Secondly, increase in stability and road ability is provided due to introduction of the automatic adjusting change of angle of rotation of steered wheels (podrulivaniye) [7–10]. Perhaps combined steering on the basis of the specified approaches [11].

Slow development of similar systems in domestic automotive industry is the reason of insufficient probes of distribution of power in the specific traffic conditions characteristic of passenger all-wheel drive cars.

The purpose of work is justification of methods of management of the curvilinear movement of the biaxial cars with various schemes of transmission providing increase in their road-holding and trajectory ability.

1. Distribution of torque between the axles

When developing mathematical model of the flat movement of the biaxial car we will accept the following assumptions:

a) the angles of withdrawal of the right and left wheels of each shaft are identical;

b) angles of rotation of steered wheels and the angles of withdrawal of wheels are small (do not exceed 10°);

c) coefficients of resistance to withdrawal of all wheels of shaft are identical.

The settlement scheme of forces operating on the car is submitted on figure 1.

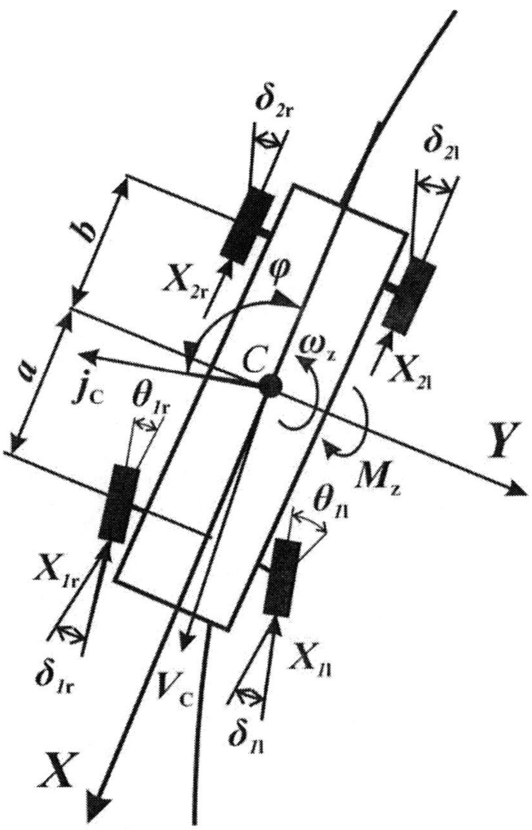

Figure 1. Scheme of forces acting on the vehicle.
C - center of масс (*CM*);
a, b - distances from the *CM* of the vehicle to the front and rear axis;
j_C - acceleration of *CM* of the vehicle;
X_{il}, X_{ir} - longitudinal forces acting on the left and right wheels from the road;
θ_{1l}, θ_{1r} - angles of rotation of the left and right wheels of the front axle;
δ_{il}, δ_{ir} - left and right wheel angles;
φ - the angle of deviation the acceleration of the *CM* from the longitudinal axis of the vehicle.

For the car with all leaders and front steered wheels to work the following differential equations (for a≈b case) are received (1)–(3):

$$\dot{\delta}_1 = \frac{V_C}{L}(\theta_{1m} + \delta_2 - \delta_1) - \frac{K_y}{V_C}\left(\frac{g}{G_C} + \frac{L^2}{4J_z}\right)\delta_1 - \frac{K_y}{V_C}\left(\frac{g}{G_C} - \frac{L^2}{4J_z}\right)\delta_2 + \frac{j_C \cos\varphi}{V_C}(\theta_{1m} - \delta_1) - \frac{\theta_{1m}}{V_C}\left(\frac{g}{G_C} + \frac{L^2}{4J_z}\right)h_f X_\Sigma; \quad (1)$$

$$\dot{\delta}_2 = \frac{V_C}{L}(\theta_{1m} + \delta_2 - \delta_1) - \frac{K_y}{V_C}\left(\frac{g}{G_C} - \frac{L^2}{4J_z}\right)\delta_1 - \frac{K_y}{V_C}\left(\frac{g}{G_C} + \frac{L^2}{4J_z}\right)\delta_2 - \frac{j_C\cos\varphi}{V_C}\delta_2$$
$$-\frac{\theta_{1m}}{V_C}\left(\frac{g}{G_C} - \frac{L^2}{4J_z}\right)h_f X_\Sigma; \tag{2}$$

$$\dot{\delta}_A = -\left(\frac{K_y}{V_C}\frac{L^2}{2J_z} + \frac{j_C\cos\varphi}{V_C}\right)\delta_A + \left(\frac{L^2}{2V_C J_z}h_f X_\Sigma - \frac{j_C\cos\varphi}{V_C}\right)\theta_{1m}, \tag{3}$$

here

L - wheelbase;

$\delta_A = \delta_2 - \delta_1$ - angle difference between the front and rear axle wheels;

$X_\Sigma = X_1 + X_2$ - full traction force;

$X_1 = X_{1l} + X_{1r}$ - front axle thrust force;

$X_2 = X_{2l} + X_{2r}$ - rear axle thrust force;

h_f – proportion of total thrust force transmitted to the front axle;

$\theta_{1m} = \frac{\theta_{1l} + \theta_{1r}}{2}$ - average steering angle;

$\delta_{im} = \frac{\delta_{il} + \delta_{ir}}{2}$ - average angle of removal of the wheels of the i-th axis.

If rear axle lagging (demolition of the front axle) see equation (4).

$$h_f = 1/(\lambda_0 + 1) \tag{4}$$

here

λ_0 - locking coefficient of central differential.

If front axle lagging (demolition of the rear axle) see equation (5).

$$h_f = \lambda_0/(\lambda_0 + 1) \tag{5}$$

At design of automatic control systems seek to provide their stability with some guarantee that changes of parameters of system in the course of its work could not result in instability of system. One of effective methods of probe of road-holding and trajectory ability of the car as multidimensional system is the frequency method which allows to carry out a priori assessment of the impact of design data of the car and parameters of its movement on its stability.

Let's consider transfer characteristic of $W_{12}(s)$ of the car with all leaders and front steered wheels (6).

$$W_{12}(s) = \frac{\delta_A(s)}{\Theta_{1cp}(s)};$$
$$s = i\omega, \tag{6}$$

here

s - Laplace operator; $i = \sqrt{-1}$ - imaginary unit; ω - circular frequency.

For our case see equation (7).

$$W_{12}(i\omega) = \frac{\left[\frac{L^2}{2V_C J_z}h_f X_\Sigma - \frac{j_C\cos\varphi}{V_C}\right]\left[\frac{K_y}{V_C}\frac{L^2}{2J_z} + \frac{j_C\cos\varphi}{V_C} - i\omega\right]}{\omega^2 + \left[\frac{K_y}{V_C}\frac{L^2}{2J_z} + \frac{j_C\cos\varphi}{V_C}\right]^2} \tag{7}$$

The amplitude of the transfer characteristic (8).

$$\left|W_{12}(\mathrm{i}\omega)\right| = \frac{\dfrac{L^2}{2V_C J_z} h_f X_\Sigma - \dfrac{j_C \cos\varphi}{V_C}}{\sqrt{\omega^2 + \left[\dfrac{K_y}{V_C}\dfrac{L^2}{2J_z} + \dfrac{j_C \cos\varphi}{V_C}\right]^2}} \tag{8}$$

Let's equate equation (8) to zero ($\left|W_{12}(\mathrm{i}\omega)\right| = 0$) that corresponds to neutral turn ability of the car, we will receive $W_{12}(\mathrm{i}\omega)$:

- if rear axle lagging (demolition of the front axle) see equation (9).

$$\frac{1}{\lambda_0 + 1} X_\Sigma = \frac{2 j_C J_z \cos\varphi}{L^2}, \text{consequently } h_f \sim \cos\varphi. \tag{9}$$

- if front axle lagging (demolition of the rear axle) see equation (10).

$$\frac{\lambda_0}{\lambda_0 + 1} X_\Sigma = \frac{2 j_C J_z \cos\varphi}{L^2}, \text{consequently } h_f \sim 1 - \cos\varphi. \tag{10}$$

From the analysis of expressions (9) and (10), it follows that for $\varphi = \pi, j_y = 0$ and $j_x \le 0$ to ensure stability, the rule $h_f = 0$ must be observed, i.e., when the vehicle slows down, all the traction force from the engine must be transmitted on the rear axle (when braking an engine, a rear-wheel drive car is most stable). Thus, the above dependencies can be considered as the algorithm of the system of the redistribution of torque between the driving axles of a biaxial vehicle.

2. Distribution of torque between the wheels of the rear axle

We again use the differential equations (1) and (2) for a car with all driving and front steered wheels (11)–(13):

$$\dot{\delta}_1 = \frac{V_C}{L}(\theta_{1m} + \delta_2 - \delta_1) - \frac{K_y}{V_C}\left(\frac{g}{G_C} + \frac{L^2}{4J_z}\right)\delta_1 - \frac{K_y}{V_C}\left(\frac{g}{G_C} - \frac{L^2}{4J_z}\right)\delta_2 + \frac{j_C \cos\varphi}{V_C}(\theta_{1m} - \delta_1)$$
$$+ \frac{L}{2J_z V_C} M_z - \frac{\theta_{1m}}{V_C}\left(\frac{g}{G_C} + \frac{L^2}{4J_z}\right) h_f X_\Sigma; \tag{11}$$

$$\dot{\delta}_2 = \frac{V_C}{L}(\theta_{1m} + \delta_2 - \delta_1) - \frac{K_y}{V_C}\left(\frac{g}{G_C} - \frac{L^2}{4J_z}\right)\delta_1 - \frac{K_y}{V_C}\left(\frac{g}{G_C} + \frac{L^2}{4J_z}\right)\delta_2 - \frac{j_C \cos\varphi}{V_C}\delta_2$$
$$- \frac{L}{2J_z V_C} M_z - \frac{\theta_{1m}}{V_C}\left(\frac{g}{G_C} - \frac{L^2}{4J_z}\right) h_f X_\Sigma; \tag{12}$$

$$\dot{\delta}_A = -\left(\frac{K_y}{V_C}\frac{L^2}{2J_z} + \frac{j_C \cos\varphi}{V_C} + \frac{1}{J_z \omega_z} M_z\right)\delta_A + \left(\frac{L^2}{2V_C J_z} h_f X_\Sigma - \frac{j_C \cos\varphi}{V_C} - \frac{1}{J_z \omega_z} M_z\right)\theta_{1m}, \tag{13}$$

here

$M_z = (1 - h_f) X_\Sigma \dfrac{\lambda - 1}{\lambda + 1} B$ - dynamic stabilization moment;

λ - wheel differential lock ratio;

B - car track width.

IOP Conf. Series: Materials Science and Engineering **820** (2020) 012008 doi:10.1088/1757-899X/820/1/012008

For the case under consideration, the transfer characteristic $W_{12}(s)$ (14).

$$W_{12}(\mathrm{i}\omega)=\frac{\left[\frac{L^2}{2V_C J_z}h_f X_\Sigma-\frac{j_C\cos\varphi}{V_C}-\frac{1}{J_z\omega_z}(1-h_f)X_\Sigma\frac{\lambda-1}{\lambda+1}B\right]}{\omega^2+\left[\frac{K_y}{V_C}\frac{L^2}{2J_z}+\frac{j_C\cos\varphi}{V_C}+\frac{1}{J_z\omega_z}(1-h_f)X_\Sigma\frac{\lambda-1}{\lambda+1}B\right]^2}\times\frac{\left[\frac{K_y}{V_C}\frac{L^2}{2J_z}+\frac{j_C\cos\varphi}{V_C}+\frac{1}{J_z\omega_z}(1-h_f)X_\Sigma\frac{\lambda-1}{\lambda+1}B-\mathrm{i}\omega\right]}{\omega^2+\left[\frac{K_y}{V_C}\frac{L^2}{2J_z}+\frac{j_C\cos\varphi}{V_C}+\frac{1}{J_z\omega_z}(1-h_f)X_\Sigma\frac{\lambda-1}{\lambda+1}B\right]^2} \quad (14)$$

Find the square of the modulus of expression (14) and equate it to zero ($\left|W_{12}(\mathrm{i}\omega)\right|^2=0$), which corresponds to the neutral understeer of the car. We get (15).

$$\left[\frac{L^2}{2V_C J_z}h_f X_\Sigma-\frac{j_C\cos\varphi}{V_C}-\frac{1}{J_z\omega_z}(1-h_f)X_\Sigma\frac{\lambda-1}{\lambda+1}B\right]^2\times\left[\left(\frac{K_y}{V_C}\frac{L^2}{2J_z}+\frac{j_C\cos\varphi}{V_C}+\frac{1}{J_z\omega_z}(1-h_f)X_\Sigma\frac{\lambda-1}{\lambda+1}B\right)^2+\omega^2\right]=0 \quad (15)$$

The second factor in brackets of expression (15) is always more than zero. Then for definition λ we will consider (16).

$$\frac{L^2}{2V_C J_z}h_f X_\Sigma-\frac{j_C\cos\varphi}{V_C}-\frac{1}{J_z\omega_z}(1-h_f)X_\Sigma\frac{\lambda-1}{\lambda+1}B=0, \quad (16)$$

whence, given the fact that $\omega_z=\frac{V_C}{L}\theta_{1m}$ and, taking $j_C=0$ (uniform motion), we obtain see equation (17).

$$\lambda=\frac{\frac{h_f L\theta_{1m}}{2(1-h_f)B}+1}{1-\frac{h_f L\theta_{1m}}{2(1-h_f)B}} \quad (17)$$

Then the expressions for the fraction of torque (of the total torque transmitted to the rear drive axle) for the running (h_{run}) and lagging (h_{lag}) wheels will have the form (18).

$$h_{\mathrm{run}}=\min\left(\frac{\lambda}{\lambda+1};\frac{1}{\lambda+1}\right);$$
$$h_{\mathrm{lag}}=\max\left(\frac{\lambda}{\lambda+1};\frac{1}{\lambda+1}\right) \quad (18)$$

3. Simulation results

The performance check of the torque distribution algorithm between the axles and the wheels of the rear axle was carried out on the example of a car with a total mass of 2400 kg with a connected front axle. The transmission scheme is presented in figure 2. Comparison is made with a rear-wheel drive car of the same mass.

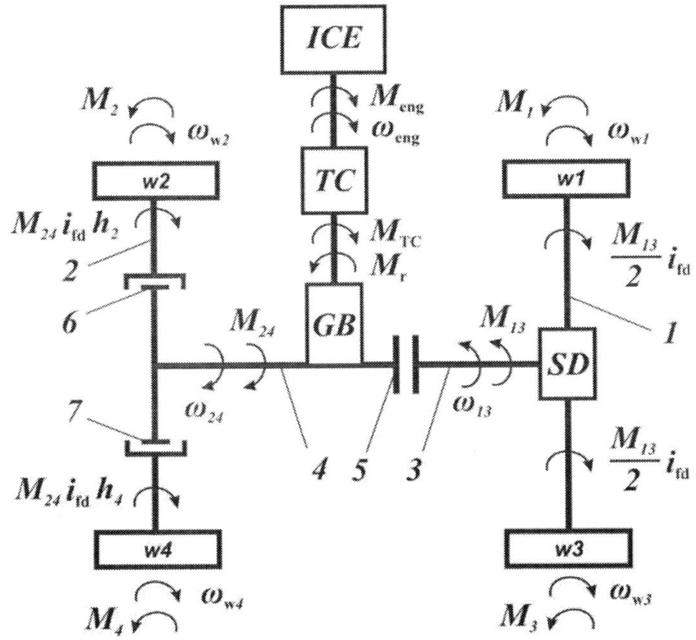

Figure 2. Car transmission scheme.
1, 2 - front and rear axles;
3, 4 - drive shafts from the gearbox to the front and rear axles;
5 - axle friction clutch;
6, 7 - friction clutches connecting the drive wheels of the rear axle;
ICE - internal combustion engine;
TC - torque converter;
GB - gearbox;
SD - symmetric cross wheel differential;
w1...w4 - wheels;
$M_1...M_4$ - rolling resistance on wheels;
$\omega_{w1}...\omega_{w4}$ - wheel rotation speeds;
M_{eng} - crankshaft torque;
ω_{eng} - crankshaft angular rotation speed;
M_{TC} - torque on the shaft of the turbine wheel torque converter;
M_{13}, M_{24} - torques on shafts 3 and 4 respectively;
ω_{13}, ω_{24} - angular speeds of rotation of shafts 3 and 4 respectively.

The transmission of a car with a connected front axle and redistribution of torque between the wheels of the rear axle is described by the following system of equations (19).

$$\begin{cases} J_w\dot{\omega}_{w1} = \dfrac{M_{13}i_{MG}}{2} - M_1 \\ J_w\dot{\omega}_{w2} = h_2M_{24}i_{MG} - M_2 \\ J_w\dot{\omega}_{w3} = \dfrac{M_{13}i_{MG}}{2} - M_3 \\ J_w\dot{\omega}_{w4} = h_4M_{24}i_{MG} - M_4 \\ J_{eng}\dot{\omega}_{eng} = i_{TC}M_{eng} - M_R \\ \dfrac{\dot{\omega}_{w1}+\dot{\omega}_{w3}}{2}i_{MG} = \dot{\omega}_{13} \\ M_{13} = h_f i_{GB}M_R \\ M_{24} = (1-h_f)i_{GB}M_R \end{cases}$$

here

J_w - the moment of inertia of the wheel relative to the axis of rotation;

i_{MG} - gear ratio main gear;

i_{TC} - gear ratio of the torque converter;

i_{GB} - gear ratio of gearbox;

h_2, h_4 - the proportion of torque transmitted to 2nd and 4th of the rear axle.

IOP Conf. Series: Materials Science and Engineering **820** (2020) 012008 doi:10.1088/1757-899X/820/1/012008

To simulate the joint operation of the engine and torque converter, a mathematical model was used, presented in [12–15].

Let us assume that the torque on the rear drive axle with permanent drive can vary in the range of 100 ... 50%, on the connected front axle - 0 ... 50% of the total torque coming from the output shaft of the gearbox.

Theoretical studies of the movement of the car using simulation mathematics were carried out. Features of the mathematical model of motion are considered in [16–21].

We study the motion on a support base "ice with snow" (with the coefficient of interaction of the mover with the support base with full slipping μ_{smax} = 0,35). Note that the term "support base" refers only to a solid non-deformable support surface. The front wheels of the car are steered. The car starts moving at a speed of V = kph. The movement was investigated at a constant level of impact on the accelerator pedal; the angle of rotation of the steering wheel changes from zero to a given value within 3 seconds and then remains unchanged.

The mode of entry into a turn and movement in a turn with a fixed radius are modeled.

The trajectory of the ice on the rear-wheel drive car when making a turn is shown in figure 3. The trajectory of the 4x4 car with a connected front axle and the redistribution of torque between the wheels of the rear axle in the same mode is shown in figure 4.

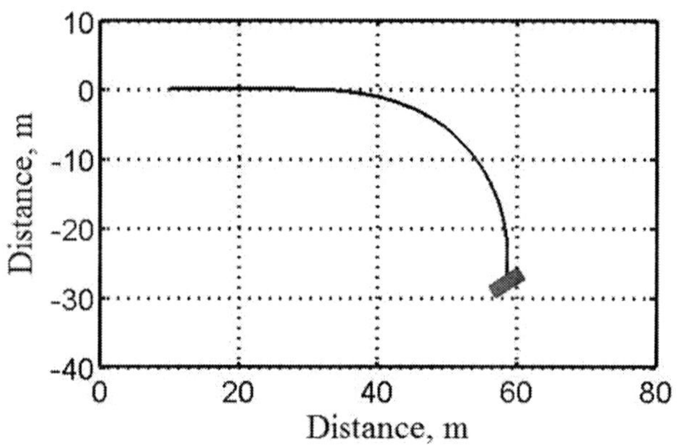

Figure 3. The trajectory of the rear wheel drive car in turn on the ice.

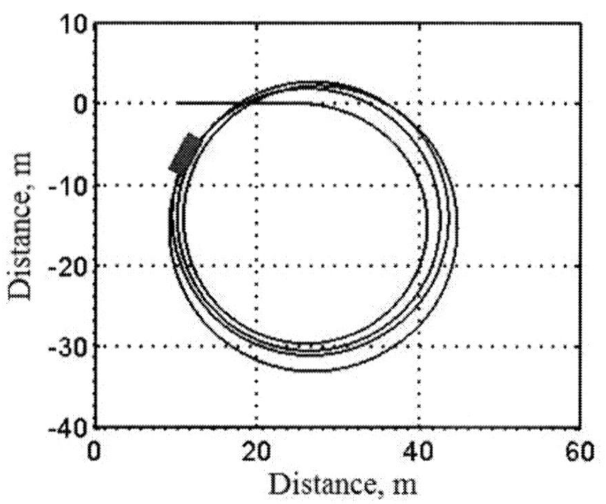

Figure 4. The trajectory of the full wheel drive car with torque vectoring in turn on the ice.

An analysis of the trajectories of the vehicle with the rear drive axle (figure 3) shows that when performing maneuvers on the support surface "ice with snow" the car loses control (skidding of the rear axle develops). When driving a car with a connected front axle and redistributing torque between the wheels of the rear axle under the same driving conditions (figure 4), the car remains manageable, small deviations from the trajectory can be compensated by steering.

Figure 5 a - c shows the time variation of the control parameters hf, h2 and h4 for a car with a connected front axle and a redistribution of torque between the wheels of the rear axle when driving in a corner.

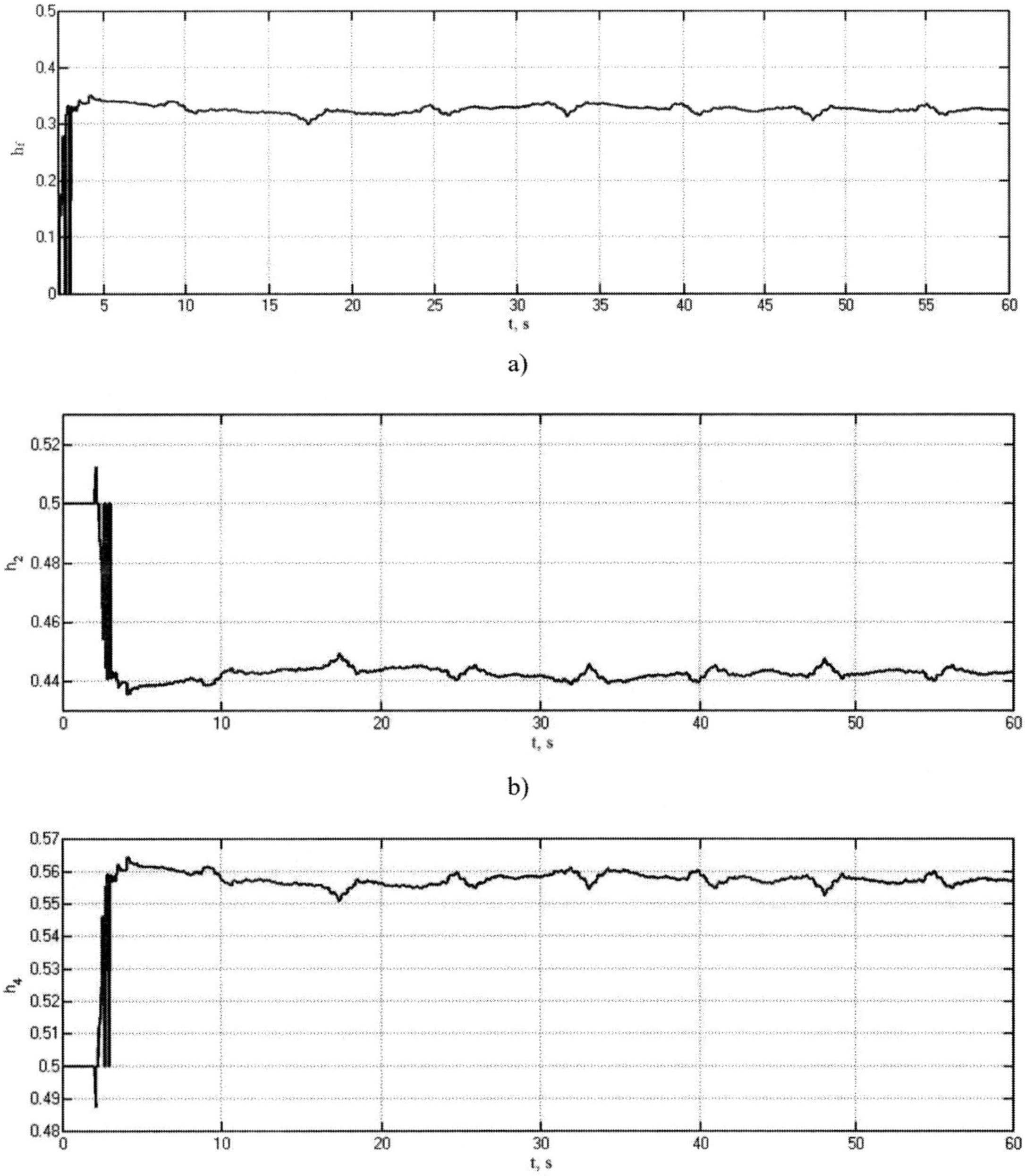

Figure 5. Parameters h_f (a), h_2 (b) and h_4 (c) for a car with a connected front axle and redistribution of torque between the wheels of the rear axle in a turn.

The nature of the changes in the parameters h_f, h_2 and h_4 (figure 5) indicates the absence of self-oscillations in the control loop, which indicates a high quality of control.

Conclusion

1. A method for increasing the stability of 4x4 4-axle vehicles with a connected front axle and torque redistribution between the wheels of the rear axle is proposed.
2. The methods and efficiency of the proposed method have been proved by simulation methods of the movement of 4x4 4-axle vehicles with a connected front axle and redistribution of torque between the wheels of the rear axle.

References

[1] Diez, D.R., Velenis, E., Tavernini, D., Smith, E.N., Siampis, E., Soltani, A. Front/Rear Axle Torque Vectoring Control for Electric Vehicles (2019) Journal of Dynamic Systems, Measurement and Control, Transactions of the ASME, 141.
[2] Wu, Z., Xie, B., Chi, R., Du, Y., Mao, E. Active modulation of torque distribution for dual-motor front- and rear-axle drive type electric vehicle based on slip ratio (2018) Nongye Gongcheng Xuebao/Transactions of the Chinese Society of Agricultural Engineering, 34 (15), pp. 66-76.
[3] Zhao, B., Xu, N., Chen, H., Guo, K., Huang, Y. Stability control of electric vehicles with in-wheel motors by considering tire slip energy (2019) Mechanical Systems and Signal Processing, 118, pp. 340-359.
[4] Xu, W., Chen, H., Zhao, H., Ren, B. Torque optimization control for electric vehicles with four in-wheel motors equipped with regenerative braking system (2019) Mechatronics, 57, pp. 95-108.
[5] Song, Y., Shu, H., Chen, X., Luo, S. Direct-yaw-moment control of four-wheel-drive electrical vehicle based on lateral tyre-road forces and sideslip angle observer (2019) IET Intelligent Transport Systems, 13 (2), pp. 356-366.
[6] Zhao, X., Guo, G. Braking torque distribution for hybrid electric vehicles based on nonlinear disturbance observer (2019) Proceedings of the Institution of Mechanical Engineers, Part D: Journal of Automobile Engineering, .
[7] Zhang, J., Zheng, H., Zhao, M. Analysis of Vehicle Steering Stability of Nonlinear Four Wheel Steering Based on Sliding Mode Control (2018) SAE Technical Papers, 2018-August (August).
[8] Ma, X., Wong, P.K., Zhao, J., Xie, Z. Cornering stability control for vehicles with active front steering system using T-S fuzzy based sliding mode control strategy (2019) Mechanical Systems and Signal Processing, 125, pp. 347-364.
[9] Hu, C., Qin, Y., Cao, H., Song, X., Jiang, K., Rath, J.J., Wei, C. Lane keeping of autonomous vehicles based on differential steering with adaptive multivariable super-twisting control (2019) Mechanical Systems and Signal Processing, 125, pp. 330-346.
[10] Chu, Z., Wu, C., Sepehri, N. Automated steering controller design for vehicle lane keeping combining linear active disturbance rejection control and quantitative feedback theory (2018) Proceedings of the Institution of Mechanical Engineers. Part I: Journal of Systems and Control Engineering, 232 (7), pp. 937-948.
[11] Mokhiamar O., Abe M. Active wheel steering and yaw moment control combination to maximize stability as well as vehicle responsiveness during quick lane change for active vehicle handling safety. // J Automobile Eng. 2002, V.216(2). – P.115–124.
[12] Cheng, G., Liu, W., Wang, H., Wang, L. Dynamic simulation of hydromechanical transmission system of tracklayer during gearshifting (2005) Zhongguo Jixie Gongcheng/China Mechanical Engineering, 16 (3), pp. 268-271.
[13] Ranogajec, V., Deur, J. An Automated Model-Order Reduction Method for Automatic Transmissions (2017) Journal of Dynamic Systems, Measurement and Control, Transactions of the ASME, 139 (7), № 071004.

[14] Lee, J.-H. Control oriented modeling of dynamic torque converter system (2017) 2017 7th International Conference on Modeling, Simulation, and Applied Optimization, ICMSAO 2017, № 7934860.
[15] Liu, C., Yan, Q., Wei, W. Three Dimensional Torque Converter Blade Modelling Based on Bezier Curves (2017) Jixie Gongcheng Xuebao/Journal of Mechanical Engineering, 53 (10), pp. 201-208.
[16] Belousov, B., Ksenevich, T.I., Vantsevich, V., Komissarov, D. 8×8 platform for studing terrain mobility and traction performance of unmanned articulated ground vehicles with steered wheels (2013) SAE Technical Papers, 9.
[17] Gorelov, V.A., Komissarov, A.I., Miroshnichenko, A.V. 8×8 wheeled vehicle modeling in a multibody dynamics simulation software (2015) Procedia Engineering, 129, pp. 300-307.
[18] Keller, A.V., Gorelov, V.A., Vdovin, D.S., Taranenko, P.A., Anchukov, V.V. Mathematical model of all-terrain truck (2015) Proceedings of the ECCOMAS Thematic Conference on Multibody Dynamics 2015, Multibody Dynamics 2015, pp. 1285-1296.
[19] Gorelov, V.A., Komissarov, A.I. Mathematical model of the straight-line rolling tire - Rigid terrain irregularities interaction (2016) Procedia Engineering, 150, pp. 1322-1328.
[20] Vol'skaya, N.S., Zhileykin, M.M., Zakharov, A.Y. Mathematical model of rolling an elastic wheel over deformable support base (2018) IOP Conference Series: Materials Science and Engineering, 315 (1), article № 012028.
[21] Wong, J.Y. Theory of Ground Vehicles / J.Y. Wong. – New York: Wiley IEEE, 2001. – 560 p.

IOP Conf. Series: Materials Science and Engineering **820** (2020) 012009 doi:10.1088/1757-899X/820/1/012009

Mathematical Modeling of a Linear Motion on a Deformable Bearing Surface of a Saddle-Type Road Train with Active Semi-Trailer Element

O Chudakov[1], V Gorelov[1] and B Padalkin[1,2]

[1]Bauman Moscow State Technical University

[2]E-mail: padalkin@bmstu.ru

Abstract. The paper introduces a mathematical modeling of linear motion on a deformable bearing surface (DBS) of a two-link road train consisting of four-axle truck with independent wheel drive and a three-axle semi-trailer. Predictive modeling based on application of pull-traction and energy efficiency obtained during the field research was used in the course of the model development to predict wheeled vehicle mover properties. Results of the theoretical research are given that relates to road train uphill motion with various wheel drives of the semi-trailer.

1. Introduction

The most useful technique to improve efficiency of transportation is to increase carrying capacity of vehicles, in particular by forming road trains. Under operational conditions typical of extractive and primary industries, the need often arises to transport cargo, heavy machinery, equipment along temporary roads or off-road (Fig. 1). At the same time, the most serious limitation for operation of 2-wheel road trains in the above industries is the low level of their cross-country performance [1–5].

a b

Fig. 1. Off-road operation of semi-trailer trucks

Development and construction of road trains with a sufficiently high level of traction dynamics is thus a pressing scientific and practical issue, which may be effectively solved by introducing an active drive of link element wheels [6–8].

By their transmission gear, driving systems of link element active axles are divided into mechanical, hydrostatic and electric transmission types. Subject to its characteristics, a mechanical transmission, traditional for commercial multiple-purpose vehicles (TVs), is hardly suitable for multi-wheel cars or all-wheel drive road trains [9]. When designing multi-wheel cars, it is assumed most rational to use electromechanical transmission [10–13]. It is expedient to design link elements with electric wheel drive for future road tractors, which already have a sufficient capacity generator installed, featuring an electromechanical transmission [14]; it is also possible to use standard algorithms for an electromechanical transmission control.

To solve the issue of increased energy efficiency of road trains under various operational conditions, designers should have at their disposal a tool for simulation experiments and methods, which allows predicting their performance subject to different designs. Nowadays, the most efficient estimation method with regard to technical solutions and vehicle mobility characteristics prediction at the design stage is a computer-aided mathematical simulation. Complexity of the task of modeling a cross-country vehicle movement is due to vehicle interaction with deforming grounds featuring high heterogeneity and complex structure, which significantly impacts a vehicle dynamics [15]. Efficiency of mathematical models of wheeled vehicles movement depends greatly on the soil characteristics applied in modeling [16–18]. In most of the known models, wheel propeller interaction with soil base is described by different empirical equations, which requires rather wide range of test data [19]. Such formalization does not always provide high reproducibility of calculation results or full-scale experiments [20].

2. Theoretical Background

In this regard, it seems appropriate to use integrated test data, obtained in the course of bench or field tests, for modeling of a wheel mover interaction with DBL, which was offered and demonstrated in [21]. Within such approach, linear rolling parameters of a wheel under different motion conditions are defined by rate of energy loss f_w(energy loss in rolling per a unit of distance covered by the wheel at a vertical unit load), free specific thrust φ (axial force applied to the rolling wheel at vertical unit load at its axis), as well as by skid coefficient:

$$S_b = \frac{\omega_k \cdot r_{ks} - V_{xk}}{\omega_k \cdot r_{ks}} = 1 - \frac{r_k}{r_{ks}},$$

where ω_k is the angular spin rate of a wheel; r_{ks} is the free roll wheel radius; r_k is the rolling radius ($r_k = V_{xk} / \omega_k$); V_{xk} is the center-of-mass velocity of a wheel in X-direction.

Experiments should reveal values in the energy balance equation, which states that energy delivered to a rolling wheel with even motion is consumed for axial force work and losses caused by interaction with DBL:

$$M_k \cdot \omega_k = P_x \cdot V_{xk} + f_w \cdot P_z \cdot V_{xk}, \tag{1}$$

where M_k is the torque moment delivered to the wheel axis; P_x is the axial force acting on the wheel axis; P_z is the vertical force acting on the wheel axis.

Subject to test, $P_x = R_x$, and $P_z = R_z$, where R_x is the force of interaction between a wheel and the bearing line; R_z is the normal response in the contact spot of the wheel and the bearing line; the equation for calculation of the specific energy losses will be as follows:

$$f_w = \frac{M_k}{P_z} \cdot \frac{\omega_k}{V_{xk}} - \frac{P_x}{P_z} = \frac{M_k}{P_z} \cdot \frac{\omega_k}{V_{xk}} - \varphi.$$

The values included in the right side of equation (1) are found in the course of experiment. We can thus obtain pull and energy $f_w = f(\varphi)$ and traction $\varphi = f(S_b)$ capacity values for a set of road conditions (Fig. 2).

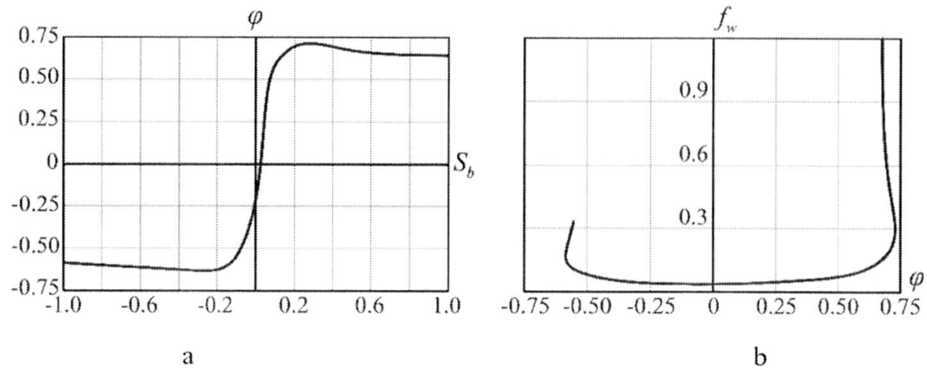

Fig. 2. Characteristic of interaction between a wheel and deforming soil: a – traction; b – pull and energy

Method of experimental characteristic application for simulation modeling of single wheel linear dynamics is described in detail in [22]. Design model of single wheel dynamics in the drive mode is given in Fig. 3 and may be described by a system of equations (2).

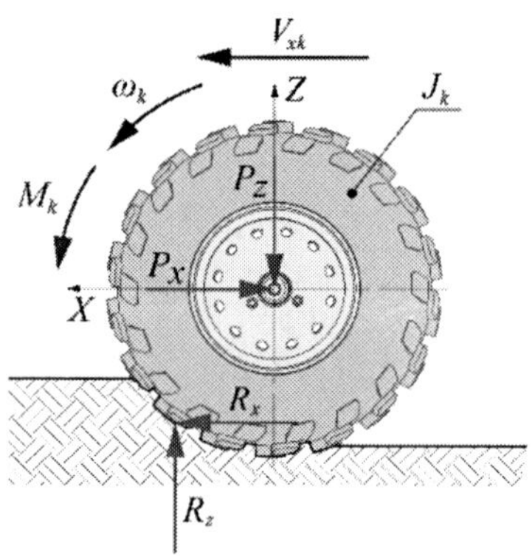

Fig. 3. Design model of wheel motion on a deformable bearing surface

$$\begin{cases} m \cdot \dot{V}_{xk} = \varphi \cdot P_z - P_x \ ; \\ J_k \cdot \dot{\omega}_k = M_k - (1 - S_b)(f_w + \varphi) \cdot P_z \cdot r_{ks} \ , \end{cases} \qquad (2)$$

where m is the weight per wheel; $\dot{V}_{xk}$ is the longitudinal acceleration of center of mass of the wheel; J_k is the spin-axis inertia; $\dot{\omega}_k$ is the angular acceleration of the wheel.

3. Mathematical Model of Road Train Motion

To solve a wide range of tasks related to dynamics of traction and cross-country performance, a mathematical model of two-link semi-trailer truck motion has been created. Development of the mathematical model included the following principal assumptions:

- Linear motion of a road train on an even deformable bearing surface is considered.
- The system is symmetric to the road train longitudinal axis.
- The wheel is considered as a hard rotating mass interacting with the bearing surface with the coefficient shown in Fig. 2.
- Soil deformation is not set through its immediate physical features, but as a specific energy loss during wheel rolling on a bearing surface.

Mathematical modeling of a semi-trailer truck movement is reviewed through the example of a 4-axle tractor truck with independent wheel drive and an active 3-axle semi-trailer. Design model of the road train link elements given in Fig. 4 is taken subject to the current task nature and is described with corresponding equations of motion (3).

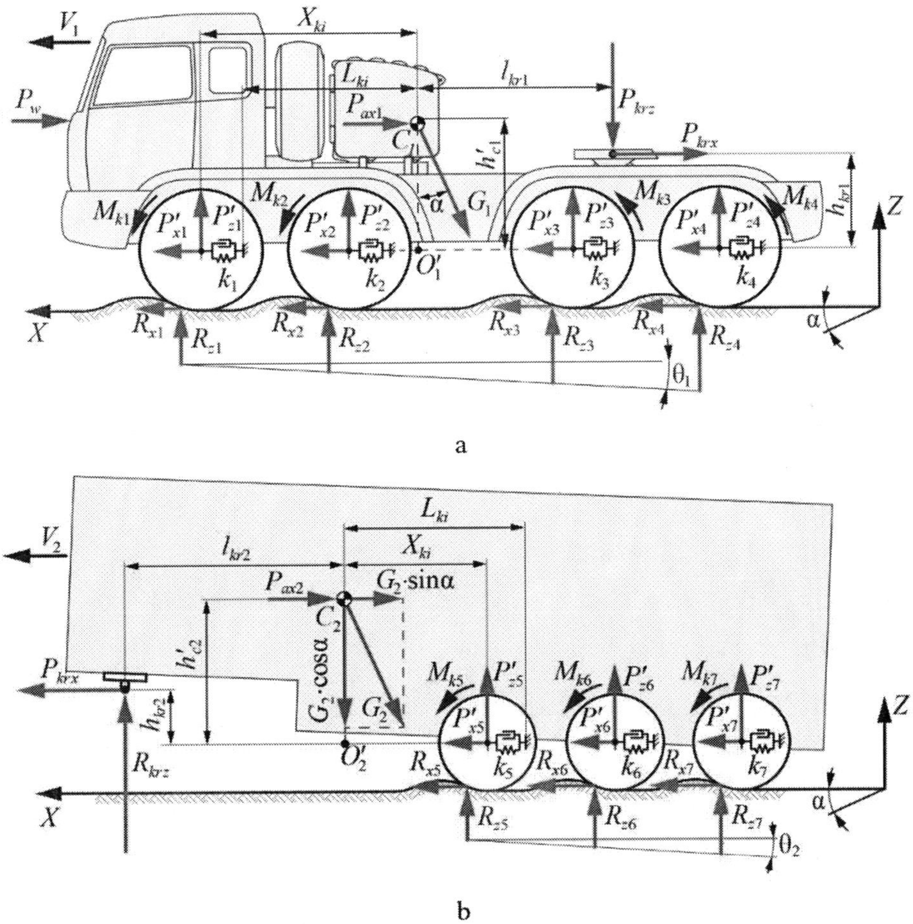

Fig. 4. Design model of tractor truck link elements motion on a deformable bearing surface: a – all-wheel drive tractor; b – active (all-wheel drive) semi-trailer

$$\begin{cases} m_1 \cdot \dot{V}_1 = 2 \cdot \sum_{i=1}^{4} P'_{xi} - G_1 \cdot \sin\alpha - P_w - P_{krx} \ ; \\ m_2 \cdot \dot{V}_2 = 2 \cdot \sum_{i=5}^{7} P'_{xi} - G_2 \cdot \sin\alpha + P_{krx} \ ; \\ J_{ki} \cdot \dot{\omega}_{ki} = M_{ki} - (1 - S_{bi}) \cdot (f_{wi} + \varphi_i) \cdot R_{zi} \cdot r_{ks} \ ; \\ m_{ki} \cdot \dot{V}_{xki} = \varphi_i \cdot R_{zi} - P'_{xi} - m_{ki} \cdot g \cdot \sin\alpha \ . \end{cases} \tag{3}$$

The following designations are introduced in the system of equations (3): i is the road train axle number; m_1, m_2 are mass values of a tractor and a semi-trailer, correspondingly; $\dot{V}_1$, $\dot{V}_2$ are the longitudinal acceleration of the center of mass of a tractor and a semi-trailer, correspondingly; P'_{xi} is the longitudinal component of force acting on a tractor body from the ith axle side; G_1, G_2 are weights of a tractor and a semi-trailer, correspondingly; P_{krx} is the longitudinal component of force in the bolster hitch; P_w is air resistance; M_{ki} is the torque of ith axle wheels; m_{ki} is the mass of ith axle wheels; g is the acceleration of gravity; α is the bearing surface slope.

Force acting on a tractor body from wheels in X-direction:

$$P'_{xi} = (X_{ki} - L_{ki}) \cdot k_p + (V_{xki} - V_1) \cdot B_p, \tag{4}$$

where k_p is the longitudinal stiffness ratio of suspension; B_p is the longitudinal damping ratio of suspension; X_{ki} is the distance between the center of mass and the ith axle; L_{ki} is the distance between the center of mass and suspension attachment points in X-direction; V_{xki} is the wheel speed in the ith axle direction; V_1 is the center-of-mass velocity of a tractor.

Normal response values from interaction of ith axle wheels and the bearing line are defined using the following equation:

$$R_{zi} = P'_{zi} + 2 \cdot m_{ki} \cdot g \cdot \cos\alpha\,, \tag{5}$$

where P'_{zi} is the normal force acting on a body from i^{th} axle wheels.

Normal response values are calculated using potential mismatch of trim angle of tractor and semi-trailer bodies (θ_1 and θ_2, correspondingly). P'_{zi} is calculated using the system of equations (6):

$$\begin{cases} 2 \cdot \sum_{i=1}^{4} P'_{zi} = \left(G_1 - 8 \cdot m_{k1} \cdot g\right) \cdot \cos\alpha + P_{krz}\ ; \\ 2 \cdot \sum_{i=1}^{4} P'_{zi} \cdot L_{ki} + \left(P_w + P_{ax1} + \left(G_1 - 8 \cdot m_{k1} \cdot g\right) \cdot \sin\alpha\right) \cdot h'_{c1} + P_{krz} \cdot l_{kr1} + P_{krx} \cdot \left(h_{kr1} - l_{kr1} \cdot \theta_1\right) = 0\ ; \\ \left(P'_{z2} - P'_{z1}\right)/k_1 = \left(L_{k1} - L_{k2}\right) \cdot \theta_1\ ; \\ \left(P'_{z3} - P'_{z1}\right)/k_1 = \left(L_{k1} - L_{k3}\right) \cdot \theta_1\ ; \\ \left(P'_{z4} - P'_{z1}\right)/k_1 = \left(L_{k1} - L_{k4}\right) \cdot \theta_1\ , \end{cases} \tag{6}$$

where h'_{c1} is the Z-distance from tractor wheel axle to the tractor center of gravity; P_{ax1} is the tractor inertia ($P_{ax1} = m_1 \cdot \dot{V}_1 \cdot h_{c1}$); l_{kr1} is the distance between the tractor center of mass and the bolster hitch; h_{kr1} is the distance between the tractor wheel axle and its fifth-wheel coupling in Z-direction.

Normal response acting on semi-trailer wheels is calculated in a similar way. Longitudinal component of the tractor and semi-trailer coupling is introduced through the force at the hook, which is calculated as follows:

$$P_{krx} = \left(X_{C1} - X_{C2}\right) \cdot k_0 + \left(V_1 - V_2\right) \cdot B_0, \tag{7}$$

where k_0 is the longitudinal stiffness ratio of the bolster hitch; B_0 is the coupler damper longitudinal resistance ration; X_{C1} and X_{C2} are a coordinate on the X-axis of tractor and semi-trailer centers of mass, correspondingly.

4. Calculation Results

The developed mathematical model is used to estimate the effect from link element wheel drive at road train uphill movement on a deformable bearing surface. The routine of the simulation experiment comprises as follows: a road train starts moving after a preliminary stop on the uphill. The experiment is considered passed if the road train has managed to build on steady speed of ascent motion.

Two-link semi-trailer truck with a total weight of 115,000 kg (max. weight of the semi-trailer is 90,000 kg) was selected as the object of study. A road train with an inactive semi-trailer was chosen to be the basic option. Alternative options differ due to their semi-trailer wheels drive. Transmission of the tractor and the semi-trailer is electromechanical, made by "in-wheel electric motor" layout. In the basic option, each wheel drive is implemented using a traction electric motor with the capacity of 60 kW. The total capacity of all traction electric motors of the tractor is thus 480 kW. The capacity of active road train traction electric motors has been selected so as to have equal specific capacity of all the options. Partial parameters of traction electric motors were obtained by additional multiplication by the capacity factor.

Three options of capacity split between the road train link elements have been reviewed: in the first option, total capacity goes to the tractor wheels (basic option); in the second and third options, part of the total capacity (25 and 50%, correspondingly) is distributed between the semi-trailer wheels. Simulation experiments were conducted for various weights of a semi-trailer. Angle of climb was defined subject to maximum possible weight of the road train under the given conditions. Moreover, maximum speed of the full weight road train was defined for climbing at different angles of the surface slope. Calculation results are given in Fig. 5 and 6.

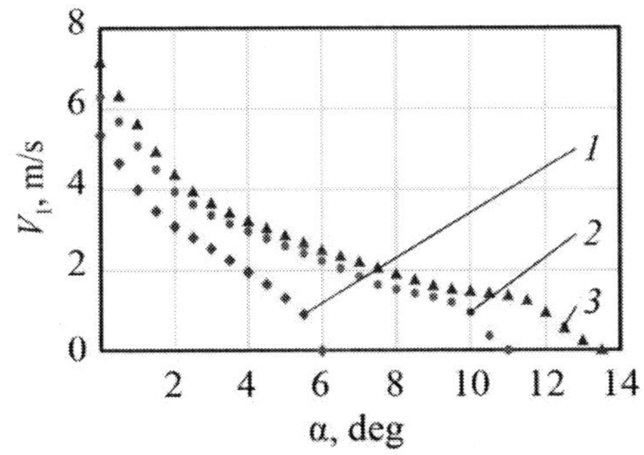

Fig. 5. Relation between the maximum transportable mass of a semi-trailer and the angle of climb for different cases of power unit capacity distribution between the link elements: 1 – basic option; 2 – 25% of the total capacity to the semi-trailer; 3 – 50% of the total capacity to the semi-trailer

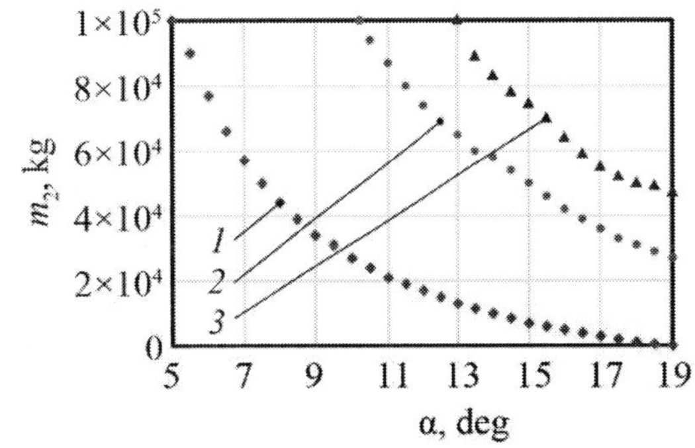

Fig. 6. Relation between the maximum speed of a road train and the angle of climb for different cases of power unit capacity distribution between the link elements: 1 – basic option; 2 – 25% of the total capacity to the semi-trailer; 3 – 50% of the total capacity to the semi-trailer

Analysis of the obtained results allowed to draw a conclusion that activation of semi-trailer wheels while moving on the analyzed type of bearing surface leads to increased angle of climb. Full-weight road train in the basic option loses its mobility at the slope of 6°, in the second case (45/20) the angle of climb increased to 11°, in the third case (30/40) – to 13.5°. Moreover, it can be seen from Fig. 6 that the basic option of a road train builds up the minimum speed in comparison with the other two cases when moving both on horizontal surface and uphill.

5. Conclusions

A model and a design program have been developed that allow studying two-link semi-trailer truck dynamics in its linear motion on deformable bearing surfaces. The developed model can be used to predict traction and dynamic properties and determine rational balance of capacity split between tractor wheels and semi-trailer wheels achieving improved energy efficiency of a road train in cross-country conditions.

Based on the conducted simulation experiments and with the help of the developed model, it has been proved that worse traction dynamic performance of motion on deformable bearing surfaces is caused by insufficient trailing weight. From the calculation results, it is seen that the basic option of a road train with its full weight can lose mobility at a longitudinal slope of 6°, which is clearly insufficient for on ground motion. The tractor itself (without useful load) is able to climb slopes of up to 19°, while being a part of an active road train, it can climb the same slope with the useful load of up to 47 tons.

Acknowledgment

The research has been conducted in the Bauman Moscow State Technical University under financial support from the Fellowship Council of the Russian Federation President (grant registration No. MD-582.2019.8).

References

[1] Diakov A S and Kotiev G O 2015 *MATEC Web of Conferences* vol **224** 02096
[2] Kotiev G O, Padalkin B V, Kartashov A B and Diakov A S 2017 *ARPN Journal of Engineering and Applied Sciences* vol **12** pp 1064–71
[3] Bazhenov E E, Buynachev S K and Chernyshev DO2019 *IOP Conf. Series: Journal of Physics: Conf. Series***1177** 012001
[4] Belousov B, Ksenevich T, Vantsevich V and Naumov S 2014*SAE Technical Paper*vol **1** 2014-01-2288
[5] Vol'skaya N S, Zhileykin M M and Zakharov A Y 2018 *IOP Conference Series: Materials Science and Engineering* vol **315** 012028
[6] Boonstra S and Rieck F 2018*World Electric Vehicle Journal* vol**9**
[7] Kurmaev R, Karpukhin K, Korkin S and Terenchenko A 2018 IEEE Transportation Electrification Conference (ITEC-India)vol **1**
[8] Usinin U, Gladyshev S, Grigoryev M, Shishkov A, Belousov E, Zhuravlev AM, Bychkov A and Sychev D 2015*SAE Technical Paper* 2015-01-2823
[9] Diba F and Esmailzadeh E 2018 *Int. Journal of Heavy Vehicle Systems* vol 25 pp 203-222
[10] Beloousov B, Ksenevich T and Izosimov D 2013 *SAE Technical Paper* 2013-01-2867
[11] Beloousov B and Ksenevich T 2013 *SAE Technical Paper*2013-01-2473
[12] Liang H, Ma Y, Zhi J, Li Y and Peng Y 2018 *Proceedings of 9th International Conference On Modelling, Identification and Control* pp 815-820
[13] Kondakov S, Pavlovskaya O, Aliukov S and Smirnov V 2019*SAE Technical Paper* 2019-01-0914
[14] Diba F and Esmailzadeh E 2013 *IEEE Transportation Electrification Conference and Expo: Components, Systems, and Power Electronics* 6574525
[15] Andreev AF and Vantsevich V V 2018 *The Dynamics of Vehicles on Roads and Tracks* vol **1** pp 149-156
[16] Ageykin Ya S 1973 Journal of Terramechanics vol **10** pp 105-111
[17] Papunin A V, Belyakov V V and Makarov V S 2018 *IOP Conference Series: Earth and Environmental Science* vol **194** 062027
[18] Shirai T and Ishigami G 2015 Journal of Terramechanics vol **62** pp 51-61
[19] Ragheb H, El-Gindy M and Kishawy H 2013*SAE Technical Paper* 2013-01-0632
[20] Gorelov V A and Komissarov A I 2016 Procedia Engineering vol **150** pp 1322-28
[21] Kotiev G O and Serebrennyy I V2008 *Herald of the Bauman Moscow State Technical University. Series Mechanical Engineering* spec. iss. pp 193–201
[22] Gorelov V A and Kotiev G O 2009 *Trudy NAMI* vol **241** pp 25-39

IOP Conf. Series: Materials Science and Engineering **820** (2020) 012010 doi:10.1088/1757-899X/820/1/012010

Intelligent system for identification of the soil parameters during the movement of the tracked vehicle

V Naumov[1], K Mashkov[1,2], K Byakov[1] and A Fedorenkov[1]

[1]Bauman Moscow State Technical University

[2]E-mail: makon111@rambler.ru

Abstract. The authors propose a schematic structure of the intelligent system for the real time identification of the soil parameters during the movement of the tracked vehicle.The system is compared with the methods of the driving conditions prediction based on the analysis of databases, contactless telemetry, terrain scanning, etc. The possible advantages and disadvantages of these methods are specified. The descriptions of the system blocks are given, the purpose and interrelation between the components are considered in detail. Requirements of the blocks and separate components of the system are specified. The concept of integration of the system for identification of the soil parameters during the movement of the tracked vehicle into an autonomous motion control system is presented.The authors offer a method for the soil characterisation without the use of databases. The method is based on the solution of the direct and inverse problems of terramechanics for the tracked autonomous mobile complexes.The article contains typical performance examples fordifferent types of soils used for the development of the algorithms of the system foridentification of the soil parameters.Different identity criteria are proposed for checking the conformity of the soil performance models to its real performance.The article contains the analysis of the technical problems to be solved during development of a system for identification of the soil parameters.

1. Introduction

At present, one of the most difficult problems to ensure an effective motion control and vehicle off-road capability is the creation of a complex of information-measuring systems capable of providing a complete picture of the road situation in real time [1, 2]. Up to the certain moment the problem had been successfully solved due to the available information about the terrain and research works on the wheel – soil [3-9] and caterpillar track – soil [10-15] interaction. The problem naturally appeared again with the appearance and development of autonomous mobile complexes (AMC), when the movement planning began to be imposed directly on the vehicle management system [16-20].

2. The basic principles of the intellectual system development

The movement planning and route selection algorithms of the AMC are largely based on the a priori data on the terrain features in the operation area. In the absence of such information, the choice of movement should be based on information from its own built-in sensors [1]. Thus, the most well-known and recommended method for identificationof the terrain current parameters is to use the GPS navigation, when the GPS tracker together with high-precision cartographic terrain data allows to predict road conditions with high accuracy. However, it is difficult for the system to operate in areas with poor radio communications and in cases where there are no terrain data or environmental conditions are not constant, such as operation in combat environment. The problem can be solved by using speed sensors, a compass and a distance sensor that measures the distance to the reference points on a map.

The following method involves the application of tele- and laser vision systems.Cameras are installed in different parts of the tracked vehicle to measure the distance to objects and identify them. The complex is also able to determine the relief of the terrain with sufficient accuracy, which allows to solve the problem of cross-country ability.

However, the above methods have a significant disadvantage. In the case of motion in non-deterministic environment (planetary surfaces, modified terrain due to natural or weather effects, combat areas, etc.) the effectiveness of such systems is minimal. This means that the solution of the problem of the terrain characterisation would likely to be in the application of the contact measuring systems. To present day, the penetration method as well as the method of the surface properties analysis with the use of dynamic effects [21] have been widely known and developed.The essence of the methods is the detectionof the force reactions of the terrain at a single test impact on the soil and empirical identification of the required parameters. However, these methods are too cumbersome, require mounting additional equipment, and are not applicable at transport speeds.

In general, it should be noted that the above methods and techniques of terrain analysis have limited application not only because of technical difficulties, but also because of the uncertainty of the received information about the bearing properties of the passable areas, when the type of soil can be identified incorrectly.This leads in some cases to the classification of the passable areas of the environment as impassable and does not allow to build trajectories of the further movement, as well as to apply adaptive control algorithms. Thus, the basis of the method for obtaining data on the soil parameters should be based on the direct contact measurement, which is carried out when the tracked vehicle passes through the area under explanation. This requires development of contact measuring instruments integrated into the existing units and assemblies of the tracked AMC.

Another promising method of soil characterization is to use neural networks for the data analysis. It is known that neural networks are good at classifying data. In our case, it is possible to classify soil types and their properties by information about loads and deformations of the ground surface. The result of this procedure is a library of solutions, which allows the neural network to choose the right type of soil and provide the specified characteristics in real time. Taking into account the modern tendencies, the method is very promising and deserves attention. In general, such principles are laid down in the classification of soil types and the creation of extended databases, which ensures the unification of management systems for this type of AMC. The disadvantage is a large amount of learning information required for correct operation of the automatic motion control system (AMCS). Taking into account the variety of soil parameters, the task becomes difficult to solve and requires the development of other approaches.

An alternative to the databases is the calculation of the soil parameters based on empirical models describing the soil – caterpillar track interaction. There exist a fair number of models describing the interaction between the caterpillar track and the ground surface. The choice of the model has a direct impact on the identified parameters of the soil, which leads to certain reactions of the control system in general.And this in turn determines whether the mission entrusted to the AMC is fulfilled. That is why it is so important to choose, and in our case to develop, a model describing the interaction of the caterpillar track with the ground, taking into account the features of the sensor system. As a result, an equally important requirement for the development of an intelligent system for identification of the soil properties during the movement of a tracked vehicle is the necessity of the obtained results verification. In addition, the terrain characterization system must sattisfy the following specific requirements:

1) Ensuring the availability of primary data.
2) High speed of primary data acquisition and processing.
3) Guaranted aquisition of the minimum information required to organize the movement process.
4) Adaptation to the specific vehicle.

5) High reliability of the sensor system; protection from external environment.

6) Capability to correct the empirical coefficients of the model of the terrain characteristics calculation in the process of movement. This is necessary to update the collected information and exclude the outdated information.

3. Intelligent system development

From the aforesaid, the following schematic structure of the intellectual system for identification of the soil parameters was offered (figure 1).

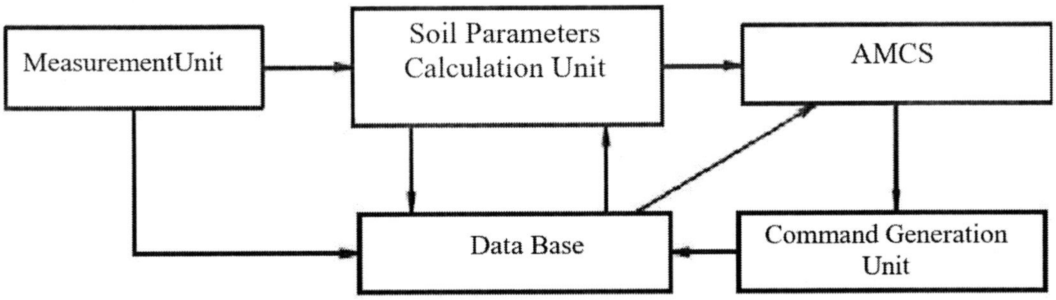

Figure 1. Schematic structure of the intelligent system for the identification of the soil parameters during the movement of the tracked vehicle

The Measurement Unit is a system of sensors integrated into the chassis of a tracked vehicle. The data obtained at this stage is then transferred to the Soil Parameters Calculation Unit. There the data is modified into a format suitable for calculation, the suspect data is filtered, and measurement errors are excluded. After that, the environmental data are calculated and checked. The calculated data are transmitted to the AMCS, where the executive commands are generated.At all the stages, the results of the operation of the units are recorded in a database. Thus, during their calculation the parameters are accumulated, classified and sorted, and AMCS actions are recorded.

3.1. Measurement Unit description

The Measurement Unit is designed to obtain the primary information about the soil mechanical properties.Figure 2 shows the schematic structure of the Measurement Unit. The information is collected through the measurement of the normal and tangential reactions that occur during the vehicle movement causing variable stressesin the soil. At the same time, the relative deformation of the soil under the next roadwheel and in the space between the roadwheels is detected. This simulates continuously changing loading of the underlying soil. The interaction between the chassis and the soil can be static when the machine «probes» the terrain at minimum speed and dynamic when the vehicle is moving at operating speed. The performance of the deformable soil will be different.This requires the calculation of parameters by using different models, which is taken into account further in the Calculation Unit. At the present time, it has been revealed that there is a great interest in the creation of distributed sensor systems to obtain the above-mentioned data, but the direction of search is still at the early stages.Along with technical difficulties, there is no methodological basis for such measurements.The basic direction of search is the creation of strain-measuring systems, as well as systems measuring the dielectric (capacitive) permeability of the environment. It is advisable to use the other types of sensors that measure additional characteristics of the soil base, such as density or humidity.

IOP Conf. Series: Materials Science and Engineering **820** (2020) 012010 doi:10.1088/1757-899X/820/1/012010

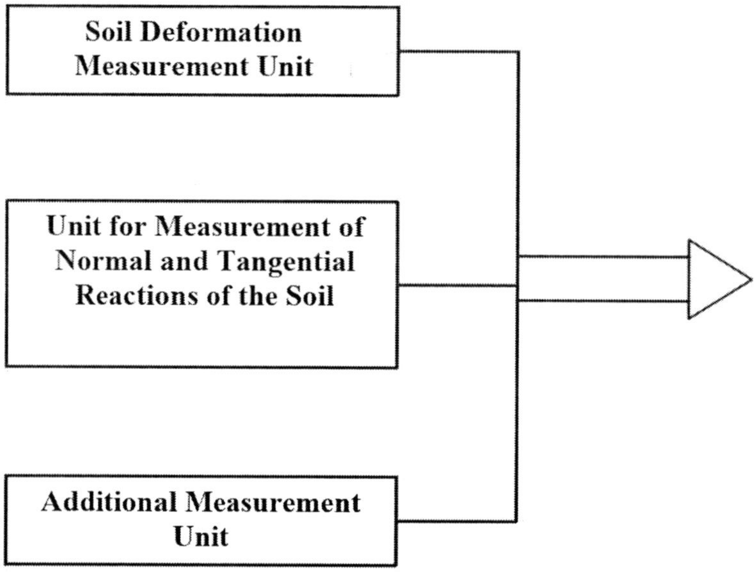

Figure 2. Schematic structure of the Measurement Unit.

3.2. Soil Parameters Calculation Unit

The data from the Measurement Unit is fed into the Simulation Unit. There it is converted into «stress-strain», «stress-time», «stress-travel distance» dependencies. The main task of the Simulation Unit is to calculate the mechanical characteristics of the terrain based on the performance of the soil. In fact, this is nothing less than the inverse task of terramechanics.

In the course of interaction of the vehicle runnig gear with the deformable soil the main mechanical characteristics are: normal sinkage, occurring under the influence of the weight of the vehicle, and shear - under the influence of longitudinal forces applied to the chassis.

Different empirical formulas can be used to describe the normal pressure – sinkage curve, for example, the Bernstein-Letoshnev formula, which is valid for the penetration of a flat plate into the soil [1]:

$$p = C_g \lambda_v^n, \tag{1}$$

p is the reaction pressure on the dF area;

λ_v is the penetrator plate sinkage;

C_g, n are soil parameters.

To describe the shear curve, it is necessary to make a distinction between situations when the chassis interacts with a hard surface, such as rocky ground, and a case of interaction with deformable soils. In the first case, the interaction of the caterpillar track is largely defined by the friction between the contact surfaces, in the second case – by the shear characteristics of the soil. Quantitative and qualitative estimates of the shear interactions are usually performed by using the Janosi-Hanamoto formula:

$$\tau = \left(c + p tg \varphi_g\right)\left(1 - e^{\frac{-|\lambda_\tau|}{\lambda_{\tau s}}}\right), \tag{2}$$

here φ_g is the internal friction angle;

c is the apparent cohesion of the soil;

$\lambda_{\tau s}$is the shear constant;

$\tau$is the shear stress of the soil;

$\lambda_\tau$is the shear of the soil.

So, the soil constants fully determine its mechanical properties.

The inverse problem of terramechanics in relation to tracked vehicles is to determine the properties of the terrain using the data obtained during the movement of the vehicle.The initial data to be entered into the solution are: stresses acting on the ground, deformations of the ground from the action of stresses, movement conditions of the vehicle. As an example, figure 3 shows the load diagrams for different types of soil, obtained by calculation.To solve the inverse problem, the researchers usually use the following approach: first, they select a ground surface with known mechanical characteristicsfor movement of the tracked vehicle. Then, the diagram of normal reactions distribution over the entire contact area of the tracks is determined.In this way, the diagram of normal reactions and the current terrain properties are matched. In the real operational conditions such a method is not used at least because it is impossible to make an unambiguous conclusion about the type of the soil according to the known loading diagrams due to different environmental conditions.The possible solution is based on the creation of empirical models of interaction that take into account the relationship between the mechanical, traction and bearing properties of the soil.

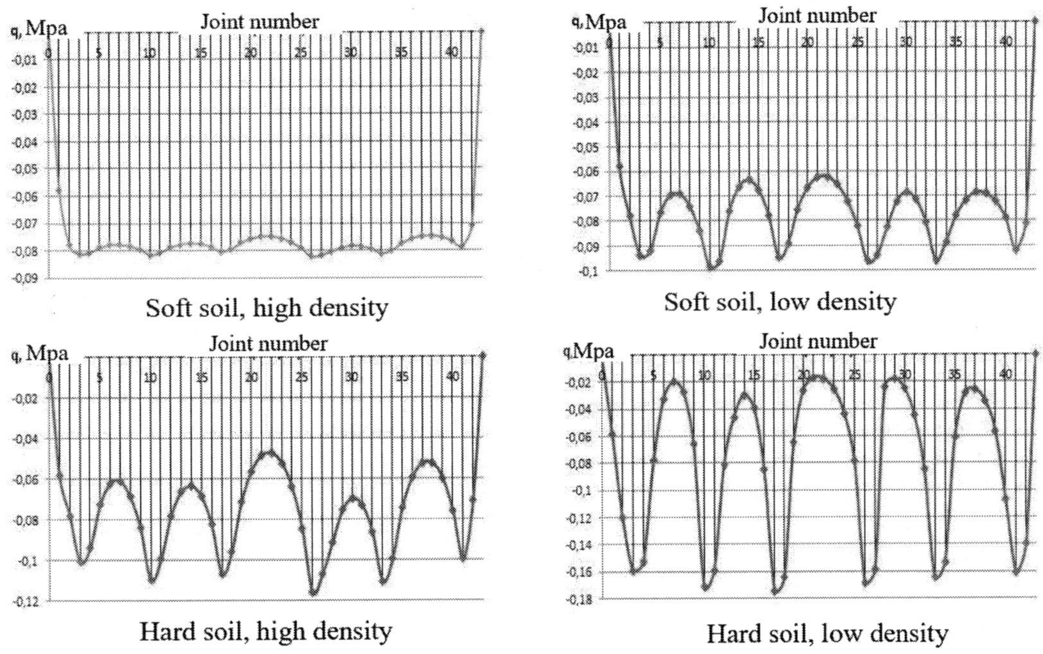

Figure 3. Different types of soil. Input data for the Simulation Unit

The calculated parameters are then passed to the Verification Unit (figure 4).

The Verification Unit is used for verificationof the calculated characteristicsagainst the actual ones.

This is achieved by calculating the normal reactions under the track and comparing the calculated and actual loading diagrams.If the diagrams match, a positive decision is made about the accuracy of the calculated values. If the result is negative, the model is recalculated using other algorithms in accordance with another interaction model. Identity criteria should be developed to verify the compliance. These can be integrated assessments, estimates of the peaks of the diagrams, or similarity assessments of individual areas.

Figure 4. Input data for the Simulation Unit

Conclusion

Evolution of the robotic transportation systems infers transition from remote control to autonomous control. For this kind control the data on the terrain properties is of crucial importance. The structural model of the proposed intelligent motion control system allows identification of the necessary parameters of the terrain. The syntesized system is adaptive due to its feedback. By comparing the characteristics of the real behavior of the terrain with the developed interaction model it becomes possible to configure the model in real time mode.This in turn leads to the flexibility of the system, which is expressed in the ability to adapt to different types of terrain, as well as in the rapid update of the internal knowledge base about the types and properties of the terrain.

Existing methods of direct contact measurement even with the correct identification of the terrain have a significant drawback.Since the measurement takes place when the first roadwheels of the tracked vehicle hit the terrain under exploration, there is a risk to lose mobility even with such a maneuver, not to mention the movement with operational speeds.The solution to this problem should be sought, most likely, in a combination of laser vision systems, GPS navigation, and the method of contact research.It is certainly more reliable, more accurate, but more expensive, energy-intensive, which significantly reduces the possible scope of application.To neutralize the disadvantages of such a system, it makes sense to identify the terrain properties at low speeds.In this case the control system will have time to make correct decisions and the risk to lose mobility is reduced to zero.

However, for this type of system to function, the following methodological and technical problems must be solved:

1) It is necessary to develop sensor systems that would allow obtaining input data that are the source for the system of determining the characteristics of the terrain in a real environment.

2) It is required to develop a device for verification of the calculated characteristics.

3) It is necessary to develop a method for obtaining load-strain relationships for a particular type of terrain.The data obtained with this method at the next calculation step will allow comparrison of the simulated behavior of the terrain with the real one and conclude whether the calculated characteristics of the terrain are correct.

4) It is necessary to develop a method for assessing the physical properties of the terrain by the mechanical characteristics of its soil.

References

[1] Naumov VN, Mashkov K Y and Byakov KE 2019Automatic determination of soil parameters by robotic vehicles *Journ. of Phys.: Conf. Ser.* **1177** Is.1

[2] Naumov V N, Mashkov K Y and Byakov KE 2019 Autonomous tracked vehicles effectiveness estimation *IOP Conf. Ser.: Mat. Sc. and Eng.* **534** N.1

[3] Klubnichkin V E, Klubnichkin EE and Naumov V N2018 Improvement of bogie tracks for forestry wheeled vehicles *IOP Conf. Ser.: Mat. Sc. and Eng.* **386**, Is. 1

[4] Naumov V N, Fedorenkov AP and Byakov K E2019The research of the elastic propulsion unit and soil contact area shapes and sizes *Journ. of Phys.: Conf. Ser.* **1177**Is.1

[5] Vol'skaya NS, Zhileykin MM and Zakharov A Y 2018 Mathematical model of rolling an elastic wheel over deformable support base *IOP Conf. Ser.: Mat. Sc. and Eng.* **315** Is. 1

[6] Agejkin Ya S and Vol'skaya N S 2004 Modeling the movement of automobile on soft soils: Problems and solutions *Avtomobil'naya Promyshlennost***10**pp 24-25
[7] Agejkin Ya S and Vol'skaya N S 2004 Specific features of wheel vehicles motion along rough dirt surfaces *Avtomobil'naya Promyshlennost***6** pp 22-24
[8] Klubnichkin E E, Klubnichkin V E and Kotiev G O 2018 Theoretical research of soil packing by timber harvester running gear *IOP Conf. Ser.: Mat. Sc. and Eng.***386**, Is. 1
[9] Horn R, Vossbrink J, Peth S and Becker S 2007 Impact of modern forest vehicles on soil physical properties *Forest Ecol. and Manag.***248**Is. 1-2 pp 56-63
[10] Alyabiev A F, Kotov AA, Klubnichkin VE, Klubnichkin EE and Bykovskiy M A 2019 Experimental determination of soil shear force by track model depending on track surface to soil friction coefficient *Journ. of Phys.: Conf. Ser.***1177**Is.1
[11] Makarov V S, Markovnina A I, Papunin A V, Belyaev A M, Goncharov KO, Klubnichkin V E, KlubnichkinE E and Belyakov V V 2019 Study of uneven surfaces distribution on forestry roads *Journ. of Phys.: Conf. Ser.***1177**Is.1
[12] Klubnichkin VE, Klubnichkin E E, Kotiev G O, Beketov S A, and Makarov V S 2018 Interaction between elements of the track ground contacting area with the soil at curvilinear motion of the timber harvesting machine *IOP Conf. Ser.: Mat. Sc. and Eng.***386**, Is. 1
[13] Kar MK 1987 Prediction of track forces in skid-steering of military tracked vehicles *Journ. of Terramech.***24** Is. 1 pp 75-84
[14] Chen W F and Liu X L 1990*Limit analysis in soil mechanics*(Elsevier; Developments in Geotechnical Engineering) p477
[15] Bodin A 1999 Development of a tracked vehicle to study the influence of vehicle parameters on tractive performance in soft terrain *Journ. of Terramech.***36**Is. 3 pp 167-181
[16] Al-Milli S, Seneviratne L D and Althoefer K 2010 Track-terrain modelling and traversability prediction for tracked vehicles on soft terrain *Journ. of Terramech.***47** Is. 3 pp 151-160
[17] Song Z, Hutangkabodee S, Zweiri Y H, Seneviratne LD and Althoefer K 2004 Identification of soil parameters for unmanned ground vehicles track-terrain interaction dynamics *Proc. of the SICE Ann. Conf.*FAI-9-3 pp 1651-1656
[18] Iagnemma K, Shibly H and DubowskyS 2002 On-line terrain parameter estimation for planetary rovers*Proc. - IEEE Int. Conf. on Robot.and Automat.***3**pp3142-3147
[19] Longoria R G, Brushaber R and Simms A 2019 An in-wheel sensor for monitoring tire-terrain interaction: Development and laboratory testing *Journ. of Terramech.***82** pp 43-52
[20] Tan C, Zweiri YH, Althoefer K and Seneviratne LD 2003 On-line soil property estimation for autonomous excavator vehicles *Proc. - IEEE Int. Conf. on Robot.and Automat***1**pp 121-126
[21] Nohse Y, Hashiguchi K, Ueno M, Shikanai T, Izumi H and Koyama F 1991 A measurement of basic mechanical quantities of off-the-road traveling performance *Journ. of Terramech.***28** Is. 4 pp 359-370

IOP Conf. Series: Materials Science and Engineering **820** (2020) 012011 doi:10.1088/1757-899X/820/1/012011

Special aspects of the test of ATV equipped with the electronic engine management system Continental M3C on a dynamometer test bench

D Ilyushin[1], Y Salikin[1], V Shekhovtsov[1], A Diakov[2] and K Evseev[2]

[1]Volgograd State Technical University, Volgograd, Russian Federation
[2]Bauman Moscow State Technical University

[3]E-mail: kb_evseev@bmstu.ru

Abstract. In this article the problems of taking characteristics of the power plant of BRP Can-Am Outlander max 570 at its loading on a dynamometer test bench DynaPack are considered. The possibilities of using input and output electrical signals of sensors and actuators of the electronic engine management system Continental M3C, including the use of third-party electronic engine management unit M74CAN cars VAZ. The schematic solutions for obtaining the value of the engine crankshaft speed of the ATV by the standard software of the DynaPack dynamometric stand are proposed. Based on the analysis of the developed circuit solutions, as well as the features of loading, power plant in general, and the engine in particular, on the dynamometer test bench DynaPack decided to use the fuel injector control signals to determine the speed of the crankshaft engine. To increase the stability of the results of the series of experiments, a method of monitoring the position of the engine throttle valve was proposed. In conclusion, the speed characteristics of the BRP Can-Am Outlander max 570 quad bike power plant at various levels of opening the throttle valve are shown.

Description of the test bench and the ATV

The dynamometer test bench Dynapack [1–13] was used (Figure 1) to estimate the power plant power parameters of the BRP Can-Am Outlander max 570 ATV in the whole operational range of engine crankshaft rotation frequency by means of obtaining its external and partial speed characteristics. In connection with the use of stepless transmission in the design of the ATV power plant at the stage of preparation of tests the problem of introduction of the value of the crankshaft rotation frequency of its engine into the software of the dynamometer test bench was solved.

The BRP Can-Am Outlander max 570 [5] ATV is equipped with a gasoline, V — shaped, two-cylinder engine with distributed fuel injection and electronic control. The engine/electronic control module (EECM) is based on an electronic control module (ECM) Continental M3C. This control unit processes the crankshaft position sensor (CSPS) signal, among other things, to determine both the crankshaft position and the speed of the crankshaft. The ECM also has a diagnostic line via the CAN bus and a closed proprietary diagnostic information transfer protocol [14–20], which includes, among other things, information on the current crankshaft speed of the engine. The direct connection of the ECM diagnostic line to the DynaPack dynamometer has not been successful. This version of the DynaPack dynamometer software does not support the operation of this diagnostic information transfer protocol.

a) b)

c)

Fig. 1. The BRP Can-Am Outlander max 570 ATV mounted on a dynamometer test bench DynaPack: a) rear view, b) front view, c) wheel connection to the loader.

Sensor equipment

The mechanism for determining the position of the engine crankshaft is shown in Figure 2 and consists of a reference disc — 1 with the number of teeth 36-2 and CSPS — 2.

Figure 1 shows the waveform of the CSPS signal. The shape of the CSPS signal is close to the sinusoidal one, with a surge of increased amplitude and period at the point of passage of two teeth. Signal frequency with CSPS at the given reference disk, without taking into account bursts, is in a range from 28 kHz to 220 kHz at speeds of rotation of a crankshaft from 800 to 6000 rpm accordingly. Possibility of signal supply with CSPS directly to the logic input of the DynaPack dynamometric stand speed control is limited by the negative component of the signal and the value of the signal amplitude, which reaches 50 V in some modes. In order to eliminate these drawbacks, the implemented scheme shown in Figure 3 was implemented.

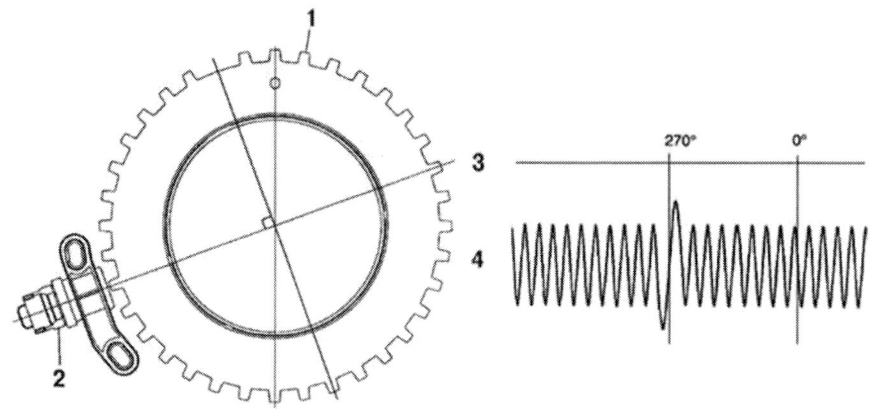

Fig. 2. Mechanism for determining the position of the engine crankshaft.

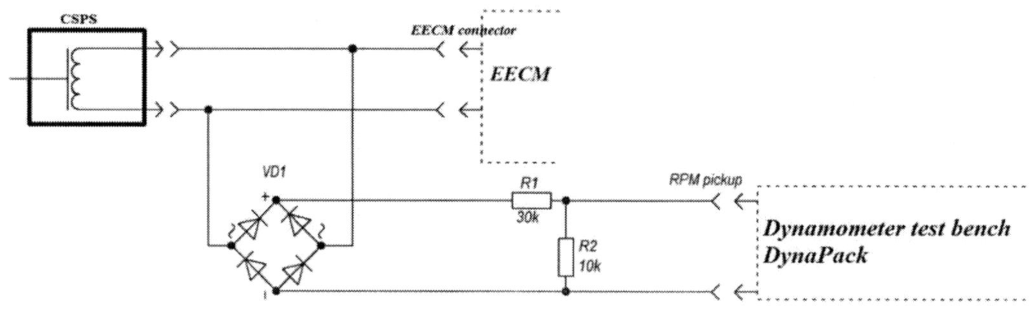

Fig. 3. Connecting the CSPS to the RPM pickup input of the DynaPack dynamometer test bench.

In this scheme, the CSPS signal is rectified by a two-period rectifier VD1, and is fed to the voltage divider assembled on the resistors R1 and R2, which has a division coefficient of 4:1. A positive signal with a safe amplitude of no more than 12.5V is fed to the logical speed monitoring input of the DynaPack dynamometer test bench. Application of the above described scheme has not yielded satisfactory results. The speed control of the DynaPack dynamometer test bench did not function adequately and the low input resistance of the circuit negatively affected the quality of signal processing by the standard ECM, which was expressed in the difficult starting of the engine.

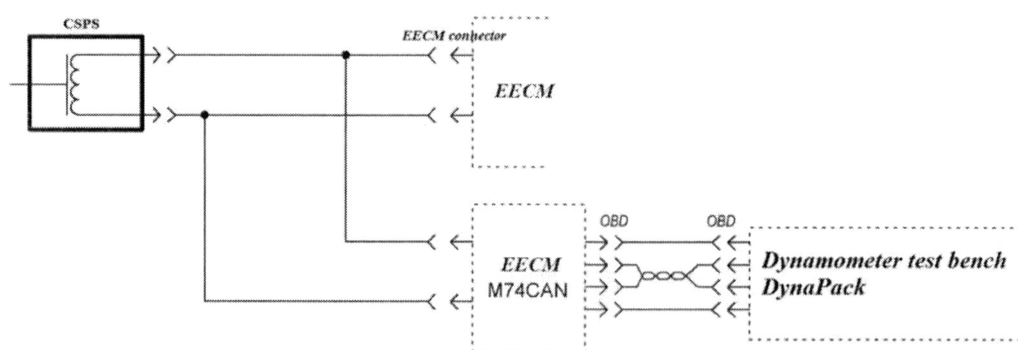

Fig. 4. Connection of CSPS to ECM M74CAN of VAZ vehicles to DynaPack dynamometer test bench by means of CAN bus.

The next way to receive the signal from CSPS was the use of a external ECM M74CAN car VAZ. In this scheme ECM M74CAN was connected to the CSPS (Figure 4) in parallel to the regular ECM as in the previous scheme, but the input resistance of the signal processing channel with CSPS ECM M74CAN has an order of magnitude greater resistance than the previous scheme, which minimizes the negative impact on the operation of the regular ECM. The ECM M74CAN processed the signal from the CSPS and transmitted the engine crankshaft speed to the DynaPack dynamometer using the CAN bus with a known diagnostic protocol.

The use of ECM M74CAN allowed the DynaPack dynamometer to obtain the value of the engine crankshaft speed that does not correspond to the actual speed, but is proportional to the actual speed with unsystematic and unpredictable obviously erroneous values. The discrepancy between the obtained and the actual speed is explained by the difference between the reference disks used to work with ECM M74CAN and M3C is expressed in a conversion factor of 1.67. In the first case the reference disc has 60-2 teeth, and in the second case it has 36-2 teeth. The discrepancy between the obtained and the real engine speed and the presence of irregular and clearly erroneous values complicated the processing of experimental data. This fact pointed to the need to apply another solution. Later it was decided not to use the signal with CSPS, due to its relatively complicated processing, and the use of the fuel injector control signal. In this control system fuel injection is a priority in the phased mode. That is, each nozzle, without taking into account transients, receives a control pulse once in two turns of the crankshaft. During transients, such as sudden acceleration or engine deceleration, it is possible to switch to asynchronous injection and completely switch off the fuel supply, which will bring significant errors in the value of the engine crankshaft speed obtained from these signals. On the dynamometer test bench DynaPack in particular, and taking into account the peculiarities of loading of the power plant in general, and the engine in particular, expressed in the absence of sharp acceleration and deceleration, and, accordingly, errors caused by these transients, it was decided to use the fuel injector control signals to determine the speed of the engine crankshaft. Figure 5 shows the scheme of the fuel injector control signal converter into the signals received by the input RPM pickup of the DynaPack dynamometer test bench with high input resistance to minimize changes in the dynamic parameters of the fuel injector and, consequently, the working process and engine parameters in particular and the power plant in general.

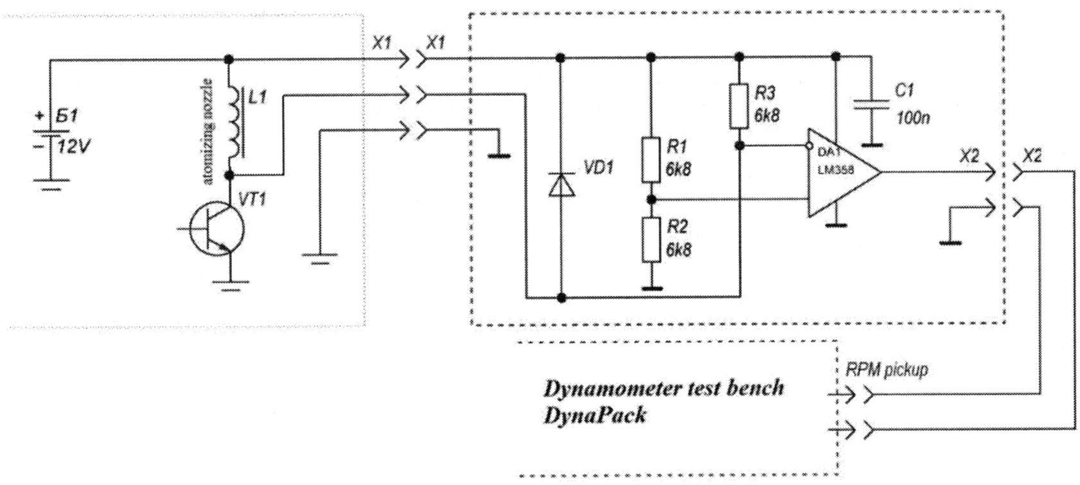

Fig. 5. Fuel Injector Signal Converter Diagram to the signals received by the RPM pickup input of the DynaPack dynamometer.

The circuit is based on a comparator made on an operating amplifier (OA) LM358. When the VT1 nozzle control transistor L1 is opened, the potential of the OA DA1 inverter input becomes less than the potential at the non-inverter input (half the supply voltage) and is close to the mass potential, resulting in switching the output to a logical unit. When the VT1 transistor is closed, the potential of the OA DA1 inverter input will be greater than the potential of the non-inverting input and the output will switch to logic zero. The VD1 diode is required to protect the OA DA1 inverter input from dangerous voltage surges when the VT1 transistor is closed. The large input impedance of the OA LM358 ensures minimal impact on the fuel injector dynamics. The operation of this circuit has been stable and found to be satisfactory.

In order to ensure the stability of the results of the series of experiments, it is advisable to control the position of the throttle valve. The standard EECM is equipped with a throttle position sensor (TPS), which is a potentiometer. The removal of the signal from the standard TPS is easy and does not require any signal level converters. Figure 6 shows the connection diagram of the adapter to remove the signal from the TPS. The adapter is designed to prevent damage to the wiring harness of the TPS to ECM connection of the EECM standard.

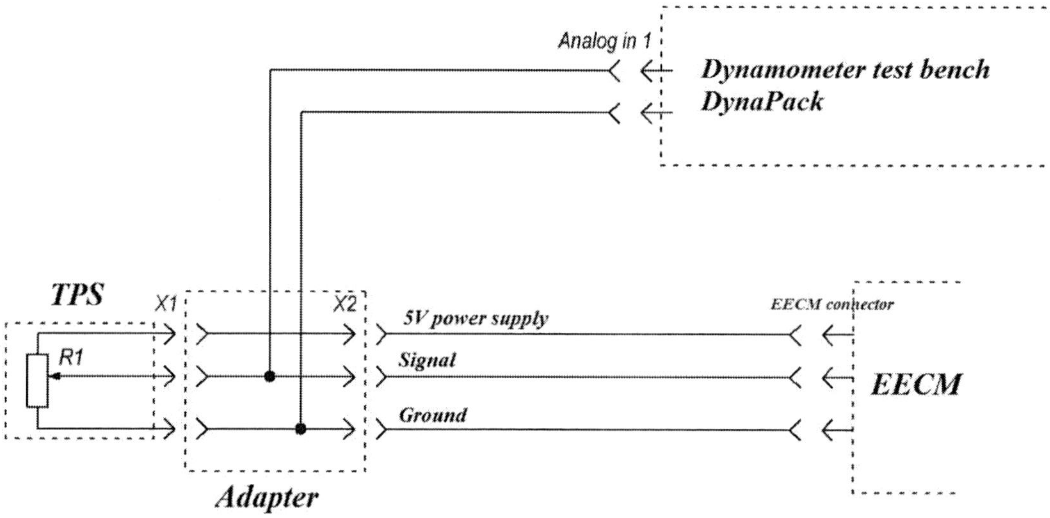

Fig. 6. The schematic diagram of the adapter to remove the signal from the TPS.

The signal of this sensor changes within the limits suitable for the stand (0-5V, taking into account the sensitivity zones). Then the signal was calibrated to display as 0-100% of the throttle opening.

Experiment results

The measures have allowed to provide stable reception by the software of dynamometric stand DynaPack of authentic value of speed of a crankshaft of the engine and an angle of opening of a throttle valve during carrying out of tests.

The result of the tests was a series of partial throttle angle increments of 10% and the external speed characteristics (Figure 7) of the power plant of the BRP Can-Am Outlander max 570 in the entire operational range of engine speeds.

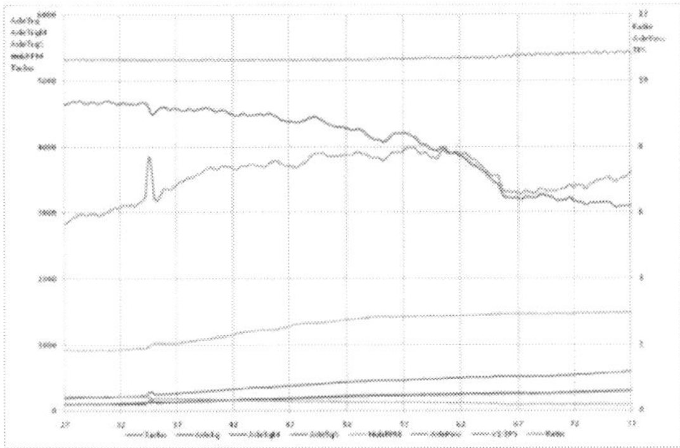

10%

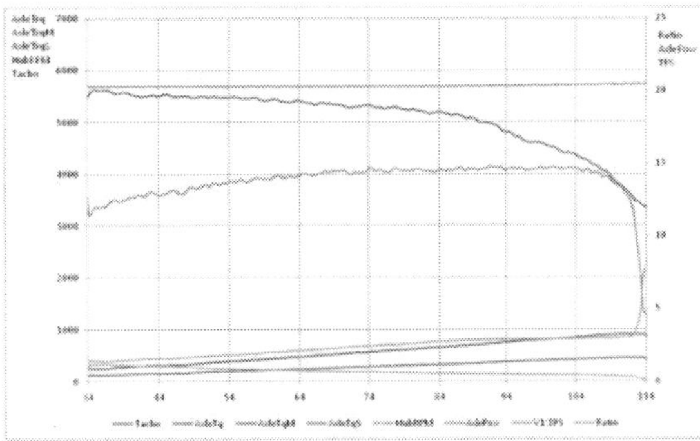

20%

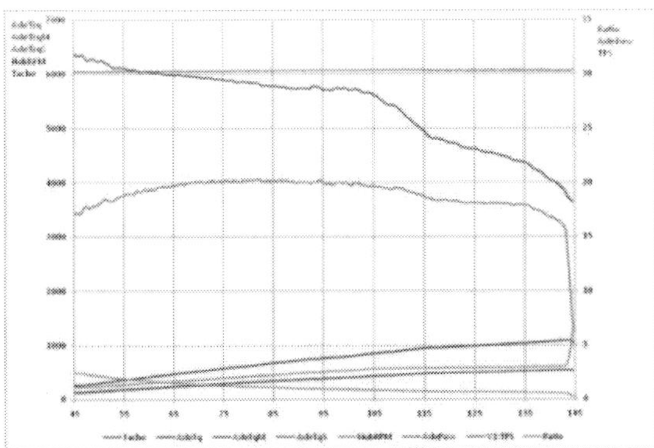

30%

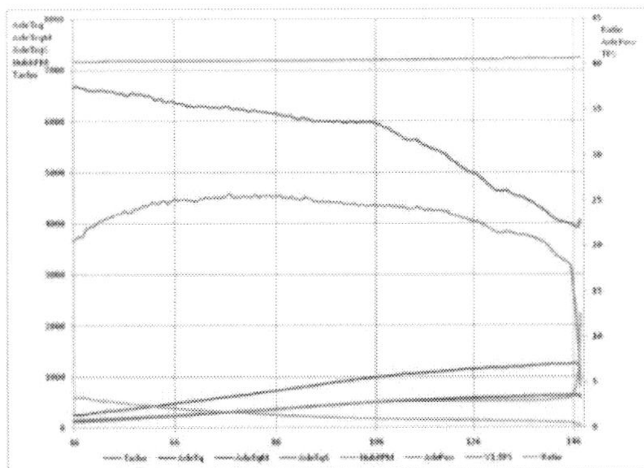

40%

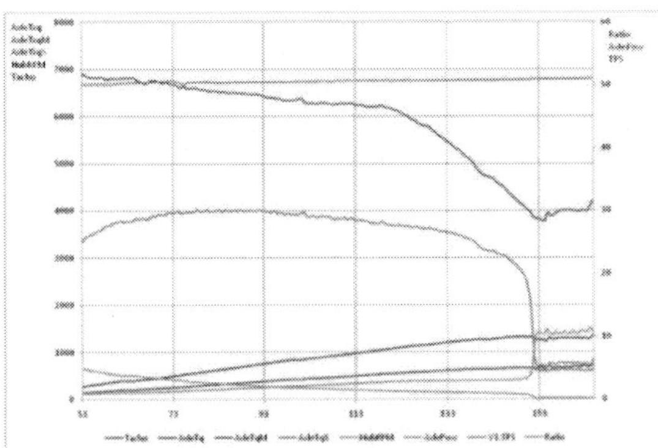

50%

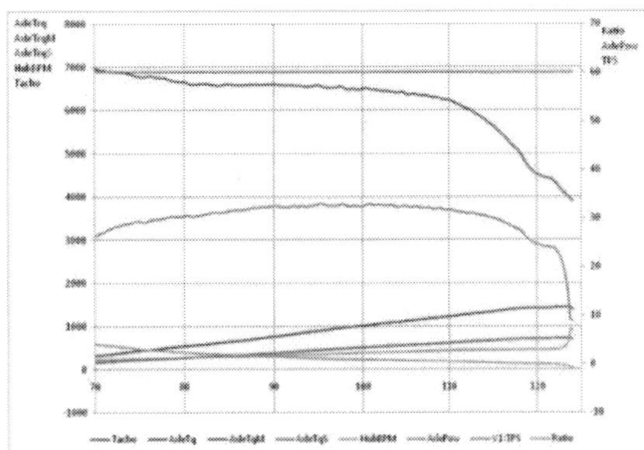

60%

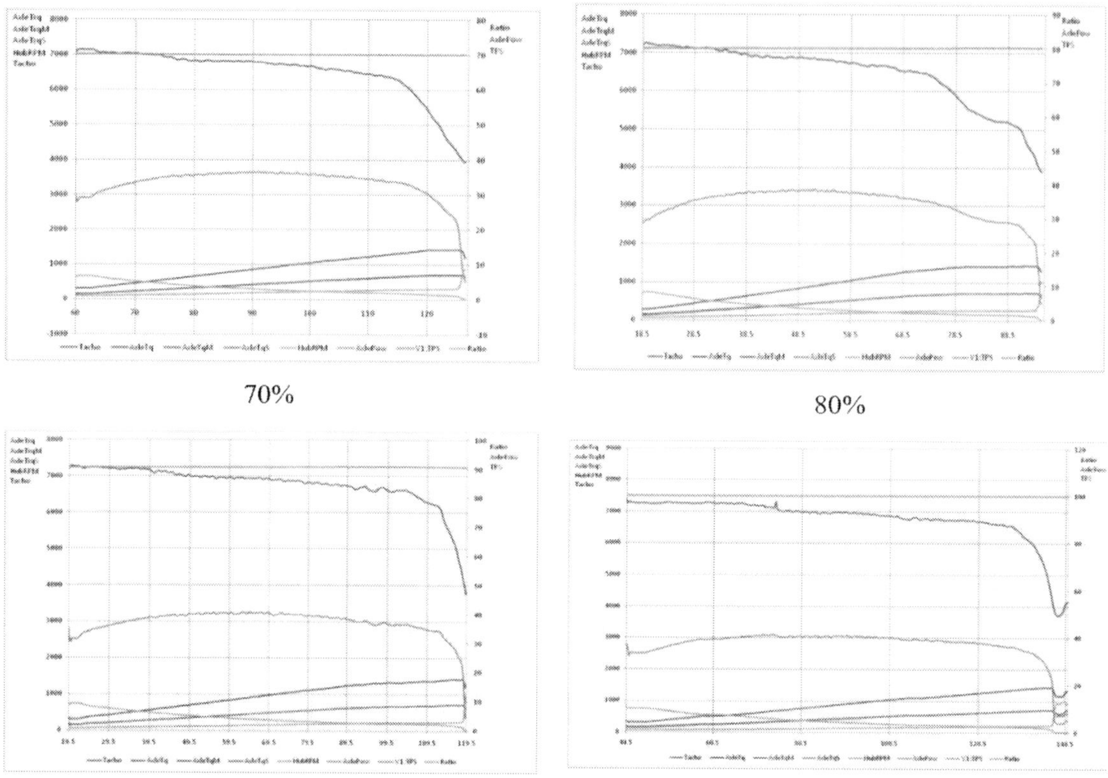

70% 80%

90% 100%

Fig. 7. Partial and external speed characteristics of BRP Can-Am Outlander max. 570 ATV

Conclusion

As a result of the experiments, stable reception of the signal by the DynaPack dynamometer bench software was provided and reliable values of the engine crankshaft speed and the throttle opening angle were obtained during the tests.

As a result of the tests, a series of partial and external high-speed characteristics of the BRP Can-Am Outlander max 570 power plant was obtained in the entire operating range of engine speeds.

It should be noted that the this work was carried out at the Bauman Moscow State Technical University, with financial support from the government in the face of the Russian Ministry of Education under the project: №14.577.21.0272. (Identification number: RFMEFI57717X0272).

References

[1] Safonov, B.A., Travnikov, A.N. Unmanned all-terrain cargo and passenger transportation system for operation conditions when automobile roads are unavailable (2019) Journal of Physics: Conference Series, 1177 (1) DOI: 10.1088/1742-6596/1177/1/012043

[2] Klubnichkin, V.E., Dyakov, A.S., Klubnichkin, E.E., Zakharov, A.Y., Vakhidov, U.S., Suchenina, A.S., Basmanov, I.V. Experimental evaluation of stability and controllability of domestic and foreign made utility terrain vehicles (2019) Journal of Physics: Conference Series, 1177 (1), DOI: 10.1088/1742-6596/1177/1/012045

[3] Klubnichkin, V.E., Dyakov, A.S., Klubnichkin, E.E., Zakharov, A.Y., Vakhidov, U.S., Suchenina, A.S., Basmanov, I.V. Experimental evaluation of speed and brake properties of domestic and foreign made utility terrain vehicles (2019) Journal of Physics: Conference Series, 1177 (1), DOI: 10.1088/1742-6596/1177/1/012048

[4] Klubnichkin, E., Klubnichkin, V. Experimental evaluation of traction and coupling properties of snowmobile motor vehicles (2018) MATEC Web of Conferences, 224, DOI: 10.1051/matecconf/201822402082

[5] Can Am Outlander https://brp-russia.blogspot.com/2016/04/brp-outlander-570-650-850-1000-2016.html

[6] Kolator, B., Janulin, M., Vrublevskyi, O. Influence of traction conditions on the power balance in the vehicle 4wd system (2018) IOP Conference Series: Materials Science and Engineering, 421 (2), . DOI: 10.1088/1757-899X/421/2/022014

[7] Irimescu, A., Mihon, L., Pădure, G. Automotive transmission efficiency measurement using a chassis dynamometer (2011) International Journal of Automotive Technology, 12 (4), pp. 555–559.

[8] A. J. Martyr M. A. Plint Engine Testing. The Design, Building, Modification and Use of Powertrain Test Facilities 2012 Elsevier Ltd 572 p.

[9] Dynapack dynamometers www.dynapack.com

[10] Barr, B.D., McKinstry, D.J., Messenger, A.S., Seidenstucker, J.M., Kallio, D.H., Hoffman, D.P., Bettig, B.P. Adaptation of four-stroke motorcycle engine to continuously variable transmission for snowmobile application (2013) SAE Technical Papers DOI: 10.4271/2003-32-0083

[11] Moskalik, A., Dekraker, P., Kargul, J., Barba, D. Vehicle Component Benchmarking Using a Chassis Dynamometer (2015) SAE International Journal of Materials and Manufacturing, 8 (3), DOI: 10.4271/2015-01-0589

[12] W Grzegoek and M Szczepka Analysis of an employment of a gear ratio rate in CVT control system 2016 IOP Conf. Ser.: Mater. Sci. Eng. 148 012007

[13] Road and Off-Road Vehicle System Dynamics Handbook Gianpiero Mastinu, Manfred Ploechl 2014 CRC Press

[14] Tyree, Z., Bridges, R.A., Combs, F.L., Moore, M.R. Exploiting the Shape of CAN Data for In-Vehicle Intrusion Detection (2019) IEEE Vehicular Technology Conference, 2018-August, paper № 8690644

[15] Perisoara, L.A., Vasile, A., Sacaleanu, D.L. Vehicles diagnosis based on LabVIEW and CAN interfaces (2018) 2017 IEEE 23rd International Symposium for Design and Technology in Electronic Packaging, SIITME 2017 — Proceedings, 2018-January, pp. 383–386

[16] Afify, M., Abuabed, A.S.A., Alsbou, N. Smart engine speed control system with ECU system interface (2018) I2MTC 2018 — 2018 IEEE International Instrumentation and Measurement Technology Conference: Discovering New Horizons in Instrumentation and Measurement, Proceedings, pp. 1–6.

[17] Kocić, J., Axmann, H. A novel solution for an ECU simulator as a key component for automated testing and verification of a vehicle diagnostic device (2018) 2017 25th Telecommunications Forum, TELFOR 2017 — Proceedings, 2017-January, pp. 1–4.

[18] Udhaya Kumar, R., Kavitha, K. Canoe tool for Ecu automated communication testing (2018) International Journal of Innovative Technology and Exploring Engineering, 8 (2S), pp. 390–392.

[19] Yang, X., Yu, Z., Xiao, M., Ji, G., Wang, Z. Automated test system design based on Tellus for in-vehicle CAN network (2015) International Congress on Ultra Modern Telecommunications and Control Systems and Workshops, 2015-January (January), paper № 7002089, pp. 118–122.

[20] Wu, Y., Chung, J.-G., Sunwoo, M.H. Design and implementation of CAN data compression algorithm (2014) Proceedings - IEEE International Symposium on Circuits and Systems, paper № 6865202, pp. 582–585.

IOP Conf. Series: Materials Science and Engineering **820** (2020) 012012 doi:10.1088/1757-899X/820/1/012012

A theoretical study on the high-speed electric tracked vehicle mobility

G Kotiev[1], B Padalkin[1], A Miroshnichenko[2], A Stadukhin[1] and B Kositsyn[1,3]

[1]Bauman Moscow State Technical University
[2]OKB "Tekhnika", 115191 Moscow, Russia

[3]E-mail: kositsyn_b@bmstu.ru

Abstract. The article deals with the problem of the characteristics selection for the electric motors of the drivetrains of the high-speed tracked vehicles with individually driven sprockets. The existing method of the tracked vehicle cornering dynamics analysis is not suitable for the analysis of the motion at the speeds near to the critical speeds in the skidding mode. The authors offer a method for improving the accuracy of the existing analytical dependencies by the use of neural network. The above-mentioned improvements provide estimation of the limit values of the required traction forces on the sides of the vehicle but they do not allow gathering statistical data on the electric motor operating conditions of short and long duration during the vehicle operation. The authors offer to solve this problem by means of a driving simulation complex. The complex provides simulation of the tracked vehicles driving along stochastic routes in real time mode and collection of the stochastic data on the electric motor loads.

Introduction

Today the electromechanical transmission for a tracked vehicle with individual driving sprockets is the most common choice for mobile robots and superlight tracked vehicles. Individual drive mode can be one of the operation modes for complex hybrid parallel electromechanical transmission. Individual drive is also a good option for articulated tracked vehicles with separately driven sections. Electric transmissions with individually driven wheels can also be used for wheeled vehicles to improve their agility.

The most heavy-duty working mode for a tracked vehicle with this type of transmission is turning. When the vehicle moves along a curved path, up to 100% of the overall thrust power could be used by the electric motor of one track, with the other electric motor switching into generating mode and creating the breaking force on the second track of the vehicle. To increase the average speed, one must provide steering of the tracked vehicle at the speeds close to the critical speeds at drifting. This demands high torque and power of the electric motor of the outside track, thus leading to increase in its mass and size. Besides, this power is not used fully during a straight-line motion.

Description of the mathematical model

Table 1 shows formulae for determination of needed power on outside track and inside track of the vehicle during a turn with radius R_f at speed V. Fig. 1 shows the corresponding schematic diagram.

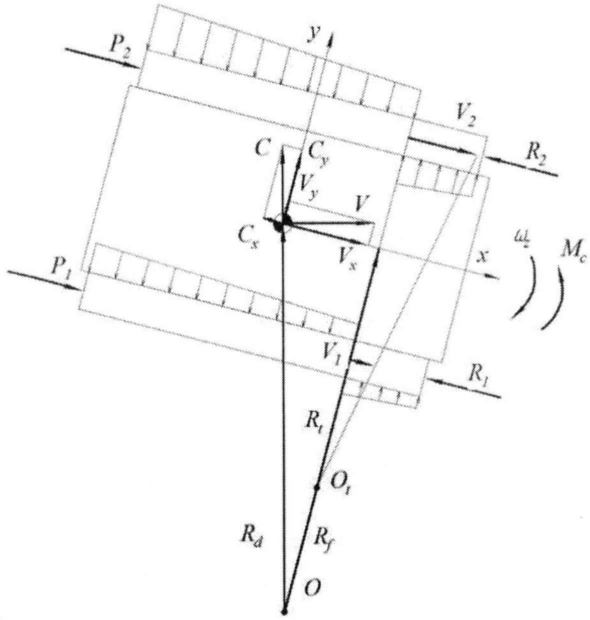

Figure 1. Forces and velocities of a tracked vehicle during a turn.

Table 1. Tracked vehicle thrust calculation formulae.

1. Theoretical turning radius R_t:	$$R_t = R_f \frac{B}{k \cdot L},$$ where B — track width; L — wheelbase length; k — proportional coefficient (individual for every vehicle) ;
2. Nikitin's formula for turning resistance μ [1]:	$$\mu = \frac{\mu_{max}}{0{,}925 + 0{,}15\frac{R_t}{B}},$$ where μ_{max} — maximum turning resistance coefficient for particular soil.
3. Critical speed before vehicle drift, V_{crit}:	$$V_{crit} = \sqrt{\mu_{\max} \cdot g \cdot R_f},$$ where g — acceleration of gravity.
4. Outside track speed V_2 and inside track speed V_1:	$$V_2 = V\left(1 + \frac{B}{2R_t}\right); \; V_1 = V\left(1 - \frac{B}{2R_t}\right),$$
5. Required thrusts on outside track P_2 and inside track P_1 [2]:	$$P_2 = \frac{m \cdot g}{2}\left(f_{gr}\left(1 + \frac{2V^2 \cdot H_z}{B \cdot R_f \cdot g}\right) + \frac{\mu \cdot L}{2B}\left(1 - \left(\frac{V}{V_{crit}}\right)^4\right)\right) + \frac{m \cdot V^4 \cdot L}{4 \cdot R_f^3 \cdot \mu \cdot g};$$

End of table 1

	$P_1 = \frac{m \cdot g}{2}\left(f_{gr}\left(1 - \frac{2V^2 \cdot H_z}{B \cdot R_f \cdot g}\right) - \frac{\mu \cdot L}{2B}\left(1 - \left(\frac{V}{V_{crit}}\right)^4\right)\right) + \frac{m \cdot V^4 \cdot L}{4 \cdot R_f^3 \cdot \mu \cdot g}$, where m – vehicle mass; f_{gr} – coefficient of resistance to straight-line motion over particular soil; H_z – vehicle center of mass height;
6. Track efficiency [3]:	$\eta_2 = 0{,}95 - 0{,}005 \cdot 3{,}6\ V_2$; $\eta_1 = 0{,}95 - 0{,}005 \cdot 3{,}6\ V_1$;
7. Required power for outside track N_2 and inside track N_1	$N_2 = P_2 \cdot V_2 / \eta_2$; $\begin{cases} N_1 = P_1 \cdot V_1 \cdot \eta_1, if\ P_1 < 0 \\ N_1 = P_1 \cdot V_1 / \eta_1, if\ P_1 > 0 \end{cases}$

As one can see in the presented formulae it is crucial to know proportional coefficient k between theoretical and true turning radius, and current amount of resistance to turning — coefficient μ. Nikitin's formula for μ was derived from a number of experiments, and the proportional coefficient k is a function of speed. Due to this fact, these formulae for stationary modes of motion are not suitable for studying critical modes close to drifting with satisfying accuracy. Besides, with this formulae it is not possible to estimate stability and control of the tracked vehicle.

Solution to the problem of unsteady movement parameters calculation could be numerical simulation using mathematical models [4, 5, 6, 7, 8, 9, 10].

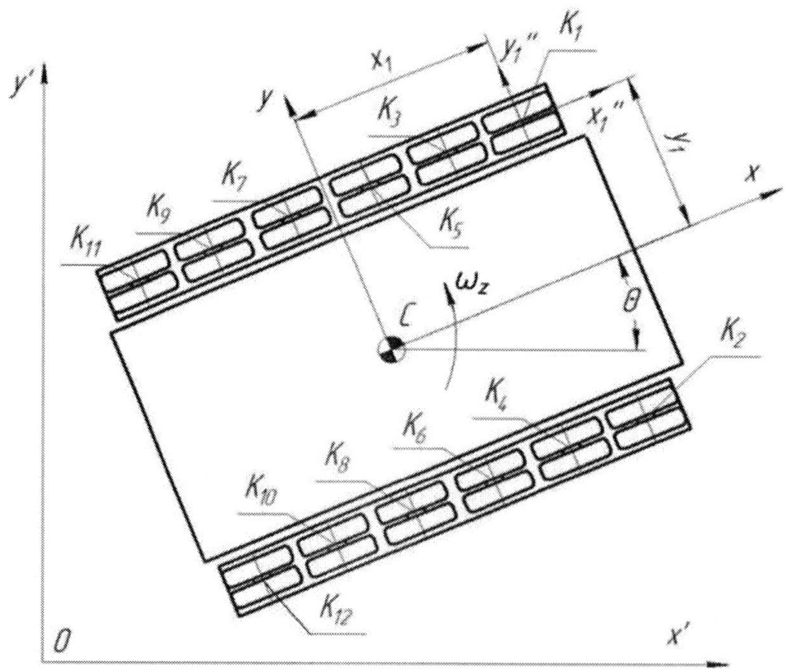

Figure 2. Planar motion of a tracked vehicle.

In this study, mathematical models [11, 12, 13, 14, 15, 16, 17] were developed where the un-steady motion of the tracked vehicle was simulated as the motion of a rigid body in horizontal plane on a planar rigid surface. Translational and rotational motion of the vehicle center of mass was considered as shown in figure 2. Translational connection of the body and the track rollers was rigid (suspension stiffness is not considered), but redistribution of normal reactions was taken into account under all track rollers due to center mass acceleration, track rollers rolling resistance and air resistance forces.

Equations of motion for this vehicle model (1):

$$\begin{cases} a_x = \dfrac{dv_x}{dt} - \omega_z v_y = \dfrac{1}{m}\left(\sum\limits_{i=1}^{n} R_{xi} - P_w\right) \\ a_y = \dfrac{dv_y}{dt} + \omega_z v_x = \dfrac{1}{m}\left(\sum\limits_{i=1}^{n} R_{yi}\right) \\ J_z \dfrac{d\omega_z}{dt} = \sum\limits_{i=1}^{n} M(R_{yi}) - \sum\limits_{i=1}^{n} M(R_{xi}) - \sum\limits_{i=1}^{n} M_{spi} \\ v_{x'} = \dfrac{dx'}{dt} = v_x \cos(\theta) - v_y \sin(\theta) \\ v_{y'} = \dfrac{dy'}{dt} = v_x \sin(\theta) + v_y \cos(\theta) \\ \omega_z = \dfrac{d\theta}{dt} \end{cases}, \tag{1}$$

where m — vehicle mass; J_z — vehicle inertia around vertical axis z; a_x, a_y — center of mass acceleration projections onto axes $x-y$; v_x, v_y — center of mass velocity projections onto axes $x-y$; $v_{x'}$, $v_{y'}$ — center of mass velocity projections onto axes $x'-y'$; $\frac{dv_x}{dt}, \frac{dv_y}{dt}$ — center of mass relative derivative of velocity projections onto axes $x-y$; ω_z — center of mass vertical rotational velocity projection onto axis z; θ — rotation angle of vehicle in ground coordinate system; x', y' — center of mass coordinates in coordinate system $x'-y'$; R_{xi} — longitudinal reaction, acting on the active part of the track under the i-th roller; R_{yi} — transversal reaction, acting on the active part of the track under the i-th roller; P_w — air resistance force, projected onto x axis of the coordinate system $x-y$;; M_{spi} — moment of resistance to turning for the part of the track under the i-th roller; n – number of rollers.

This system of equations is suitable for vehicle's center of mass accelerations determination using forces and moments acting on the vehicle.

Forces under a track roller in the active part of the track are determined from the friction ellipse [18, 19, 20, 21]. Surface reaction force in the active part of the track is directed opposite to the slip velocity.

Shown mathematical model improves analytical steady mode formulae for a tracked vehicle in the case of unsteady critical motion close to drifting. Nikitin's formula for resistance to turning and proportional coefficient k were replaced by a neural network that estimates modeling data for the tracked vehicle with different L/B range during a turn for a particular range of speeds and turning radii. Comparison of the analytical solution and numerical modelling is shown in figure 3. All calculation data are given for GM-569 tracked vehicle.

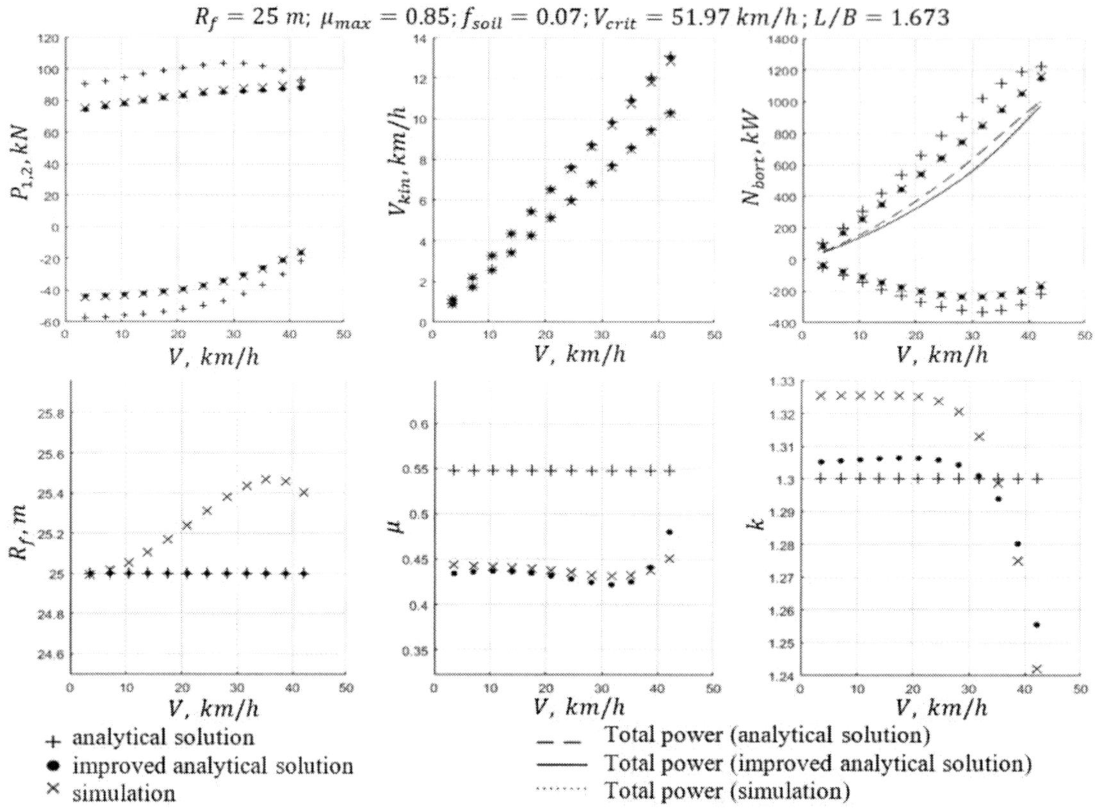

Figure 3. Calculation results for GM-569 tracked vehicle during a turn.

Resistance to turning in the mathematical model:

$$\mu = \frac{4\left(\left|\sum_{i=1}^{n} M\left(R_{xi}\right)\right| + \left|\sum_{i=1}^{n} M\left(R_{zi} f_{zp}\right)\right| - \frac{mV^2 H_z}{R_f} f_{gr}\right)}{mg \cdot L\left(1-\left(\frac{V}{V_{crit}}\right)^4\right)}, \tag{2}$$

where R_{zi} — normal reaction in the active part of the track under the i-th roller.

Simulation results estimate maximum required moments and power on driving wheels (figure 4), and allow estimation of the required thrust for maximum agility of the vehicle (turning at forward speed $0{,}8V_{crit}$), as shown in figure 5.

However, presented model does not estimate stochastic data on the short-term and long term modes of electric motor duty cycles and therefore it is not suitable for modelling vehicles with individual drive of all rollers and wheels.

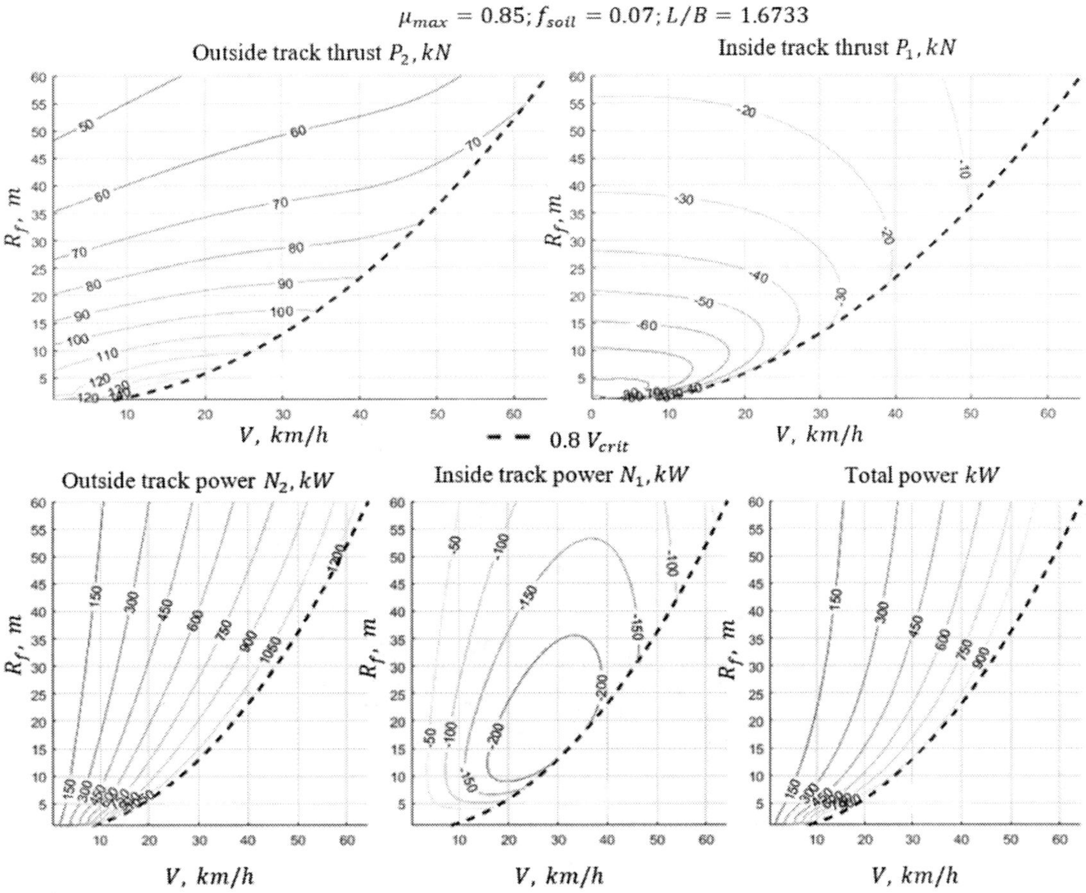

Figure 4. Estimation of required power and thrust/braking forces on each side of the vehicle

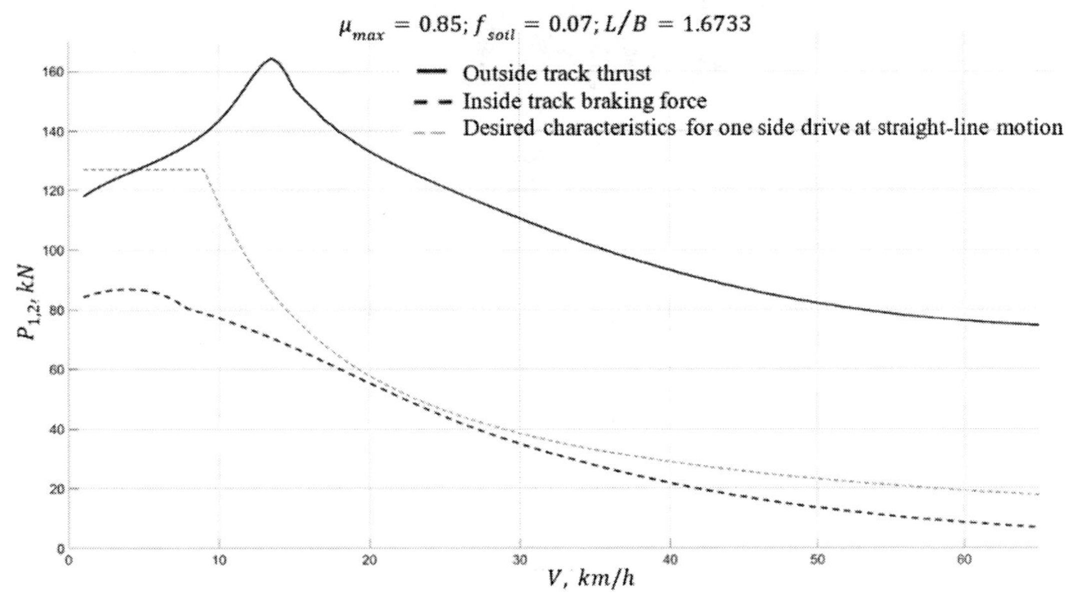

IOP Conf. Series: Materials Science and Engineering **820** (2020) 012012 doi:10.1088/1757-899X/820/1/012012

Figure 5. Required vehicle thrust performance curve for obtaining agility when turning at forward speed of $0,8V_{crit}$.

Designing of the driving simulator

To solve this problem the authors designed a driving simulator based on the presented mathematical model. The simulator is capable of real-time simulation and allows studying vehicle reaction to human driver's commands while simulating motion over different surfaces built on the base of the stochastic data on their properties, trajectory curvature, coefficient of friction, and coefficient of resistance to straight-line motion.

In order to maintain stable driving the authors developed a controller allowing connection of the driver's commands (accelerator and brake pedal positions, steering wheel angle) with control parameters of the electric motor.

Figure 6. Driving simulator workspace.

Gathered vehicle motion statistics with information on electric motors work state allows to estimate the duty cycles of the motors (including short-term overloads), required zone of motor's high efficiency, maximum efficiency curve for desired average vehicle speed, an also to formulate requirements for the cooling system of the motors.

Figure 7A shows a prototype tracked vehicle built at Bauman Moscow State Technical University. The prototype features diagonal driving wheels. Due to this fact, it is possible to study different types of electric transmissions. The prototype is equipped with complex measuring system capable to store data from:

- torques on the drive shafts on both sides of the vehicle;
- rotational speeds of the driving wheels;
- longitudinal and transversal vehicle accelerations;

- rotational speed of the vehicle about its vertical axis.

An example of the experimental measurements is shown in figure 8.

Figure 7. Prototype tracked vehicle (m=3000 kg; L/B=1,78).

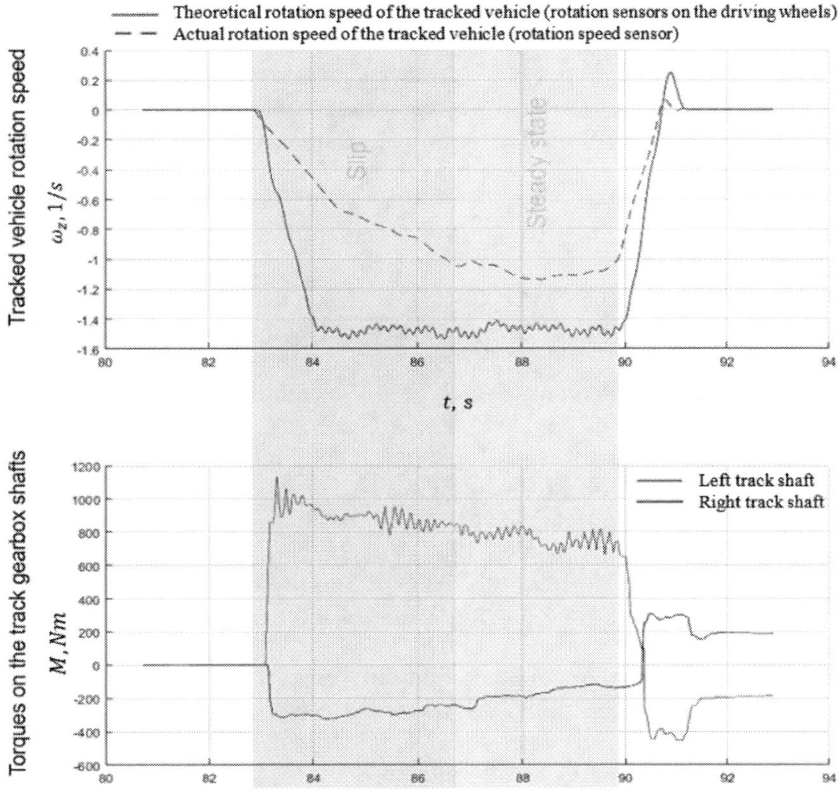

Figure 8. Turning around with one track stopped. Experimental measurements.

Conclusion

As was determined in performed research, the transmission with individual drive of the driving wheels has significant disadvantages. At present, engineers of Bauman Moscow State Technical University are studying different types of electromechanical transmissions (figure 9) including the ones with individual drive of all tracked wheels.

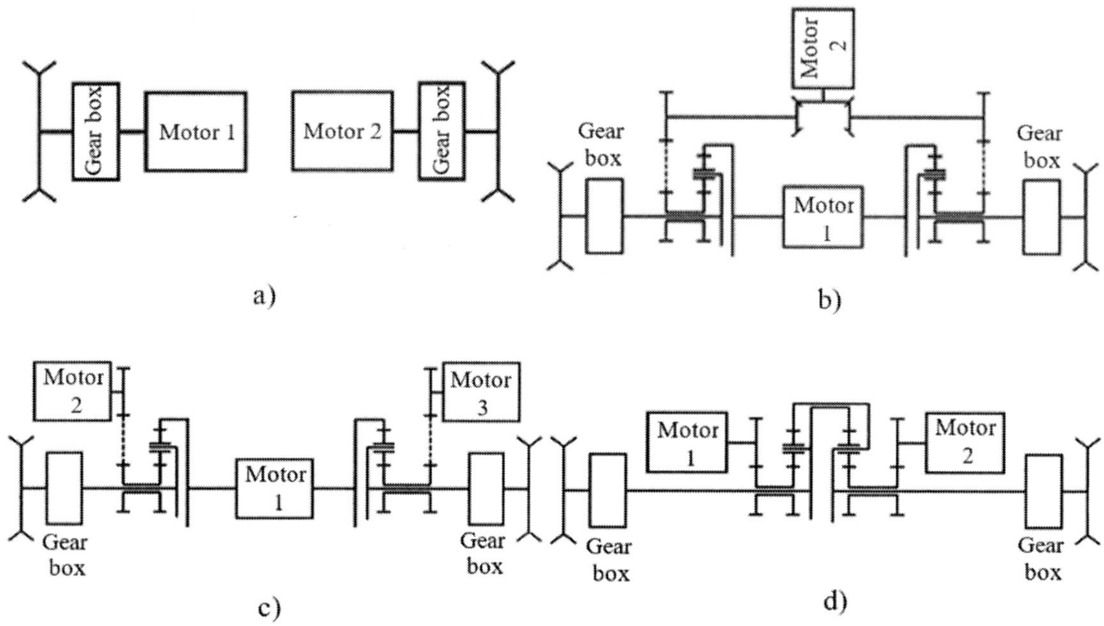

Figure 9. Types of electromechanical transmissions:
a) — two side electric motors; b) — dual way electromechanical transmission; c) — three way electromechanical transmission; d) — electromechanical transmission with central planetary gear.

References

[1] Nikitin A.O. Teoriya tanka. M.: Tipografiya Voennoj ordena Lenina akademii bronetankovyh vojsk, 1962. 584 s.

[2] CHobitok V.A. Teoriya dvizheniya tankov i BMP. M.: Voenizdat, 1984. 264 s.

[3] Zabavnikov N.A. Osnovy teorii transportnyh gusenichnyh mashin. M.: «Mashinostroenie», 1975, 448 s.

[4] Kitano M., Jyozaki H. A Theoretical Analysis of Steerability of Tracked Vehicle // Journal of Terramechanics. 1976. Vol. 13. No. 4. PP. 241–258.

[5] Kitano M., Kuma M. An Analysis of Horizontal Plane Motion of Tracked Vehicles. Journal of Terramechanics. 1977. Vol. 14. No. 4. PP. 211–226.

[6] Kitano M., Watanabe K., Takaba Y., Togo K. Lane-Change Maneuver of High Speed Tracked Vehicles. Journal of terramechanics. 1988. Vol. 25. No. 2. PP. 91–102.

[7] Baladi G.Y., Rohani B. A Mathematical Model of Terrain-Vehicle Interaction for Predicting the Steering Performance of Track-Laying Vehicles. Proc. 6th ISTVS, Austria. 1978. Vol. III. PP. 959–990.

[8] Baladi G.Y., Rohani B. Analysis of Steerability of Tracked Vehicles. Theoretical Predictions vs Field Measurements. Proc. 7th ISTVS, Canada. 1981. Vol. III. PP. 1175–1220.

[9] Dhir A., Sankar S. Analitical model for dynamic simulation of off-road tracked vehicles // Vehicle system dynamics. 1994. N.24. PP. 26–49.

[10] Dhir A., Sankar S. Analitical wheel model for dynamic simulation of off-road tracked vehicles // Vehicle system dynamics. 1997. N.27. PP. 37–63.

[11] Gorelov V.A., Kotiev G.O., Miroshnichenko A.V. Matematicheskoe mo-delirovanie rabochih processov kolesnyh i gusenichnyh mashin s ehlektromekhanicheskimi transmissiyami // Razrabotka i ispol'zova-nie ehlektricheskih transmissij dlya obrazcov vooruzheniya i

IOP Conf. Series: Materials Science and Engineering **820** (2020) 012012 doi:10.1088/1757-899X/820/1/012012

voennoj tekhniki: Sbornik statej nauchno-prakticheskoj konferencii OAO «VNIITransmash». 2016. S. 76–86.

[12] Kotiev G.O., Padalkin B.V., Kartashov A.B., Diakov A.S. Designs and development of Russian scientific schools in the field of cross-country ground vehicles building // ARPN Journal of Engineering and Applied Sciences. 2017. V.12, N.4. PP 1064–1071.

[13] Kotiev G.O., Butarovich D.O., Kositsyn B.B. Energy efficient motion control of the electric bus on route // IOP Conference Series: Materials Science and Engineering. 2018. V. 315, N.1. DOI: 10.1088/1757-899X/315/1/012014.

[14] Zhileykin M.M., Kotiev G.O., Nagatsev M.V. Synthesis of the adaptive continuous system for the multi-axle wheeled vehicle body oscillation damping // IOP Conference Series: Materials Science and Engineering. 2018. V. 315, N.1. DOI: 10.1088/1757-899X/315/1/012031.

[15] Gorelov V.A., Komissarov A.I., Miroshnichenko A.V. 8×8 wheeled vehicle modeling in a multibody dynamics simulation software // PROCEDIA ENGINEERING Elsevier BV. 2015. V.129. PP.300-307. DOI: 10.1016/j.proeng.2015.12.066

[16] Keller A.V., Gorelov V.A., Anchukov, V.V. Modeling truck driveline dynamic loads at differential locking unit engagement // Procedia Engineering. 2015. V.129. P. 280–287. DOI: 10.1016/j.proeng.2015.12.063.

[17] Keller, A.V., Gorelov, V.A., Vdovin, D.S., Taranenko, P.A., Anchukov, V.V. Mathematical model of all-terrain truck // Proceedings of the ECCOMAS Thematic Conference on Multibody Dynamics 2015, Multibody Dynamics 2015. PP. 1285–1296.

[18] Pavlov V.V. Teoriya povorota gusenichnyh transportnyh mashin. Uchebnoe posobie / MADI(TU). — M.,2000. — 224 s.

[19] Beketov S.A. Teoriya upravlyaemogo dvizheniya gusenichnyh mashin. — Moskva: Izdatel'stvo MGTU im. N.EH. Baumana, 2017. — 125 s.

[20] Antonov A.S., Blagonravov A.I., Binovich YA.E., i dr. Pod ředakciej Kristi M.K. Tanki osnovy teorii i rascheta. — Moskva/Leningrad: Glavnaya redakciya mashinostroitel'noj i avtotraktornoj literatu-ry, 1937. 436 s.

[21] Opejko F.A. EHksperimental'noe issledovanie anizotropnogo tre-niya // MIMEHSKH: Sb. nauchno-tekhnicheskih rabot. M.: Sovetskaya nauka, 1952. S. 57–64.

Using strain gages to assess the bearing properties of underlying surface with a robot crawler

K Mashkov[1] and T Pozdnyakov [1,2]

[1]Bauman Moscow State Technical University

[2]E-mail: t.pozdnyakov@bmstu.ru

Abstract. The advent of military robotic systems has raised requirements for control systems, especially independent motion control, in particular, lateral possibility and flotation when moving in partially defined or undefined environment and in the presence of external perturbations. This paper proposes approaches to determine bearing reactions under the dynamic interaction of RV running gear with the bearing surface. It demonstrates a set of gages built into the tracked running gear allowing to evaluate the underlying surface properties. It proposes a computer-aided engineering system based on the finite element method to identify the most efficient locations to install strain gages.
Keywords. Robotic systems, flotation, finite element method, strain gage

1. Introduction

Advancement of robotic vehicles (RV) is directly related to the development of the necessary artificial intelligence technologies [1–3].

In terms of control type, RV systems can be divided into three groups:

• remotely controlled
• remotely controlled with autonomous motion elements
• autonomous.

As the necessary artificial intelligence and autonomous robotics technologies are developed, remote controlled robot vehicles become more and more capable gradually moving away from operator control to partial autonomy in performing the overall task.

The third group of vehicles is of particular interest for science. This is the class of robotic vehicles capable of unmanned motion and performing various kinds of tasks in partially defined or undefined environment and in the presence of external perturbations.

The automatic motion control system (AMCS) shall ensure lateral passability and flotation of RV, including:

• integrated processing of information from onboard sensors, machine vision and navigation systems with reference to the digital terrain map
• automatic real-time trajectory (path) planning (correction)
• path testing including RV obstacle avoidance and control automation at all stages of movement.

The control system shall ensure lateral passability and flotation of RV.

The problems of lateral passability of autonomous robots are mainly solved today based on machine vision systems [4–8]. However, the full-scale problem of RV autonomous control requires the design and implementation of a flotation system.

One way to determine the flotation is mathematical simulation of the robotic vehicle behaviour in specific soil conditions. The main task of this subsystem is the dynamic simulation of robot movement based on measured inputs (angular velocity and torque of traction wheels) during cross-country movement of the robot, as well as specific soil-propulsor interaction. To do this, the subsystem shall solve the inverse problem of determining road soil conditions and adjusting the control action on actuators based on dynamic effects of the subgrade on the running gear.

Mathematical modelling of RV as a control object is quite challenging. Firstly, RV is a complex mechanical system consisting of a large number of mechanically coupled subsystems (body and attachments, wheeled or tracked running gear with a transmission, cushioning subsystem, etc.). Secondly, RV often moves on rough terrain replete with a variety of obstacles: downgrade and upgrade, uphill slopes and ravines; in these conditions, the mathematical model must simulate RV movement in 3D. Thirdly, RV movement involves slipping and skidding associated with soil deformation.

Information on subgrade soil and nature of soil-propulsor interaction is obtained using methods based on different physical tactile, acoustic principles on the basis of a complex machine vision system according to stresses arising in the propulsor and running gear as a whole and other [9–20].

Previously, the most common methods were undertaken using strain-gage tracks, but went nowhere due to the imperfection of technologies at that time. Today the development of microprocessor technology and AMCS requirements make the use of strain-gage tracks feasible [21].

It should be noted that the problem at hand is easily solved at the design stage for light tracked RVs equipped with radiotransparent solid rubber or polyurethane crawler tracks (Fig. 1 and 2), when the strain gage is baked into the track body.

Installing strain gages on medium- and heavy-weight vehicles is more challenging. It is particularly difficult to install gages in the track pin.

Some track pin designs enable the installation of strain gages, in particular, when it has a hollow center as in Fig. 3.

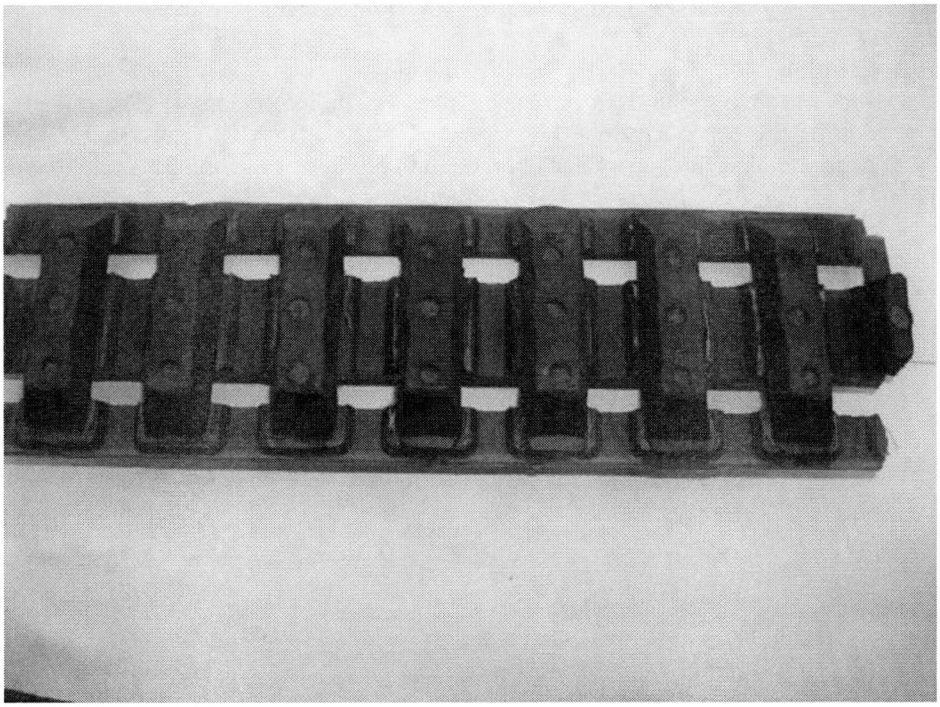

Fig. 1. RV track link weighing 200 kg with removable grouser

Fig. 2. RV track link weighing 700 kg with a metal liner

Fig. 3. Track pin with a hollow center

2. Model Description

When tracked RV is moving, the track pin experiences tensile loads (Fig. 4). These loads are defined as static tension of the track envelope and dynamic effects from RV movement. The maximum efficiency of gages can be achieved by installing them in regions of the highest strain.

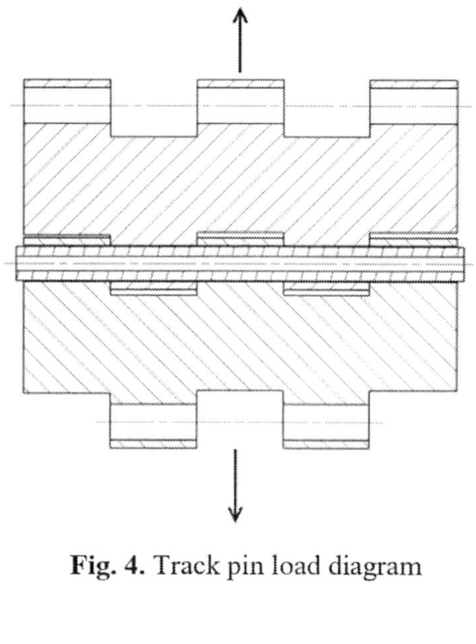

Fig. 4. Track pin load diagram

IOP Conf. Series: Materials Science and Engineering **820** (2020) 012013 doi:10.1088/1757-899X/820/1/012013

This paper proposes to identify such regions using mathematical simulation based on the finite element method (FEM) [22–23]. FEM is currently widely used for research and design. The advantages of FEM are minimum costs and time to obtain results of the most stressed regions compared to a series of experiments, but FEM-based mathematical simulation can not replace experimental testing. This is because the mathematical model can be highly simplified as compared to the real system, and simulation results will not fully reflect the behaviour of a real object. Experimental tests confirm the correctness of the chosen mathematical model.

Fig. 5 and 6 show a model of RV track pin.

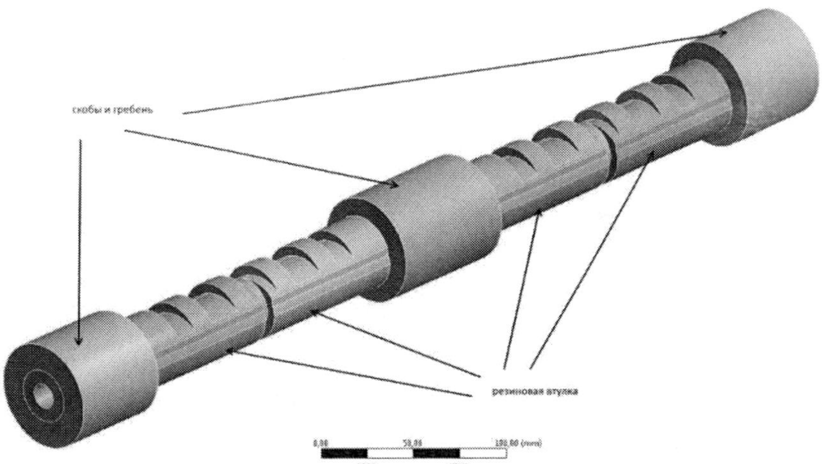

Fig. 5. Track pin model

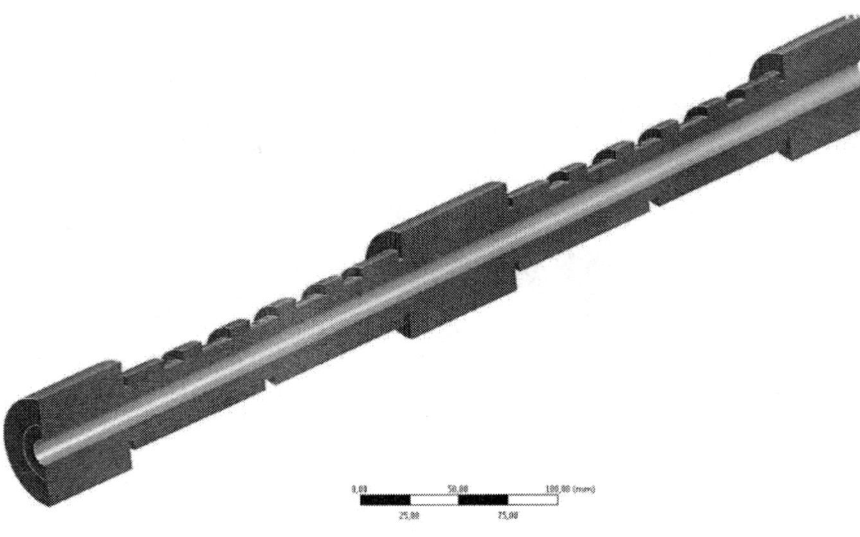

Fig. 6. Track pin model (cross-section)

3. Simulation Results

It is convenient to use equivalent stress to identify the most stressed regions. Von Mises equivalent stress is conventionally used for plastic materials:

$$\sigma_{экв} = \frac{1}{\sqrt{2}}\sqrt{(\sigma_x - \sigma_y)^2 + (\sigma_y - \sigma_z)^2 + (\sigma_z - \sigma_x)^2 + 6(\tau_{yz} + \tau_{zy} + \tau_{yx})^2}$$

Fig. 7–9 show the distribution of displacement and equivalent stress in the pin during its operation as part of the tracked running gear.

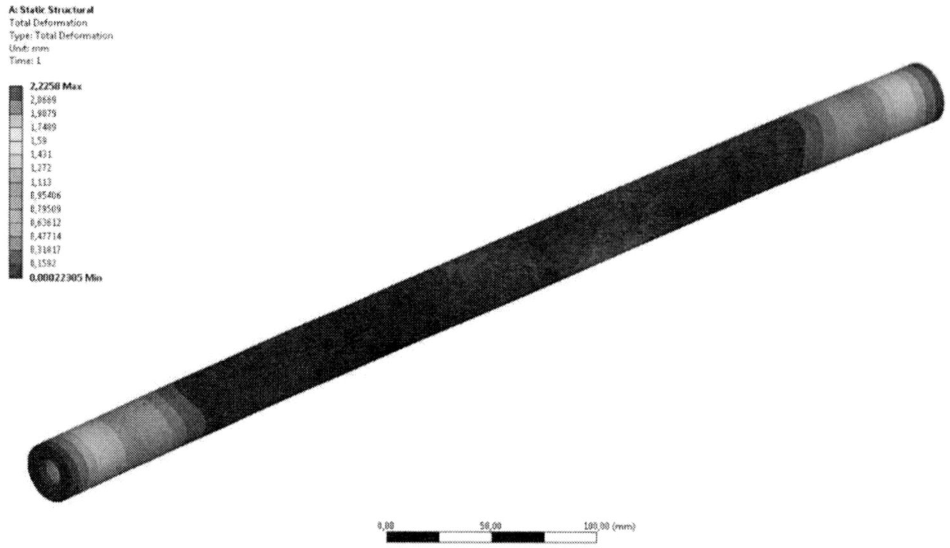

Fig. 7. Displacement distribution, mm

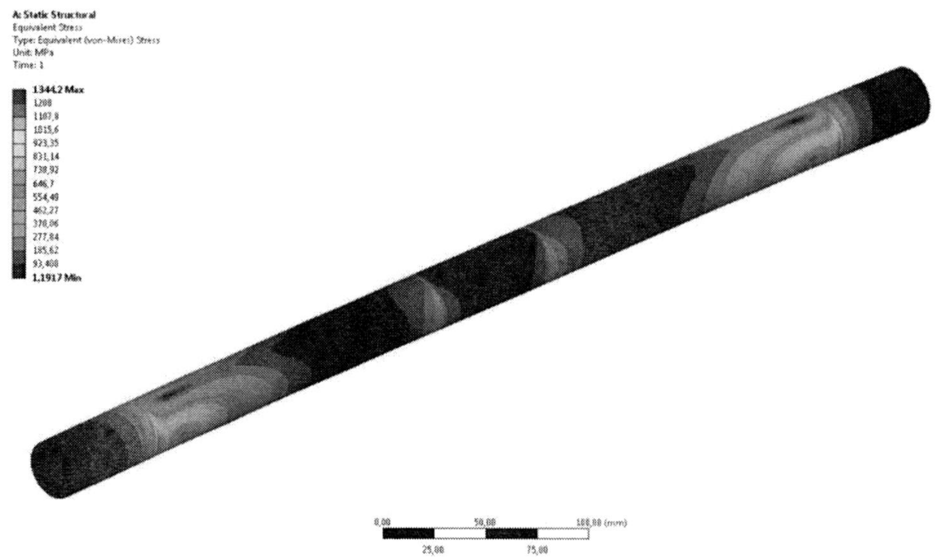

Fig. 8. Equivalent stress distribution, MPa

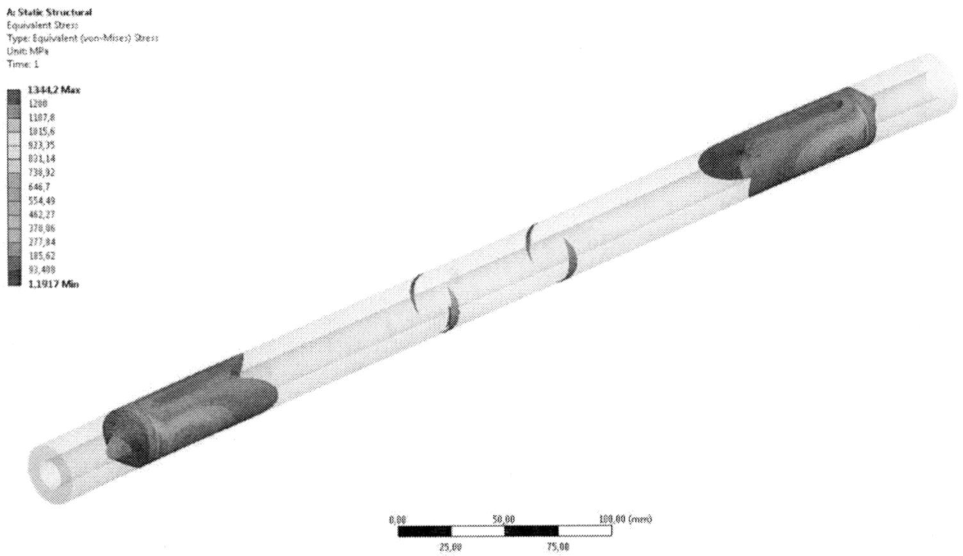

Fig. 9. Distribution of equivalent stress over 300 MPa, MPa

4. Results and Discussion

The analysis of simulated results allows us to conclude that the greatest displacement occurs in the pin interface with the track clamp.

It is most expedient to use an annular strain gage to assess pin loads. Its schematic design is shown in Fig. 10.

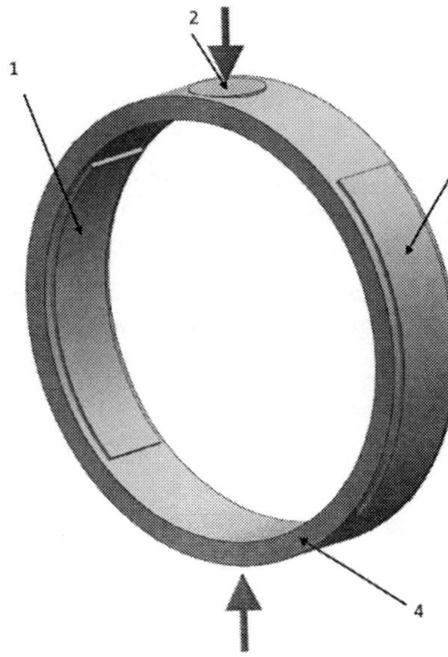

Fig. 10. Strain gage:
1, 3 — foil or wire resistor; 2 — load application pad; 4 — elastic member.

The annular strain gage provides data on the loading of RV track link.

Therefore, modern computer-aided engineering systems based on the finite element method allow to identify the most efficient locations to install gages. Computer-aided engineering systems enable us to select and design a gage, which in turn will allow necessary data to be obtained for AMCS.

References

[1] Naumov, V.N., Mashkov, K.Y., Byakov, K.E. Automatic determination of soil parameters by robotic vehicles (2019) Journal of Physics: Conference Series, 1177 (1)

[2] Ojeda, L., Borenstein, J., Witus, G., Karlsen, R. Terrain characterization and classification with a mobile robot (2006) Journal of Field Robotics, 23 (2), pp. 103-122

[3] Mashkov, K., Rubtsov, V., Rubtsov, I. Development of robotics technologies in agriculture (2018) MATEC Web of Conferences, 224

[4] Ostroukh, A., Vakhrushev, O., Maikov, K., Kolbasin, A. Combined identification method by reconstruction and analysis of face 3D structure (2019) ARPN Journal of Engineering and Applied Sciences, 14 (7), pp. 1385-1388.

[5] Naumov, V.N., Fedorenkov, A.P., Byakov, K.E. The research of the elastic propulsion unit and soil contact area shapes and sizes (2019) Journal of Physics: Conference Series, 1177 (1)

[6] Sarach E, Kotiev G and Beketov S 2018 Methods for road microprofile statistical data transformation *MATEC Web of Conf.* **224**

[7] Hutangkabodee, S. Soil parameter identification and driving force prediction for wheel-terrain interaction / Suksun Hutangkabodee, Yahya Zweiri, Lakmal Seneviratne, Kaspar Althoefer // International Journal of Advanced Robotic Systems. — 2008. — vol. 5. — No. 4. — pp. 425-432.

[8] Hutangkabodee, S. Traversability prediction for unmanned ground vehicles based on identified soil parameters / Suksun Hutangkabodee, Yahya H. Zweiri, Lakmal D. Seneviratne, Kaspar Althoefer // 16th IFAC World Congress: Conference Proceedings — Czech Republic, 2005. — No. 16. — p.1.

[9] Lee D H, Kim S K, Kim C S and Huh K S 2014 Development of an autonomous braking system using the predicted stopping distance Int. Journ. of Aut. Techn. 15 (2) pp. 341-346

[10] Jin P F, Kim J H and Kim J H. 2015 Design of unmanned vehicle advanced braking system using smart motor 12th Int. Conf. on Ubiquit. Rob. and Amb. Intell. pp 167-171

[11] Hsu J Y, Ku K L, Jhang T K, Lin H P and Yeh C J 2017 Integration and implementation of a lane keeping system with vehicle dynamics control Int. Autom. Contr.l Conf. 2017 November pp 1-6

[12] Lee D H, Kim S K, Kim C S and Huh K S 2014 Development of an autonomous braking system using the predicted stopping distance Int. Journ. of Aut. Techn. 15 (2) pp. 341-346

[13] Cerone V, Chinu A and Regruto D 2002 Experimental results in vision-based lane keeping for highway vehicles *Proc. of the Amer. Contr. Conf.* **2** pp 869-874

[14] Kirovskii O M and Gorelov V A 2019 Driver assistance systems: analysis, tests and the safety case. ISO 26262 and ISO PAS 21448 *IOP Conf. Ser.: Mat. Scien. and Eng.* **534(1)**

[15] Rao S J, Seitz T, Lanka V R R and Forkenbrock G 2019 Analysis and mathematical modeling of car-following behavior of automated vehicles for safety evaluation *SAE Techn. Pap.* **2019-April**

[16] Tominaga K, Takeuchi Y, Tomoki U, Kameoka S, Kitano H, Quirynen R, Berntorp K and Cairano S 2019 GNSS based lane keeping assist system via model predictive control SAE Tech. Pap. **2019 April**

[17] Cerone V, Milanese M and Regruto, D 2009 Combined automatic lane-keeping and driver's steering through a 2-DOF control strategy *IEEE Transact. on Contr. Syst. Techn.* **17 (1)** pp 135-142

[18] Beglerovic H, Schloemicher T, Metzner S and Horn M 2018 Deep learning applied to scenario classification for lane-keep-assist systems *Appl. Scienc. (Switzerland)* **8(12)**

[19] Stallkamp J, Schlipsing M, Salmen J and Igel C 2012 Man vs. computer: Benchmarking machine learning algorithms for traffic sign recognition *Neur. Netw.* **32** pp 323-332

[20] Sermanet P and Lecun Y 2011 Traffic sign recognition with multi-scale convolutional networks *Proc. of the Int. Joint Conf. on Neur. Netw.* pp 2809-2813

[21] Littell, Justin D. (2011), "Large Field Photogrammetry Techniques in Aircraft and Spacecraft Impact Testing", Dynamic Behavior of Materials, Volume 1, Conference Proceedings of the Society for Experimental Mechanics Series, Springer New York, pp. 55-67

[22] O.C. Zienkiewicz, R.L. Taylor The Finite Element Method Published August 28th 2000 by Butterworth-Heinemann

[23] William McGuire, Richard H. Gallagher, Ronald D. Ziemian Matrix Structural Analysis, 2nd Edition Published July 16th 1999 by John Wiley & Sons Inc

IOP Conf. Series: Materials Science and Engineering **820** (2020) 012014 doi:10.1088/1757-899X/820/1/012014

The stress-strain state and optimization of the cross section of the wheel rim tubeless truck tire

I Balabin[1], M Lyamin[2], I Chabunin[3,4] and V Ryazantsev[1]

[1]Bauman Moscow State Technical University
[2]Russia 41800 Moscow region, Dmitrov, Kosmonavtov str. 33, Branch of the state budgetary educational institution of higher education of the Moscow region "Dubna University" — Dmitrov Institute of continuing education, Department of " Road transport and construction of buildings and structures"
[3]Russia 109380 Moscow, Golovacheva street 2, Moscow Higher All-Arms Command School, Department of General professional disciplines

[4]E-mail: tchabunin@rambler.ru

Abstract. This article is devoted to the issues of strength and reliability of the wheel, as well as minimizing its weight, which allows to improve traction and brake dynamics, smooth running, fuel efficiency, etc. The solution of the problem of determining the stress-strain state of the automobile wheel with fifteen-degree landing shelves is presented. In the article it is offered to equalize the stresses arising at internal and external surfaces of a rim of a variable thickness, at the expense of variation of thickness of a rim of a car wheel in various zones.

Introduction

As you know, the wheel is one of the most important elements of the car, which has a direct impact on its basic functional qualities, including reliability, safety in operation [1–5].

Despite the availability of a sufficient number of publications related to wheel problems, the issues of strength and reliability of wheel rims remain largely unresolved. Naturally, the requirements of ensuring the guarantee of strength and reliability of the wheel, minimizing its weight can be fully satisfied only with the use of scientifically sound methods of calculation used in the design of such a responsible and important unit of the car. The absence of such methods to date has led to the need to provide for obviously excessive reserves of strength, eliminating the risk of destruction, but inevitably causing overloading of the wheel with all the ensuing negative consequences, which worsened the traction and braking dynamics, smoothness, fuel efficiency, etc., and also, given the mass production of wheel products, led to the production of a giant overrun of scarce metal products, estimated in thousands of tons annually [6–9].

Currently, the efforts of specialists and scientists of the scientific school under the guidance of doctor of technical Sciences, Professor Balabin IV developed a set of methods of calculation and design of automobile wheels, allowing you to create structures that have the necessary strength at a minimum weight and inertial characteristics.

Problem statement

As you know, the dominant power factor for the rim of a vehicle wheel on which to base the calculation of the stress-strain state (VAT), the latter is the internal air pressure in the tire, the value of which shall be such that with a reserve to counter any (of course, real) force; otherwise, it will cause the loss of bearing capacity of tires with all the negative consequences. In this regard, the prediction of the contact interaction diagram of the tire and the rim of the car wheel, as a necessary input data, is an urgent task. In addition, the plot will not only determine the value of the rim, but also to check the reliability of the tire landing on the rim, excluding its rotation relative to the rim from the action of the moment due to extreme traction or braking forces in the spot of contact with the tire bearing surface.

Consider the problem of determining the stress-strain state of the rim of the automobile wheel with fifteen-degree landing shelves, widely used at present for trucks and buses.

The solution of such a problem only using the finite element method (FEM) is extremely difficult, since the tire is a shell with a complex generatrix consisting of various parts (frame, breaker, protector, etc.), firmly connected to each other; its modeling is a rather complex and time — consuming task.

We divide this task into two stages: the initial stage - the construction of diagrams of the force interaction of the tire with the rim, the second-the definition of the RIM VAT. Thus for the second step you can enter a simplification, since the junction of the rim with the wheel disc, the rigidity of the shell, at least an order of magnitude greater stiffness of the other zones, it is possible to consider only half of the rim, for which the boundary conditions are zero linear and angular movement at one end of the shell, at the location of the disk, and free the other end of the shell.

Decision

Consider the real diagram of the force interaction between the tire and the rim, obtained by measuring the system based on the electro-hydraulic sensor (Fig. 1) [10]. For Fig.2, and is represented by half the wheel rim with fifteen-degree landing shelves.

We divide the profile of the rim part AB, experiencing variable pressure, into two parts-AC and CB, as shown in Fig.2, a. Place the beginning of the z coordinate used to describe the right side of the plot in section C, located under the side ring of the tire and corresponding to the maximum pressure, and direct it along the rim sweep. For a positive take the direction of reference to the section B.

To construct the right side of the power plot, there are three boundary conditions for the two extreme points (Fig. 2, a): at $z = b$ $p = 0$; at $z = 0$ $p = p_{\max}$; at $z = 0$ $\frac{dp(z)}{dz} = 0$, since at this point there is an extremum of the total force plot.

Since there are three boundary conditions, we use the equation:

$$p(z) = \left(Az^2 + Bz + C\right)^n, \tag{1}$$

where A, B, C and n are the coefficients to be determined;

After its solution, the equation of the right part (2) of the power plot takes the form:

$$p(z) = p_{\max}\sqrt{1 - \frac{z^2}{b^2}}, \tag{2}$$

As shown by the analysis of the experimentally obtained diagrams of the force interaction between the tire and the rim with fifteen-degree landing shelves, the most acceptable, that is, at the same time having sufficient accuracy and ease of calculation, to describe the left side of the power diagram is the power dependence of the form:

$$p(z_1) = A \cdot e^{-B \cdot z_1^n}, \tag{3}$$

where A, B, n are the coefficients to be determined;

IOP Conf. Series: Materials Science and Engineering **820** (2020) 012014 doi:10.1088/1757-899X/820/1/012014

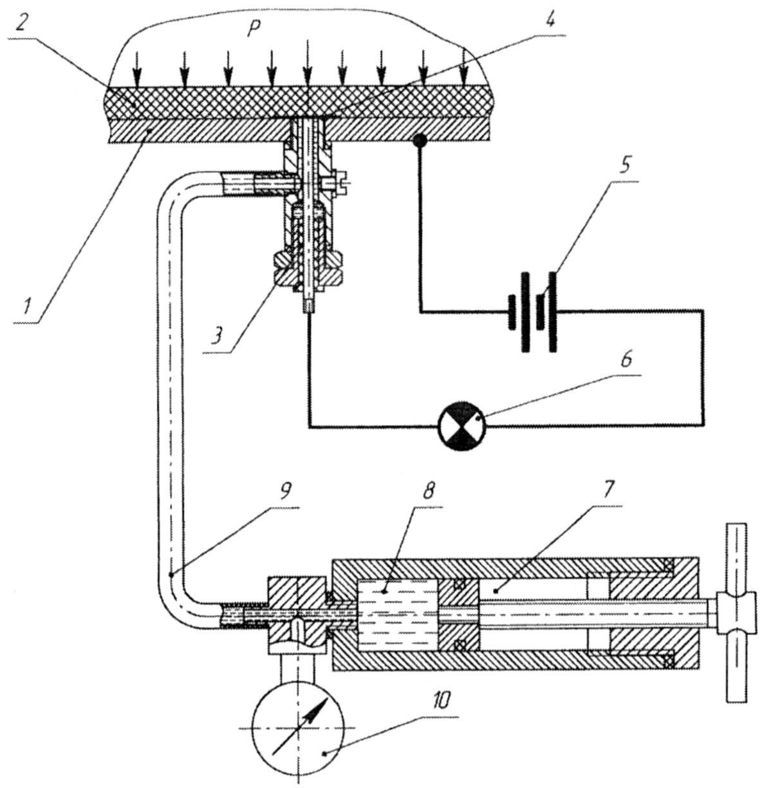

Figure 1. Diagram of the measuring installation on the basis of electrohydrostatic, where 1 — rim; 2 — tire; 3 — electrohydraulic sensor; 4 — conductive layer; 5 — electric power supply; 6 — indicator lamp; 7 — hydraulic cylinder; 8 — working fluid; 9 — hydraulic slang; 10 — pressure gauge.

After solving the equation describing the left part of the power plot has the form:

$$p(z) = p_{\max} \cdot e^{-10\left(\frac{z_1}{c}\right)^{3,065}}. \tag{4}$$

From the obtained dependences it can be seen that the maximum pressure is unknown. It is determined by comparing the magnitude of the axial force acting on the considered part of the rim AB (except for the friction force) [11]:

$$T = \pi p\left(R_1^2 - R_2^2\right), \tag{5}$$

where p is the tire air pressure;

R_1 — radius of zero curvature of the tire;

R_2 — radius of the cylindrical part of the rim

and the axial force in fractions obtained from the analytical expressions for pressure (4) and (6).

Sections AG and GC (Fig.2, a) are a conical shell, the section CB is divided into sections CD, DE, ED, representing, respectively, a conical shell and shells of negative and positive Gaussian curvature. The values perceived by each section of the axial force T denote respectively by T_{AG}, T_{GC}, T_{CD}, T_{DE}, T_{EB}. Their sum will be equal to the total axial force T (excluding friction force T_{mp}) in fractions $p_{\max}$. Equating it to the value of the force T, obtained by dependence (7) (minus T_{mp}), determine the value $p_{\max}$.

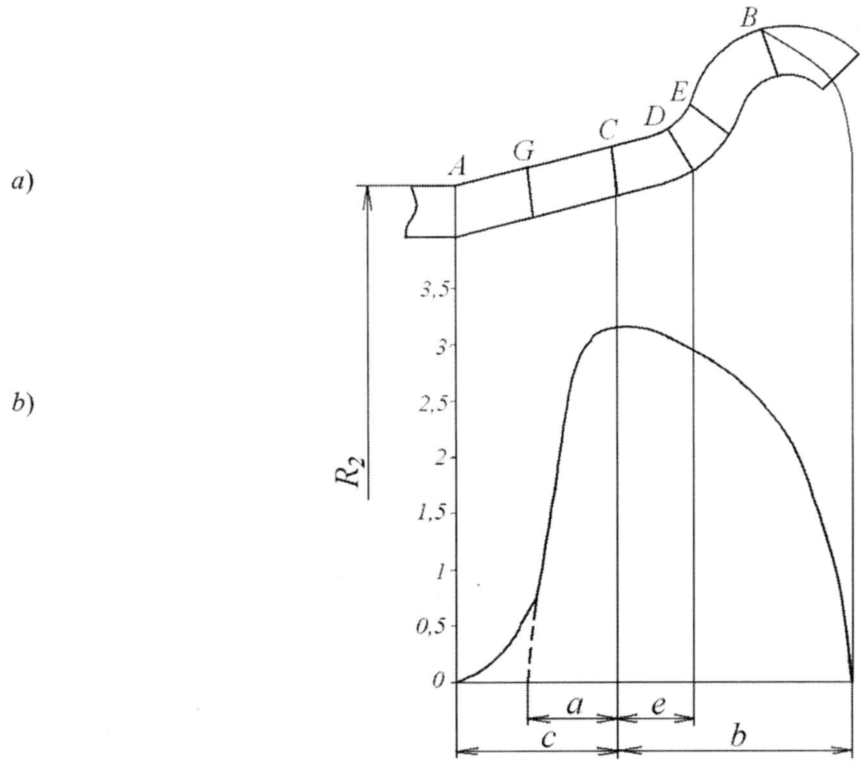

Figure 2. Diagram of the force interaction of tubeless tires and rims with fifteen-degree landing shelves

Consider, for example, a section of CD. Let us denote its current radius by R, as shown in Fig. 3, and take it as an argument to the load balancing function $p(z)$. The expression for determining the axial force acting on the elementary ring of this section will be as follows:

$$dT_{CD} = p(z) \cdot 2\pi R \cdot \sin\alpha \cdot dz, \tag{6}$$

where α is the angle of the landing shelf.

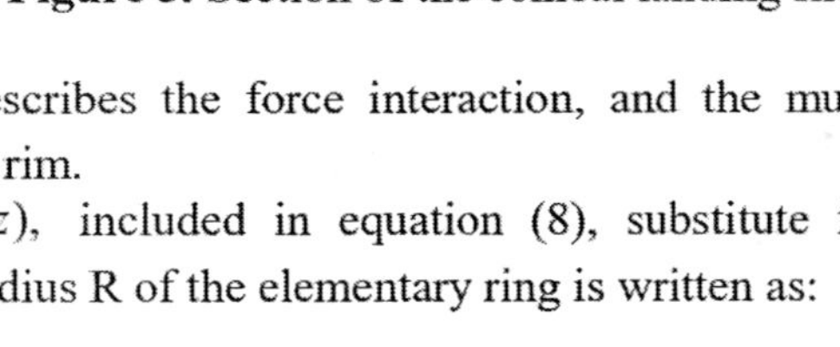

Figure 3. Section of the conical landing shelf

The multiplier $p(z)$ describes the force interaction, and the multiplier $2\pi R \cdot \sin\alpha \cdot dz$ — the geometric parameters of the rim.

The expression for $p(z)$, included in equation (8), substitute from expression (4), and the expression for the current radius R of the elementary ring is written as:

$$R = r + \left(z + \frac{dz}{2}\right)\sin\alpha. \tag{7}$$

IOP Conf. Series: Materials Science and Engineering **820** (2020) 012014 doi:10.1088/1757-899X/820/1/012014

Therefore, the axial force acting on the entire section of the CD will be written as:

$$T_{CD} = 2\pi\int_0^d p_{\max}\sqrt{1-\frac{z^2}{b^2}}\left(r+\left(z+\frac{dz}{2}\right)\sin\alpha\right)\sin\alpha dz. \tag{8}$$

Thus, the unknown quantity is also the maximum pressure $p_{\max}$.

Similarly, determining the axial forces for other sites and summing the results, we obtain an expression for the total axial force T (excluding friction force T_{mp}) in fractions $p_{\max}$.

Equating the resulting expression for T in fractions $p_{\max}$ to the value of T, obtained by the formula (7) (minus T_{mp}), determine the value $p_{\max}$ and then the dependencies (4) and (6) plot the force interaction between the tire and the rim, as shown in Fig. 2,b.

The first stage of calculation is completed.

The stress-strain state of the wheel rim can then be determined. To do this, we use the finite element method, in particular, the widely used Nastran software complex.

Since the rim of the wheel with fifteen-degree landing shelves is a thin-walled shell, the plate shell element, which perceives membrane, shear, transverse, bending loads, was used as a finite element [12].

Results

The results of calculation in the form of distribution fields of equivalent stresses arising at the inner and outer surfaces of the rim of constant thickness and obtained using the Nastran software package are presented in Fig. 4, a and Fig. 4, b respectively. As seen in Fig. 4, the stresses are distributed extremely unevenly. The most loaded sections are the landing shelf and the transition zone of the Assembly stream in the cylindrical part. The peak of stress falls on the last. Stresses in all other areas are within normal limits. The least is at the bottom of the mounting Creek.

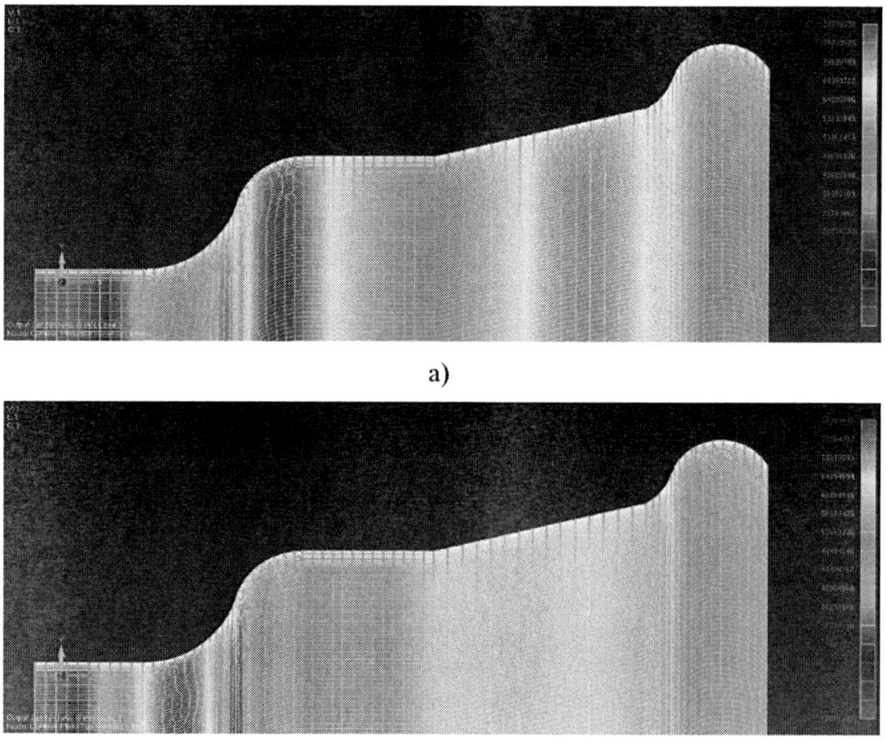

a)

b)

Figure 4. Equivalent stresses in the wheel rim of constant thickness

Thus, it is possible to reduce the weight of the rim by equalizing the stresses by varying its thickness in different zones; the outer surface of the rim adjacent to the tire should remain unchanged. The initial design of the rim of constant thickness is shown in Fig. 5 dashed line.

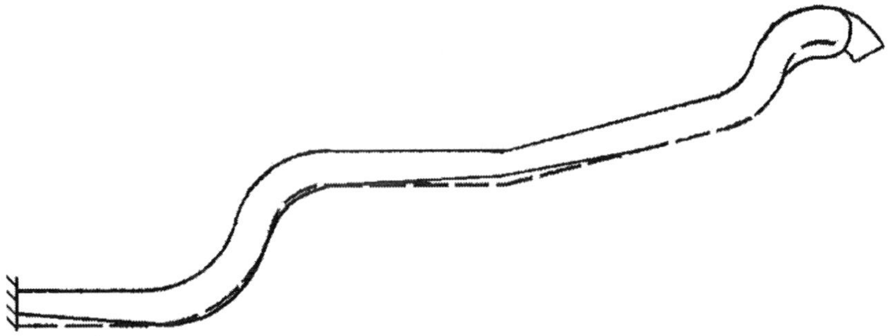

Figure 5. Rim Profiles of constant and variable thickness

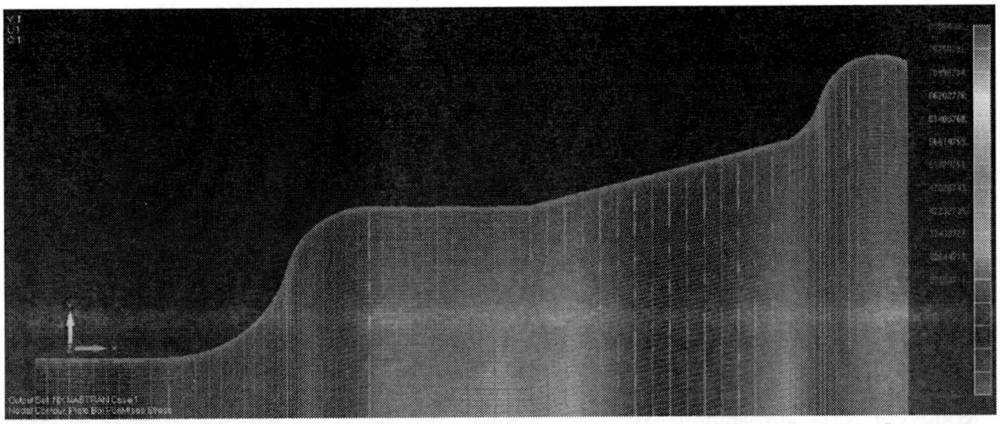

a)

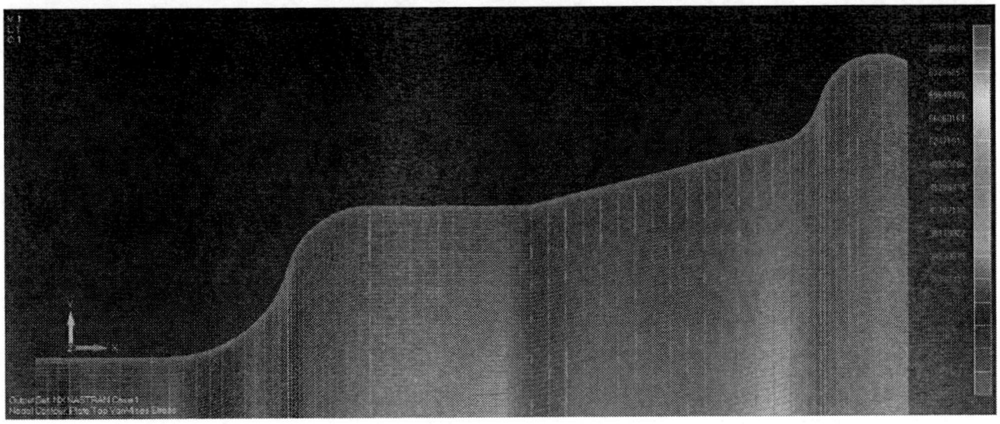

b)

Figure 6. Equivalent stresses in the wheel rim of variable thickness

First, to protect the rim from accidental side impacts, reduce its width, thereby increasing the overlap of the rim with the sides of the tire, but compensating for the decrease in stiffness by increasing its thickness, as shown in Fig. 5. Secondly, slightly increase the thickness in the most intense areas of the rim-radius transitions in the wall of the stream. Due to these measures, it is possible to reduce the thickness of less loaded zones, in particular, the bottom of the stream, the cylindrical part located between the stream and the landing shelf, and partially the conical landing shelf. The proposed rim design is shown in Fig. 5 solid line.

First, to protect the rim from accidental side impacts, reduce its width, thereby increasing the overlap of the rim with the sides of the tire, but compensating for the decrease in stiffness by increasing its thickness, as shown in Fig. 5. Secondly, slightly increase the thickness in the most intense areas of the rim-radius transitions in the wall of the stream. Due to these measures, it is possible to reduce the thickness of less loaded zones, in particular, the bottom of the stream, the cylindrical part located between the stream and the landing shelf, and partially the conical landing shelf. The proposed rim design is shown in Fig. 5 solid line.

The distributions of equivalent stresses occurring at the inner and outer surfaces of the rim of variable thickness are shown in Fig. 6, a and Fig. 6, b respectively.

Summary

As a result of the optimization of the voltage, as follows from the comparison Fig. 4 and 6 are distributed along the generatrix more evenly, the equal strength of the rim increases, while the maximum value of equivalent stresses in the proposed design not only does not exceed the maximum stresses arising in the original version, but on the contrary, decreases by about 7%, despite the fact that the mass has decreased by about 12 %, which on the scale of mass production will give significant savings of metal.

References

[1] Kotiev G. O., Miroshnichenko A. V., Stadukhin A. A., B. B. Kositsyn Estimating operation modes for the individual wheel electric drive of the all-wheel drive vehicle with the use of the driving simulator. Volume 534, Issue 1, 12 June 2019, the International Automobile Scientific Forum: Intelligent Transport System Technologies and Components, IASF 2018

[2] Evseev, K. B., Kartashov, A. B., Dashtiev, Z. I., Pozdeev, A. V. Analysis viscoelastic properties of fiber-reinforced composite spring for the all-terrain vehicle, MATEC Web of Conferences 224, 02039 (2018).

[3] Kotiev G. O., Miroshnichenko A. V., Stadukhin A. A., Kositsyn B. B. Determination of mechanical characteristics of high-speed tracked vehicles traction motor with individual drive wheels. Journal of Physics: Conference Series. Volume 1177, Issue 1, 10 April 2019, No. 012058, 3rd Science and Technology Seminar on Mobility of Transport and Technological Machines, MTTM 2018; Nizhny Novgorod; Russian Federation; 28 August 2018

[4] Kotiev G. O., Butarovich D. O., B. B. Kositsyn Energy efficient motion control of the electric bus on route. Volume 315, Issue 1, 22 February 2018, 2017 International Automobile Scientific Forum, IASF 2017

[5] Gorelov, V. A., Komissarov, A. I., Miroshnichenko, A. V. 8×8 wheeled vehicle modeling in a multibody dynamics simulation software (2015) Procedia Engineering, 129, pp. 300–307.

[6] Klubnichkin, V. E., Klubnichkin, E. E., Kotiev G. O., Beketov, S. A., Makarov, S. V. Interaction between elements of the track ground contacting area with the soil at curvilinear motion of the timber harvest-ing machine (2018) IOP Conference Series: Materials Science and Engineering, 386 (1), article number 012016.

[7] V. V. Novikov, V. A. Pozdeev, A. S. Diakov, Research and testing complex for analysis of vehicle suspension units, Procedia Engineering, Vol. 129, pp. 465–470 (2015) DOI: 10.1016/j.proeng.2015.12.153

[8] Kotiev G. O., Padalkin, B. V., Kartashov, B. A., Diakov, A. S, Designs and development of Russian scientific schools in the field of cross-country ground vehicles building, ARPN Journal of Engineering and Applied Science, Vol. 12, Issue 4, 2017, pp. 1064–1071

[9] Novikov, V. V., Pozdeev, A. V., Diakov, A. S., Research and testing complex for analysis of vehicle suspension units, Procedia Engineering, 129, pp. 465–470

[10] Balabin, I. V., Zaitsev S. A. Chabunin I. S., Lyamin M. S., Lyamin S. N. The measuring system of the force interaction between rim and tire car wheels // 18 Symposium "Problems of tires and rubber-cord composites". M.-15 - 19 October 2007., pp. 61-72.
[11] Biderman V. L. et al. Car tires. — M.: Mashinostroenie, 1963. — 384c.
[12] Rychkov S. P. modeling of structures in Femap with NX Nastran. — Moscow: DMK Press, 2013. — 784 p.

IOP Conf. Series: Materials Science and Engineering **820** (2020) 012015 doi:10.1088/1757-899X/820/1/012015

Simulation model of a transport vehicle with a fixed-ratio transmission and a flywheel energy storage in case of random external action

V Korsunskiy[1,2]

[1]Bauman Moscow State Technical University

[2]E-mail: vakors@bmstu.ru

Annotation. The purpose of this study is to evaluate the dynamic qualities and efficiency of a heavy-duty transport vehicle equipped with a mechanical transmission, using a combined power plant (CPP) with a flywheel energy storage (FES), establish an acceptable level of power reduction of the main internal combustion engine (ICE), determine the range of installed power of the FES drive and its average value of efficiency, as well as the calculation of the level of required energy capacity of the flywheel when driving in various types of terrain. A universal technique for simulating statistical modeling of external disturbances acting on the transport vehicle has been developed. Using the developed methodology, the vehicles' characteristics have been compared according to the criterion of average speed and an assessment of the efficiency of the use of CPP has been carried out. The results obtained allow us to carry out a comparative analysis of vehicles equipped with a conventional or combined power plant with a random external action of the road, as well as optimize their parameters.

Introduction (Immediacy of the problem)

The feasibility of using a combined power plant (CPP) on transport vehicles (TVs), which uses a flywheel energy storage (FES) as an additional energy source, has been proved by a number of scientific studies: [1-7]. An example of a practical use of a device that accumulates the braking energy of a vehicle and develops great power during acceleration using a fast-rotating flywheel is a system developed by Flybrid Systems for racing cars [8].

To use CPP with FES in TVs of various weight categories, the Oak Ridge National Laboratory (ORNL) conducted a comparative analysis of their characteristics and designs [9]. Tests of heavy-duty vehicles equipped with CPP [9], conducted by a number of leading US and Western European companies, made it possible to determine the level of energy accumulated (delivered) by the flywheel equal to 2.0 kWh (7.2 MJ) with a peak (short-term) power developed by the FES drive of 150 ~ 200 kW. In addition, to date, extensive experience has been gained in the countries of the European Union in operating fast-rotating flywheels as part of CPPs used for heavy-duty urban vehicles, for example, buses with ICEs and various electric vehicles: [10-13].

Statement of the problem, assumptions, design formulas and limitations

The purpose of this study is to evaluate the dynamic qualities and efficiency of heavy-duty TVs equipped with CPP and mechanical transmission, determine the acceptable level of power reduction of the main engine while maintaining the average speed of the vehicle, calculate the required energy capacity of the flywheel, the drive's installed power and its average value of efficiency when driving at random terrain.

The external environment action on TV is represented by a vector $\bar{F}$ with random components possessing the properties of stationarity and ergodicity

$$\bar{F} = \left(f_{\text{gr}}, \alpha, \phi_{\text{ad}}, \mu_{\text{max}}, h, k_{\text{r}}\right)^{\text{T}}, (1)$$

where f_{gr} is the coefficient of ground resistance to the rectilinear movement of the vehicle; α is slope angle of the terrain profile; φ_{ad} is adhesion factor; μ_{max} is maximum coefficient of resistance to rotation of the vehicle; h is the height of the irregularities of the micro profile of the path; k_{r} is road curvature of the path.

An accurate description of the processes occurring in the mathematical model elements describing the transient motion of TV with CPP, considered, for example, in [14], complicates the modeling algorithm. The difficulty of solving the problem using computer due to the complexity of solving a system of nonlinear differential equations when calculating one random implementation of the vector $\bar{F}$ minimizes the total number of possible implementations.

The level of formalization of a system of differential equations describing the motion of the TV with CPP can be various. In simulation modeling, the models that are implemented in the form of difference equations are most often used.

In relative form, the formalized motion equation of the center of mass of TV with CPP appears as follows:

$$\frac{d\bar{V}_c}{dt} = \frac{g}{\delta}\left(f_{\text{dv}} \pm f_{\text{m}} - f_{\text{c}}\right), (2)$$

where $\bar{V}_{\text{c}}$ is the velocity vector of the center of mass of TV; t is time; f_{dv} is specific thrust of ICE realized in the soil, taking into account losses in the transmission and propulsion of the TV; f_{m} is the specific traction force developed by the FES in the soil, taking into account losses in the drive [15], the transmission (in case the drive is connected to the drive wheels through the transmission) and the propulsion; f_{c} is total coefficient of resistance during curvilinear motion of the TV; δ is coefficient of conditional mass increment of the TV; g is gravity acceleration.A negative value f_{m} means that the FES is in charge mode.

The total coefficient of resistance to the rectilinear motion of the TV is determined by the known relationship

$$f_{\text{st}} = f_{\text{gr}}(\cos\alpha + \sin\alpha). (3)$$

When driving along a random path, the external action exerted on the TV, determined by the right-hand side of equation (2), is taken into account by the relationship of the total coefficient of resistance to movement of the vehicle f_{st} and travelS. For this purpose, it is necessary to set the distribution function of the coefficient f_{st} along the path, i.e. $\Phi_s(f_{\text{st}})$. Graphs $\Phi_s(f_{\text{st}})$ for two types of terrain are shown in Fig. 1.

To simplify the solution of the problem, we assume that the components $f_{\text{gr}i}$ and $k_{\text{r}i}$ of the random vector $\bar{F}$ specified by formula (1) have constant values on the intervals of the path ΔS .The value ΔS determines the degree of detail of the road.The choice of a small value ΔS makes it possible to take into account disturbances that affect the dynamic loading of the transmission units and the vehicle's cushioning system more than the average speed.

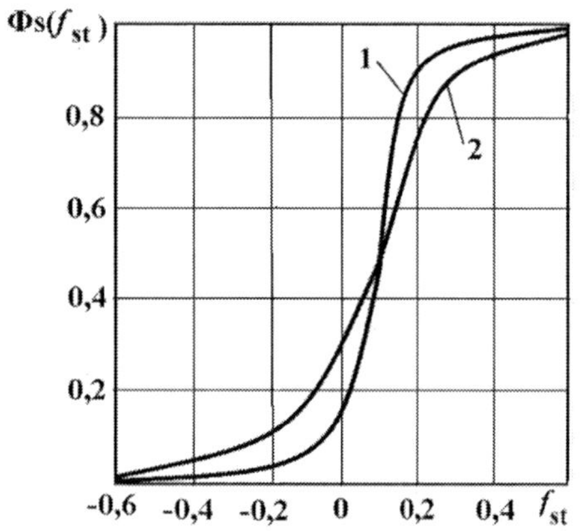

Figure 1. Distribution functions of the total coefficient of resistance to rectilinear motion:
1 - mildterrain; 2 - ruggedterrain.

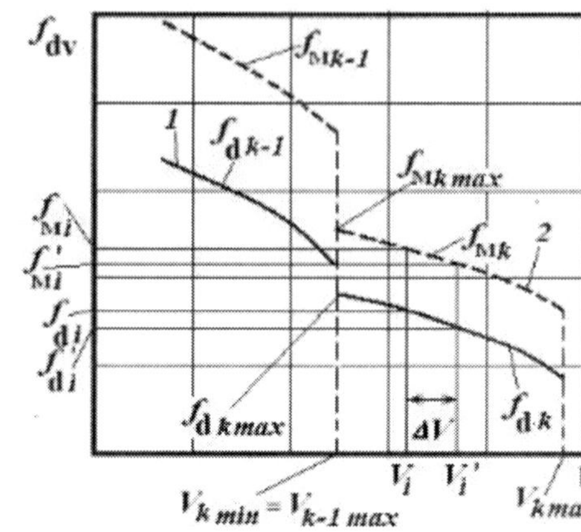

Figure 2. Traction characteristic of the vehicle with a mechanical transmission:1 - with ICE; 2 - with CPP.

According to equation (2), it is possible to determine the acceleration of TV with CPP on each section ΔS of the road using the traction characteristic (Fig. 2).

To construct the traction characteristic of the TV with a general power plant (GPP), it is necessary to have an experimental external characteristic of the ICE (i.e., the relationship between the torque and speed and control parameter), the values of the general gear ratios i_o of the transmission in various gears in the gear box, the drive wheel radius R_{dw}, and also calculated or experimental relationships of the overall efficiency η_o or losses vs load and speed of the vehicle [16].For vehicles with CPP, you also need to know the maximum power transmitted by the FES drive, as well as the relationship of the losses in the drive vs the load and speed.

The critical speed at which the sidewise skidding of the vehicle begins is found from the formula

$$V_{cr} = V_c \leq \sqrt{\frac{g\mu_{max}}{k}}, \quad (4)$$

where$k$ is the real curvature of the TV movement, which differs from the road curvature k_r and depends, in particular, on the type of steering mechanism.

The traction characteristic of TV with GPP is based on the experimental points obtained, for example, using spline interpolation.

A similar characteristic for TV with CPP differs by a constant value within the speed change in the k-th gear in the box by $f_{mk\,max}$ (see Fig. 2).The limit value $f_{mk\,max}$ is limited by the strength of the hydrostatic (or power transmission) and mechanical branches of the FES drive, i.e. maximum torque (or maximum power) transmitted by the drive.In the case of using frameless double-sided synchronous DC motors in the adjustable drive branch, power losses can be taken into account according to the method described in [17].

In the calculations, the assumption is made that switching to the highest gear occurs when the TV reaches the maximum speed value in this gear. That is, with $V_{ci} = V_{k-1\,max}$, the gear box changes from (k-1)-th to k-th gear (see Fig. 2).

In contrast to that proposed in [18], the control algorithm of the vehicle is chosen so that when switching from the (k-1)-th to k-th gear, the following conditions are satisfied:

$$\left.\begin{array}{l} V_{k-1\max} < V_{\text{perm}i}, \\ f_{\text{d}k\max} + f_{\text{m}k\max} \geq f_{ci}(S_i) \text{ when } \Delta S < S_i < \Delta S + \Delta L, \\ E_{\text{m}i} = E_{\text{m}\max}, \end{array}\right\} \quad (5)$$

where $V_{k-1\max}$ is the maximum speed of the car in (k-1)-th gear in the gearbox;

$V_{\text{perm}i}$ – permissible speed of the vehicle on the i-th section of the road;

$f_{\text{d}k\max}$ and $f_{mk\max}$ – maximum values of specific thrust forces developed, respectively, by ICE and FES on propulsion devices when gearbox is engaged in the k-th gear (see Fig. 2);

S_i and ΔL - the current value of the path and the section of the road visually assessed by the driver;

$E_{\text{m}i}$ and $E_{\text{m}\max}$ – the current and maximum energy value (maximum energy capacity) of the flywheel energy storage.

The permissible speed of the vehicle on the i-th section of the road is determined by the formula

$$V_{\text{perm}i} = \min(V_{\text{cr}i}, V_{\text{cush}i}), \quad (6)$$

where $V_{\text{cr}i}$ is the critical speed of the TV in the i-th section of the road, determined by formula (4) for random values $\mu_{\max i}$ generated by the computer and real curvature k_i and their distribution functions along the path); V_{cushi} - TV speed on the i-th section of the road, limited by the cushioning system and determined for random values h_i generated by the computer and obtained using the speed characteristics of the cushioning system $h(V)$ and the given distribution function $\Phi_s(h)$ for height of the irregularities along the path [19].

Switching from the k-th gear to the $(k-1)$-th gear occurs when the TV reaches the minimum speed in this gear $V_{k\min} = V_{k-1\max}$ in the following cases (Fig. 2):

a) when the minimum speed in the k-th gear is greater than the permissible speed, i.e. at $V_{k\min} > V_{\text{perm}i}$;

b) under condition $f_{\text{st}k\max} + f_{\text{m}k\max} < f_{ci}$.

Since the traction characteristic of the vehicle (Fig. 2) is non-linear in nature, in order to take the movement equally variable, it is necessary to refine the current values $f_{\text{st}ki}$ and $f_{\text{m}ki}$ even when the vehicle passes sections of the road less than the value set at the beginning of the calculation of the constant value ΔS.

The above assumptions allow us to proceed to the determination of the vehicle with CPP parameters without solving a system of nonlinear differential equations.

The movement of the vehicle with CPP at a speed V_i when the k–th gear is engaged on the i–th road section will be described in a generic form by the system:

$$\left\{\begin{array}{l} \dfrac{dV_{ci}}{dt} = \dfrac{g}{\delta_k}(f_{\text{dv}\,ki} \pm f_{\text{m}\,ki} - f_{ci}), \\ \Delta f_{\text{dv}i} = \psi_1(\text{M}_{\text{dw}i}; n_{\text{dw}i}; u_{\text{dv}}), \\ \Delta f_{\text{m}i} = \psi_2(\text{M}_{\text{m}i}; n_{\text{dv}i}; u_{\text{fw}}), \\ \Delta E_{\text{m}i} = \psi_3(f_{\text{m}i}; \Delta f_{\text{m}i}; n_{\text{dv}i}; u_{\text{fw}}), \\ V_{ci} \leq V_{\text{perm}i}, \end{array}\right. \quad (7)$$

where δ_k is the coefficient of the conditional increment of the vehicle mass in the$k$–th gear; $\Delta f_{дi}$ and Δf_{mi} - total losses in the transmission and propulsion, as well as losses in the FES drive; M_{mi} and M_{dwi} are the torques on the FES drive shaft and on the driving wheels; n_{dvi} and n_{dwi} - rotational speed of the ICE shaft and driving wheels; ΔE_{mi} - change in FES energy on the i-th road section; u_{dv} and u_{fw} - control parameters of the ICE and FES drive.

Quantitative values of Δf_{dvi} and Δf_{mi} depend on the type of transmission and drive FES. For comparative calculations, the values of Δf_{dvi} and Δf_{mi} are assumed constant on each gear in the gearbox.

Analysis of the simulation results. Quantitative calculations were obtained for a vehicle with a general ICE, as well as for a TV with a CPP with a mechanical transmission, the kinematic scheme and numerical characteristics of which are given in [20].

The maximum speed of TV adopted in the calculations was V_{max} = 16.91 m/s (61 km/h).In the calculations, the section of the road ΔL estimated by the driver ranged from 20 to 100 m.

The relative energy capacity E_o of the flywheel energy storage was determined by the formula

$$E_o = {E_{m\,max}}/{E_{v\,max}}, \quad (8)$$

where $E_{v\,max}$ is the kinetic energy of the TV at maximum speed.

The maximum installed power of the FES hydrostatic-mechanical drive adopted in the calculations was 258 kW.

Figure 3 shows the relationships of the TV average speed V_{av} and the ICE load factor $\bar{z}_{dv}$ vs the relative energy capacity E_o determined by the formula (8).

Figure 4 shows the curves of changes in the vehicle average speed V_{av} depending on the limitation $V_{perm} = const$.

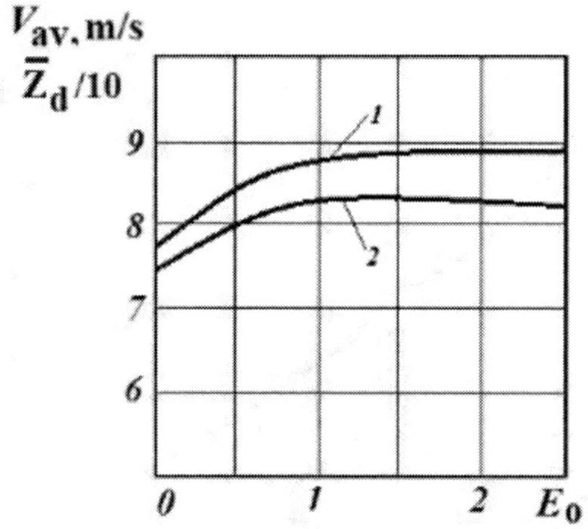

Figure 3. 1 – ICE loading $\bar{z}_{dv}$; 2 – TV average speed V_{av}.

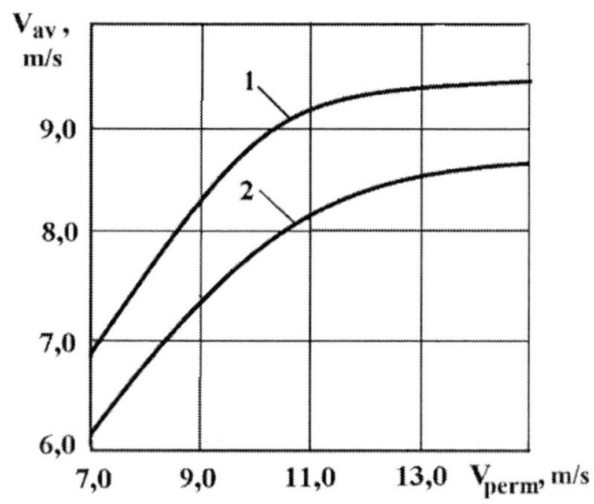

Figure 4. Graphs of the TV average speed for various values of the speed limit: 1 - TV with CPP; 2 - TV with ICE

Figure 5 illustrates graphs of changes in the speed of TV with CPP, TV with a conventional ICE, the current energy value of the flywheel energy storage E_m and the total coefficient of resistance to rectilinear movement f_{st} when it is linearly interpolated along the path S (a fragment of unsteady TV movement over mild terrain with a length of 1500 m with mean value along the path $m_s(f_{st}) = 0.104$ and dispersion $D_s(f_{st}) = 0.00728$). The relative energy capacity of the flywheel was $E_o = 1$ ($E_{m\,max} =$ = 5.71 MJ) with the mean value of the FES drive efficiency along the path $m_s(\eta_m) = 0.8$. At the beginning of the TV with CPP movement, the flywheel was not charged.

To determine the possibility of reducing the installed power of ICE in the case of using CPP, it is necessary to carry out a comparative calculation of vehicles with CPP and conventional ICE with the equality of their average speeds (i.e., at $V_{av\,CPP} = V_{av\,CICE}$).

Fig. 6 shows the curves of changes V_{cp} depending on E_o for various values of installed power and the mean values of the flywheel drive efficiency $m_s(\eta_m)$ along the way when the vehicle is moving in mild terrain. The value of ICE free power in the calculation was 383 kW (80% of the nominal). An analysis of the curves shows that only with high mean values of the drive efficiency ($m_s(\eta_m)$= 0.9) and with the installation power of the FES drive $N_{M\,max}$ equal to 440 kW, a flywheel with a relative energy capacity $E_o = 1.5$ (i.e., at E_{mmax}= 2.475 kW•h) provides the same average speeds for a conventional vehicle with a nominal ICE power of 485 kW and for a TV with CPP with ICE power reduced by 20%.

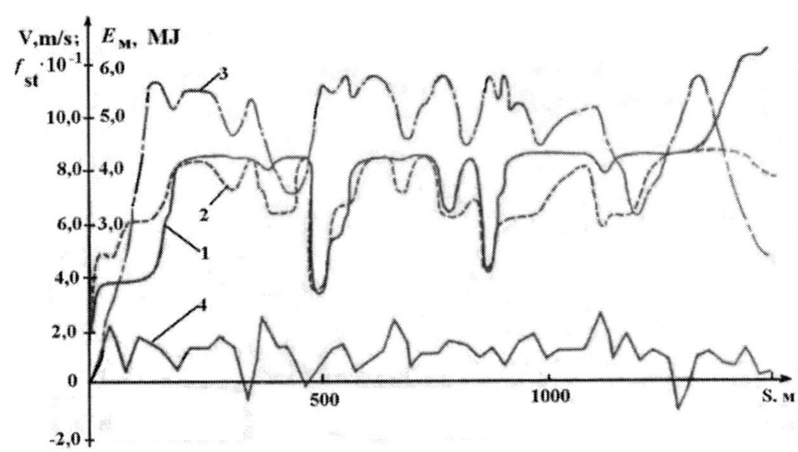

Figure 5. Graphs of changes in the TV parameters along the path: 1 - speed of TV with CPP; 2 - speed of TV with ICE; 3 - FES energy; 4 - coefficient of total resistance to rectilinear motion

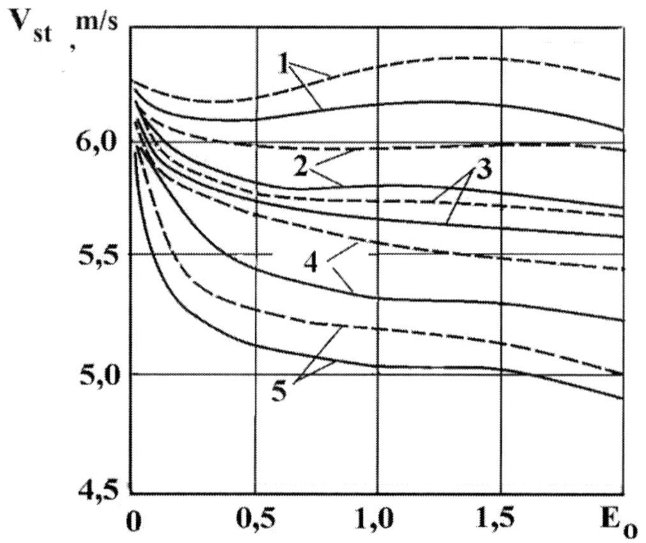

Figure 6. Curves for different values of the installed power of the FES drive: 220 kW - solid lines; 440 kW - dotted lines. 1 – $m_s(\eta_m)$=0.9; 2 – $m_s(\eta_m)$=0.8; 3 – $m_s(\eta_m)$=0.7; 4 – $m_s(\eta_m)$== 0.6; 5 – $m_s(\eta_m)$=0.5.

Conclusions

The performed statistical analysis of the possibility of using CPP in TVs with a fixed-ratio transmission allows us to draw the following conclusions:

The use of CPP containing ICE and flywheel energy storage (FES), allows to increase the average speed of the TV by an average of 10~20% depending on the mean values of the FES drive efficiency, its installed power and the type of terrain (road).

Direct speed limitations when driving on the road allow us to get an additional 5% increase in average speed due to energy recovery during TV braking.

While maintaining the same average speed of a vehicle moving on mild terrain, it is possible to use ICE with a power reduced by 10% of the nominal value for TV with CPP. In this case, the maximum energy capacity of the FES should be at least 2 kW•h, and the mean value of the FES drive efficiency along the path $m_s(\eta_m)$ should be at least 0.8.

A decrease in the installation power of the main ICE of the vehicle will, among other things, contribute to increased fuel economy.

References

[1] Genta G. Kinetic energy storage: theory and practice of advance flywheel systems. Butterworth–Heinemann Ltd, 1985. – 374 p.
[2] Hebner R., Beno J., Walls W. Flywheel batteries come around again. IEEE Spectrum, 2002, vol. 39, no. 4, pp. 46–51.
[3] The design of an engine-flywheel hybrid drive system for a passenger car / Schilke N., DeHart A., Hewko L. et al. Proceedings of The Institution of Mechanical Engineers, Part D: Journal of Transport Engineering. 1986, vol. 200 (44), pp. 231-248.
[4] Kharade R. Regenerative Braking in Automobiles. International Journal of Engineering and Techniques, 2017, vol. 3, issue 6, pp.216–221.
[5] Thoolen F. Development of an advanced high speed flywheel energy storage system.Eindhoven TechnischeUniversiteit, Eindhoven, 1993. doi: 10.6100/IR406829.
[6] Flywheel Energy Storage for Automotive Applications / M. Hedlund, J. Lundin, J. Santiago et al. Energies, 2015, no 8, pp. 10636–10663. doi:10.3390/en81010636.
[7] Dhand A., Pullen K. Review of flywheel based internal combustion engine hybrid vehicles. International Journal of Automotive Technology, 2013, no 10, vol.14, issue 5, pp. 797–804.

[8] Tinham B. Breakthrough in Ricardo Kinergy high-speed flywheel technology. Transport Engineer. 23.08.2011. Available at: URL: http://www.transportengineer.org.uk/transport-engineer-news/breakthrough-ricardo-kinergy-flywheel-technology/36277/ (accessed 31 May 2019).
[9] Hansen J., O'Kain D. An Assessment of Flywheel High Power Energy Storage Technology for Hybrid vehicles. Reports. December 2011, ORNL/TV-2010-280.
[10] Principle, Design and Experimental Validation of a Flywheel-Battery Hybrid Source for Heavy-Duty Electric Vehicles / O. Briat, J.M. Vinassa, W. Lajnefet al. IET Electric Power Applications, vol. 1, issue 5, September 2007, pp.665–674. doi:10.1049/iet-epa:20060458.
[11] Low Cost Flywheel Energy Storage for a Fuel Cell Powered Transit Bus / C. Hearn, M. Flynn, M. Lewis et al. Vehicle Power and Propulsion Conference, Sept. 9 – 12, 2007. VPPC 2007. IEEE. doi: 10.1109/VPPC.2007.4544239.
[12] Ogasa M. Application of energy storage technologies for electric railway vehicles – examples with hybrid electric railway vehicles. IEEJ Transactions on Electrical and Electronic Engineering, 2010, no. 5(3), pp. 304–311. doi:10.1002/tee.20534.
[13] Model-based comparison of hybrid propulsion systems for railway diesel multiple units / Schmid S., Ebrahimi K., Pezouvanis A. et al. International Journal of Rail Transportation, 2017, vol. 6, issue 1, pp. 16–37.
[14] Korsunskiy V.A. [Mathematical model of hydrostatic-mechanical drive for the flywheel energy storage]. Trudy NAMI, 2017, no. 4 (271), pp. 38–45. (In Russian).
[15] Abrahamsson J., Bernhoff H. Magnetic bearings in kinetic energy storage systems for vehicular applications. Journal of Electrical Systems. 2011, vol. 7, no. 2, pp. 225-236.
[16] Kotiev G.O., Miroshnichenko A.V., Stadukhin A.A., Kositsyn B.B. / Determination of mechanical characteristics of high-speed tracked vehicles traction motor with individual drive wheels (2019). Journal of Physics: Conference Series, 1177 (1), article № 012058.
[17] Analysis on operational power and eddy current losses for applying coreless double-sided permanent magnet synchronous motor/generator to high-power flywheel energy storage system / Jang S.M., Park Ji.H., You D.J. et al.JournalofAppliedPhysics.2009. vol. 105, no 7. pp. 07F116.
[18] Shen S.; Veldpaus F.E. Analysis and control of a flywheel hybrid vehicular powertrain. IEEE Transaction on Control Systems Technology, 2004, vol. 12, pp. 645–660.
[19] Methods for road microprofile statistical data transformation / Evgeniy Sarach, George Kotiev and Sergey Beketov. MATEC Web of Conferences. 2018, v.224, article № 04009.
[20] Korsunskiy V.A. Mathematical model of transport vehicle with combined power installation, flywheel energy storage and stepped transmission. Journal of advanced research in technical science, 2019, no. 14, vol. 2, pp. 78–88. https://doi.org/10.26160/2474-5901-2019-14-190-196.

IOP Conf. Series: Materials Science and Engineering **820** (2020) 012016 doi:10.1088/1757-899X/820/1/012016

Scenario-based definition of technical safety requirements for autonomous road vehicles

O Kirovsky [1,3] and K Byakov[2]

[1]ZF North America, 34605 W 12 Mile Rd, MI 48033, Farmington Hills, USA
[2]Bauman Moscow State Technical University

[3]E-mail: Oleg.Kirovskii@trw.com

Abstract. The article provides a brief overview of the standards ISO 26262 and ISO 21448 and the safety life cycle defined therein. A method for technical requirements development based on road situation statistics is proposed. Possibilities for improvement of safety life cycle for autonomous vehicles using traffic data are considered. An outlook towards the implementation of traffic scenarios into the safety lifecycle is briefly described.

1. Introduction

The automotive industry is undergoing perhaps the most important changes since Henry Ford launched the conveyor belt. Over the last hundred years, engineers have been working to improve the vehicles driven by humans. Today, the direction of development is changing. More and more systems now take over the functions previously performed by human drivers. These systems are called Driver Assistance Systems (DAS). Already developed and launched in the series are systems like automatic emergency braking (AEB) [1–4], lane keep assist (LKA) [5–9], signage assistance systems (SAS) [10–15] and many others [16–18].

However, car manufacturers are not going to limit themselves and focus on driver assistance systems only. For example, Tesla Motors (USA) developed the “Autopilot” system (Figure 1A). According to the company, this system can fully take over the control of the vehicle in some cases. Other car manufacturers are developing similar systems [19–22].

The main barrier to the widespread introduction of autonomous driving technologies is posed by safety issues. Tesla Motors’ cars with activated Autopilot system have already caused a number of accidents (Figure 1B). Similar incidents have also occurred with vehicles of other manufacturers (e.g. Uber). According to the Fatality Analysis Reporting System (FARS), seven people have already died in autonomous car accidents. The indicator of the number of fatalities per 100 km driven is higher for Tesla cars than the average for cars of the same class ($p<0.05$). Investigations of the various accidents confirm the potential danger of autonomous driving.

It is believed that the methods of dealing with the threats related to the operation of vehicles on public roads are different for autonomous and non-autonomous systems.

IOP Conf. Series: Materials Science and Engineering **820** (2020) 012016 doi:10.1088/1757-899X/820/1/012016

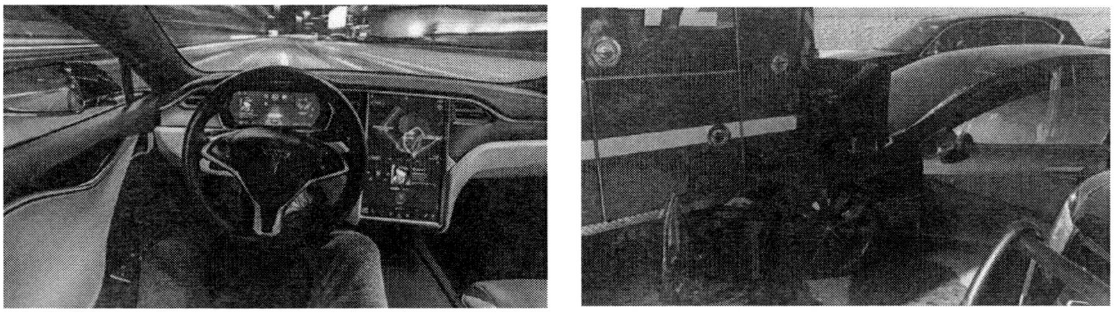

A B

Figure 1. Tesla Model S with activated Autopilot
A — Functioning “Autopilot” system
B — Autopilot-induced collision with a fire vehicle (Culver City, California, USA, January 2018)

2. Autonomous systems: Methods for safety lifecycle

The safety methods used in the development of vehicle systems cannot fully protect passengers, drivers and other road users from hazards associated with autonomous vehicles on public roads. ISO 26262 is the standard for the vehicle systems safety. It does not cover hazards not related directly to system component failures.

The newly developed ISO 21448 "Safety of the Intended Functionality" deals with hazards caused by environmental factors. Such factors include e.g. the illumination that affects the functionality of systems based on camera. In addition, it also covers unintentional errors committed by the driver or operator of the vehicle. The distinction between ISO 26262 and ISO 21448 is shown in Figure 2.

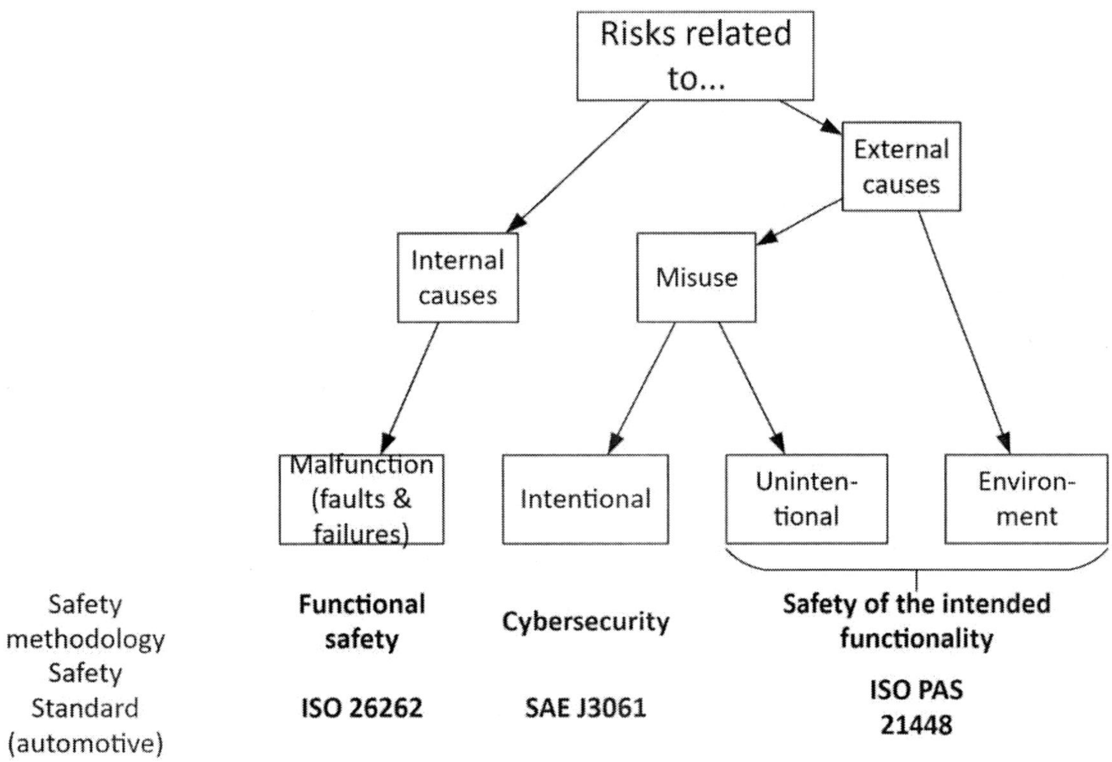

Figure 2. Methodologies to ensure the safety of vehicle systems [23, 24]

The number of failure modes of the system components is finite. At the same time, the number of different traffic situations, taking into account different variants of parameterization, is infinite. Therefore, ISO 26262 methods (inductive and deductive system analysis, testing and validation of specifications) do not fully cover the safety needs of AD and ADAS systems. Comparison of safety life cycles according to ISO 26262 and ISO 21448 is presented in [23, 24]. There one can also find a definition of the scenario for the description of the traffic situation. In the next section we will consider the differences between the hazard analysis for the systems of zero level of autonomy and subsequent levels.

3. Methods

3.1. Hazard analysis

According to ISO 26262 and ISO 21448, a hazard is a potential source of risk. The purpose of the hazard analysis and risk assessment (HARA) is to identify and categorize the hazards associated with the operation of the system under analysis, as well as to determine the safety goals required to deal with those threats.

3.1.1. Non-autonomous systems. The HARA for non-autonomous systems in accordance with ISO 26262-3 is carried out according to the scheme presented in Figure 3.

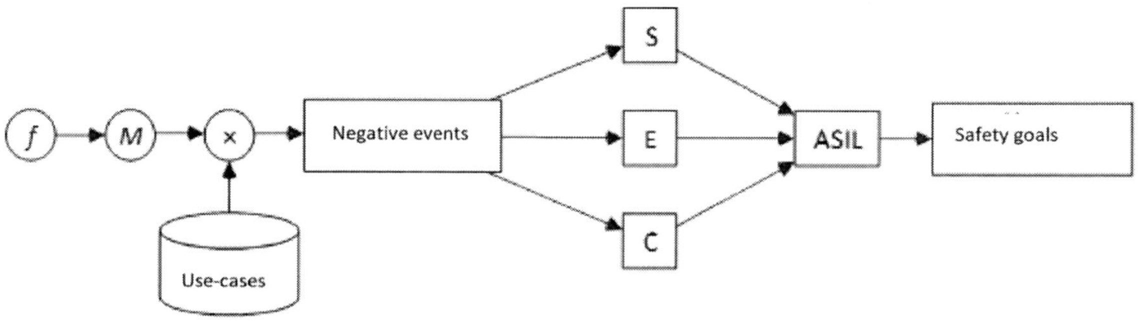

Figure 3. Hazard analysis for non-autonomous systems

Here f is the list of functions of the system under study, M is the list of functional failures (malfunctions). By analyzing the behavior of a car with different types of failures in different scenarios, possible negative consequences are determined. This is followed by a classification of these consequences by severity (S), the time that the driver and passengers spend in the state where the consequences may occur (exposure time, E), and the driver's ability to avoid the consequences (C). The table from ISO 26262-3 is then used to determine an ASIL – the safety integrity level suitable to this failure in the selected scenario. The result of the analysis is a list of safety targets with corresponding ASIL levels. Scenarios are described here in natural language, classifications S, E and C are estimated.

3.1.2. Additional aspects for autonomous systems. For hazards caused by component failures in autonomous systems, the HARA is performed in the same way as for non-autonomous systems (see Figure 2). Due to the more diverse nature of the events that can cause autonomous systems to fail, classification by S, E, and C parameters, as well as finding a common ASIL level, is meaningless. Instead, the probability of consequences of various degrees of severity is determined, which can then be compared to accident statistics. If the overall probability of severe consequences does not increase due to the operation of an autonomous system, the system remains safe [23, 24].

To determine the probability of the occurrence of certain negative consequences, Monte Carlo simulation based on parameterized scenarios is used.

3.2. Parameterization of scenarios

According to ISO GTC 21448, the following information is included in the scenario definition:

- Scene (information about nearby objects)
  - o Dynamic elements (other cars, pedestrians, cyclists, etc.)
  - o Static elements (road infrastructure, trees, etc.)
- Actions and events occurring during the scenario performance
- Objectives and indicators defining the safety of the scenario, quality of driving, compliance with the requirements of the road traffic regulations and other evaluation factors (KPI)

Static and dynamic elements of the scene are determined by numerical parameters. These parameters cannot be specified unambiguously, they have some probability distribution. For example, let's consider a scenario where two cars follow each other along a straight road section (Figure 4). This scenario is important for designing an autonomous emergency braking system (AEB).

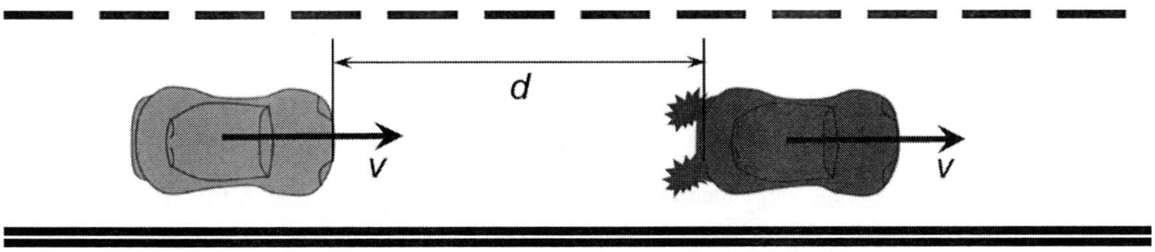

Figure 4. Scenario «car following»

Figure 4 shows the following scenario. On a straight stretch of road, two cars follow each other at a distance *d* in one direction and at the same speed *v*. The front car is equipped with an AEB system. At a certain point in time t_0, the AEB system is activated in the front car. This results in a maximum negative acceleration braking *a*. The negative consequence of this situation is that the rear car is hitting the vehicle in front of it because the driver of the rear car may not be able to respond to a sudden change in speed. The result of such collision may be human casualties.

Let's assume that the maximum braking acceleration of both cars is the same. The second car starts to brake in time τ after the first one because of the delay in the driver's reaction. After braking starts, the distance between the two vehicles is determined by the formula:

$$d(t) = d - \frac{a\tau^2}{2} - a\tau t \quad (1)$$

The collision time (i.e. the moment when *d(t)* becomes zero) is determined by the formula:

$$t_c = \frac{d - a\tau^2/2}{a\tau} \quad (2)$$

Vehicle speed difference at the moment of collision $\Delta v = a\tau$. The difference in speed determines the severity of injuries to drivers and passengers of both vehicles during the collision.

3.3. *Probabilistic modelling*

From the traffic data [25], the distributions of drivers' response time to sudden changes in situation (τ) as well as the time distance $d_t = d / v$ are known. The corresponding distributions are shown in Figure 5.

Negative acceleration *a* was assumed to be constant and equal to 9 m/s^2.

A Monte Carlo simulation was performed using the Markov property, i.e. each subsequent pair of random variables did not depend on the previous results. In addition, random d_t and τ values were considered as independent.

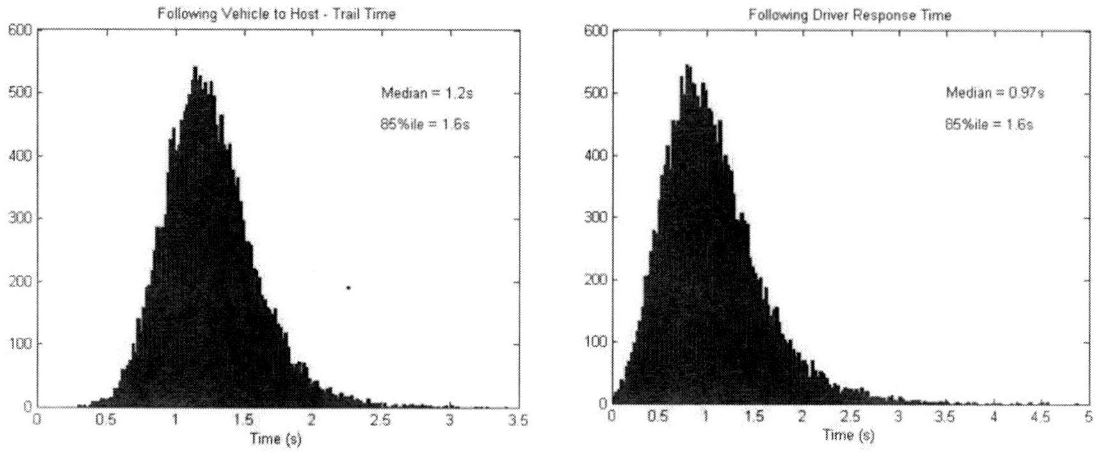

A. Trail Distance d_t B. Reaction time τ

Figure 5. Distributions of the model parameters

4. Results

As a result of modeling the dependence between initial velocity and probability of realization of various negative consequences has been received. Avoidance of a collision or a non-injury collision (S0 according to ISO 26262 classification), minor (S1), moderate (S2), or severe (up to fatal) injuries were chosen as possible implementations. The result is shown graphically in Figure 6.

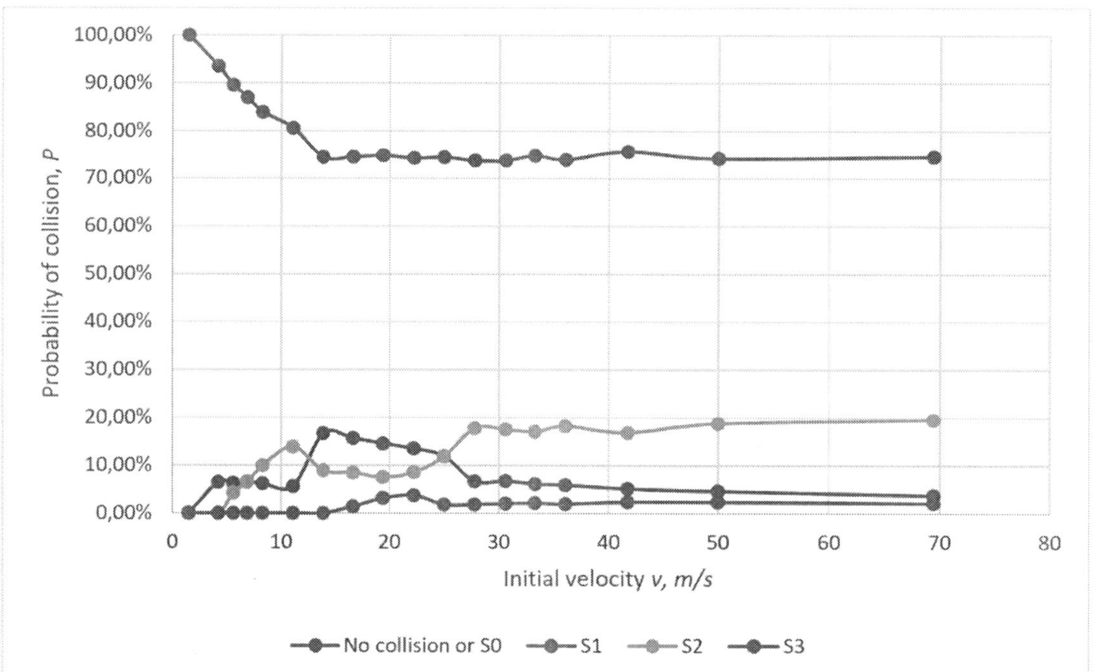

Figure 6. Probability of various negative consequences, modeling results (p<0,001)

By using the risk graph from ISO 26262-3, it can be established that the required level of integrated safety in this scenario is ASIL B. There are various ways to reduce this level. One possibility is to reduce the maximum speed at which the AET system is active to 20 m/s. Another way is to change the system so that the AEB system is deactivated if a vehicle equipped with the system is followed by another vehicle at a dangerous distance. This will require the installation of additional sensors and software components.

These results characterize vehicles driven by human drivers. They can be used as a benchmark for writing specifications for stand-alone vehicles. It can be claimed that in order to lower the injuries and fatalities in the considered scenario, an autonomous vehicle would need to have a lower response time than the driver. If it is not possible to do in case the situation corresponding to the considered scenario is detected, the autonomous vehicle will have to initiate an evasive maneuver. The development of such a maneuver will require consideration of additional scenarios and creating simulation similar to that discussed in this paper.

5. Conclusions

The article describes a method for elicitation of requirements for autonomous vehicle control systems based on parameterized traffic scenarios. As an example, the "car following" scenario (Figure 4) was chosen. This scenario was parameterized using two values - trail distance d_t and driver reaction time τ. Based on the available statistics, probability distributions were created for both parameters (Figure 5). This was followed by a Monte Carlo simulation, which resulted in a probability of various possible negative consequences in the "car-following" scenario (Figure 6).

It should be noted that in reality, d_t and τ parameters seem to be dependent on each other and on the speed. Drivers with longer reaction times (e.g. people of advanced age) would be more likely to maintain a longer trailing distance than drivers showing good reaction time (e.g. young sports car drivers). In addition, when the flow rate is reduced, the distance between vehicles on the road is usually shortened. Consequently, simulation methods need to be improved to take these dependencies into account.

Scenario, parametrization, creation of parameter distribution and modeling in this paper were performed using MathWorks MATLAB and Microsoft Excel software products. As the level of autonomy of the systems under consideration increases, the number of scenarios to be simulated grows rapidly [23, 24]. Further it seems necessary to use formal or semi-formal language for scenario description together with an interpreter capable to create necessary models and distributions in an automated mode in a form acceptable for using in simulation software.

References

[1] Miholjcic D, Fabbroni M and Robinson R 2019 A Study of the Performance of Automatic Emergency Braking (AEB) Systems Equipped on Passenger Vehicles for Model Years 2013 to 2018 *SAE Tech. Pap.* **2019 April**

[2] Ariyanto M, Haryadi G D, Munadi M, Ismail R and Hendra Z 2019 Development of Low-Cost Autonomous Emergency Braking System (AEBS) for an Electric Car *Proc. — 2018 5th Int. Conf. on Elec. Vehic. Techn.* pp 167–171

[3] Lee D H, Kim S K, Kim C S and Huh K S 2014 Development of an autonomous braking system using the predicted stopping distance *Int. Journ. of Aut. Techn.* **15 (2)** pp. 341–346

[4] Jin P F, Kim J H and Kim J H. 2015 Design of unmanned vehicle advanced braking system using smart motor *12th Int. Conf. on Ubiquit. Rob. and Amb. Intell.* pp 167–171

[5] Tominaga K, Takeuchi Y, Tomoki U, Kameoka S, Kitano H, Quirynen R, Berntorp K and Cairano S 2019 GNSS based lane keeping assist system via model predictive control SAE Tech. Pap. **2019 April**

[6] Cerone V, Chinu A and Regruto D 2002 Experimental results in vision-based lane keeping for highway vehicles *Proc. of the Amer. Contr. Conf.* **2** pp 869–874

[7] Cerone V, Milanese M and Regruto, D 2009 Combined automatic lane-keeping and driver's steering through a 2-DOF control strategy *IEEE Transact. on Contr. Syst. Techn.* **17 (1)** pp 135–142

[8] Beglerovic H, Schloemicher T, Metzner S and Horn M 2018 Deep learning applied to scenario classification for lane-keep-assist systems *Appl. Scienc. (Switzerland)* **8(12)**

[9] Hsu J Y, Ku K L, Jhang T K, Lin H P and Yeh C J 2017 Integration and implementation of a lane keeping system with vehicle dynamics control *Int. Autom. Contr.l Conf.* **2017 November** pp 1–6
[10] Stallkamp J, Schlipsing M, Salmen J and Igel C 2012 Man vs. computer: Benchmarking machine learning algorithms for traffic sign recognition *Neur. Netw.* **32** pp 323–332
[11] Sermanet P and Lecun Y 2011 Traffic sign recognition with multi-scale convolutional networks *Proc. of the Int. Joint Conf. on Neur. Netw.* pp 2809–2813
[12] Bahlmann C, Zhu Y, Ramesh V, Pellkofer M and Koehler T 2005 A system for traffic sign detection, tracking, and recognition using color, shape, and motion information *IEEE Intel. Vehicl. Symp., Proc.* pp 255–260
[13] Maldonado-Bascón S, Lafuente-Arroyo S, Gil-Jiménez P, Gómez-Moreno H and López-Ferreras F 2007 Road-sign detection and recognition based on support vector machines *IEEE Trans. on Intel. Transp. Syst.* **8(2)** pp 264–278
[14] Shadeed W G, Abu-Al-Nadi D I and Mismar M J 2003 Road traffic sign detection in color images *Proc. of the IEEE Int. Conf. on Electr., Circ., and Syst.* **2** pp 890–893
[15] Loy G and Barnes N 2004 Fast shape-based road sign detection for a driver assistance system *IEEE/RSJ Int. Conf. on Intel. Robots and Syst.* **1** pp 70–75
[16] Kotiev G O, Butarovich D O and Kositsyn B B 2018 Energy efficient motion control of the electric bus on route *IOP Conf. Ser.: Mat. Scien. and Eng.* **315(1)**
[17] Medvedev V, Butarovich D and Smirnov A 2019 Prototyping of the remote monitoring system of battery condition of the electric vehicle *Journ. of Phys.: Conf.e Ser.* **1177(1)**
[18] Sarach E, Kotiev G and Beketov S 2018 Methods for road microprofile statistical data transformation *MATEC Web of Conf.* **224**
[19] Rao S J, Seitz T, Lanka V R R and Forkenbrock G 2019 Analysis and mathematical modeling of car-following behavior of automated vehicles for safety evaluation *SAE Techn. Pap.* **2019-April**
[20] Yakusheva N, Proletarsky A and Basarab M 2018 Pedestrian-Vehicle Collision Avoidance Strategy for NLOS Conditions *26th Telecom. For. — Proc.*
[21] Munir F, Azam S, Hussain M I, Sheri A M, Jeon M 2018 Autonomous vehicle: The architecture aspect of self driving car *Proc. — IEEE INFOCOM 2018* **April** pp 72–80
[22] Osipkov V, Ksenevich T I, Belousov B, Karasev O, Sokolov S, Rudynets M, Skalyga M, Naumov S, Stepnov V and Klimachkova A 2016 Intelligent Transport Systems: Revolutionary Threats and Evolutionary Solutions *SAE Techn. Papers* **April**
[23] Kirovskii O M and Gorelov V A 2019 Driver assistance systems: analysis, tests and the safety case. ISO 26262 and ISO PAS 21448 *IOP Conf. Ser.: Mat. Scien. and Eng.* **534(1)**
[24] Kirovskii O 2019 Determination of Validation Testing Scenarios for an ADAS Functionality: Case Study *SAE Techn. Pap.* 2019-April
[25] Christopher Kober 2018 Stochastische Verkehrsflusssimulation auf Basis von Fahrerverhaltensmodellen zur Absicherung automatisierter Fahrfunktionen *Springer Vieweg,* (Stuttgart) p 162

IOP Conf. Series: Materials Science and Engineering **820** (2020) 012017 doi:10.1088/1757-899X/820/1/012017

The method of increasing the stability of trailer-trucks in case of emergency braking in a turn and emergency failure of the trailer brake system

M Zhileykin[1] and G Skotnikov[1,2]

[1]Bauman Moscow State Technical University

[2]E-mail: skotnikov.g@yandex.ru

Abstract. One of the most important properties of semi-trailer trucks are stability and steerability. This is because with an increase in driving speeds, these properties largely determine the safety of vehicle operation. Improving the semi-trailer trucks safety is all the more relevant because the consequences of road accidents involving semi-trailer trucks are the most severe and cause enormous damage. One of the most common causes of accidents is the loss of stability during emergency braking in a turn, especially in the event of a failure of trailer braking system. The goal of this research is to develop principles for improving the stability of semi-trailer trucks movement during emergency braking in a turn and failure of trailer braking system. A method has been developed to increase semi-trailer trucks stability during emergency braking in a turn and the failure of trailer braking system, which allows to maintain the trajectory stability of the semi-trailer truck and to avoid accidents with severe consequences. Using simulation methods, the effectiveness of the proposed method is proved.

Introduction

The widespread use of trailer-trucks in the transportation structure is determined by the need for transportation of heavy bulky goods, the need to ensure low specific pressures on the road while increasing the carrying capacity of vehicles and increasing operating speeds [1].

One of the most important properties of semi-trailer trucks are stability and steerability. This is because with an increase in driving speeds, these properties largely determine the safety of vehicle operation. Improving the semi-trailer trucks safety is all the more relevant because the consequences of road accidents involving semi-trailer trucks are the most severe and cause enormous damage.

One of the most common causes of accidents is the loss of stability during emergency braking in a turn, especially in the event of a failure of trailer braking system.

Due to its relevance, a large number of publications are devoted to the issues of ensuring maneuverability, steerability, and stability in the design of trailer trucks [2-11].

Literary sources are mainly devoted to the development of algorithms for the operation of dynamic stabilization systems of separately moving vehicles either by methods of creating stabilizing moments by braking individual wheels [12-15], or by steering [16-19]. As regards the stabilization of the movement of trailers and semi-trailers, it is usually a question of increasing overturning stability [20], as well as equipping trailer-trucks with anti-lock brake systems.

The goal of this research is to develop principles for improving the stability of semi-trailer trucks movement during emergency braking in a turn and failure of trailer braking system.

IOP Conf. Series: Materials Science and Engineering **820** (2020) 012017 doi:10.1088/1757-899X/820/1/012017

Substantiation of the principles of stabilization of trailer-trucks movement during emergency braking in a turn and emergency failure of trailer braking system

In case of emergency braking when trailer braking system is failed, a characteristic picture of the loss of trailer-truck stability is the deviation of the truck from a given trajectory due to the fact that heavier trailer push the truck to an adjacent lane. An emergency can be prevented by applying a range of measures.

1. Preventive reduction in the speed of a trailer-truck in a turn if it is close to critical values. This is achieved by reducing the engine power demand and braking the wheels of the trailer.

2. Emergency braking mode leads to a redistribution of normal reactions between the axles of the truck. The front wheels are loaded with additional vertical force. This circumstance, as well as the operation of the anti-lock braking system, which prevents the wheels from locking by the braking torque, leads to an increase in the grip of these wheels. These facts allows to apply a corrective change in the angle of rotation of the steered wheels (steering), which will help to keep the truck on the path set by the driver.

A simplified mathematical model of trailer-truck planar motion will be consider to compose the equations of trailer-truck motion. The model has several assumptions:

a) the slip angles of right and left wheels of each axis are the same;

b) the steering angles of the wheels and wheel slip angles are small, i.e. do not exceed 10°;

c) slip resistance coefficients for all wheels of the axle are the same.

The design diagram of the forces acting on the trailer-truck in curved motion is shown in Fig. 1.

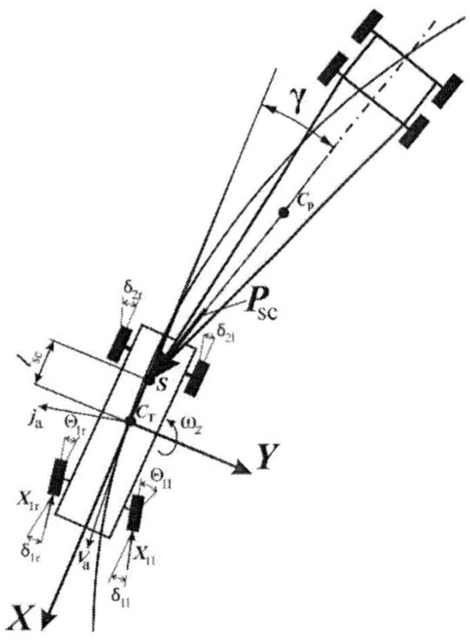

Fig. 1. Scheme of forces acting on the trailer-truck: *X*, *Y* – coordinate axes associated with the truck center of mass; C_T, C_P – centers of mass of the truck and semi-trailer; *S* – the fifth wheel center; l_{sc} – distance from the center of mass of the truck to the fifth wheel center; j_a – acceleration of the center of mass of the truck; V_a – linear velocity of center of mass of the truck; ω_z – the angular velocity of rotation of the truck around a vertical axis passing through its center of mass; P_{sc} – force acting at the fifth wheel center from the side of the semi-trailer to the truck during braking; X_{1l}, X_{1r} — longitudinal forces acting on the left and right wheels of the first axle of the truck from the road; Θ_{1l}, Θ_{1r} – steering angles of left and right wheels of the first axle of the truck; δ_{1l}, δ_{1r} – slip angles of left and right wheels of the first axle of the truck; δ_{2l}, δ_{2r} – slip angles of left and right wheels of the second axle of the truck; γ – jackknifing angle of the trailer-truck

Consider the movement of a truck in braking mode, replacing the effect of the semi-trailer with the force P_{sc} applied to the fifth wheel center on the truck, acting along the longitudinal axis of the semi-trailer For a rear drive two-axle vehicle the following differential equations were obtained in [21]:

$$\dot{\delta}_1 = \frac{V_a}{L}(\Theta_{1sr} + \delta_2 - \delta_1) - \frac{K_y}{V_a}\left(\frac{g}{G_a} + \frac{L^2}{4J_z}\right)\delta_1 - \frac{K_y}{V_a}\left(\frac{g}{G_a} - \frac{L^2}{4J_z}\right)\delta_2 +$$

$$+\frac{j_a}{V_a}(\Theta_{1sr} - \delta_1) + \frac{g}{G_a V_a}P_y + \frac{L}{2J_z V_a}M_z - \frac{\Theta_{1sr}}{V_a}\left(\frac{g}{G_a} + \frac{L^2}{4J_z}\right)X_1;$$

$$\dot{\delta}_2 = \frac{V_a}{L}(\Theta_{1sr} + \delta_2 - \delta_1) - \frac{K_y}{V_a}\left(\frac{g}{G_a} - \frac{L^2}{4J_z}\right)\delta_1 - \frac{K_y}{V_a}\left(\frac{g}{G_a} + \frac{L^2}{4J_z}\right)\delta_2 -$$

$$-\frac{j_a}{V_a}\delta_2 + \frac{g}{G_a V_a}P_y - \frac{L}{2J_z V_a}M_z - \frac{\Theta_{1sr}}{V_a}\left(\frac{g}{G_a} - \frac{L^2}{4J_z}\right)X_1;$$

$$M_z = P_{sc}\sin(\gamma)l_{sc}; V_a = \frac{L\omega_z}{\Theta_{1cp} + \delta_2 - \delta_1}; P_y = P_{sc}\sin(\gamma)$$

$$\dot{\delta}_A = -\left(\frac{K_y}{V_a}\frac{L^2}{2J_z} + \frac{j_a}{V_a} + P_y\sin(\gamma)l_{sc}\frac{1}{J_z\omega_z}\right)\delta_A - \left(P_y\sin(\gamma)l_{sc}\frac{1}{J_z\omega_z} - \frac{L^2}{2V_aJ_z}X_1 + \frac{j_a}{V_a}\right)\Theta_{1sr}, \quad (1)$$

where $\delta_A = \delta_2 - \delta_1$ – difference between rear and front slip angels; $X_1 = X_{1l} + X_{2r}$ – full force on the front axle of the truck; K_y – total (for the vehicle axis) coefficient of resistance to tire retraction; $\Theta_{1sr} = \frac{\Theta_{1l} + \Theta_{1r}}{2}$ – average steering angle; $\delta_i = \frac{\delta_{il} + \delta_{ir}}{2}, i = 1,2$ – average slip angle of *i*-truck axle; G_a – truck weight; L – truck wheelbase; J_z – the moment of inertia of the truck relative to the vertical axis passing through its center of mass.

As known, difference between rear and front slip angels $\delta_A = \delta_2 - \delta_1$ characterizes its turnability: $\delta_A = 0$ – neutral; $\delta_A > 0$ – oversteer; $\delta_A < 0$ – understeer. In case of loss of stability by a trailer-truck (jackknifing) during braking, the truck acquires oversteer to stabilize the truck, neutral steering is required.

For disturbed motion, equation (1) takes the form

$$\Delta\dot{\delta}_A = -\left(\frac{K_y}{V_a}\frac{L^2}{2J_z} + \frac{j_a}{V_a} + P_y\sin(\gamma)l_{sc}\frac{1}{J_z\omega_z}\right)\Delta\delta_A - \left(P_y\sin(\gamma)l_{sc}\frac{1}{J_z\omega_z} - \frac{L^2}{2V_aJ_z}X_1 + \frac{j_a}{V_a}\right)\Delta\Theta_{1sr}, \quad (2)$$

where the symbol Δ denotes the increment of the value over the time Δt.

Equation (2) has the form

$$\dot{X}(t) = A(t)X(t) + R(t)U(t), \quad (3)$$

where $X(t) = \Delta\delta_A$ – state variable; $U(t) = \Delta\Theta_{1sr}$ – control action;

$$A(t) = -\left(\frac{K_y}{V_a}\frac{L^2}{2J_z} + \frac{j_a}{V_a} + P_y\sin(\gamma)l_{sc}\frac{1}{J_z\omega_z}\right);\ R(t) = -\left(P_y\sin(\gamma)l_{sc}\frac{1}{J_z\omega_z} - \frac{L^2}{2V_aJ_z}X_1 + \frac{j_a}{V_a}\right).$$

The control *U(t)* for obvious physical reasons is limited, i.e. belonging to a closed set:

$$|U(t)| \le U_{max}.$$

Lyapunov's studying of the stability of movement, which allows one to judge the properties of perturbed movements, indicates the path to the rational design of regulators. Therefore, the Lyapunov function method is widely used directly for the synthesis of the so-called stabilizing controls, i.e. providing stability of the designed dynamic system. The idea of a general approach to finding stabilizing control for objects (3) with a control action *U(t)* is as follows.

Let some scalar function *V (X)* be given, which can be considered as a measure of the deviation of a moving object (3) from the steady state. In this case, the steady state is determined by a trivial solution: *X=0*. In this case, a positive definite Euclidean norm of the vector ||X|| or the square of its value will be introduced as the Lyapunov function *V(X)*

$$V(X) = \Delta\dot{\delta}_A^2 \tag{4}$$

The role of the control system is to reduce the distance between the moving object (3) and the steady-state value *X=0*.

The direct Lyapunov method reduces the problem of studying the stability of system (3) to studying the properties of the function *V(X)* and its first derivative, which is calculated by the formula

$$\frac{dV(X)}{dt} = \frac{\partial V}{\partial X} AX = -2\Delta\delta_A^2 \left(\frac{K_y}{V_a} \frac{L^2}{2J_z} + \frac{j_a}{V_a} + P_y \sin(\gamma) l_{sc} \frac{1}{J_z \omega_z} \right) \tag{5}$$

To find the optimal control vector *U(t)*, we use the method of optimal damping of transients V.I. Zubova [22]:

$$U(t) = -U_{max} sign\left(\frac{\partial V}{\partial X} R(t) \right) \tag{6}$$

The magnitude order of the summands of equation (5) and their signs based on the positive directions of vectors adopted in Figure 1 are shown in table 1.

Table 1. The order of magnitude of the terms of equation (5) and their signs

№	Summand	Sign	Magnitude order
1	$\frac{K_y}{V_a} \frac{L^2}{2J_z}$	>0	$\approx 10^3 \ldots 10^4$
2	$\frac{j_a}{V_a}$	<0	$\approx 10^1$
3	$P_y \sin(\gamma) l_{sc} \frac{1}{J_z \omega_z}$	<0	$\approx 10^2$

Therefore, the expression in brackets of equation (5) is always greater than zero. Then $\frac{dV(X)}{dt} < 0$, and the function $V(X)$ is a Lyapunov function. Then the optimal control (6) takes the form

$$U(t) = -U_{max} sign\left[-2D\Delta\delta_A\right]$$
$$D = \left(P_y \sin(\gamma) l_{sc} \frac{1}{J_z \omega_z} - \frac{L^2}{2V_a J_z} X_1 + \frac{j_a}{V_a} \right) \tag{7}$$

Because $\frac{L^2}{2V_a J_z} X_1 > 0$, then $D < 0$, and expression (7) can be written

$$U(t) = -U_{max} sign(\Delta\delta_A) \tag{8}$$

A control restriction is introduced $U_{max} = \left|\Delta\gamma(t) = \gamma(t) - \gamma_0\right|$, where $\gamma(t)$ – actual hitch angle, γ_0 – value of hitch angle when the driver presses the brake pedal.

Since the difference between slip angels of the axes $\Delta\delta_A$ cannot be measured in practice, it is necessary to replace this variable with another one that has the same sign when jackknifing is occur. As such a variable, the hitch angle *γ(t)* can be used. Then the finally optimal control (8) will have the form

$$U(t) = \Delta\Theta_{1sr}(t) = -\left|\Delta\gamma(t)\right| sign\left[\gamma(t)\right] \quad (9)$$

The research of the effectiveness of the proposed algorithm for the stabilization of the trailer-truck with the help of active steering when braking by simulation methods

As a criterion for the effectiveness of the algorithm of the stabilization system of the trailer-truck during braking, the absence of the road train from the lane was taken.

Description of the research object and the conditions of simulation tests

Consider the movement of a trailer-truck with a gross weight of 36,000 kg, consisting of a truck and a semi-trailer. 4x2 truck has double-tire rear wheels.

The simulation model of the trailer-truck takes into account the displacement of the vertical axis of the fifth wheel to the right by 0.03 m with respect to the longitudinal axis of symmetry of the truck. The braking systems of the truck and semi-trailer are equipped with an anti-lock system that prevents the wheels from locking during braking.

Theoretical studies were conducted using simulation methods to confirm the effectiveness of the proposed algorithm of the stabilization system of the trailer-truck. Features of the mathematical model of the trailer-truck motion are considered in [23].

The motion on the "dry asphalt" rigid flat road (tire grip coefficient is $\mu smax = 0.7$) was investigated.

The simulated maneuver consisted of the following parts: entrance to the turn, the movement in the turn with a constant angle of the steered wheels and an increasing heading speed not exceeding critical speed according to the rollover condition, emergency braking. At the same time, the braking torque of 0.7 from the maximum value was realized at the wheels of the left side of the semi-trailer, and 0.1 of the maximum value of the braking moment at the wheels of the right side.

Research of the movement of trailer-truck not equipped with a stabilization system

In Figure 2 shows the final position of the trailer-truck, not equipped with a stabilization system, after stopping. Fig. 3 shows the time-history of the hitch angle γ.

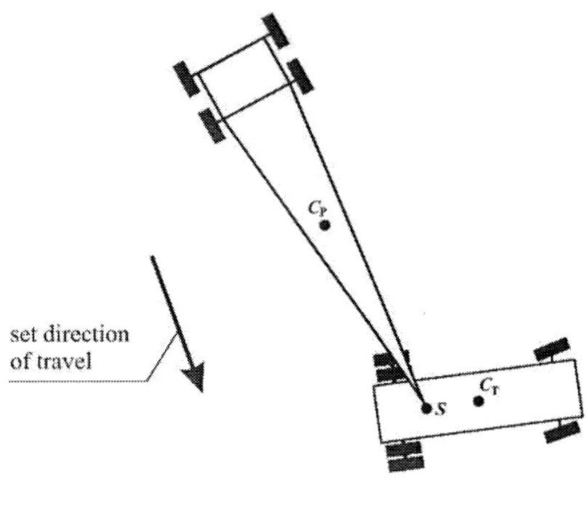

Fig. 2. Final position of the trailer-truck, not equipped with a stabilization system, after emergency braking in a turn

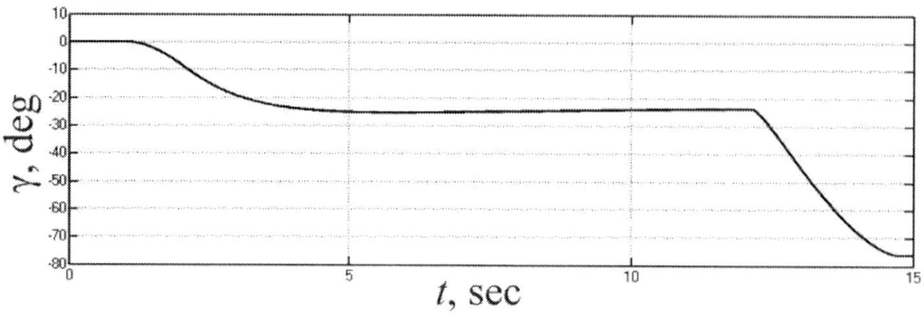

Fig. 3. Time-history of the hitch angle

Figure 2 and 3 show that during emergency braking in a turn during an emergency failure of the braking system of a semi-trailer truck, deviations from a given trajectory are significant. The hitch angle reaches 75^0.

Research of the movement of trailer-truck equipped with a stabilization system

In Figure 4 shows the final position of the trailer-truck, equipped with a stabilization system, after stopping. Fig. 5 shows the time-history of the hitch angle γ.

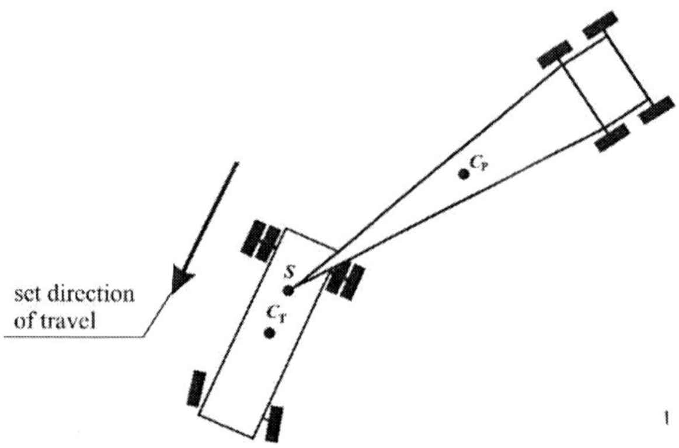

Fig. 4. Final position of the trailer-truck, equipped with a stabilization system, after emergency braking in a turn

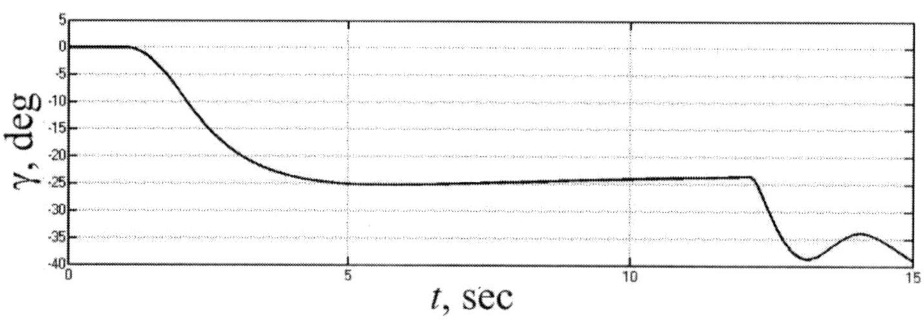

Fig. 5. Time-history of the hitch angle

From the presented simulation results (Fig. 4-5), it is seen that during emergency braking in a turn during an emergency failure of the trailer braking system, the developed method of stabilizing the movement is effective. In no case did the road train leave its lane; the hitch angle is reduced by 49%.

Conclusion

1. A method has been developed to increase the stability of trailer-truck during emergency braking in a turn and failure of the braking system of the trailer, which allows to maintain the trajectory stability of the trailer-truck and to avoid accidents with severe consequences.
2. The effectiveness of the proposed method for increasing trailer-truck stability during emergency braking in a turn and failure of the braking system of trailer has been proved using simulation methods.

References

[1] Keller, A.V., Gorelov, V.A., Vdovin, D.S., Taranenko, P.A., Anchukov, V.V. Mathematical model of all-terrain truck (2015) Proceedings of the ECCOMAS Thematic Conference on Multibody Dynamics 2015, Multibody Dynamics 2015, pp. 1285-1296.
[2] Skotnikov, G.I., Jileykin, M.M., Komissarov, A.I. Increasing the stability of the articulated lorry at braking by locking the fifth wheel coupling (2018) IOP Conference Series: Materials Science and Engineering, 315 (1), article № 012027.
[3] ASHLEY Liston. Jackknife stability of articulated tractor semitrailer vehicles with high-output brakes and jackknife detection on low coefficient surfaces[D]. Columbus, USA: the Ohio State University, 2003.
[4] Richard W. Radlinski, "The anatomy of a tractor trailer jackknife", President Radlinski & Associates, Inc, http://www.raibrakes.com
[5] ZONG Changful, ZHU Tianjun, WANG Chang, and LIU Haizhen 2012 Multi-objective Stability Control Algorithm of Heavy Tractor Semi-trailer Based on Differential Braking CHINESE JOURNAL OF MECHANICAL ENGINEERING Vol. 25, No. 1, 2012 DOI: 10.3901/CJME.2012.01.088
[6] Mohamed Abd Elazeem Elhemly, Mohamed Abd Elghany Fayed, Ali Abd Elghany Elmaihy 2013 Tractor–semitrailer jackknifing elimination using semitrailer differential braking technique, Int. J. Heavy Vehicle Systems, Vol. 20, No. 1, 2013
[7] Azad, N.L., Khajepour, A. and McPhee, J. 2009 A survey of stability enhancement strategies for articulated steer vehicles, International Journal of Heavy Vehicle Systems, Vol. 16, Nos. 1–2, pp.26–48
[8] Elhemly, M.A.E. and Fayed, M.A.G. 2011 'Simulation of tractor semitrailer manoeuvre at high speed using MATLAB/SIMULINK', Int. J. Heavy Vehicle Systems, Vol. 18, No. 4, pp.341–358.
[9] Kaneko, T. and Kageyama, I. 2003 A study on the braking stability of articulated heavy vehicles, JSAE Review, Vol. 24, pp.157–164.
[10] Palkovics, L. and El-Gindy, M. (1994) 'Design of an active unilateral brake control system for five axle tractor – semitrailer based on sensitivity analysis', Vehicle System Dynamics: International Journal of Vehicle Mechanics and Mobility, Vol. 24, No. 10, pp.725–758.
[11] Bouteldja M Cerezo V 2011 Jackknifing warning for articulated vehicles based on a detection and prediction system (Indianapolis: Submitted to the 3rd International Conference on Road Safety and Simulation, September 14-16, 2011) http://onlinepubs.trb.org/onlinepubs/conferences/2011/RSS/3/Bouteldja,M.pdf
[12] Karogal I., Ayalew B. Independent Torque Distribution Strategies for Vehicle Stability Control, in World Congress of the Society of Automotive Engineers. 2009, SAE, Inc.: Detroit, MI.
[13] Osborn R.P. Shim T. Independent Control of All-Wheel-Drive Torque Distribution. SAE paper number 2004-01-2052, 2004.
[14] Belousov, B.N., Merkulov, I.V., Fedotov, I.V. Synthesis of dynamic control system for active mountings of multiaxis ATM(2004) Avtomobil'naya Promyshlennost, (4), pp. 15-17.
[15] Belousov, B.N., Kupreyanov, A.A., Leksin, K.G., Popov, S.D., Shelomkov, S.A.Control algorithm of ATM motor-wheels (2003) Avtomobil'naya Promyshlennost, (4), pp. 15-18.
[16] Mammar S., Baghdassarian V.B. Two-degree-of-freedom formulation of vehicle handling improvement by active steering. // Proc. Amer. Contr. Conf., 2000, V. – P.105–109.

IOP Conf. Series: Materials Science and Engineering **820** (2020) 012017 doi:10.1088/1757-899X/820/1/012017

[17] Rodrigues A.O. Evaluation of an active steering system. Master's degree project / A.O. Rodrigues.– Sweden 2004. (http://www.s3.kth.se/~kallej/grad_students/rodriguez_orozco_thesis04.pdf).
[18] Mokhiamar O., Abe M. Active wheel steering and yaw moment control combination to maximize stability as well as vehicle responsiveness during quick lane change for active vehicle handling safety. // J Automobile Eng. 2002, V.216(2). – P.115–124.
[19] Beloousov, B., Ksenevich, T.I., Vantsevich, V., Komissarov, D. 8×8 platform for studing terrain mobility and traction performance of unmanned articulated ground vehicles with steered wheels (2013) SAE Technical Papers, 9.
[20] Sampson D.J. Active Roll Control of Articulated Heave Vehicles. Dissertation submitted to the University of Cambridge for the Degree of Doctor of Philosophy. – United Kingdom, Cambridge University Engineering Department. – September 2000.
[21] Litvinov A.S. Ustojchivost' i upravlyaemost' avtomobilya. – Moscow: Mashinostroenie, 1971. – 416 p.
[22] Optimal'noe upravlenie elektrotekhnicheskimi ob"ektami / V.I.Lovchakov, B.V.Suhinin, V.V. Surkov – Tula: TulGU, 2005. – 149 p.
[23] ZHilejkin M.M. Teoreticheskie osnovy povysheniya pokazatelej ustojchivosti i upravlyaemosti kolesnyh mashin na baze metodov nechetkoj logiki / Moscow: BMSTU, 2016. – 238 p.

IOP Conf. Series: Materials Science and Engineering **820** (2020) 012018 doi:10.1088/1757-899X/820/1/012018

Analysis of ATV transmission operation according to the results of tests on a dynamometer test bench

M Lyashenko[1], P Potapov[1], A Dolotov[1], A Diakov[2], K Evseev[2,3] and A Zverev[2]

[1]Volgograd State Technical University, Volgograd, Russian Federation
[2]Bauman Moscow State Technical University

[3]Email: kb_evseev@bmstu.ru

Abstract. The article describes the features and methods of testing the ATV on a hub dynamometer test bench. The graphs of the test results are presented, and the graphs analyzes, aimed at researching the operation of its engine-transmission bench, is carried out. The change of the gear ratio at different throttle positions is specific. This specific operation characteristic of the V-belt variator used in ATV transmission as a gearboxes found.

Introduction

All-terrain vehicles are quite interesting, however, a specific type of transport. Their range of application is quite wide, but at the same time, it is limited by off-road use. Their operating conditions determine the features of their design, in general, transmissions and powertrain systems [1-6].

There are several common schemes and layouts of such vehicles today. The most common is a scheme with a low-volume two-cylinder four-stroke gasoline engine and continuously variable transmission, which includes a transfer case with a reduction gear. V-belt variator is used as a gearbox. In addition, the transmission is all-wheel drive, usually without interaxial differential (sometimes without interwheel). This choice is due to a number of objective reasons [1-4]:

-the engine has sufficient power in its compact size, but the maximum power develops at high revolutions and the torque value is relatively low, especially in the low revolutions [1-6];

- the use of a stepless transmission makes it possible to simplify the overall management of the vehicle, as it is an automatic transmission, which makes it easy to control the traction on off-road conditions, especially in the light of these engine features. -

- V-belt variator with rubber belt and mechanical control (centrifugal regulator) allows simplifying the design and providing ease of maintenance with acceptable reliability. At the same time, such a gearbox has a high mechanical efficiency [1-6];

- the use of a transfer case with a reduction gear is conditioned by the above mentioned features of the engine, namely low torque, which is a significant disadvantage in off-road conditions. In addition, vehicles can still tow trailers, which means that the trailer's thrust-to-weight ratio should be high enough [1-6];

- all-wheel drive without axle differential (with rigid front axle connection) does not show any characteristic drawbacks (decrease of machine stability, deterioration of turning ability, increased tire wear, additional loading of transmission with circulating power) on soft soils and in mud, but, on the contrary, increases traction possibilities [1-6];

The design of most of the elements of ATV transmission is already well known and therefore their work does not cause much interest. However, the V-belt variator in this transmission is interesting, because the nature of its work significantly affects the work of the entire powertrain system, and the principle of its operation makes it difficult to analyze - it is necessary to know the law of the centrifugal regulator, which is possible only after individual tests of the unit, or having the manufacturer's data. Accordingly, carrying out tests, which will allow studying the work of the transmission with a wedge-belt variator and applying the obtained characteristics to improve the design of the variator, or to synthesize the control program, is an urgent task.

Test equipment

ATV is used as a prototype Can-Am Outlander 570 [7,8], which fully corresponds to the abovementioned design features.

For a long time, we have been using dynamometer test benches for testing complete vehicles, which allow us to investigate the operation of powertrain systems of vehicles as a whole and do not require dismantling the units [9-14]. These stands can be of different types - drum (roller) or hub stands. This paper describes the tests carried out with the help of a step dynamometer stand Dynapack Daqplus 43located in the laboratory of the Department of Transport Machines and Engines of Volgograd State Technical University.This test bench allows to test vehicles with power up to 2000 hp and torque up to 7000 Nm (all-wheel drive) [15].

The use of a wheelbase eliminates the influence of tyre contact with drum tyres, which is a relevant issue in the case of the vehicle in question, as it is equipped with advanced off-road tread and high profile wheels. In addition, the stand provides a high repeatability and accuracy of results.

The vehicle hubs are connected to the stand loaders by means of special adapters for testing (Figure 1).

Fig. 1. ATV mounted on the stand: 1 - loader of the stand, 2 - adapter

Test methods

The usual test procedure involves entering the value of the transmission ratio into the test bench control program, where the tests will be conducted. As a result, precise measurement of torque and power on wheels can be made in different loading modes (e.g. external engine speed response or simulation of road conditions). However, the tests described were intended to evaluate the reactions of a motor transmission system to loading under conditions of constant change in the gear ratio by a variable speed gearbox. In this regard, the program was introduced some "starting" value of the gear ratio, which, in fact, did not correspond to the reality. In order to calculate the current gear ratio, it was necessary to have a wheel speed signal (the default data are recorded by the bench) and the motor speed. The last signal could be received in two ways: by sending a signal to the input of the bench using the OBD protocol, or by sending a signal to the special input of the stand RPM Pickup from any sensor of the ATV. It turned out that the used ATV does not have a standard OBD connector and the corresponding protocol, an attempt to use the CAN-bus on the corresponding connector pins was impossible. The RPM Pickup input requires a 0-5V, 0-20V rectangular pulse signal. The only available source of this signal was the fuel injector control wire, which had to be used with a special converter. This allowed for synchronous recording of the engine speed during the tests. In addition, an adapter was used to feed the standard throttle position sensor signal to the test bench.

In view of the availability of these sensor signals, it was decided to use the following test methods:

- acceleration of the ATV to maximum speed at a given constant throttle position;
- increasing the load on the wheels with the help of the stand until the minimum steady speed of the ATV engine is reached (the load was increased with some experimentally selected step with holding it for a few seconds at each step to achieve the established mode if necessary, at a sufficiently small step it was not necessary) [16-18];

Test results

During each test, engine speed, axle wheel speed, throttle position, axle wheel torque (two axles in all-wheel drive mode), wheel power were recorded. Then the received data have been transferred in the tabular editor and processed.

Examples of the dependencies obtained are given below. The curve designations on the graphs are as follows:

Tacho–engine speed by sensor, rpm;
HubRpm - wheel spins on the stand, rpm
AxleTrq– wheel torque (total), Nm;
AxleTrqM - torque on the rear wheels (total for the axle), Nm;
AxleTrqS -torque on front wheels (total for axle), Nm;
AxlePow - wheel power, hp.;
V1:TPS - throttle opening degree by sensor, %;
Ratio - current gear ratio.
time - test time, sec.

Figures 2-4 show the graphs of the above mentioned values recorded by the time of measurement.

All diagrams have the following characteristic features that are the peculiarity of the loading process: the time scale starts from the moment of reaching the specified position of the throttle valve; it was not possible to reach the minimum revolutions (they would have to correspond to the moment of opening of the centrifugal regulator - about 1700 rpm), but due to the discrepancy of the ratio of transmission of the stand control program it was wrong to think that the engine had reached the minimum stable revolutions (about 600 rpm) and discharged the load. Therefore, the minimum speed zone was left out of the test.

The graphs (Figures 2-4) show the graphical dependencies for opening the throttle valve 60, 80, 100%.

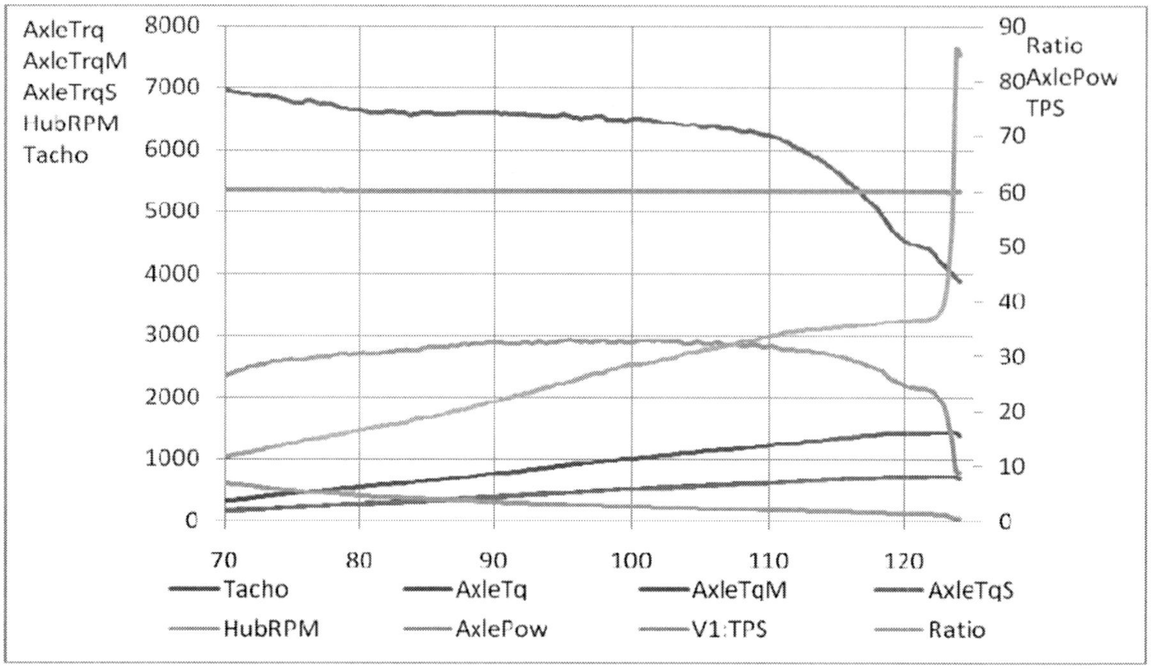

Fig. 2. Test results at 60% throttle opening

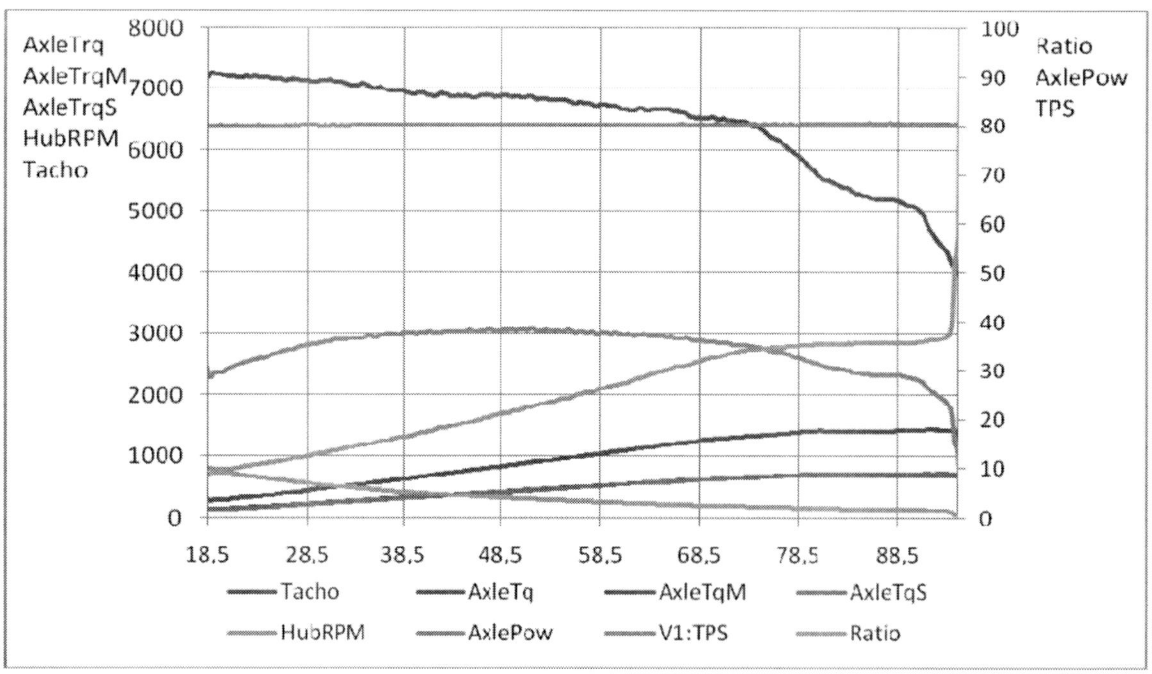

Fig. 3. Test results at 80% throttle opening

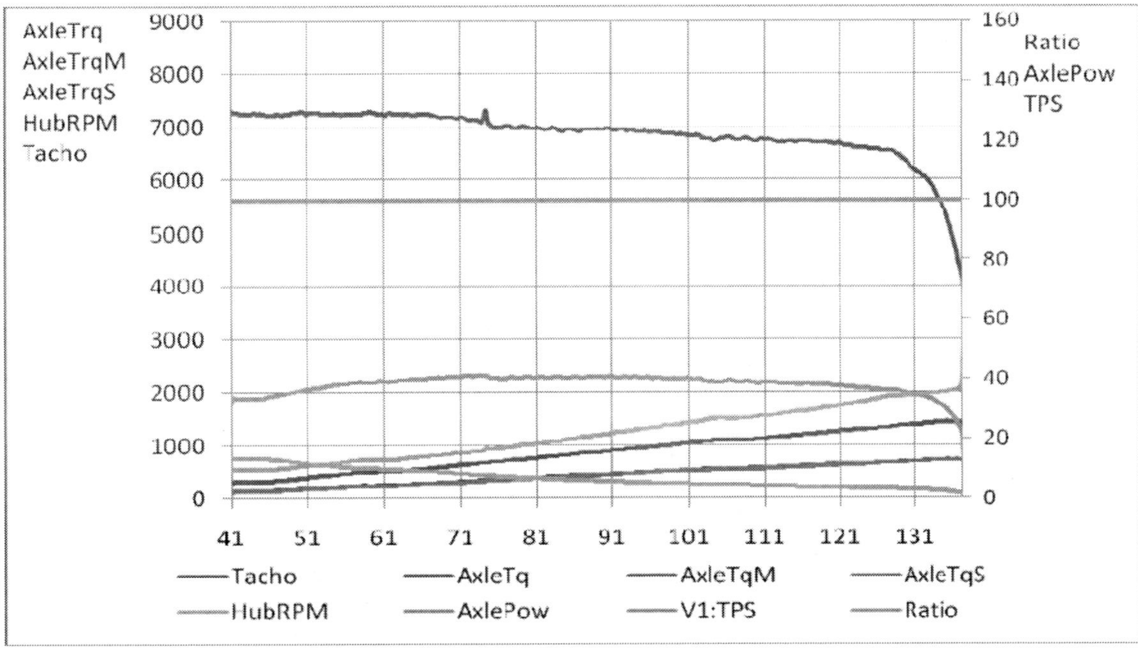

Fig. 4. Test results at 100% throttle opening

All the graphs show a generally similar pattern of change in the performance of the powertrain system: it is clear that the overall ratio changes almost linearly throughout most of the test, with the values of the ratio close to the different degrees of opening the throttle, the nature of the reduction in engine speed is almost identical. The maximum gear ratio is about 80, the minimum is about 10.

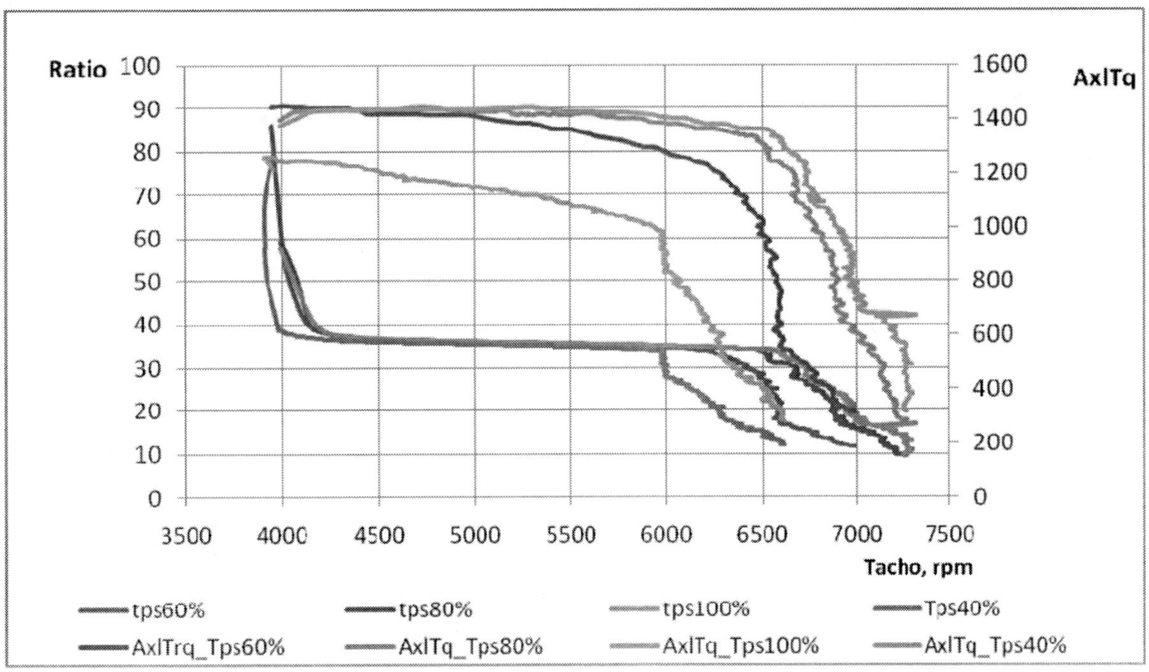

Fig. 5. Dependence of transmission ratio and total torque on engine speed

Figure 5 shows that the nature of the change in the total gear ratio of the transmission is the same for all the throttle positions - the main section corresponds to a linear increase (in fact, almost constant value) and short growth areas. Moreover, the areas of rapid growth of the ratio at different degrees of opening of the throttle valve correspond to different engine revolutions. So, for TPS 60, 80, 100 the growth area at low revolutions practically coincides, and at high revolutions the shift on 250 revolutions is observed. Thus for TPS 40% both growth areas differ - the beginning of a growth area at high revolutions is shifted already on 500 revolutions/min from TPS100%, and at low revolutions - on 250 revolutions/min. However, the linear section coincides in size with other cases. Similar character is observed at other degrees of opening of the throttle valve - the smaller the opening, the earlier the switching begins on revolutions and the greater the minimum gear ratio.

The change in torque also corresponds to the change in the gear ratio, but the graphs show that the maximum torque at TPS 60,80,100 is the same at 4250-5000 rpm. With a TPS of 60%, the maximum torque is less and is achieved at lower speeds. The torque increase in the gear ratio corresponds to the moment when the engine starts to reduce the speed and passes the maximum torque zone.

This pattern of change suggests that the V-belt variable speed gear ratio is adjusted not only in terms of drive sheave speed (motor speed), but also in terms of motor torque (torque response) and is likely to depend on the cam profile of the driven pulley.

Conclusion

The conducted researches have shown the possibility of testing the ATV on a stepped dynamometer stand according to the given method, but have revealed several features that should be paid attention to during further researches: for example, achievement of the minimum engine revolutions at a high transmission ratio and peculiarities of operation of the program of stand control, which does not allow to introduce too high value of the transmission ratio; transient processes in changing the gear ratio, which require the selection of the incremental load step and the stabilization of the load at each step

The specific character of the change in the transmission ratio of stepless transmission, which depends not only on the speed of rotation, but also on the load factor, is revealed. The assumed characteristic of the centrifugal regulator of the V-belt variable speed gearbox should be a curve that characterizes the change in the position of the weights depending on the input shaft revolutions. This curve should be common for all modes of motor operation [19]. However, from the graph analysis, the actual characteristic of the change in gear ratio (which depends on the characteristic of the regulator) by engine speed is a group of curves that have the same nature of change, but different values depending on the position of the throttle valve.

It should be noted that the this work was carried out at the Bauman Moscow State Technical University, with financial support from the government in the face of the Russian Ministry of Education under the project: №14.577.21.0272. (Identification number: RFMEFI57717X0272)

References

[1] Safonov, B.A., Travnikov, A.N. Unmanned all-terrain cargo and passenger transportation system for operation conditions when automobile roads are unavailable (2019) Journal of Physics: Conference Series, 1177 (1) DOI: 10.1088/1742-6596/1177/1/012043

[2] Klubnichkin, V.E., Dyakov, A.S., Klubnichkin, E.E., Zakharov, A.Y., Vakhidov, U.S., Suchenina, A.S., Basmanov, I.V. Experimental evaluation of stability and controllability of domestic and foreign made utility terrainvehicles (2019) Journal of Physics: Conference Series, 1177 (1), DOI: 10.1088/1742-6596/1177/1/012045

[3] Klubnichkin, V.E., Dyakov, A.S., Klubnichkin, E.E., Zakharov, A.Y., Vakhidov, U.S., Suchenina, A.S., Basmanov, I.V. Experimental evaluation of speed and brake properties of domestic and foreign made utility terrain vehicles (2019) Journal of Physics: Conference Series, 1177 (1), DOI: 10.1088/1742-6596/1177/1/012048

[4] Klubnichkin, E., Klubnichkin, V. Experimental evaluation of traction and coupling properties of snowmobile motor vehicles (2018) MATEC Web of Conferences, 224, DOI: 10.1051/matecconf/201822402082

[5] Vora, K.C., Agrewale, M.R.B., Desai, M.M., Mishra, H., Narkar, O. Development, analysis and testing of an electric all-terrain vehicle (2018) 2017 IEEE Transportation Electrification Conference, ITEC-India 2017, 2018-January, pp. 1-6.

[6] Lee, T., Cho, S., Lee, S., Roh, C., Kang, Y. Design of an autonomous antarctic exploration platformsthrough manipulations of All-Terrain Vehicles (2016) 2016 IEEE Transportation Electrification Conference and Expo, Asia-Pacific, ITEC Asia-Pacific DOI: 10.1109/ITEC-AP.2016.7512963

[7] CAN-AM ATVhttps://brp-world.com/en/brand.action?brandName=canamatv

[8] CanAmOutlanderhttps://brp-russia.blogspot.com/2016/04/brp-outlander-570-650-850-1000-2016.html

[9] Kolator, B., Janulin, M., Vrublevskyi, O. Influence of traction conditions on the power balance in the vehicle 4wd system (2018) IOP Conference Series: Materials Science and Engineering, 421 (2), .DOI: 10.1088/1757-899X/421/2/022014

[10] Kolator, B., Janulin, M. Methodology for measuring car traction parameters (2018) MATEC Web of Conferences, 182, DOI: 10.1051/matecconf/201818202001

[11] Yang, Z., Deng, B., Deng, M., Huang, S. An overview of chassis dynamometer in the testing of vehicle emission (2018) MATEC Web of Conferences, 175, DOI: 10.1051/matecconf/201817502015

[12] Mohamed, E.S., Khalil, M.I., Abouel-Seoud, S.A. Assessment of manual, automatic and continuously variable transmission powertrains for gasoline engine powered midsize saloon vehicle (2015) International Journal of Vehicle Structures and Systems, 7 (3), pp. 85-91 DOI: 10.4273/ijvss.7.3.01

[13] Irimescu, A., Mihon, L., Pădure, G. Automotive transmission efficiency measurement using a chassis dynamometer (2011) International Journal of Automotive Technology, 12 (4), pp. 555-559.

[14] A. J. Martyr M. A. Plint Engine Testing. The Design, Building, Modification and Use of Powertrain Test Facilities2012 Elsevier Ltd 572 p.

[15] Dynapack dynamometers www.dynapack.com

[16] Barr, B.D., McKinstry, D.J., Messenger, A.S., Seidenstucker, J.M., Kallio, D.H., Hoffman, D.P., Bettig,B.P. Adaptation of four-stroke motorcycle engine to continuously variable transmission for snowmobile application (2013) SAE Technical Papers DOI: 10.4271/2003-32-0083

[17] Moskalik, A., Dekraker, P., Kargul, J., Barba, D. Vehicle Component Benchmarking Using a Chassis Dynamometer (2015) SAE International Journal of Materials and Manufacturing, 8 (3), DOI: 10.4271/2015-01-0589

[18] W Grzegoek and M Szczepka Analysis of an employment of a gear ratio rate in CVT control system2016 IOP Conf. Ser.: Mater. Sci. Eng. 148 012007

[19] William C. Orthwein Clutches and Brakes. Design and Selection.Second Edition 2004 Marcel Dekker 337 p.

Verification of mathematical model in curvilinear motion of the car with AWD system

G Ankinovich[1], A Verzhbitsky[1] and A Antonyan[1,2]

[1]Bauman Moscow State Technical University

[2]E-mail: antonyan.akop@yandex.ru

Abstract. In the vehicles development also developed a variety of control systems. However, in their development it is necessary to have a mathematical model of the control object, that is, the vehicle model for the study of electronic control systems. Mathematical modeling is the most effective and promising method of studying the control object behavior. It is possible to carry out vehicle mathematical modeling and its components with sufficient high accuracy and compliance with its real prototype. But there are cases that it is not possible to simulate in detail some phenomena of the control object. In order to verify the accuracy of the compiled model, model verification of the developed vehicle is carried out. The basis used the KIA Sorento. It has AWD based on front wheel drive. This vehicle, equipped with measuring and recording devices, which carried out on the roads of the testing ground (FSUE "NAMI"). The work focuses on modelling the simulation of transmission units in the vehicle. The systems of equations of the engine, gearbox, transfer case, main gears and elastic drive shafts are written. Races on curvilinear trajectories were carried out. It also presents results comparison of the real vehicle movement and the results carried out from modeling in MATLAB&Simulink. The accuracy of the developed mathematical model and admissibility in further application in the development of control systems are proved.

1. Introduction

Vehicles with AWD have recently become more popular. The transmission, which provides traction on all wheels, is found not only on SUVs, but also on crossovers and sedans. It contains a transfer case, and drives axles to front and rear axis, giving an increase in weight and mechanical loss. But, despite this, the frequent use of such a transmission is explained by improvements in the vehicle dynamic qualities, increased driving safety, and increased vehicle stability when passing curved road sections, as well as increased cross-country. Mathematical modeling [1–5] is used for these purposes. Mathematical modeling makes it possible to simulate the vehicle movement in critical situations [6–11]. The developed mathematical model must be verified to ensure the correctness of the mathematical description of the vehicle body, components and assemblies.

1.1 Targets

Vehicle mathematical model development, with AWD based on front wheel drive, carrying out the experiment and checking the accuracy of the developed mathematical model.

2. Vehicle motion mathematical model

The paper focuses on the transmission units modeling. When developing a mathematical model of vehicle dimensional movement the following assumptions are made: the masses of unsprung elements are given to the wheels axes, and the sprung ones – to the body system. The road is assumed to be undeformed. Assumptions for transmission and driveline units, the following: the gear box elements, transfer case elements, the final drive elements are absolutely rigid. Elastic driveline transversal shafts are accepted. The model of wheel interaction with the basic basis is developed on methods used in works [12–20].

2.1 Description of the mathematical model of the car

The basis is a passenger car with an AWD based on front wheel drive. Figure 1 shows a diagram of the forces acting on the vehicle:

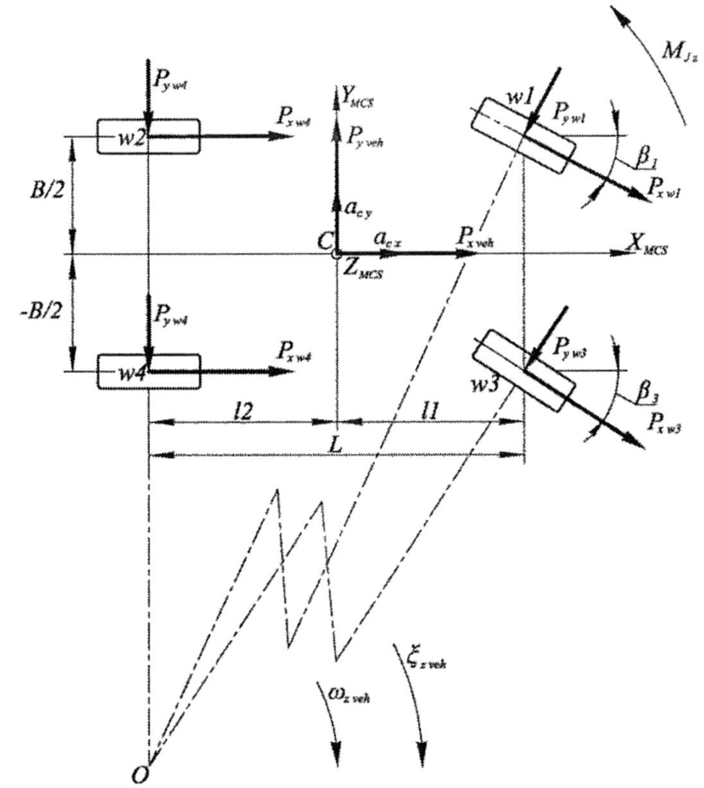

Figure 1. Vehicle design model:
B – wheel track; L – wheelbase; C – center of mass; $l1$ – distance between center of mass front axle; $l2$ – distance between center of mass rear axle; X_{MCS}, Y_{MCS}, Z_{MCS} – axes X, Y, Z movable coordinate system (MCS); $w1$ – wheel 1; $w2$ – wheel 2; $w3$ – wheel 3; $w4$ – wheel 4; β_1 – wheel 1 steering angle; β_3 – wheel 3 steering angle; a_{cx} – acceleration vector projection on the X-axis MCS; a_{cy} – acceleration vector projection on the Y-axis MCS; P_{xw1} – wheel 1 force projection on the X_{MCS}-axis; P_{xw2} – wheel 2 force projection on the X_{MCS}-axis; P_{xw3} – wheel 3 force projection on the X_{MCS}-axis; P_{xw4} – wheel 4 force projection on the X_{MCS}-axis; $P_{y\,w1}$ – wheel 1 force projection on the Y_{MCS}-axis; P_{yw2} – wheel 2 force projection on the Y_{MCS}-axis; P_{yw3} – wheel 3 force projection on the Y_{MCS}-axis; P_{yw4} – wheel 4 force projection on the Y_{MCS}-axis; $P_{x\,veh}$ – inertial force projection on the X_{MCS}-axis; $P_{y\,veh}$ – inertial force projection on the Y_{MCS}-axis; M_{Jz} – body inertial torque around the vertical axis; $\omega_{z\,veh}$ – body angular velocity; $\xi_{z\,veh}$ – body angular acceleration.

The vehicle motion is described by an equations system based on theorem about the change in the quantity of motion and the angular momentum [1–4]:

$$\begin{cases} ma_{cx} + m\left(\omega_y v_{cz} - \omega_z v_{cy}\right) = G_x + F_x + \sum\limits_{j=1}^{2N} R_j^x; \\ ma_{cy} + m\left(\omega_z v_{cx} - \omega_x v_{cz}\right) = G_y + F_y + \sum\limits_{j=1}^{2N} R_j^y; \\ ma_{cz} + m\left(\omega_x v_{cy} - \omega_y v_{cx}\right) = G_z + F_z + \sum\limits_{j=1}^{2N} R_j^z; \\ J_x\dot{\omega}_x - \omega_y\omega_z\left(J_z - J_y\right) = M_x\left(F\right) + \sum\limits_{j=1}^{2N} M_x\left[R_j\right]; \\ J_y\dot{\omega}_y - \omega_z\omega_x\left(J_x - J_z\right) = M_y\left(F\right) + \sum\limits_{j=1}^{2N} M_y\left[R_j\right]; \\ J_z\dot{\omega}_z - \omega_x\omega_y\left(J_x - J_y\right) = M_z\left(F\right) + \sum\limits_{j=1}^{2N} M_z\left[R_j\right] + \sum\limits_{j=1}^{2N} M_{tr\,j} \end{cases} \quad (1)$$

m, vehicle mass, kg; a_{cx}, a_{cy}, a_{cz}, acceleration vector projections on the X, Y,Z-axis movable coordinate system (MCS), m/s^2; v_{cx}, v_{cy}, v_{cz}, speed vector projections on the X, Y, Z-axis in MCS, m/s; ω_x, ω_y, ω_z angular speed vector projections on the X, Y, Z-axis in MCS, rad/s; $\dot{\omega}_x$, $\dot{\omega}_y$, $\dot{\omega}_z$, angular acceleration vector projections on the X, Y, Z-axis in MCS, rad/s^2; J_x, J_y, J_z, body inertias around the X, Y, Z-axis in MCS, $kg \cdot m^2$; G_x, G_y, G_z, gravity force projections on the X, Y, Z-axis in MCS, N; F_x, F_y, F_z, outside force projections on the X, Y, Z-axis, N; R_j^x, R_j^y, R_j^z, reactive force projections on the X, Y, Z-axis, N; $M_{tr\,j}$, wheel turn resistance torque, $N \cdot m$; j, number of axes.

2.2 Transmission and driveline mathematical model description

Also, a transmission and driveline mathematical model, in which each transmission and driveline unit are described by own system of equations. The design model of the transmission and drive line are shown in figure 2:

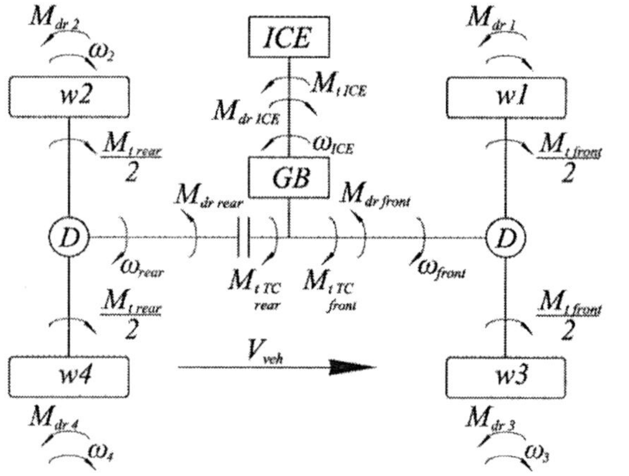

Figure 2. Transmission and driveline design model:
ICE – internal combustion engine;
GB – gearbox;
D – differential;
1 – front left wheel;
2 – rear left wheel;
3 – front right wheel;
4 – rear right wheel

Crankshaft equation:

$$J_{ICE} \cdot \dot{\omega}_{ICE} = M_{tICE} - M_{drICE} \tag{2}$$

J_{ICE}, moving parts inertia of the engine, $kg \cdot m^2$; $\dot{\omega}_{ICE}$, angular engine shaft acceleration, rad / s^2; M_{tICE}, M_{drICE}, torque and drag torque on the motor shaft, $N \cdot m$.

Gearbox equation system:

$$\begin{cases} \dot{\omega}_{GB} = i_{GB} \cdot \dot{\omega}_{ICE}; \\ J_{GB} \cdot \dot{\omega}_{GB} = M_{tGB} - M_{dr\,GB} \end{cases} \tag{3}$$

J_{GB}, moving parts inertia of the gearbox, $kg \cdot m^2$; $\dot{\omega}_{GB}$, gearbox output shaft angular acceleration, rad / s^2; i_{GB}, gearbox ratio; M_{tGB}, gear box output shaft torque, which is $M_{tGB} \cdot i_{GB}$, $N \cdot m$; M_{drGB}, gearbox output shaft drag torque, $N \cdot m$.

Transfer case equation system:

$$\begin{cases} J_{shaft} \cdot \dot{\omega}_{front} = M_{in} - \dfrac{M_{cl}}{2} - M^{dr}_{front}; \\ J_{shaft} \cdot \dot{\omega}_{rear} = \dfrac{M_{cl}}{2} - M^{dr}_{rear}; \\ \dot{\omega}_{input} = i_{TC} \cdot \dot{\omega}_{front} \end{cases} \tag{4}$$

J_{shaft}, front and rear output drive shafts inertia, $kg \cdot m^2$; $\dot{\omega}_{front}$, $\dot{\omega}_{rear}$, front and rear output shafts angular accelerations, rad / s^2; M^{dr}_{front}, M^{dr}_{rear}, front and rear shafts drag torques, $N \cdot m$.

Final drive equation system:

$$\begin{cases} J_w \cdot \dot{\omega}_{w\,right} = i_{FD} \cdot \dfrac{M_{t\,input}}{2} - M^{dr}_{right}; \\ J_w \cdot \dot{\omega}_{w\,left} = i_{FD} \cdot \dfrac{M_{t\,input}}{2} - M^{dr}_{left}; \\ \dot{\omega}_{input} = i_{FD} \cdot \dfrac{\dot{\omega}_{w\,right} + \dot{\omega}_{w\,left}}{2} \end{cases} \tag{5}$$

J_w, wheel inertia, $kg \cdot m^2$; $\dot{\omega}_{wright}$, $\dot{\omega}_{wleft}$, right and left wheel angular accelerations, rad / s^2; i_{FD}, final drive ratio; M^{dr}_{right}, M^{dr}_{left}, right and left wheel drag torques, $N \cdot m$.

Elastic driveshaft transversal equation system:

$$\begin{cases} J_{input} \cdot \dot{\omega}_{input} = M_{t\,input} - c \cdot (\varphi_{in} - \varphi_{out}) - b \cdot (\dot{\varphi}_{in} - \dot{\varphi}_{out}); \\ J_{output} \cdot \dot{\omega}_{output} = c \cdot (\varphi_{in} - \varphi_{out}) + b \cdot (\dot{\varphi}_{in} - \dot{\varphi}_{out}) - M_{dr\,output} \end{cases} \tag{6}$$

J_{input}, J_{output}, reduced inertia of input and output drive shaft sides, $kg \cdot m^2$; $\dot{\omega}_{input}$, $\dot{\omega}_{output}$, angular accelerations of input and output drive shaft sides, rad / s^2; M_{tinput}, $M_{droutput}$, input torque and output drag torque drive shaft sides, $N \cdot m$; c, stiffness coefficient, $N \cdot m / rad$; b, damping coefficient, $N \cdot m / rad / s$; φ_{in}, φ_{out}, torsion angles of input and output drives haft sides, rad; $\dot{\varphi}_{in}$, $\dot{\varphi}_{out}$, torsion velocity of input and output drive shaft sides, rad / s.

IOP Conf. Series: Materials Science and Engineering **820** (2020) 012019 doi:10.1088/1757-899X/820/1/012019

3. Experimental object description

During the tests, the Kia Sorento was used (figure 3).

Figure 3. During the test «TurnR_{turn} = 35m»

3.1 Test description

The experiments were on the testing ground roads (FSUE "NAMI") with the drivers assistance of the motor vehicles research center for testing and refining.

During the tests, the following motion parameters were recorded: longitudinal and transverse acceleration, vehicle speed, yaw rate.

The vehicle motion on the support base "asphalt" (the coefficient of wheel interaction with the support base at full slipping $\mu_{s\,max} = 0,9$) is investigated. Figures 4–6 show the trajectory and parameters of the vehicle when driving in test "lane change S = 20 m".

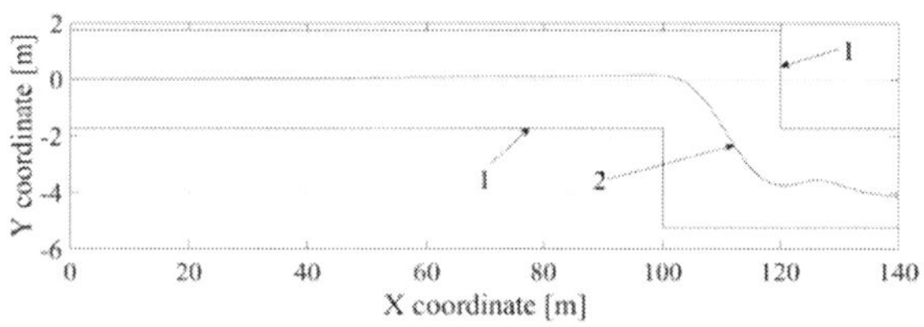

Figure 4. Vehicle trajectory:
1 –test border;
2 – center mass trajectory

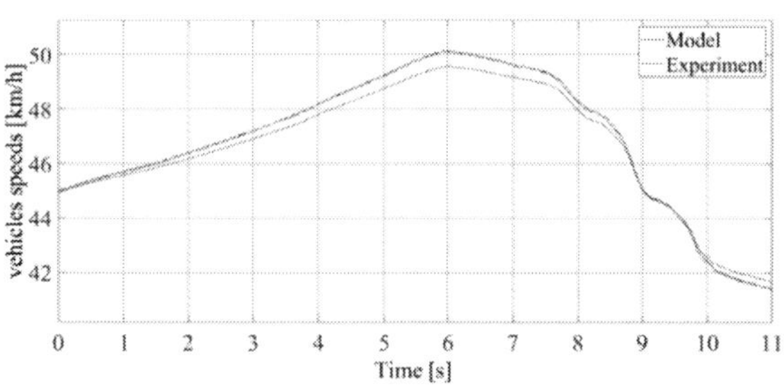

Figure 5. Comparison of longitudinal vehicles speeds

IOP Conf. Series: Materials Science and Engineering **820** (2020) 012019 doi:10.1088/1757-899X/820/1/012019

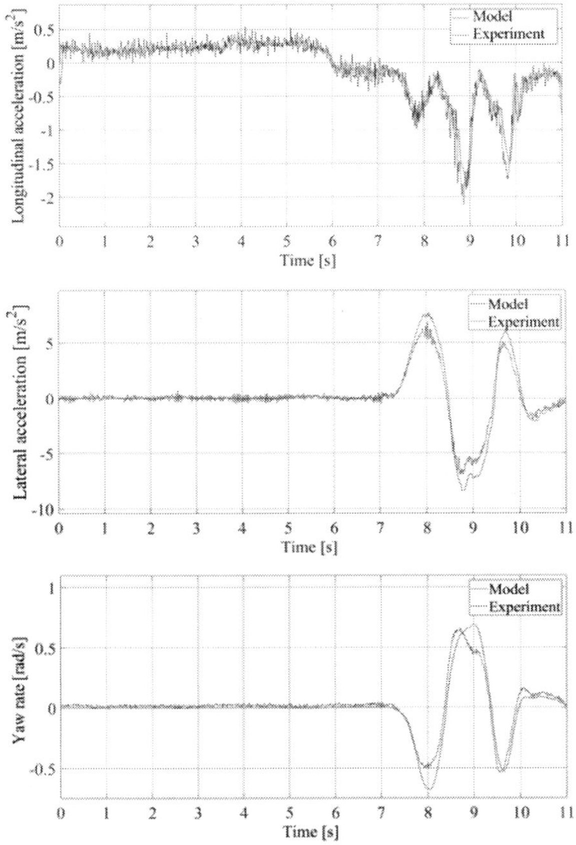

Figure 6. Comparison of longitudinal and lateral vehicle accelerations and yaw rates

Then were tests «TurnR$_{turn}$ = 35 m» (figures 7–9).

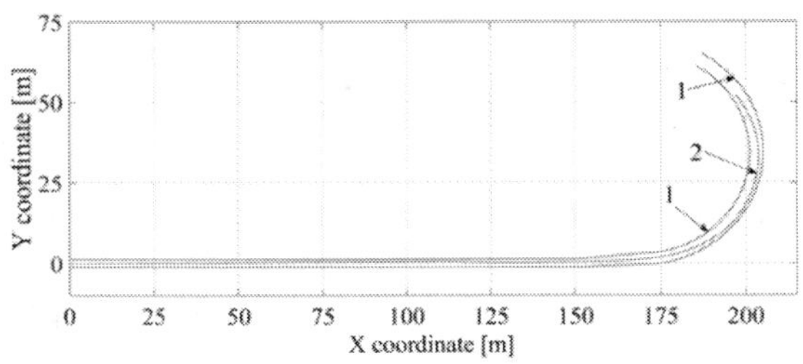

Figure 7. Vehicle trajectory:
1 – test border;
2 – center mass trajectory

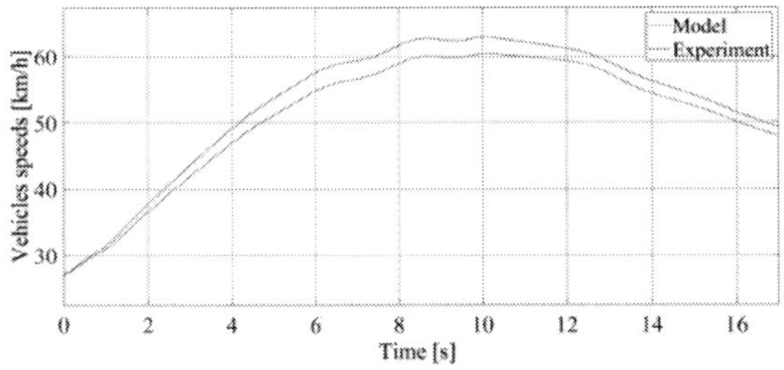

Figure 8. Comparison of longitudinal vehicles speeds

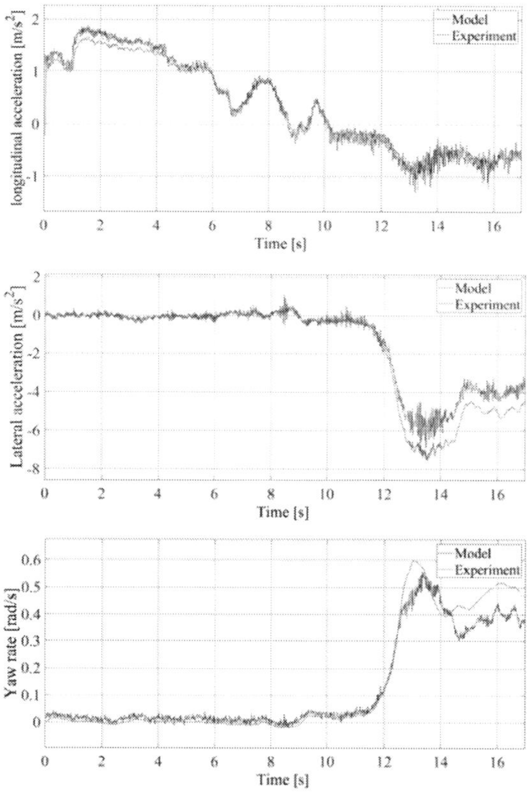

Figure 9. Comparison of longitudinal and lateral vehicle accelerations and yaw rates

4. Results

A mathematical model of the vehicle motion is developed, as well as an experiment in the simulation environment in the MATLAB&Simulink. The simulation and comparison of the results showed good convergence in all parameters, which indicates the accuracy of the developed mathematical model and the possibility of further application.

References

[1] RajeshRajamani2012 *VehicleDynamics and Control Second edition*, 496 p
[2] Reza N Jasar*Vehicle Dynamics, Theory and Application*, 1015 p, 2008
[3] Guzzella L and Sciaretta A2013*Vehicle Propulsion Systems. Introduction to Modeling and Optimization Third edition*, 409p
[4] Waschl H, Kolmanovsky I,Steinbuch M and Luigi del Re 2014*Optimization and Optimal Control in Automotive Systems*, 326p
[5] Popp K, Schnielen W 2010*Ground Vehicle Dynamics*, 348p
[6] Odenthal D, Bunte T, Ackermann J 1999Nonlinear steering and braking control for vehicle rollover avoidance(InProceedingsofEuropeanControlConference,Karlsruhe,Germany)
[7] BelousovB N, Merkulov I V, Fedotov I V2004 Synthesis of dynamic control system for active mountings of multiaxis(ATM Avtomobil'nayaPromyshlennost, (4)), pp. 15-17.
[8] Belousov B, Demik V, Kozlova A, Ksenevich T, Naumov S, Medvedev E, Lyushnin S,Kuzminkov K 2018The schematic diagrams of actuators of an mechatronic wheel steering system(FISITA World Automotive Congress)
[9] Belousov B N, Boldorev A G 2005All-wheel steering with autonomous drive (Avtomobil'nayaPromyshlennost, (11)), pp. 16-17.
[10] Belousov B N, Boldorev A G 2005 System for automatic control of steering for wheel automobile(Avtomobil'nayaPromyshlennost, (9)), pp. 15-18.

[11] Ergin, A.A., Kolomejtseva, M.B., Kotiev, G.O. 2004Antiblocking control system of the brake drive of automobile wheel (PriboryiSistemyUpravleniya, (9)), pp. 11-13.

[12] Gorelov, V.A., Komissarov, A.I. Mathematical model of the straight-line rolling tire - Rigid terrain irregularities interaction (2016) Procedia Engineering, 150, pp 1322-1328

[13] Sarach E, Kotiev G, Beketov S 2018 Methods for road microprofile statistical data transformation (MATEC Web of Conferences), 224, article № 04009

[14] Vol'skaya N S, Zhileykin, M M , Zakharov, A Y 2018 Mathematical model of rolling an elastic wheel over deformable support base(IOP Conference Series: Materials Science and Engineering), 315 (1), article № 012028

[15] Beloousov B, Ksenevich T, Vantsevich, V Komissarov D 2013 8×8 platform for studying terrain mobility and traction performance of unmanned articulated ground vehicles with steered wheels (SAE Technical Papers), 9

[16] Klubnichkin V E, Klubnichkin E E, Kotiev G O, Beketov S A, Makarov V S2013 Interaction between elements of the track ground contacting area with the soil at curvilinear motion of the timber harvesting machine (IOP Conference Series: Materials Science and Engineering, 386 (1)), article № 012016

[17] Belousov B N, Shelomkov S A, Ksenevich T I, Kupreyanov A A 2009 Experimental verification of a mathematical model of the wheel-supporting surface interaction during nonstationary rolling motion (Journal of Machinery Manufacture and Reliability, 38 (5)), pp. 501-505

[18] Gorelov V A, Komissarov A I, Miroshnichenko A V 2015 8×8 wheeled vehicle modeling in multibody dynamics simulation software (Procedia Engineering), 129, pp. 300-307.

[19] Zhileykin M M, Kotiev G O, Nagatsev M V2018 Comparative analysis of the operation efficiency of the continuous and relay control systems of a multi-axle wheeled vehicle suspension (IOP Conference Series: Materials Science and Engineering, 315 (1)), article № 012030

[20] Zhileykin M M, Kotiev G O, Nagatsev M V 2018 Synthesis of the adaptive continuous system for the multi-axle wheeled vehicle body oscillation damping (IOP Conference Series: Materials Science and Engineering, 315 (1)), article № 012031

IOP Conf. Series: Materials Science and Engineering **820** (2020) 012020 doi:10.1088/1757-899X/820/1/012020

Mathematical model of air suspension with internal throttling of working fluid

K Chernyshov[1], V Novikov[1], A Pozdeev[1,3], I Ryabov[1], A Diakov[2], K Evseev[2] and D Chumakov[1]

[1]Volgograd State Technical University, Volgograd, Russian Federation
[2]Bauman Moscow State Technical University

[3]E-mail: pozdeev.vstu@gmail.com

Abstract. The article describes a mathematical model of a two-mass vibrating system with an air suspension consisting of a two-chamber air compressor, the working and auxiliary chambers of which are connected to each other through an air throttle. The results of the research have been used to construct the amplitude-frequency characteristics of vertical displacements and accelerations of the suspended mass, and conclusions have been drawn.

1. Introduction
One of the ways to provide damping in air-sprung suspensions is the use of internal throttling of working fluid in air-springs [1-22]. To determine the efficiency of internal throttling in the air suspension was developed a model of two-mass oscillating system, which simulates the motion on the road profile of the vehicle with air springs, which provides internal throttling, based on the theory of viscous flow of gas through the hole.

2. The description of the mathematical model
Figure 1 shows the scheme of a two-mass oscillating system that simulates the motion of a vehicle with a two-chamber air spring by the road profile. Internal throttling of gas is provided by a throttle hole made in the separating wall between the main and additional chambers of the spring.

The equations of system oscillations:

$$\begin{cases} M\ddot{z}_1 + k\left(\dot{z}_1 - \dot{z}_2\right) - F_{\text{spr}} + Mg = 0, \\ m_t\ddot{z}_2 + k_t\left(\dot{z}_2 - \dot{q}\right) + c_t\left(z_2 - q\right) - k\left(\dot{z}_1 - \dot{z}_2\right) + F_{\text{spr}} - Mg = 0. \end{cases} \quad (1)$$

Where $F_{\text{spr}} = F_1$ – elasticity force depending on the pressure in the main chamber; g – the acceleration of free fall. The elastic force is a function of time:

$$F_{\text{spr}}(t) = F_1(t) = \left(\frac{l_1(t)}{z_1 - z_2 + l_{10}}\right)^n \cdot (Mg + p_{\text{atm}}S) - p_{\text{atm}}S, \quad (2)$$

where p_{atm} – atmospheric pressure; n – polytrope index.

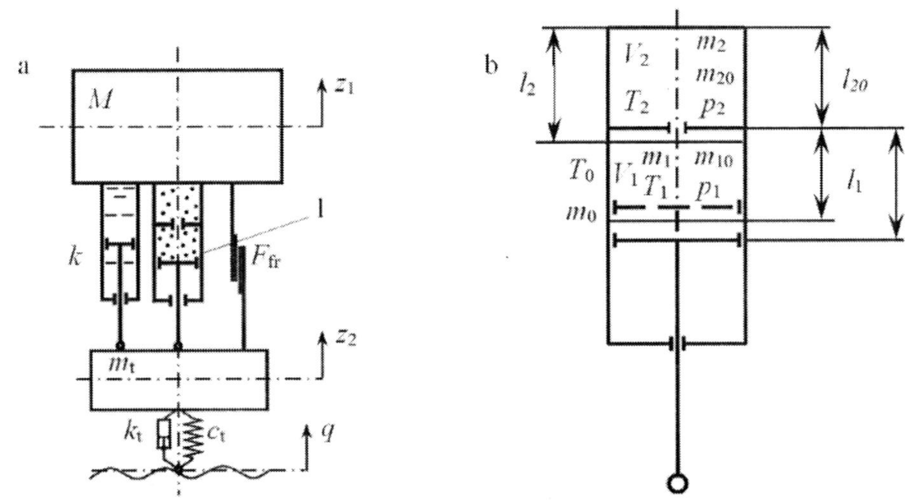

Figure 1. Design model of air suspension with internal throttling of working fluid: a - scheme of oscillating system with air suspension; b - scheme of air spring with internal throttling; 1 - air spring with internal throttling; M – sprung weight; m_t – unsprung weight; k – suspension damping coefficient; F_{fr} – dry friction force in the suspension; c_t – tyre stiffness; k_t – tyre damping coefficient; l_{10}andl_{20} – the initial height of the gas column in the main chamber and the additional chamber under static load when the gas mass flows out of it (into it)Q; $l_1(t)$ and$l_2(t)$ – the current height of the gas column in the main and additional chambers under static load;$p_1(t)$, $p_2(t)$ – current pressure in the main and additional chambers; $l_0 = l_{10} + l_{20} = l_1(t) + l_2(t)$ – total height of the gas column; $V_{10} = l_{10} \cdot S$, $V_{20} = l_{20} \cdot S$corresponding volumes of main and additional chambers under static load; S– piston area; $V_1(t) = (l_{10} + z - q) \cdot S$, $V_2(t) = V_{20} = l_{20}(t) \cdot S$ – the corresponding volume of cameras at the current load; m_{10}иm_{20} – Initial gas mass in the main and additional chambers; $m_0 = m_{10} + m_{20}$ – gas mass in the spring; T_0 – Initial gas temperature in the main and additional chambers; $m_1(t)$, $T_1(t)$ and$m_2(t)$, $T_2(t)$ – current weight and temperature of gas in the main and additional chambers; p_{10}andp_{20} – initial pressure in the main and additional chambers under static load; q – vertical displacement of the road profile;z_1 – vertical displacement of the sprung mass; z_2 – vertical displacement of unsprung mass.

The elasticity force of the gas in the additional chamber:

$$F_2(t) = \frac{l_2^n(t)}{l_{20}^n} \cdot (Mg + p_{atm} S) - p_{atm} S. \quad (3)$$

Because $l_{10} - l_1(t) = l_2(t) - l_{20}$, then

$$F_2(t) = \left(\frac{l_0 - l_1(t)}{l_0 - l_{10}} \right)^n \cdot (Mg + p_{atm} S) - p_{atm} S. \quad (4)$$

Pressure in the main volume:

$$p_1(t) = \frac{F_1(t)}{S} = \left(\frac{l_1(t)}{z_1 - z_2 + l_{10}} \right)^n \cdot \left(\frac{Mg}{S} + p_{atm} \right) - p_{atm}. \quad (5)$$

Pressure in additional volume:

$$p_2(t) = \frac{F_2(t)}{S} = \left(\frac{l_0 - l_1(t)}{l_0 - l_{10}} \right)^n \cdot \left(\frac{Mg}{S} + p_{atm} \right) - p_{atm}. \quad (6)$$

The formula for mass flow of gas from the pressure chamber is known p_1 into a pressure chamber p_2 through a hole in the area f_0 [23, 24]:

$$Q=\alpha_g f_0\left(\frac{p_2}{p_1}\right)^{\frac{1}{n}}\sqrt{\frac{2n}{n-1}\cdot\frac{p_1 m_1}{V_1}}\sqrt{1-\left(\frac{p_2}{p_1}\right)^{\frac{n-1}{n}}}\,t,(7)$$

where α_g – gas consumption factor, depending on the shape and length of the hole; f_0 – the area of the throttle hole. Accordingly, the mass flow rate of gas from the pressure chamber p_2 into a pressure chamber p_1:

$$Q=\alpha_g f_0\left(\frac{p_1}{p_2}\right)^{\frac{1}{n}}\sqrt{\frac{2n}{n-1}\cdot\frac{p_2 m_2}{V_2}}\sqrt{1-\left(\frac{p_1}{p_2}\right)^{\frac{n-1}{n}}}\,t,(8)$$

As the pressure, volume and mass of the leaked gas changes, the gas flow rate through the hole per unit of time also changes.

With $\Delta p=p_1(t)-p_2(t)>0$ the gas flows from the main chamber to the secondary chamber.

Then, taking into account the formula (7), we will get:

$$dQ=\alpha_g f_0\left(\frac{p_2(t)}{p_1(t)}\right)^{\frac{1}{n}}\sqrt{\frac{2n}{n-1}\cdot\frac{p_1(t)m_1(t)}{V_1(t)}}\sqrt{1-\left(\frac{p_2(t)}{p_1(t)}\right)^{\frac{n-1}{n}}}\,dt.(9)$$

The mass of the gas flowing through it changes with time according to the laws:

$$Q(t)=m_{10}-m_1(t).(10)$$

As at a static suspension position the relation of current mass to initial mass in the main chamber is equal to the relation of current height of gas column in this chamber to initial height, we have:

$$m_1(t)=m_{10}\cdot\frac{l_1(t)}{l_{10}}.(11)$$

Considering that the current volume of gas in the main chamber $V_1(t)=(l_{10}+z-q)\cdot S,$ and the formula (11), we will get:

$$dQ=d\left(m_{10}-m_{10}\cdot\frac{l_1(t)}{l_{10}}\right)=\alpha_g f_0\left(\frac{p_2(t)}{p_1(t)}\right)^{\frac{1}{n}}\cdot\sqrt{\frac{2n}{n-1}\cdot\frac{p_1(t)\cdot m_{10}\cdot l_1(t)}{(l_{10}+z_1-z_2)\cdot l_{10}\cdot S}}\cdot\sqrt{1-\left(\frac{p_2(t)}{p_1(t)}\right)^{\frac{n-1}{n}}}\,dt.(12)$$

Then:

$$\frac{dl_1(t)}{dt}=-\alpha_g f_0\cdot\sqrt{\frac{l_{10}}{m_{10}\cdot S}}\cdot\left(\frac{p_2(t)}{p_1(t)}\right)^{\frac{1}{n}}\cdot\sqrt{\frac{2n}{n-1}\cdot\frac{p_1(t)\cdot l_1(t)}{l_{10}+z_1-z_2}}\cdot\sqrt{1-\left(\frac{p_2(t)}{p_1(t)}\right)^{\frac{n-1}{n}}}.(13)$$

Use the Clapeyron-Mendeleyev formula to describe the initial state of the working fluid in the main chamber:

$$p_{01}V_{10}=\frac{m_{10}}{\mu}RT_0,(14)$$

where μ – molar gas mass.

Initial gas pressure is provided by the mass of the suspended load and the piston area:

$$p_{01} = \frac{Mg}{S}. (15)$$

From here on out:

$$\frac{l_{10}}{m_{10}} = \frac{RT_0}{\mu Mg}. (16)$$

Use the formula (16) in the equation (13):

$$\frac{dl_1(t)}{dt} = -\alpha_g f_0 \cdot \sqrt{\frac{RT_0}{\mu MgS}} \cdot \left(\frac{p_2(t)}{p_1(t)}\right)^{\frac{1}{n}} \cdot \sqrt{\frac{2n}{n-1} \cdot \frac{p_1(t) \cdot l_1(t)}{l_{10} + z_1 - z_2}} \cdot \sqrt{1 - \left(\frac{p_2(t)}{p_1(t)}\right)^{\frac{n-1}{n}}}. (17)$$

With $\Delta p = p_1(t) - p_2(t) < 0$ the gas flows from the additional chamber to the main chamber. Taking into account the formula (8), we get:

$$dQ = \alpha_g f_0 \cdot \left(\frac{p_1(t)}{p_2(t)}\right)^{\frac{1}{n}} \cdot \sqrt{\frac{2n}{n-1} \cdot \frac{p_2(t) m_2(t)}{V_2(t)}} \cdot \sqrt{1 - \left(\frac{p_1(t)}{p_2(t)}\right)^{\frac{n-1}{n}}} dt. (18)$$

The mass of the leaked gas changes with time according to the law:

$$Q(t) = m_{20} - m_2(t). (19)$$

Since the current mass of gas in the additional chamber:

$$m_2(t) = m_{10} + m_{20} - m_1(t) = m_0 - m_1(t), (20)$$

additional chamber volume

$$V_2(t) = l_{20} \cdot S, (21)$$

then

$$dQ = d\left(m_{10} \cdot \frac{l_1(t)}{l_{10}} - m_{10}\right) = \alpha_g f_0 \left(\frac{p_1(t)}{p_2(t)}\right)^{\frac{1}{n}} \cdot \sqrt{\frac{2n}{n-1} \cdot \frac{p_2(t) \cdot (m_0 l_{10} - m_{10} \cdot l_1(t))}{l_{10} \cdot l_{20} \cdot S}} \cdot \sqrt{1 - \left(\frac{p_1(t)}{p_2(t)}\right)^{\frac{n-1}{n}}} dt, (22)$$

where

$$\frac{dl_1(t)}{dt} = \alpha_g f_0 \cdot \frac{l_{10}}{m_{10}} \cdot \left(\frac{p_1(t)}{p_2(t)}\right)^{\frac{1}{n}} \cdot \sqrt{\frac{2n}{n-1} \cdot \frac{p_2(t) \cdot (m_0 l_{10} - m_{10} \cdot l_1(t))}{l_{10} \cdot l_{20} \cdot S}} \cdot \sqrt{1 - \left(\frac{p_1(t)}{p_2(t)}\right)^{\frac{n-1}{n}}}. (23)$$

As at a static suspension position the relation of total mass of gas in a suspension to initial mass of gas in the basic chamber is equal to the relation of initial total height of a column of gas to initial height of a column of gas in the basic chamber, we have:

$$m_0 = m_{10} \cdot \frac{l_0}{l_{10}}. (24)$$

When

$$\frac{dl_1(t)}{dt}=\alpha_g f_0\cdot\sqrt{\frac{l_{10}}{m_{10}\cdot S}}\cdot\left(\frac{p_1(t)}{p_2(t)}\right)^{\frac{1}{n}}\cdot\sqrt{\frac{2n}{n-1}\cdot\frac{p_2(t)\cdot(l_0-l_1(t))}{l_{20}}}\sqrt{1-\left(\frac{p_1(t)}{p_2(t)}\right)^{\frac{n-1}{n}}}.\ (25)$$

Use formula (16) in equation (25):

$$\frac{dl_1(t)}{dt}=\alpha_g f_0\cdot\sqrt{\frac{RT_0}{\mu MgS}}\cdot\left(\frac{p_1(t)}{p_2(t)}\right)^{\frac{1}{n}}\cdot\sqrt{\frac{2n}{n-1}\cdot\frac{p_2(t)(l_0-l_1(t))}{l_0-l_{10}}}\cdot\sqrt{1-\left(\frac{p_1(t)}{p_2(t)}\right)^{\frac{n-1}{n}}}.\ (26)$$

If the ratio of pressure in the chambers $\frac{p_2}{p_1}$ (with $\Delta p=p_1(t)-p_2(t)>0$) or $\frac{p_1}{p_2}$ (then $\Delta p=p_1(t)-p_2(t)<0$) will reach a critical value $\left(\frac{2}{n+1}\right)^{\frac{n}{n-1}}$, gas consumption becomes constant:

$$Q_{\max}=\alpha_g f_0\cdot\left(\frac{2}{n+1}\right)^{\frac{n}{n-1}}\cdot\sqrt{\frac{2n}{n-1}\cdot\frac{p_1 m_1}{V_1}}\cdot t\ (27)$$

Then $\Delta p=p_1(t)-p_2(t)>0$ and

$$Q_{\max}=\alpha_g f_0\cdot\left(\frac{2}{n+1}\right)^{\frac{n}{n-1}}\sqrt{\frac{2n}{n-1}\cdot\frac{p_2 m_2}{V_2}}\cdot t\ (28)$$

Then $\Delta p=p_1(t)-p_2(t)<0$.

When we make transformations similar to the previous ones, we will get:

$$\frac{dl_1(t)}{dt}=-f_0\cdot\sqrt{\frac{RT_0}{\mu MgS}}\cdot\left(\frac{2}{n+1}\right)^{\frac{n}{n-1}}\cdot\sqrt{\frac{2n}{n-1}\cdot\frac{p_1(t)\cdot l_1(t)}{l_{10}+z-q}}\ (29)$$

Then $\Delta p=p_1(t)-p_2(t)>0$ and

$$\frac{dl_1(t)}{dt}=f_0\cdot\sqrt{\frac{RT_0}{\mu MgS}}\cdot\left(\frac{2}{n+1}\right)^{\frac{n}{n-1}}\cdot\sqrt{\frac{2n}{n-1}\cdot\frac{p_2(t)(l_0-l_1(t))}{l_0-l_{10}}}\ (30)$$

then $\Delta p=p_1(t)-p_2(t)<0$.

As a result, the following suspension working conditions were obtained.

The dynamics of the system is described by equations (1). Suspensionelastic force:

$$F_{spr}=F_1(t)=p_1(t)S.\ (31)$$

The pressure is mainly determined by formula (5), the pressure in the additional volume is determined by formula (6).

When $\Delta p=p_1(t)-p_2(t)>0$ gas column height change l_1 is determined by the equation (17), if $\frac{p_2}{p_1}>\left(\frac{2}{n+1}\right)^{\frac{n}{n-1}}$ and by the equation(29), if $\frac{p_2}{p_1}\le\left(\frac{2}{n+1}\right)^{\frac{n}{n-1}}$.

When $\Delta p = p_1(t) - p_2(t) < 0$ gas column height change l_1 is determined by the equation(26),if $\frac{p_1}{p_2} > \left(\frac{2}{n+1}\right)^{\frac{n}{n-1}}$ and by the equation (30), if $\frac{p_1}{p_2} \le \left(\frac{2}{n+1}\right)^{\frac{n}{n-1}}$.

3. Analysis of calculation results by mathematical model

For research of the mathematical model the program in language Pascal ABC has been written which results of calculations have shown that vibrating system can have two low-frequency and two high-frequency resonances (an example in figure 2). The first low-frequency resonance corresponds to 1 Hz, the second low-frequency - 2.3 Hz, the first high-frequency - 8.8 Hz, the second high-frequency - 9.8 Hz. These results were obtained at the following initial data: the volume of the main chamber - 3.1 liters; the volume of the additional chamber - 15.5 liters; static pressure - 0.13 MPa; spring piston area – 450 cm^2; unsprung weight – 125kg; tyre rigidity – 300000 N/m.

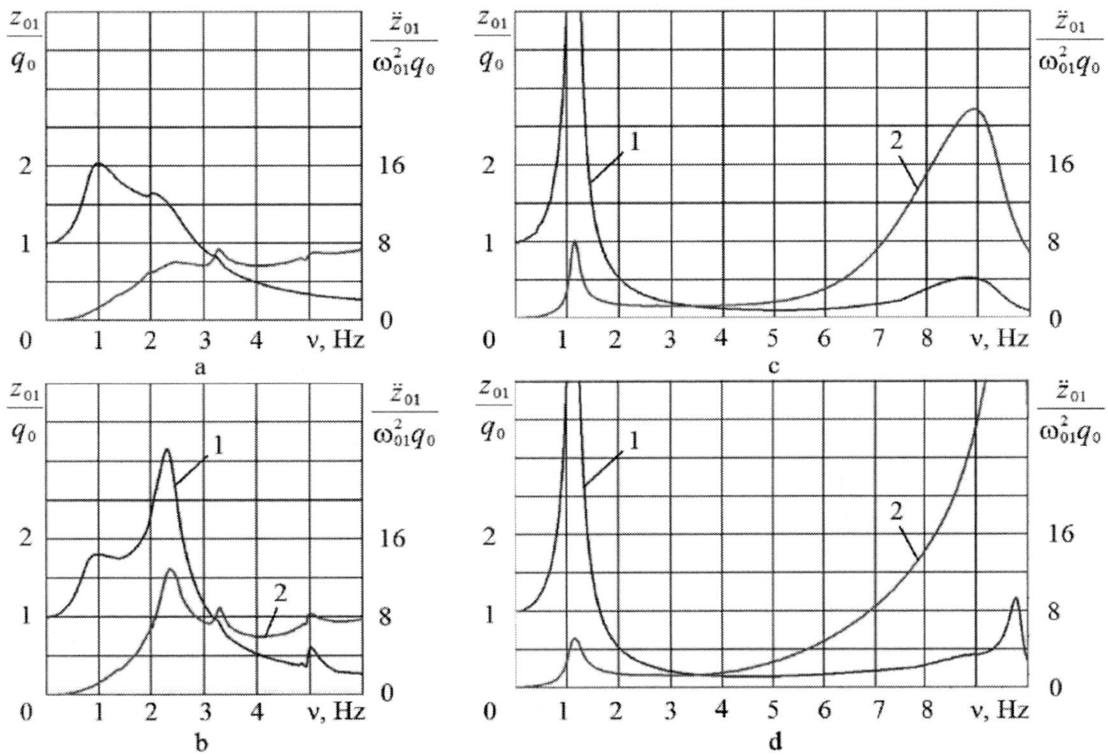

Figure 2. Amplitude-frequency characteristics of movements (1) and accelerations (2) of the suspended mass$M$ = 0,5 tonon the air suspension with internal air throttling: z_{01}/q_0 – the relative amplitude of the movements of the suspended mass; $\ddot{z}_{01}/\omega_{01}^2 q_0$ – relative amplitude of acceleration; ν – disturbance frequency: a – q_0 = 5 mm; d_{thr} = 2 mm; b– q_0 = 8 mm; d_{thr} = 2 mm; c – q_0 = 5 mm; d_{thr} = 5 mm; d – q_0 = 5 mm; d_{thr} = 4 mm

4. Conclusions

- The developed mathematical model allows to define the degree of influence of parameters of a two-chamber air suspension with internal throttling of working fluid on its vibration protection properties.

- Depending on the modes of vibration in the suspension it is possible to achieve 4 resonances, at two of which it is possible to achieve the minimum amplitudes of movement and acceleration of the body: the first low-frequency and the first high-frequency.
- The amplitudes are most affected by the ratio of additional volume to main volume and throttle diameter.

It should be noted that the this work was carried out at the Bauman Moscow State Technical University, with financial support from the government in the face of the Russian Ministry of Education under the project: №14.577.21.0272. (Identification number: RFMEFI57717X0272)

References

[1] Bachrach B I, Rivin E 1983 Analysis of a damped pneumatic spring *Journal of Sound and Vibration*vol 86 **2** pp 191-197
[2] Kornhauser A A 1994 Dynamic modeling of gas springs *Transactions of the ASME (Journal of Dynamic Systems, Measurement and Control)* **116** pp 414-418
[3] Erin C, Wilson B, Zapfe J 1998 An improved model of a pneumatic vibration isolator: theory and experiment *Journal of Sound and Vibration* vol 218 **1** pp 81-101
[4] Toyofuko K, Yamada C, Kagawa T, Fujita T 1999 Study on dynamic characteristic analysis of air spring with auxiliary chamber *JSAE Review* **20** pp 349-355
[5] Quaglia G, Sorli M 2001 Air suspension dimensionless analysis and design procedure *Vehicle System Dynamics* **35** pp 817-829
[6] Deo H, Suh N P 2006 Pneumatic suspension system with independent control of damping, stiffness and ride-height *Proceedings of ICAD2006 (4th International Conference on Axiomatic Design, Firenze)* **22** pp 1-6
[7] Gavriloski V, Jovanova J 2006 Dynamic behaviour of an air spring element *Machines, Technologies, Materials* pp 24-27
[8] Docquier N, Fisette P, Jeanmart H 2007 Multiphysic modeling of railway vehicles equipped with pneumatic suspensions *Journal of Vehicle System Dynamics* **45(6)** pp 505-524
[9] Kim H S, Lee H C 2007 Asynchronous and synchronous load leveling compensation algorithm in airspring suspension *Proceedings of the International Conference on Control, Automation and Systems* pp 367-372
[10] Lee J-H, Kim K-J 2007 Modeling of nonlinear complex stiffness of dual-chamber pneumatic spring for precision vibration isolations *Journal of Sound and Vibration* **301** pp 909-926
[11] Sugahara Y, Takigami T, Kazato A 2007 Suppressing vertical vibration in railway vehicles through air spring damping control *Journal of System Design and Dynamics* vol 1 **2** pp 212-223
[12] Jun K J et al. 2008 Prediction of fatigue life and estimation of its reliability on the parts of an air suspension system *International Journal of Automotive Technology* vol 9 **6** pp 741-747
[13] Nieto A J, Morales A L, Gonzalez A, Chicharro J M, Pintado P 2008 An analytical model of pneumatic suspensions based on an experimental characterization *Journal of Sound and Vibration* **313** pp 290-307
[14] Wolf-Monheim F, Frantzen M, Seemann M, Wilmes M 2008 Modeling, testing and correlation of interlinked air suspension systems for premium vehicle platforms *F2008-SC-040* pp 18-29
[15] Berg M 2009 A three-dimensional air spring model with friction and orifice damping *Journal of Vehicle System Dynamics* **33** pp 528-539
[16] Khamitov R N, Aver'yanov G S, Korchagin A B 2009 Pneumatic Shock Absorber with an Active Damping System *Russian Engineering Research*vol 29 **9** pp 871-873
[17] Sayyaadi H, Shokouhi N 2009 A new model in rail-vehicles dynamics considering non-linear suspension components behavior *International Journal of Mechanical Sciences* pp 1016-1035
[18] Moon Jun-Hee, Lee Bong-Gu 2010 Modeling and sensitivity analysis of a pneumatic vibration isolation system with two air chambers *Mechanism and Machine Theory* **45** pp 1828-1850

[19] Pu H et al. 2011 Modeling and analysis of dual-chamber pneumatic spring with adjustable damping for precision vibration isolation *Journal of Sound and Vibration* **03** pp 1016-1021

[20] Novikov V V, Ryabov I M, Pozdeev A V, Chernyshov K V, Chumakov D A 2018 Experimental study of air suspension with combined damping*Truck: transport complex, special equipment* (Moscow, Russia) **9** pp 3-7

[21] Pozdeev A V, Chumakov D A, Novikov V V, Golyatkin I A, Chernyshov K V, Gavrilov A E, Leonard A V 2019Thermographic bench tests of rubber-cord pneumatic spring *IOP Conference Series: Materials Science and Engineering (ToPME, Moscow, IMASH RAS)* **489** 9 p

[22] Novikov V V, Chumakov D A, Golyatkin I A 2019 Experimental Modular Unit for Testing Nodes of Air Suspension with Combined Amortization *Proceedings of the 4th International Conference on Industrial Engineering (ICIE 2018)* (Moscow, Russia) ed A A Radionov (Lecture Notes in Mechanical Engineering (LNME) Springer) pp 385-392

[23] Shumsky K P1974 Vacuum apparatus and instruments for chemical engineering *(Moscow, Mechanical engineering)* 576 p

[24] Prandtl L 2000 Hydroaeromechanics (Izhevsk, “Regular and chemical dynamics”) 576 p

IOP Conf. Series: Materials Science and Engineering **820** (2020) 012021 doi:10.1088/1757-899X/820/1/012021

Results of theoretical and experimental research of pneumatic suspension with internal throttling of working fluid

K Chernyshov[1], V Novikov[1], A Pozdeev[1,3], I Ryabov[1], A Diakov[2], K Evseev[2] and D Chumakov[1]

[1]Volgograd State Technical University, Volgograd, Russian Federation
[2]Bauman Moscow State Technical University

[3]E-mail: pozdeev.vstu@gmail.com

Abstract. The article presents the mathematical model and experimental device of the air suspension installed on the dynamic stand, which imitates the work of the unsupported vehicle suspension. The air suspension is presented in the form of a two-chamber air compressor, the working and auxiliary cavities of which are communicated to each other through a throttle with a check valve open during compression. The design and experimental amplitude-frequency characteristics of scales of vertical movements of the suspended mass are presented.

1. Introduction

The mathematical modeling and bench tests are used to study the vibro-protective properties of the air suspension systems of motor vehicles, which allow to check the adequacy of the mathematical model and to reveal the peculiarities of the work of the real suspension systems [1-21]. Special and universal benches are used for testing, but they can make errors in the work of real systems, for example, in the VolgGTU at the department of "Automatic installations" there is a unique test bench BISS-ITW for the study of unsupported vibrating systems [1-3]. The peculiarity of the stand is the presence of a movable crosshead with weights imitating the suspended mass of the car, moved relatively stationary base of the stand with little friction $F_{fr.add}$, which is absent when the real car is moving along the roughness of the road.

In order to take into account the effect of additional friction on the work of the unsupported air suspension installed on the stand, a calculation scheme (Figure 1) and a mathematical model have been developed:

$$\begin{cases} M\ddot{z}_1 + k\left(\dot{z}_1 - \dot{z}_2\right) - F_{spr} + Mg + F_{fr}\,\mathrm{sgn}(\dot{z}_1 - \dot{z}_2) + F_{fr.add}\,\mathrm{sgn}\,\dot{z}_1 = 0, \\ m_t\ddot{z}_2 + k_t\left(\dot{z}_2 - \dot{q}\right) + c_t\left(z_2 - q\right) - k\left(\dot{z}_1 - \dot{z}_2\right) + F_{spr} - Mg - F_{fr}\,\mathrm{sgn}(\dot{z}_1 - \dot{z}_2) - F_{fr.add}\,\mathrm{sgn}\,\dot{z}_1 = 0. \end{cases} \quad (1)$$

Where $F_{spr} = F_1$ – the elasticity force, which depends on the pressure p_1 in the main chamber of the rubber-corded air springs (RCS); g – freefall acceleration; M and m_t – suspended and unsprung mass, k – damping factor, which takes into account the internal losses in RCS; F_{fr} – dry friction force in the air suspension; $F_{fr.add}$ – additional dry friction force between the suspended mass and the base of the bench; k_t and c_t – damping factor and tyre rigidity.

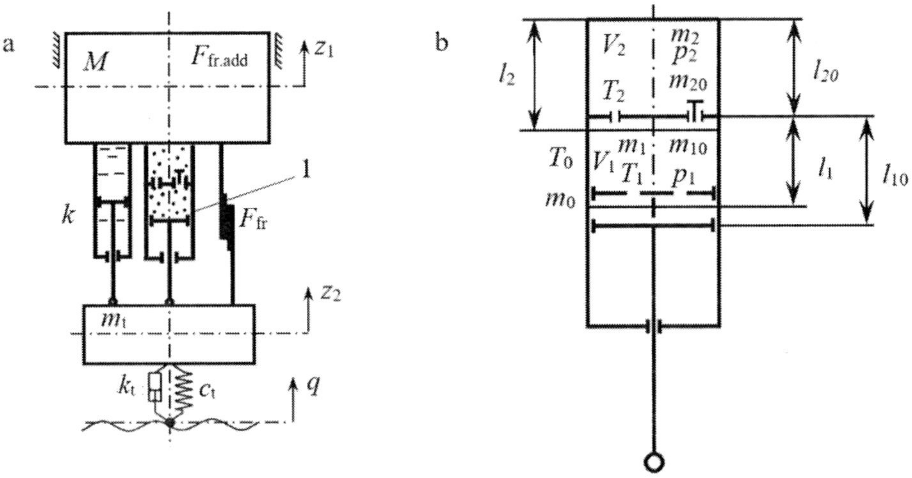

Figure 1. Design model of air suspension with internal throttling of working fluid: a - scheme of oscillating system with air suspension; b - scheme of air spring with internal throttling; 1 - air spring with internal throttling; M – sprung weight; m_t – unsprung weight; k – suspension damping coefficient; F_{fr} – dry friction force in the suspension; c_t – tyre stiffness; k_t – tyre damping coefficient; l_{10} and l_{20} – the initial height of the gas column in the main chamber and the additional chamber under static load when the gas mass flows out of it (into it) Q; $l_1(t)$ and $l_2(t)$ – the current height of the gas column in the main and additional chambers under static load; $p_1(t)$, $p_2(t)$ – current pressure in the main and additional chambers; $l_0 = l_{10} + l_{20} = l_1(t) + l_2(t)$ – total height of the gas column; $V_{10} = l_{10} \cdot S$, $V_{20} = l_{20} \cdot S$corresponding volumes of main and additional chambers under static load; S– piston area; $V_1(t) = (l_{10} + z - q) \cdot S$, $V_2(t) = V_{20} = l_{20}(t) \cdot S$ – the corresponding volume of cameras at the current load; m_{10} и m_{20} – Initial gas mass in the main and additional chambers; $m_0 = m_{10} + m_{20}$ – gas mass in the spring; T_0 – Initial gas temperature in the main and additional chambers; $m_1(t)$, $T_1(t)$ and $m_2(t)$, $T_2(t)$ – current weight and temperature of gas in the main and additional chambers; p_{10} and p_{20} – initial pressure in the main and additional chambers under static load; q – vertical displacement of the road profile; z_1 – vertical displacement of the sprung mass; z_2 – vertical displacement of unsprung mass.

Pressure in the main volume:

$$p_1(t) = \left(\frac{l_1(t)}{z_1 - z_2 + l_{10}}\right)^n \cdot \left(\frac{Mg}{S} + p_{\text{atm}}\right) - p_{\text{atm}}, \tag{2}$$

Pressure in additional volume:

$$p_2(t) = \left(\frac{l_0 - l_1(t)}{l_0 - l_{10}}\right)^n \cdot \left(\frac{Mg}{S} + p_{\text{atm}}\right) - p_{\text{atm}}, \tag{3}$$

where p_{atm} – atmospheric pressure; n – polytrope index.

The suspension elasticity:

$$F_{\text{spr}} = F_1(t) = p_1(t) S. \tag{4}$$

With $\Delta p = p_1(t) - p_2(t) > 0$ the change in the height of the gas column l_1 is determined by the equation:

$$\frac{dl_1(t)}{dt} = -\alpha_g f_0 \cdot \sqrt{\frac{RT_0}{\mu MgS}} \cdot \left(\frac{p_2(t)}{p_1(t)}\right)^{\frac{1}{n}} \cdot \sqrt{\frac{2n}{n-1} \cdot \frac{p_1(t) \cdot l_1(t)}{l_{10} + z_1 - z_2}} \cdot \sqrt{1 - \left(\frac{p_2(t)}{p_1(t)}\right)^{\frac{n-1}{n}}}, \tag{5}$$

if $\frac{p_2}{p_1} > \left(\frac{2}{n+1}\right)^{\frac{n}{n-1}}$ and equation

$$\frac{dl_1(t)}{dt} = -f_0 \cdot \sqrt{\frac{RT_0}{\mu MgS}} \cdot \left(\frac{2}{n+1}\right)^{\frac{n}{n-1}} \cdot \sqrt{\frac{2n}{n-1} \cdot \frac{p_1(t) \cdot l_1(t)}{l_{10} + z - q}}, \tag{6}$$

if $\frac{p_2}{p_1} \le \left(\frac{2}{n+1}\right)^{\frac{n}{n-1}}$.

With $\Delta p = p_1(t) - p_2(t) < 0$ the change in the height of the gas column l_1 is determined by the equation:

$$\frac{dl_1(t)}{dt} = \alpha_g f_0 \cdot \sqrt{\frac{RT_0}{\mu MgS}} \cdot \left(\frac{p_1(t)}{p_2(t)}\right)^{\frac{1}{n}} \cdot \sqrt{\frac{2n}{n-1} \cdot \frac{p_2(t)(l_0 - l_1(t))}{l_0 - l_{10}}} \cdot \sqrt{1 - \left(\frac{p_1(t)}{p_2(t)}\right)^{\frac{n-1}{n}}}, \tag{7}$$

if $\frac{p_1}{p_2} > \left(\frac{2}{n+1}\right)^{\frac{n}{n-1}}$ and equation

$$\frac{dl_1(t)}{dt} = f_0 \cdot \sqrt{\frac{RT_0}{\mu MgS}} \cdot \left(\frac{2}{n+1}\right)^{\frac{n}{n-1}} \cdot \sqrt{\frac{2n}{n-1} \cdot \frac{p_2(t)(l_0 - l_1(t))}{l_0 - l_{10}}}, \tag{8}$$

if $\frac{p_1}{p_2} \le \left(\frac{2}{n+1}\right)^{\frac{n}{n-1}}$.

Taking into account the operation of the non-return valve in equations (5), (6), (7) and (8) of area f_0, the following conditions apply:

$$f_0 = \begin{cases} f_1 + f_2 & if\ z_1 > z_2\ and\ p_1 > p_2 \\ f_1 & if\ z_1 \le z_2\ and\ p_1 \le p_2 \end{cases} \tag{9}$$

where f_1 – throttle area; f_2– check valve area.

2. Description of the experimental installation

The experimental installation of a double-chamber pneumatic compressor with a rubber-corded casing on the basis of a dynamic stand with a servo-hydraulic drive of the Indian company BISS-ITW (Fig. 2) was assembled to check the adequacy of the mathematical model.

The test air compressor 1 is installed on the test bench between the support plate 3 of the hydraulic impulse 2 and the installation channels of the suspended mass 5. Test pneumo compressor contains a rubber-cord case of hose type 1 of grade VL 260-340, hollow piston 4 and the top cover with a filling connection. The cavity A of the rubber cord sheath 1 is communicated with the cavity B of the piston 4 through a pneumatic damper, the body 6 of which is mounted on the axis on the top end of the piston 4. Inside housing 6 there is a throttle that constantly communicates the cavities A and B with each other, two rows of radial holes and an elastic check valve that covers these holes in the course of the rebound and communicates the cavity A with the cavity B of the piston 4 only in the course of compression. The upper end of piston 4 is equipped with a buffer of maximum compression stroke 8.

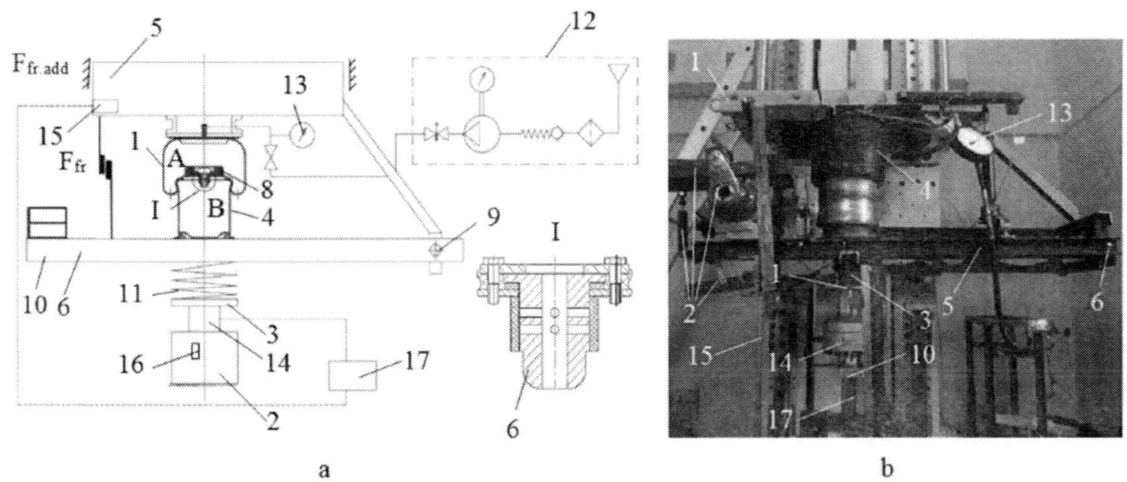

Figure 2. Experimental installation of air suspension on the bench: a - scheme; b - general view; 1 - rubber-cord cover of pneumo compressors; 2 - hydraulic impulse with support plate 3; 4 - hollow piston of pneumo compressors with a buffer of the maximum course of compression 8; 5 - suspended weight; 6 - pneumatic damper with a throttle and a check valve; 7 - directing lever with a swinging axis 9; 10 - unsprung weight; 11 - Spring with rigidity imitating wheel tire; 12 - Compressor unit; 13 - Pressure gauge; 14 - Force sensor; 15 - Movement sensor and speed of the suspended mass; 16 - Position sensor; 17 - Microcontroller; A - Rubber-cord shell cavity Pneumatic springs (main chamber); B - Pneumatic springs piston cavity (additional chamber)

The volume of the air springs above the piston in static position V_{st} = 12 l, the volume of the extended hollow piston V_p = 7,6 l. Height of the air springs in static position H_{st} = 450mm. Static pressure in the air compressor p_{st} = 1.2 kG/cm^2 (0.12 MPa). Throttle diameter d = 6,7mm, $F_{fr.add}$ = 10 N.

3. Bench testing methodology

The test methodology was as follows.

During the tests, the air compressor had a working volume of 19.6 liters. This volume of communication with the hollow piston cavity, the volume of 7.6 liters, was carried out through a throttle or through a throttle with a check valve open during compression.

During the internal throttling study, a damping unit is installed in the piston of the air springs. After the pneumo compressor is set in a static position, corresponding to its height of 450 mm, by supplying the pneumatic compressor with air from the external compressor. At the same time, the working pressure inside the air compressor is controlled by a pressure gauge, and the height of the air compressor is fixed visually by a measuring scale.

Next, the control panel of the stand is used to set the displacement of the hydraulic impulse rod according to the harmonic law with the kinematic disturbance amplitude q_0 = 12 mm at frequencies, Hz: 0.5; 0.75; 0.8; 0.9; 1.0; 1.25; 1.5; 1.75; 1.8; 1.9; 2.0; 2.2; 2.3; 2.5; 3.0; 4.0; 5.0; 6.0; 7.0; 8.0. The selected pitch and frequency range allow to determine the resonant frequencies of the vibrations of the suspended and unsprung masses. Harmonic profile is chosen as the heaviest mode of suspension loading, causing steady forced oscillations, and its parameters are close to the profile of a heavily worn and broken road.

4. Analysis of the results of theoretical and experimental research of air suspension with internal throttling

The results of calculation and experiment have been obtained the amplitude-frequency characteristics (AFC) for the suspended mass installed on a pneumatic compressor with internal throttling. The throttling system was made in the form of a throttle with a hole diameter of 6.7 mm and a check valve, which is made in the form of 12 holes with a diameter of 2 mm and operates on the course of compression.

Figure 3 shows the dependence of the amplitudes of the resonant vibrations of the suspended mass on the ratio of the volumes of additional and main chambers at a constant diameter of the throttle hole. As can be seen from the figure, with the increase in the ratio of the vibration amplitude is reduced in both the first and second resonances.

Figure 4 shows the calculated dependencies of the amplitudes of the resonant vibrations of the suspended mass on the diameter of the throttle hole with a constant ratio of the additional volume to the main volume. The figure shows that for each resonance there is an optimal diameter of the throttle opening, at which the amplitude of the vibrations of the suspended mass is minimal.

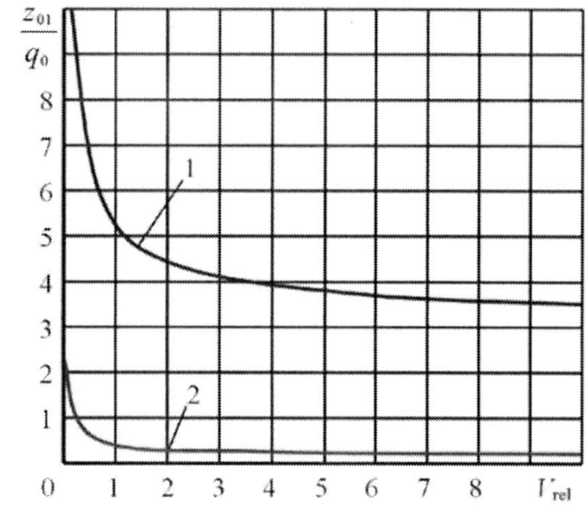

Figure 3. Dependence of the vibration amplitudes of the suspended mass at the first (1) and second (2) resonances on the ratio of the volumes of additional and main chambers with d_{thr} = 4 mm

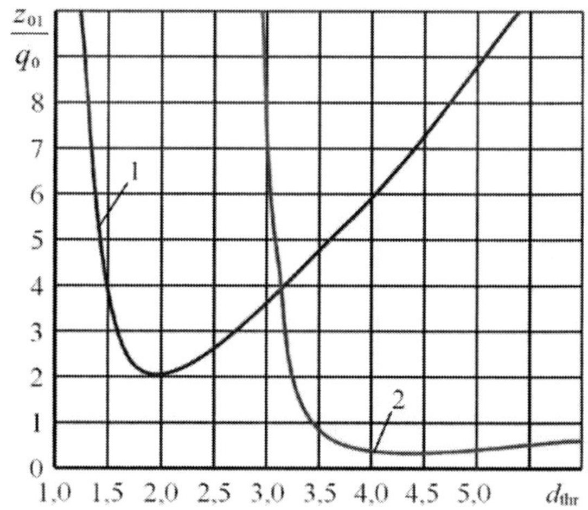

Figure 4. Dependence of the vibration amplitudes of the suspended mass at the first (1) and second (2) resonances on the throttle diameter with V_{rel}= 5/1

Figures 5 and 6 show the experimental and calculated frequency response of the absolute displacements of the suspended mass on the air suspension with the throttling system, made in the form of a throttle hole and a check valve opened on the course of compression, obtained at the following parameters: d_{thr} = 6,7 mm and V_{RCS} = 12l; V_p = 7,6 l; H_{st} = 450 mm; m_t= 128kg; c_t = 300 kN/m; k = 1,3; T_0 = 323 K; p_{atm} = 0,1 MPa; g = 9,81 m/s^2; D_p = 0,25 m; R = 297, vibration amplitudes q = 3 and 5 mm.

It can be seen from the graphs that dry friction between the suspended mass and the base of the stand leads to a decrease in the amplitude of low-frequency resonance by 13...47% for kinematic perturbation q = 3 and 5 mm.

5. Conclusions

- The analysis of the calculated and experimental results shows that the mathematical model quite adequately describes the investigated oscillating system with air suspension and air damping system in the form of a throttle with a check valve.
- The presence of even a small dry friction leads to a decrease in the amplitude of low-frequency resonance by 13...47%.
- The amplitude of oscillations decreases in both the first and second resonances as the ratio of additional and basic volumes increases.
- For each resonance, there is an optimum throttle diameter at which the amplitude of the suspended mass is minimized.

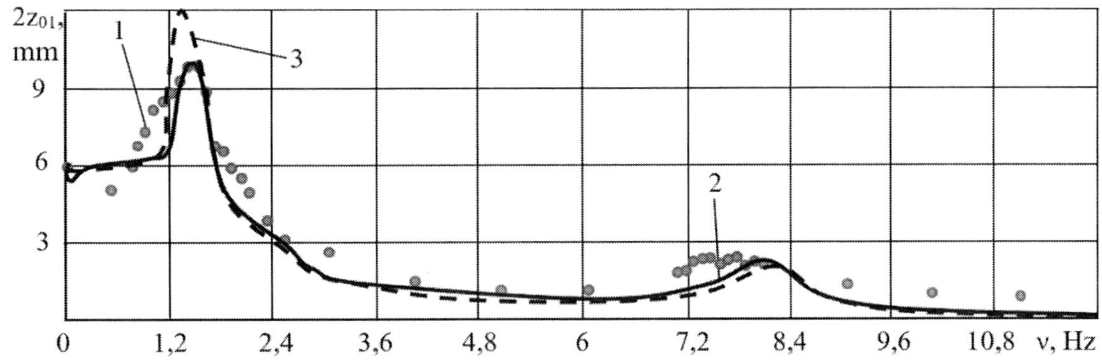

Figure 5. AFC of the vertical displacements of the suspended mass M = 0.5 t on the air suspension with internal air throttling at the kinematic perturbation amplitude q = 3 mm, 1 - experiment; 2 - calculation taking into account additional friction; 3 - calculation without taking into account additional friction; F_{fr} = 68 N; $F_{fr.add}$ = 10 N; k_d= 590 N·s/m

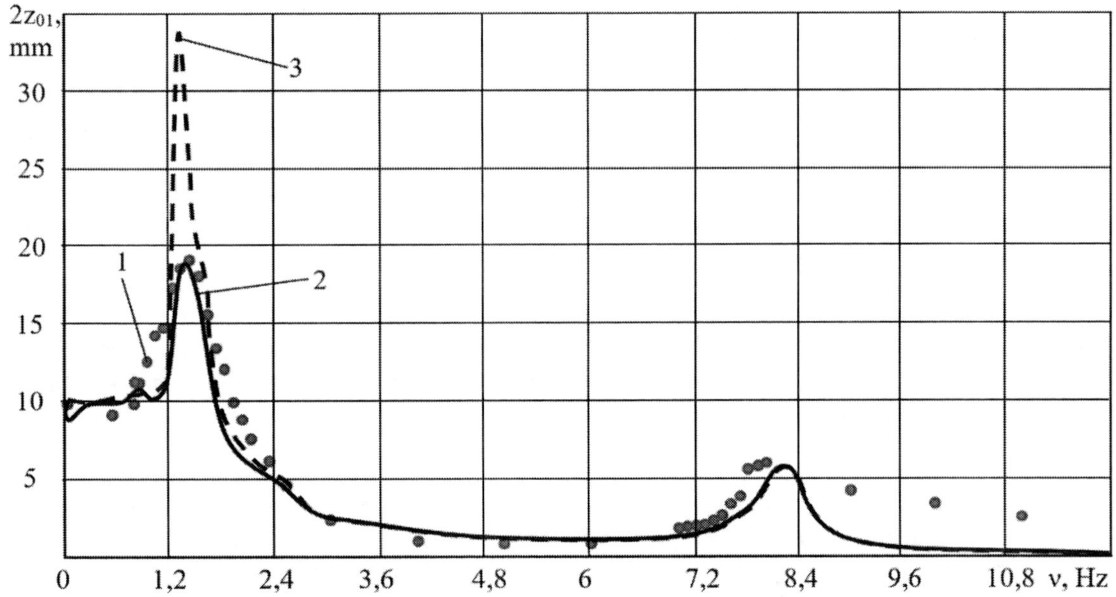

Figure 6. AFC of the vertical displacements of the suspended mass M = 0.5 t on the air suspension with internal air throttling at the kinematic perturbation amplitude q = 5 mm, 1 - experiment; 2 - calculation taking into account additional friction; 3 - calculation without taking into account additional friction; F_{fr}= 107 N; $F_{fr.add}$= 20 N; k_d= 300 N·s/m

It should be noted that the this work was carried out at the Bauman Moscow State Technical University, with financial support from the government in the face of the Russian Ministry of Education under the project: №14.577.21.0272. (Identification number: RFMEFI57717X0272)

References

[1] Novikov V V, Pozdeev A V, Diakov A S 2015 Research and testing complex for analysis of vehicle suspension units *Procedia Engineering* (*International Conference on Industrial Engineering (ICIE-2015)* vol 129) ed A A Radionov (Elsevier publishing) pp 465-470

[2] Novikov V V, Chumakov D A, Golyatkin I A 2019 Experimental Modular Unit for Testing Nodes of Air Suspension with Combined Amortization *Proceedings of the 4th International Conference on Industrial Engineering (ICIE 2018)* (Moscow, Russia) ed A A Radionov (Lecture Notes in Mechanical Engineering (LNME) Springer) pp 385-392

[3] Pozdeev A V, Chumakov D A, Novikov V V, Golyatkin I A, Chernyshov K V, Gavrilov A E, Leonard A V 2019Thermographic bench tests of rubber-cord pneumatic spring *IOP Conference Series: Materials Science and Engineering (ToPME, Moscow, IMASH RAS)* **4899** p

[4] Bachrach B I, Rivin E 1983 Analysis of a damped pneumatic spring *Journal of Sound and Vibration*vol 86 **2** pp 191-197

[5] Kornhauser A A 1994 Dynamic modeling of gas springs *Transactions of the ASME (Journal of Dynamic Systems, Measurement and Control)***116** pp 414-418

[6] Erin C, Wilson B, Zapfe J 1998 An improved model of a pneumatic vibration isolator: theory and experiment *Journal of Sound and Vibration* vol 218**1** pp 81-101

[7] Toyofuko K, Yamada C, Kagawa T, Fujita T 1999 Study on dynamic characteristic analysis of air spring with auxiliary chamber *JSAE Review***20** pp 349-355

[8] Quaglia G, Sorli M 2001 Air suspension dimensionless analysis and design procedure *Vehicle System Dynamics***35** pp 817-829

[9] Deo H, Suh N P 2006 Pneumatic suspension system with independent control of damping, stiffness and ride-height *Proceedings of ICAD2006 (4th International Conference on Axiomatic Design, Firenze)***22** pp 1-6

[10] Gavriloski V, Jovanova J 2006 Dynamic behaviour of an air spring element *Machines, Technologies, Materials* pp 24-27

[11] Docquier N, Fisette P, Jeanmart H 2007 Multiphysic modeling of railway vehicles equipped with pneumatic suspensions *Journal of Vehicle System Dynamics***45(6)** pp 505-524

[12] Kim H S, Lee HC 2007 Asynchronous and synchronous load leveling compensation algorithm in airspring suspension *Proceedings of the International Conference on Control, Automation and Systems* pp 367-372

[13] Lee J-H, Kim K-J 2007 Modeling of nonlinear complex stiffness of dual-chamber pneumatic spring for precision vibration isolations *Journal of Sound and Vibration***301** pp 909-926

[14] Jun K J et al. 2008 Prediction of fatigue life and estimation of its reliability on the parts of an air suspension system *International Journal of Automotive Technology*vol 9 **6** pp 741-747

[15] Nieto AJ, Morales AL, Gonzalez A, Chicharro JM, Pintado P 2008 An analytical model of pneumatic suspensions based on an experimental characterization *Journal of Sound and Vibration***313** pp 290-307

[16] Wolf-Monheim F, Frantzen M, Seemann M, Wilmes M 2008 Modeling, testing and correlation of interlinked air suspension systems for premium vehicle platforms *F2008-SC-040* pp 18-29

[17] Berg M 2009 A three-dimensional air spring model with friction and orifice damping *Journal of Vehicle System Dynamics***33** pp 528-539

[18] Khamitov R N, Aver'yanov G S, Korchagin A B 2009 Pneumatic Shock Absorber with an Active Damping System *Russian Engineering Research*vol 29 **9** pp 871-873

[19] Sayyaadi H, Shokouhi N 2009 A new model in rail-vehicles dynamics considering non-linear suspension components behavior*International Journal of Mechanical Sciences* pp 1016-1035

[20] Moon Jun-Hee, Lee Bong-Gu 2010 Modeling and sensitivity analysis of a pneumatic vibration isolation system with two air chambers *Mechanism and Machine Theory* **45** pp 1828-1850

[21] Pu H et al. 2011 Modeling and analysis of dual-chamber pneumatic spring with adjustable damping for precision vibration isolation *Journal of Sound and Vibration***03** pp 1016-1021

IOP Conf. Series: Materials Science and Engineering **820** (2020) 012022 doi:10.1088/1757-899X/820/1/012022

Calculated and experimental tests of dynamic vibration isolators for use in the suspension system of the traction vehicle cabin

V Shekhovtsov[1,3], M Lyashenko[1], P Potapov[1], A Diakov[2] and K Evseev[2]

[1]Volgograd State Technical University, Volgograd, Russian Federation
[2]Bauman Moscow State Technical University

[3]E-mail: shehovtsov@vstu.ru

Abstract. The article gives a brief description of operational vibration effects on the tractor cab, protective qualities of standard rubber monoblock vibration isolators used in the systems of cabin suspension of domestic tracked tractors. The principle of operation and advantages of dynamic vibration dampers are described. The description of methods and results of calculation and experimental studies of vibration insulating properties of standard and dynamic vibration isolators is given.

Introduction

The analysis of literature sources [in particular, 1] shows that caterpillar agricultural tractors on ploughing, cultivation and sowing, i.e. on the basic agrotechnical operations have a narrow bandwidth spectrum of traction frequencies, in which pronounced peaks in the frequency ranges of 3-3.5 Hz, 10-13 Hz, 14-16 Hz, 18-20 Hz and 28-32 Hz are observed. The spectral density of traction resistance and spectral density of acceleration of the frame and cab characterize different, but interrelated vibrational processes in the dynamic system of the tractor unit. Loads at these frequencies are transferred from the running system to the frame through the suspension of the chassis and then to the floor of the cab through the suspension system. The task of the suspension system is to provide maximum vibration isolation of the operator's workplace from these loads.

The systems of cabin suspension of domestic tracked tractors in many cases still use rubber monoblock vibration isolators [2, 3, 4]. Four such vibration isolators are installed between the frame and the cab floor. The free running of the cab during the vibrations from the above loads is a few millimeters [3, 4, 5], which is enough to protect the operator in high-frequency vibrations, but not enough in low and medium frequency vibrations. There is widespread evidence in the literature that the elastic damping properties of such vibration isolators do not provide a quality protection of the cab from vibrations. The analysis of sources [5–8] shows that during the operation of tractor aggregates low-frequency vibrations from the tractor frame pass through the cab suspension practically without damping, and at some frequencies even with amplification.

Dynamic dampers [8–14] are successfully used in mechanical engineering to dampen narrowband vibrations. The authors have analyzed the possibilities and prospects of using such dynamic vibration isolators in the cabin suspension system. They include additional masses associated with the main oscillating mass of the system through elastic damping elements. The size of the additional masses of the damper and the elastic and damping properties of its elements in the design are calculated in such a way as to provide the most effective damping of the vibrations of the base mass at certain frequencies.

The vibration isolator [11], the scheme of which is shown in Fig. 1, provides damping of cabin vibrations with three main frequencies of the spectrum of vibration [13, 14]. It is possible to distinguish 4 partial systems in this scheme (together with the suspended cabin). The first is suspension mass 1 with an elastic damping element 2, the second is mobile mass 3 with elements 2 and 4, the third is mobile mass 5 with elements 4 and 6, the fourth is mobile mass 7 with elements 6 and 8. This dynamic system has three own (fo_1, fo_2 and fo_3) and two partial (fp_1, fp_2 and fp_3) frequencies of vibration, which generally differ from each other.

The following example shows how the vibration isolator works. Let's assume that the partial frequency fp_1 has a partial system with elastic damping element 2 and suspended mass 1, and the partial frequency fp_2 - a partial system with element 4 and mobile mass 3. According to the theory of oscillation, under the action of vibration loads from one of these frequencies, the mobile mass, the partial frequency of which is equal to the frequency of action, and the suspended mass remains practically stationary, makes vibration with high amplitude [13, 14]. That is, in the case of action on the side of the base 9 on the mobile mass 3 through the elements 4, 6 and 8 with mobile masses 5 and 7 perturbations with a frequency of fp_2, the amplitude of vibrations of the suspended object 1 will be insignificant, and the amplitude of vibrations of the mobile mass 3 — a significant [10].

At the same time, the mobile mass 3, which vibrates with a large amplitude, acts on the suspended mass 1 force, which is almost equal in amplitude and in phase opposite to the external force of the support base 9. Thus, the dynamic damper compensates the effect of external force on the suspended mass 1.

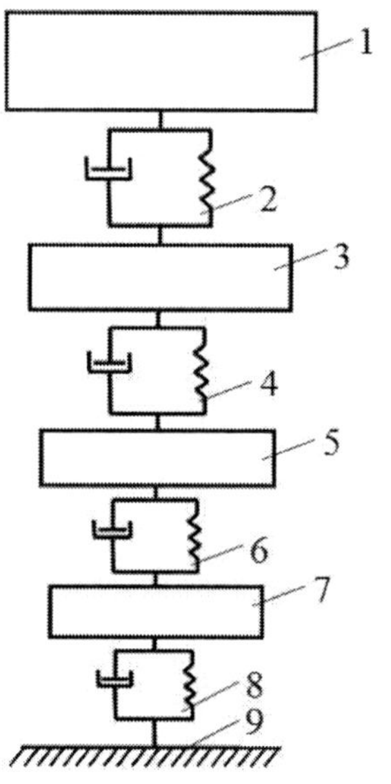

Fig. 1. Scheme of a four-mass dynamic vibration isolator

In the second, third and fourth partial systems the values of mobile masses and rigidity of elastic damping elements are selected in such a way that the partial frequencies fp_1, fp_2 and fp_3 of these systems correspond to the three basic frequencies fo_1, fo_2 and fo_3 from the spectrum of frequencies of operational influences on the cabin from the side of the base 9. Then under the action of disturbances on the side of the base 9 with the first basic frequency fo_1 the mobile mass 3 will make vibrations with a significant amplitude, and the amplitudes of vibrations of mobile masses 5 and 7, as well as the suspended mass 1 (which is important!) will be minimal. Accordingly, under the action of disturbances with the second main frequency fo_2 with a significant amplitude the mobile mass 5 will oscillate, and the mobile masses 3, 7 and the suspended mass 1 will make insignificant oscillatory movements; under the action of disturbances with the third main frequency fo_3 with a significant amplitude the mobile mass 7 will oscillate, and the mobile masses 3, 5 and the suspended mass 1 will be almost motionless.

Selection of dynamic damper parameters and methods of research

It is necessary to know the inertial and elastic damping parameters of the elements of each circuit in order to effectively dampen vibrations in given ranges. For this purpose the program [10, 12] based on MatLab package is created. It allows changing these parameters with a given step and for each variant partial frequencies and basic frequencies of the operational spectrum were compared. Thus, the size of mobile masses and elastic course of elements is limited by values of 20 kg and 40 mm. In addition, the value of the dynamic coupling coefficient of mass oscillations is determined γ [15]. The lower the coefficient, the better, i.e. the specified frequency of vibration loading "works out" one mass, while the others fluctuate significantly less. As a result of modeling, close correspondence of partial and main frequencies of vibrations is received on the second and third frequencies, on the first one it is not received as the value of one of masses thus should be more than 20 kg. As a result, 25 constructively realizable variants were chosen [10], parameters of the best 2 of them are given in Table 1.

Table 1.

№	$fp1$	fp_2	fp_3	fp_4	fo_1	fo_2	fo_3	fo_4	m_1	m_2	m_3	m_4	c_1	c_2	c_3	c_4	γ
1	8	13	30	161	1,22	13,02	30,02	163,62	7	6	2	200	22154	87697	73650	2021295	0,3577
25	7	15	30	101	1,82	15,02	30,02	95,82	18	14	20	200	47532	130194	2773677	955378	0,5582

Table 1 has the next indications: fp_i — partial frequency of i-mass, Hz; fo_i — i-th basic frequency of the spectrum of operational impacts, Hz; m_i — i-th mobile mass; c_i — stiffness of i-th elastic element of vibration isolator, N/m; γ — coefficient of dynamic connection of mass oscillations

Calculated researches of efficiency of dynamic dampers

The models of standard and dynamic vibration isolators (Fig. 2) loaded with their cabin weight have been created and their operation at different frequencies has been studied [10] to check the effectiveness of such dampers in the "Universal Mechanism" package.

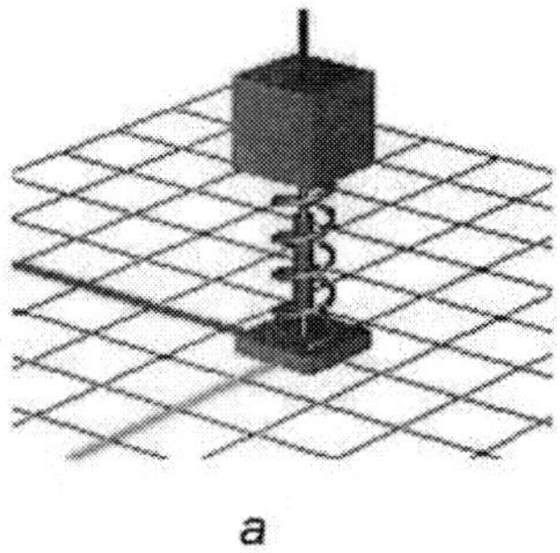

a

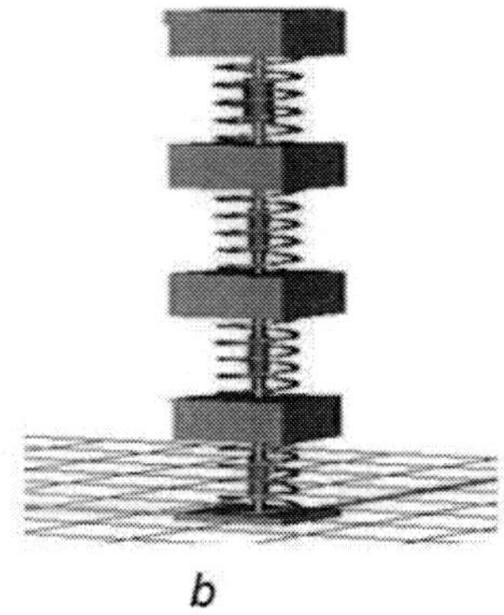

b

Fig. 2. Dynamic models of vibration isolators: a — standard; b — dynamic

Our study showed that the standard vibration isolator in the frequency range from 1 to 13 Hz does not provide damping, and in most cases even enhances the vibratory processes (Table 2). Only starting from 30 Hz is it possible to reduce vibration movements, speeds and accelerations. Both dynamic dampers from 5 Hz onwards provide significantly better vibration protection. So, for example, in the second variant (Table 3) at frequencies of 2, 5, 9, 13 and 30 Hz acceleration decreases accordingly by 20, 92, 96, 96 and 99 %.

Table 2.

f, Hz	A_{frame}, mm	A_c, mm	V_{frame}, mm/sec	V_c, mm/sec	a_{farme}, mm/sec^2	a_c, mm/sec^2
1	0,3	0,3	2	2	13	13
2	0,3	0,3	3,8	4	47	52
5	0,3	0,34	9	11	290	340
9	0,3	0,5	15	30	800	1600
13	0,3	1,2	21	100	1600	8000
30	0,3	0,05	40	12	8000	2000

Table 2 has the next indications: f — frequency of influences, Hz; A_{frame} — amplitude of the base, i.e. tractor frame, mm; A_c — amplitude of the cabin, i.e. the suspended mass, mm; V_{frame} и V_c — speed of movement of the frame and cabin, mm/sec; a_{frame} и a_c — acceleration of movement of the frame and cabin, mm/sec^2.

Table 3

f	$A_{осн}$	A_{m1}	A_{m2}	A_{m3}	A_{m4}	V_{frame}	V_{m1}	V_{m2}	V_{m3}	V_{m4}	a_{frame}	a_{m1}	a_{m2}	a_{m3}	a_{m4}
1	1,0	0,7	0,7	0,7	0,65	0,6	0,8	0,8	0,8	0,8	5,5	5,5	5,5	5,5	5,5
2	1,0	0,4	0,6	0,6	0,6	1,25	6	8	8	9	15	75	105	105	105
5	1,0	0,04	0,35	0,35	0,4	3	0,6	0,4	0,5	0,6	100	17	13	15	18
9	1,0	0,07	0,03	0,05	0,15	5,5	2	0,04	0,15	0,4	5,5	2	0,04	0,15	0,4
13	1,0	0,1	0,2	0,12	0,13	7	8	0,6	0,3	0,8	600	600	55	25	55
30	1,0	0,15	0,19	0,017	0,007	17	4	6	0,15	0,045	3200	560	790	26	38

Table 3 has the next indications: f — frequency of influences, Hz; A_{mi}, mm, V_{mi}, mm/sec, and a_{mi}, mm/sec^2 — amplitude, speed and acceleration of vibrations of the i-th mobile mass.

Thus, the calculated researches have shown that the offered dynamic damper in comparison with the standard vibroisolator possesses essentially better vibroisolating qualities.

Experimental studies of dynamic dampers

An experimental installation (Fig. 3, a), simulating the dynamic system of one point of suspension of the cabin [6, 7] for experimental testing of the vibration insulating properties of the suspension with dynamic vibration isolators, and for the calculated test — its dynamic model (Fig. 3, b) [10]. The vibrations in the range of 0-20 Hz are caused by an inertial loader. This loader is connected to a swinging bench lever that simulates a cab floor. During the tests at each excitation frequency, movements, speeds and acceleration of the cabin floor and moving masses were measured and recorded using SVAN equipment, on the basis of which a series of amplitude-frequency characteristics of vertical and longitudinal-angle accelerations of the cabin and seat without dynamic dampers and with dampers was constructed.

Since dynamic vibration isolator provides the most effective damping of vibrations when the natural frequency of vibration of its mass coincides with its own frequency of vibration of the suspended mass and with the frequency of external influence, the main experimental and calculated studies of the effectiveness of vibration isolators are carried out on such resonance modes, and video recording of processes was also carried out. Analysis of these records shows that in the resonance mode, vibroisolator mass and cabin mass vibrations occur in the antiphase (Fig. 4). Due to this, the dynamic damper reduces the vibration load on the cab. This is confirmed [10] by the comparison of the amplitude-frequency characteristics of the cabin vibrating systems without a dynamic damper (Fig. 5) and with the damper adjusted to the own frequency of the cabin vibration (Fig. 6).

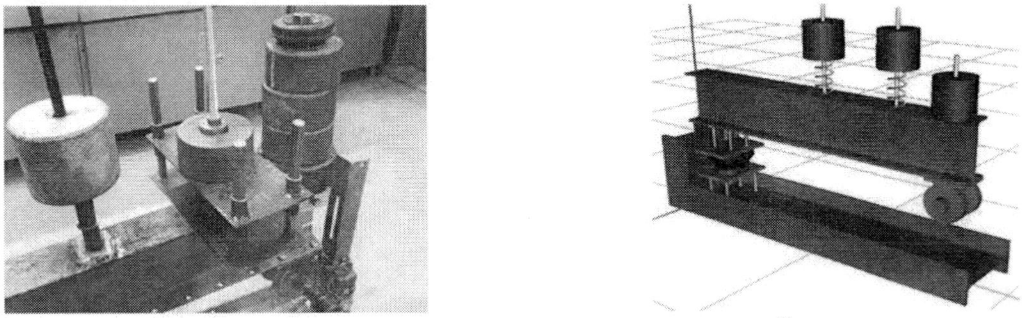

a b

Fig. 3. a — experimental installation; b — dynamic model

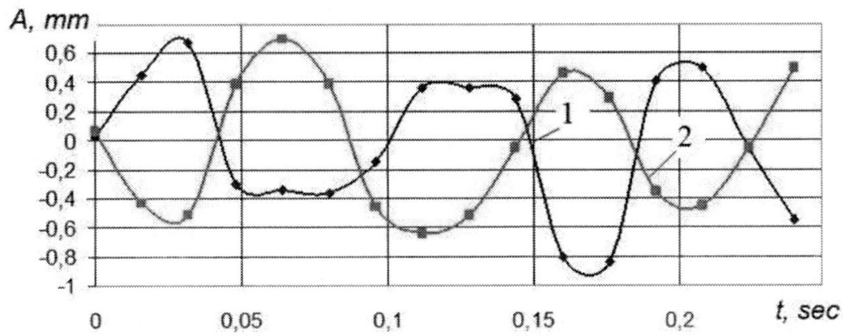

Fig. 4. Recording of movements in the cabin resonance mode (1) and vibration isolator weight (2) in the counter-phase.

The analysis of the amplitude-frequency response shows that, according to the calculated data, the use of dynamic dampers in the resonance mode reduces the vertical acceleration of the cabin by 40% (Fig. 5), and in accordance with the experimental data — by 49% (Fig. 6).

Dynamic dampers with these parameters are introduced into the model of cabin suspension and the same complex of researches as with standard vibration isolators is executed [10]. For comparison, some of the oscillograms for standard and dynamic vibration isolators are shown in Fig. 7, and the diagrams for standard vibration isolators are indicated by the number 1, for dynamic — 2.

Comparison of the whole set of the obtained oscillograms shows that [10], that the installation of dynamic cabin vibration isolators in all cases of motion considered, vertical and longitudinal-angle acceleration of the cabin and seat are reduced (Fig. 8). At the same time:

– the vertical accelerations of the seat at the frequency of 2 Hz decrease by 1.5 times, at the frequency of 3 Hz — by 3.5 times, at the frequency of 10 Hz — by 4 times;

– the vertical acceleration of the cabin at 4 Hz decreases 1.7 times, at 7 Hz — 2.5 times, at 11 Hz — 8.4 times, at 17 Hz - 9.6 times, at 18 Hz — 10 times;

– the longitudinal-angular acceleration of the cabin and seat at the frequency of 3 Hz is reduced by 2.5 times, at the frequency of 5 Hz — by 2.8 times, at the frequency of 11 Hz — by 3.6 times, at the frequency of 14 Hz — by 4.9 times, at the frequency of 17 Hz — by 8 times.

The analysis of sources [16]–[26] shows that during the operation of tractor aggregates low-frequency vibrations from the tractor frame pass through the cab suspension practically without damping, and at some frequencies even with amplification.

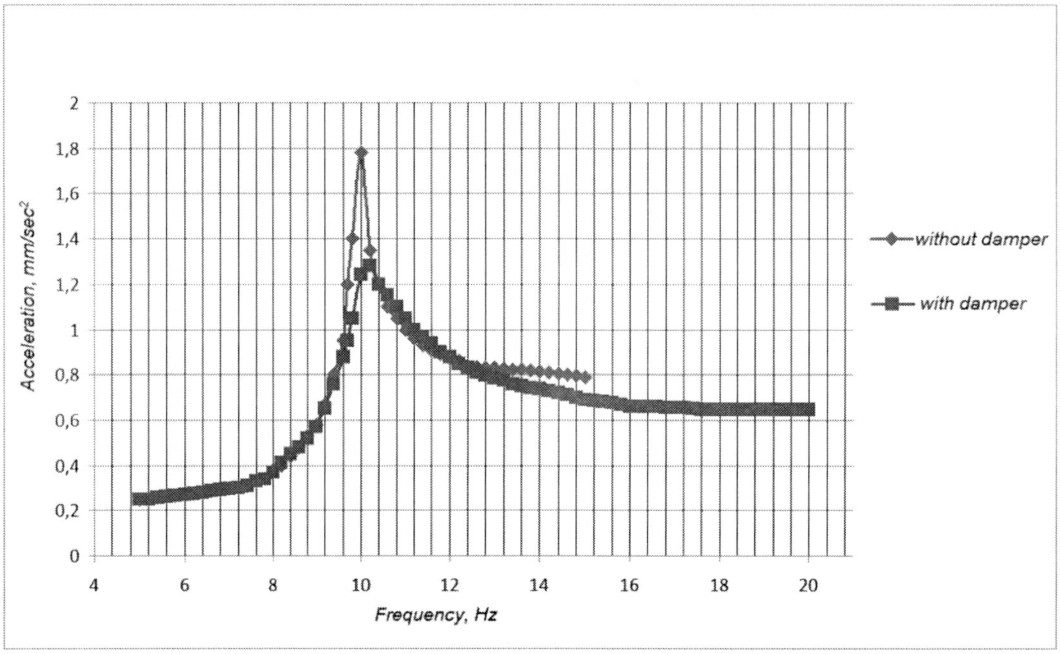

Fig. 5. Calculated amplitude-frequency characteristics of the system without dynamic damper and with damper

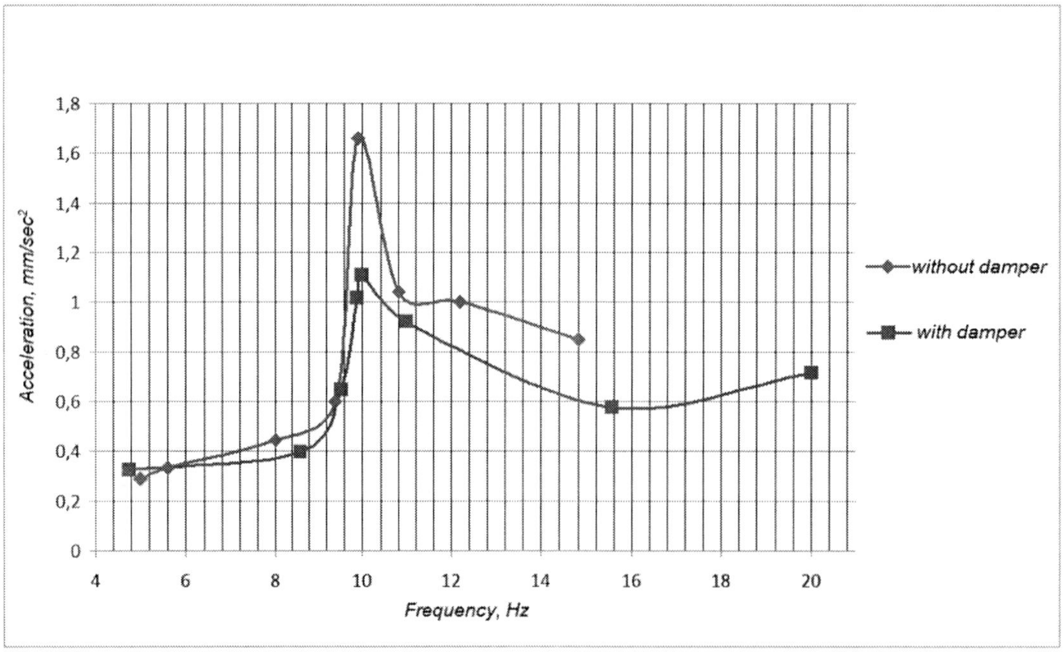

Fig. 6. Experimental amplitude-frequency characteristics of the system without dynamic damper and with damper

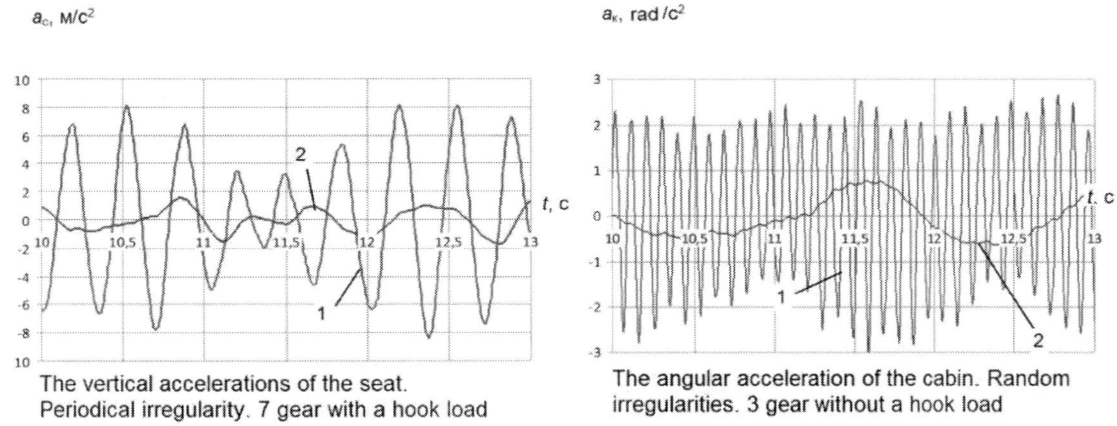

Fig. 7. Examples of compared oscillograms

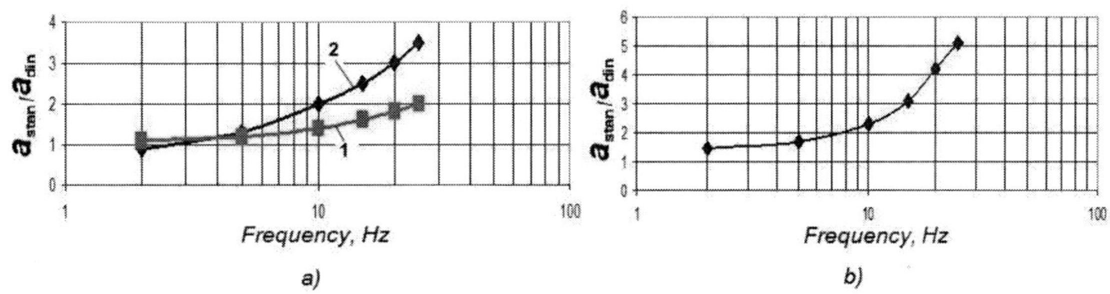

Fig. 8. Decrease in amplitudes: a — vertical accelerations of the seat (1) and cabin (2); b — longitudinal-angle accelerations of the cabin with the seat

Conclusions

Thus, both calculated and experimental researches show that vibro-protective properties of tractor cab suspension system with dynamic dampers appear to be essentially the best in the range of frequencies of operational influences. Practically on all modes of movement of the tractor unit this system provides better vibration protection of the operator's workplace than the system with standard rubber vibration isolators.

It should be noted that the this work was carried out at the Bauman Moscow State Technical University, with financial support from the government in the face of the Russian Ministry of Education under the project: №14.577.21.0272. (Identification number: RFMEFI57717X0272)

References

[1] Kuznetsov, N.G. Stabilization of operation modes of high-speed machine-and-tractor units. Monograph / N.G. Kuznetsov; Volgogr. — Volgograd, 2006. — 426 p.

[2] Technical solutions of the elastic damping devices of the tractor-rail cabin suspension / V.V. Shekhovtsov, M.V. Lyashenko, V.P. Shevchuk, A.V. Pobedin, N.P. Sokolov-Dobrev, K.V. Shekhovtsov // International research journal. — 2013. —№ 7 (ч. 2). — P. 122–125.

[3] *Modeling, simulation, and validation of a pendulum-pounding tuned mass damper for vibration control Structural Control and Health Monitoring*/Wang, W.a, Hua, X.aEmail Author, Chen, Z.a, Wang, X.b, Song, G.c Volume 26, Issue 4, April 2019.

[4] *Prediction of the vibration characteristics for wheeled tractor with suspended driver seat including air spring and MR damper* Zheng, E.Email Author, Fan, Y., Zhu, R., Zhu, Y.,

Xian, J. /Journal of Mechanical Science and Technology Volume 30, Issue 9, 1 September 2016, Pages 4143–4156.

[5] Experimental determination of the characteristics of the tractor cab vibration isolators / V.V. Shekhovtsov, M.V. Lyashenko, V.P. Shevchuk, N.P. Sokolov-Dobrev, K.V. Shekhovtsov // International research journal. — 2013. — № 7 (ч. 2). — P. 118–122.

[6] Lyashenko, M.V. Laboratory installation for testing vibration isolators / M.V. Lyashenko, A.V. Pobedin, K.V. Shekhovtsov // Vestnik of the Academy of Military Sciences. — 2011. No. 2 (special issue). — P. 270–274.

[7] Tractor cabin vibration isolator test bench equipment / A.V. Pobedin, M.V. Lyashenko, K.V. Shehovtsov, Z.A. Gojaev // Tractors and agricultural machinery. — 2012. — № 7. — P. 43–48.

[8] Evaluation of the vibration-protection properties of the standard and dynamic cabin suspension systems in accordance with the requirements of the standards / K.V. Shehovtsov, A.V. Pobedin, N.P. Sokolov-Dobrev, M.V. Lyashenko, V.V. Shekhovtsov // Izvestia VolgGTU. Terrestrial Trans-Port Systems" series. Issue. 7: Interuniversity collection of scientific articles / VolgGTU. — Volgograd, 2013. — № 21 (124). — C. 53–55.

[9] Voloshin, Yu.V. Application of the suspension systems in the foreign tractors // Tractors and agricultural machines. — 2000, — № 2. — P. 36.

[10] Shekhovtsov, K.V. Reduction of the level of vibration load of the tractor operator's workplace due to the use of dynamic vibration dampers in the system of cabin suspension: PhD. — Volgograd, VolgTU, 2014. — 17 p.

[11] The utility model 136110 RUS, F16F3/087 / Vibroisolator of the vehicle cabin / K.V. Shehovtsov, N.S. Sokolov-Dobrev, A.V. Pobedin, V.P. Shevchuk, M.V. Lyashenko, V.V. Shekhovtsov; Volgogr. — 2013.

[12] Shekhovtsov, K.V. Tractor cabins suspension using the dynamic vibration dampers / Shekhotsov, K.V.; Pobedin, A.V.; Sokolov-Dobrev, N.S.; Shekhovtsov, V.V. // Izvestia VolgGTU. Ground transportation systems series. Issue. 6: Interuniversity collection of scientific station / VolgGTU. — Volgograd, 2013. — № 10. — P. 43–46.

[13] Karamyshkin, V.V. Dynamic damping of vibrations. — L.: Mechanical engineering, Leningrad region, 1988. — P. 105

[14] Korenev B.G., Reznikov L.M. Dynamic vibration dampers: Theory and technical applications. — Moscow: Nauka, 1988. — P. 302

[15] Influence of elements dynamic cohesiveness in power shafting on torsional vibrations spreading and dynamic equality of reducible model V.V. Shekhovtsov, N.S. Sokolov-Dobrev, M.V. Lyashenko, P.V. Potapov, K.V. Shekhovtsov, A.V. Kalmykov // Mechanika (Kaunas). — 2014. — Vol. 20, No. 2. — P. 190–196.

[16] *Application and development challenges of dynamic damper in cabin booming noise elimination* Cheni, R.K.Email Author, Jain, C.P.Email Author, Muthiah, R.Email Author, Gomatam, S. /SAE Technical Papers Volume 1, 2014 SAE 2014 World Congress and Exhibition; Detroit, MI; United States; 8 April 2014 до 10 April 2014.

[17] *Analysis of driving seat vibrations in high forward speed tractors* Servadio, P. , Marsili, A. , Belfiore, N.P. (2007) Biosystems Engineering.

[18] Deprez, K., Moshou, D., Anthonis, J., De Baerdemaeker, J., Ramon, H. *Improvement of vibrational comfort on agricultural vehicles by passive and semi-active cabin suspensions (2005) Computers and Electronics in Agriculture*, 49 (3), pp. 431–440. doi: 10.1016/j.compag.2005.08.009.

[19] Zehsaz, M., Sadeghi, M.H., Ettefagh, M.M., Shams, F. *Tractor cabin's passive suspension parameters optimization via experimental and numerical methods* (2011) Journal of Terramechanics, 48 (6), pp. 439–450. doi: 10.1016/j.jterra.2011.09.005.

[20] *Whole body vibration exposure during rotary soil tillage operation: The relative importance of tractor velocity, draft and soil tillage depth International* Singh, A., Nawayseh, N.b, Singh, L.P.a, Singh, S.a, Singh, H.c /Journal of Automotive and Mechanical Engineering Volume 15, Issue 4, 1 December 2018, Pages 5927–5940.

[21] *Tractor cabin's passive suspension parameters optimization via experimental and numerical methods* Journal of Terramechanics Volume 48, Issue 6, December 2011, Pages 439–450 Zehsaz, M., Sadeghi, M.H., Ettefagh, M.M., Shams, F.

[22] V.V. Novikov, A.V. Pozdeev, A.S. Diakov, Research and testing complex for analysis of vehicle suspension units, Procedia Engineering, Vol. 129, pp. 465–470 (2015) DOI: 10.1016/j.proeng.2015.12.153

[23] Evseev, K.B., Kartashov, A.B., Dashtiev, I.Z., Pozdeev, A.V. Analysis viscoelastic properties of fiber-reinforced composite spring for the all-terrain vehicle, MATEC Web of Conferences 224, 02039 (2018).

[24] Diakov, A.S., Pozdeev, A.V., Novikov, V.V. The main directions of the development of snowmobiles in the Russian Federation (2018) MATEC Web of Conferences, 224, № 02080.

[25] E. Klubnichkin, A.S. Dyakov, E.E. Klubnichkin, A.Yu. Zakharov, U. Sh. Vakhidov, A.S. Suchenina and I.V. Basmanov. Experimental evaluation of stability and controllability of domestic and foreign made utility terrain vehicles. IOP Conf. Series: Journal of Physics: Conf. Series 1177 (2019) 012045. DOI:10.1088/1742-6596/1177/1/012045.

[26] V.E. Klubnichkin, A.S. Dyakov, E.E. Klubnichkin, A.Yu. Zakharov, U.Sh. Vakhidov , A.S. Suchenina and I.V. Basmanov. Experimental evaluation of speed and brake properties of domestic and foreign made utility terrain vehicles. IOP Conf. Series: Journal of Physics: Conf. Series 1177 (2019) 012048. DOI:10.1088/1742-6596/1177/1/012048.

IOP Conf. Series: Materials Science and Engineering **820** (2020) 012023 doi:10.1088/1757-899X/820/1/012023

Increasing the smoothness of the course of the forage harvester by optimizing the mass-dimensional and inertial parameters of its body

M Zhileykin[1] **and P Sirotin**[2,3]

[1] Bauman Moscow State Technical University
[2]Platov South Russian State Polytechnic University (Novocherkassk, Rostov region, Russia)

[3]E-mail: spv_61@mail.ru

The scientific-methodical substantiation and practical recommendations on stabilizing the movement of forage harvesters are proposed. Calculation of optimal mass and inertial parameters of the hull, as well as elastic properties of tires as the main element of the combine suspension system is theoretically grounded. The developed technique on the example of the combine prototype shows the possibility of calculating the optimal stiffness of tyres and weight of additional counterweights installed in the base of the machine to ensure the equality of natural frequencies of front and rear axle oscillations on elastic tyres. The modeling of the combine's movement on the subsoil road and asphalt concrete highway with different speeds has been carried out on the basis of mathematical and imitation model of the combine developed by the authors earlier. In the order to evaluate the efficiency of the proposed method of improving the smoothness of running the comparison of peak and mean square values of the pitch angle of the hull, as well as the levels of vertical vibroaccelerations in the cabin on the floor under the seat of the operator of the combine in its basic design and improved version due to the use of counterweights of a given mass and optimized stiffness of tires. A simulation was used to perform an octave analysis of the workplace vibration load. It is shown that optimization of mass and inertial parameters of the body, as well as elastic properties of the combine harvester tires on the main driving modes provides improvement of comfort of the operator's working place, especially in the most dangerous for human frequency range. The driving modes are defined, on which the best effect of rework is achieved. The stabilization process is described. The conclusions on the direction of further research are presented.

Introduction

The existing system of classification of motor transport, self-propelled vehicles of agricultural, road construction, military and other purposes assumes their division not only by their purpose, design and layout, but also predetermines the approaches and methods of design [1, 2]. Until recently, the self-propelled combine harvester was considered by designers and marketers, first of all, as a technological machine, and in this connection in the process of its creation they tried to provide a competitive level of only technological properties. As a result, they received a self-propelled machine, effective in terms of technological indicators, but having an extremely low level of transport and operational properties, due to a significant imbalance of mass and size indicators, as well as the lack of a suspension system. In this connection at operation of combines it is noticed: reconsolidation of agrophones because of the excessive weight which is falling on wheels of the forward axis; separation of wheels from a basic surface at movement with speed more than 8-10 km/h and emergency braking; galloping of the case on elastic tyres with formation of considerable dynamic loadings on the case, and also high vibronic loading of a workplace of the operator [3-10]. Especially such problems are shown at operation of forage harvesters (FH).

Purpose of the study

The listed disadvantages are determined, first of all, by mass and inertial parameters of the body, as well as by unbalanced level of elastic dissipative properties of tyres as the only element of the body suspension system. Despite this, the designers did not take into account the stiffness of the tires, assigning the pressure in them based on the recommendations of manufacturers, which in turn are guided by the requirements of reliability and provide a given pressure in the contact spot [11-13]. However, recently combine manufacturers began to consider the properties of tires as a factor of smooth running control. For example, the CLAAS FH model JAGUAR 900 (Fig. 1 a) is equipped with a tire pressure control system, which allows for the change of the area of the wheel's spot contact with the bearing surface, the reduction of the rolling resistance of the wheels when driving on hard roads and the rigidity of the tires in order to control the smooth running parameters.

Another obvious and easiest way to reduce the dynamic load on the combine's body is to level the weights on the axles by means of additional counterweights mounted on the rear of the FH. The weight of the counterweights is chosen on the basis of the designers' recommendations or after a set of experimental measurements on the prototypes of combines with different variants of the attached implements.

Despite the fact that such a solution is currently used by all major manufacturers of FH, however, to date there is no information on the application of specific techniques and practical recommendations for the implementation of these technical solutions. The lack of methodology and practical recommendations for selecting the parameters of counterweights and stiffness of elastic tyres predetermined the purpose of this work.

Materials and methods

The study was carried out with the help of the mathematical and corresponding simulation model FH RSM 2650, developed on the basis of the results obtained in the works [14-19]. The model describes the dynamics of the combine harvester motion taking into account its weight and dimensions, elastic and viscous bonds, peculiarities of the interaction of the propulsor with the roughness of the support base and other parameters. It is assumed that the masses of the harvester's suspension elements are reduced to the load-bearing system, and the support base is supposed to be non-deformable. When modeling the dynamics of the harvester's movement its body was considered as absolutely rigid structure. The relationship between the kinematic parameters and external disturbances is described by means of differential equations, which make up the mathematical model of the combine's motion. The modeling is carried out on an uneven non-deformable support base of the type "asphalt concrete highway" and "dirt road", to describe the profile of which used the correlative functions given in the well-known works [20-21].

The study of the distribution patterns of mass and inertial pairs-meters was carried out on the basis of the method of conservative dynamic systems with many related degrees of freedom [22].

As efficiency indicators of the proposed method are accepted: vertical vibroacceleration at the operator's workplace in the first five octaves; standard deviation φsd and peak value of longitudinal inclination of the combine harvester body φmax. The simulation measurements of vertical vibroaccelerations in the operator's cabin on the floor under the driver's seat are carried out in accordance with GOST 12.1.012.

Results and discussion

The FH dynamic model is based on the type of "bicycle" calculation scheme on elastic tires (Fig. 1) to evaluate the effectiveness of the application and to develop practical recommendations for the method of motion stabilization by adjusting the tire stiffness and application of counterweights. Such representation of the FH model is possible because of the lack of connection between transverse angular oscillations of the FH body and other types of its motion [2, 22].

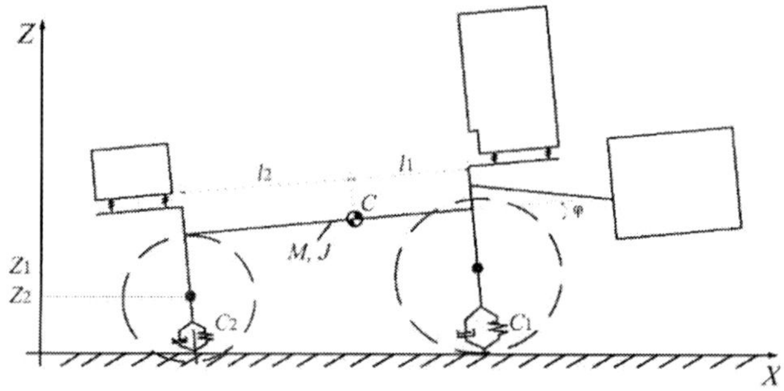

Fig. 1. Calculation diagram of the forage harvester on elastic tyres:
C – center of mass of the combine; X, Z – axes of stationary coordinate system; C_1, C_2 – tire stiffness of front and rear axles respectively; l_1, l_2 – distance from the center of masses of the combine to the front and rear axes respectively; z_1, z_2 – vertical coordinates of points of fixing the wheels of front and rear axes respectively; φ –angle of longitudinal inclination of the combine body; M, J – total mass and the main moment of inertia relative to the transverse axis of the combine

According to Fig. 1 of the equation of motion of the combine on elastic tires we write as

$$\begin{cases} M\ddot{z}_c + (C_1 + C_2)z_c + (C_1 l_1 - C_2 l_2)\varphi = 0; \\ J\ddot{\varphi} + (C_1 l_1 - C_2 l_2)z_c + (C_1 l_1^2 + C_2 l_2^2)\varphi = 0, \end{cases} \quad (1)$$

where z_c – vertical movement of the combine's centre of gravity; $\ddot{z}_c$ – acceleration of the combine's centre of gravity in the vertical direction.

We enter the symbols:

$$\omega_1^2 = \frac{C_1 + C_2}{M}; \omega_2^2 = \frac{C_1 l_1^2 + C_2 l_2^2}{J}; \frac{d}{dt} = p; \frac{d^2}{dt^2} = p^2,$$

where ω_1 – natural frequency of vertical oscillations of the enclosure; ω – natural frequency of longitudinal and angular vibrations of the enclosure.

The system identifier (1) can be recorded as [22]

$$\begin{vmatrix} p^2 + \omega_1^2 & \alpha\omega_1\omega_2 \\ \alpha\omega_1\omega_2 & p^2 + \omega_2^2 \end{vmatrix} = 0;$$

$$p^4 + (\omega_1^2 + \omega_2^2)p^2 + \omega_1^2\omega_2^2(1 - \alpha^2) = 0; \quad (2)$$

$$\alpha^2 = \frac{(C_1 l_1 - C_2 l_2)^2}{(C_1 + C_2)(C_1 l_1^2 + C_2 l_2^2)},$$

where α – the bond coefficient.

Then the roots of equation (2) can be defined as

$$\left.\begin{matrix}\Omega_1^2\\ \Omega_2^2\end{matrix}\right\} = \frac{\omega_1^2+\omega_2^2 \pm \sqrt{\left(\omega_1^2-\omega_2^2\right)^2+4\alpha^2\omega_1^2\omega_2^2}}{2},$$

where Ω_1, Ω_2 – natural angular frequencies of vertical and longitudinal angular oscillations respectively.

It has been shown in [22] that, for biaxial wheeled machines, an improvement in smooth running can be achieved by ensuring the equality of ω_1 and ω_2. In order for the two natural frequencies of the oscillating system (1) to coincide with some optimum frequency, the condition must be fulfilled $\Omega_1^2 = \Omega_2^2$. It follows from this that

$$C_1 l_1 = C_2 l_2; \tag{3}$$

$$\frac{C_1 l_1^2 + C_2 l_2^2}{J} = \frac{C_1 + C_2}{M}. \tag{4}$$

Therefore, the geometric l_1, l_2, inertial M, J and stiffness parameters C_1, C_2 should be in a certain ratio in order to minimize the system oscillations (1) caused by the influence of the supporting base, in certain expressions (3) and (4). Expressions (3) and (4) allow optimizing mass and dimensional properties and inertial parameters of FH to provide a given level of vertical oscillations and minimize longitudinal-angle oscillations.

It is often difficult to maintain these ratios during design. The tyre rigidity can be changed by changing the air pressure in the tyres. In order to ensure smooth running, FH tyres are better suited for optimising the properties of the front axle tyres, as they are larger than the rear axle tyres and therefore have a larger stiffness control range.

The expression (3) can then be rewritten as

$$C_1 = \frac{l_2}{l_1} C_2. \tag{5}$$

You can use an additional mass m_n in the form of counterweights located on the feeder to bring geometric and inertial parameters, which are determined by the weight and mutual arrangement of the combine harvester's aggregates, in accordance with expression (4). In this case, the center of mass of the combine with the counterweights is shifted to the rear axle at a distance of x (Fig. 2).

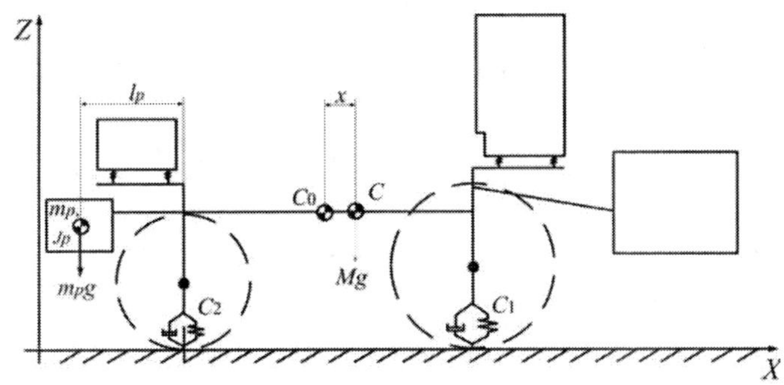

Fig. 2. Calculation diagram of the forage harvester with counterweights:
C_o – center of mass of the combine with counterweights; l_p – distance from the rear axis of the combine to the center of masses of counterweights; m_p,J_p – mass and the main moment of inertia of counterweights relative to the transverse axis of coordinates

According to the calculation scheme in Fig. 2, the distance x is defined as

$$x = \frac{m_p}{M + m_p}\left(l_2 + l_p\right). \tag{6}$$

If we subtract expression (5) from (4), we shall receive, subject to (6)

$$\frac{\left(l_2 - x\right)\left(l_2 + x + 1\right)}{J_o + m_p\left(l_2 - x + l_p\right)^2} = \frac{L}{\left(l_1 + x\right)\left(M + m_p\right)};$$

$$J_o = J + J_p; L = l_1 + l_2,$$

where J_o – main moment of inertia with respect to the transverse axis of the combine with counterweights.

For simplification of the further transformations we will accept without essential loss of accuracy $l_n \approx 0$. Finally, we obtain a quadratic equation to determine the optimal m_p:

$$am_p^2 + bm_p + c = 0;$$

$$a = 2l_1l_2 + l_2 + Ll_2 + l_1^2 + l_1;$$

$$b = l_1^2 + 2Ml_1l_2 + Ml_2 + 2l_1^2M + 2Ml_1 - LJ_o - LMl_2; \tag{7}$$

$$c = l_1^2M^2 + l_1M^2 - LJ_oM,$$

where a, b, c – quadratic equation coefficients.

Solving equation (7) by known methods, it is possible to determine the optimal value of m_p taking into account the rigidity of tyres, mass and geometric characteristics of the body. On the basis of expression (7) it is established that for FH RSM 2650 the optimal m_p is 1450 kg, and the stiffness of front wheels should be reduced from 3125000 to 1856335 N/m. The stiffness of the rear axle tyres did not change, in all cases it corresponded to the level of production tyres of the company "Mitas" with the dimension of 21.3R24 and was 1230769 N/m.

According to the results of simulation it has been established that the vibration load of the operator's workplace on the improved version of FH in comparison with the basic version is reduced by 1÷7 dB, and the greatest effect is achieved in the most dangerous for the human body 3, 4 and 5 octaves (Fig. 3). In the first and second octaves it is possible to increase the active vibration level up to 3 dB, which is associated with some increase in the swinging of FH due to a decrease in the rigidity of the front axle tires and increase in the body weight due to counterweights. However, the permissible levels of exposure do not exceed the standard requirements.

As you can see from the fig. 4, FH refinement by the proposed method resulted in some increase of φsd and φ_{max} at the speed of 5 km/h. This is due to the fact that the roughness of the supporting surface forms power disturbances with a frequency close to the frequency of the body's own vibrations on elastic tyres. It should be taken into account that swaying is also facilitated by a reduction in the stiffness of the front axle tires, as the more solid body of the combine presses the softer tire into the larger foamy ground. At 10 km/h, there has been an improvement in running smoothness: φsdhas been reduced by 0.2 ° and φ_{max} by 1.2°.

In the case of traffic without an adapter on the asphalt concrete highway (mode "long-distance transport"), the level of vibration load of the operator's workplace FH as a whole increases with the increase in speed from 10 to 35 km / h (Fig. 5). Reduced front wheel stiffness and counterweights at speeds of up to 30 km/h result in some increase of vibration at the workplace only in the first octave (Fig. 5a-c), and in more dangerous from the point of view of labor protection third, fourth and fifth vibration loadings decrease is achieved. The greatest effect of finishing the combine is manifested in the third octave, which achieves a reduction in vibration from 2.5 to 8 dB (Fig. 5a-c). In the "long-distance transport" mode, the greatest result of the harvester's reworking was found at a speed of 20 km/h. At the maximum speed, the vibration level at the operator's workplace of the reworked combine has decreased noticeably only at the first octave (Fig. 5g) due to the reduction of galloping of the combine's body as a whole (Fig. 6). At 35 km/h, the level of vibration load in the cabin of the reworked combine is practically the same as in the basic version.

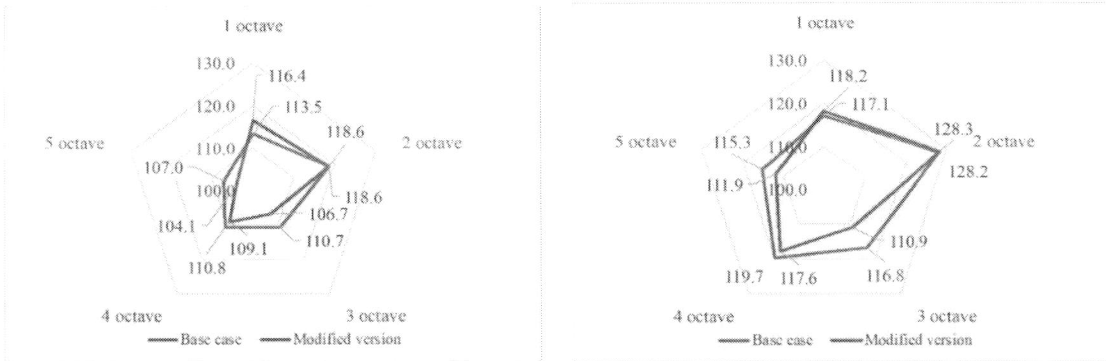

a) b)

Fig. 3. Current levels of vertical accelerations in the operator's cab for combine harvesters with an adapter when driving on a dirt road at 5 (a) and 10 (b) km/h

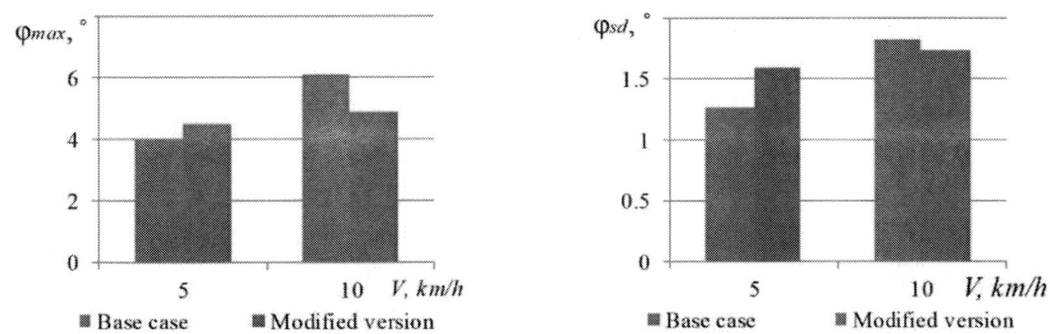

a) b)

Fig. 4. Comparison of peak (a) and standard pitch values of the hull (b) FH with the adapter when driving on a dirt road

When the combine moves along the asphalt without adapter, the finish of the combine significantly improves the smoothness of travel at a speed of 10 and 35 km/h (Fig. 6). However, at a speed of 20-30 km/h, the values of φsd and φ_{max} increased, primarily due to the prevalence of low-frequency hull oscillations (see Fig. 5b, d). Especially the smooth running parameters decrease at the speed of 30 km/h. It is obvious that at low speed the effect is achieved due to smoothening of vibrations by softer tires, and at the maximum because of the increase in the mass of the body sprung on the tires. At a speed of 10 km/h, the reduction of φsd is due to the smoothing of vibrations by softer tyres, but for the same reason the level of φ_{max} increases.

Thus, a set of measures has been developed to improve the design of the FH, consisting in changing the stiffness of the front axle tires and attaching additional counterweights to the rear of the combine. In contrast to the known methods of stabilization of self-propelled machines, the developed method involves bringing the parameters of the combine's oscillating system to a certain ratio, which ensures the minimization of vibration, primarily by narrowing the bandwidth of the body's own vibrations.

IOP Conf. Series: Materials Science and Engineering **820** (2020) 012023 doi:10.1088/1757-899X/820/1/012023

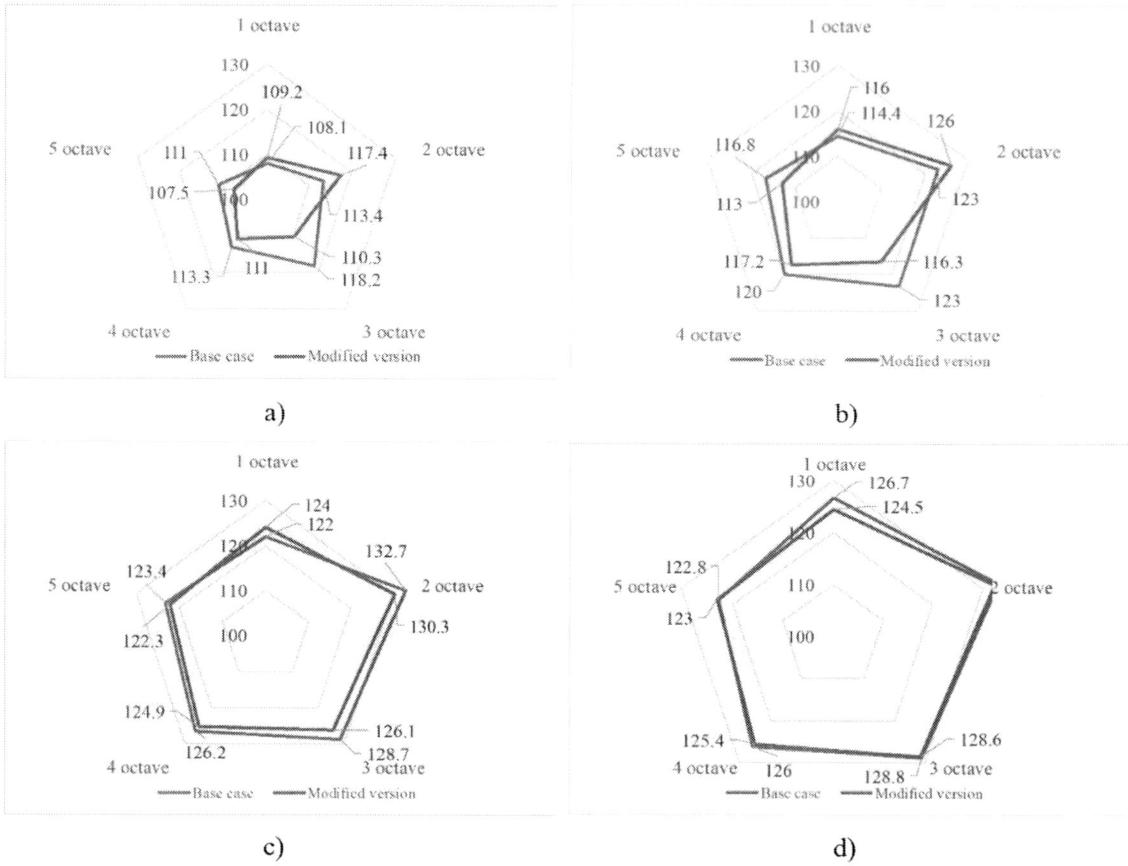

Fig. 5. Comparison of the current levels of vertical accelerations in the operator's cab for combine harvesters without an adapter when driving on an asphalt-concrete road at a speed of 10 (a), 20 (b), 30 (c) and 35 km/h (d)

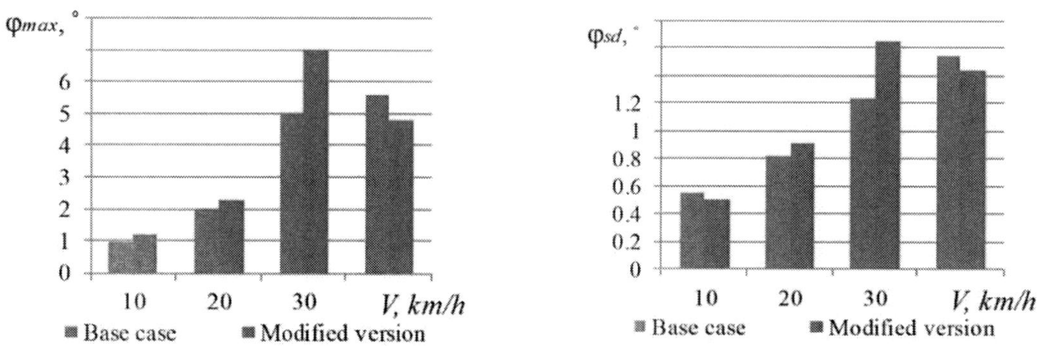

a) b)

Fig. 6. Comparison of peak (a) and standard pitch values of the hull (b) FH without adapter when driving on an asphalt-concrete road

Conclusion

1. When the combine is in transport mode, its reworking is efficient by optimizing the weight and inertia of the machine and helps to reduce the vibration load of the FH operator's workplace.

2. The optimization of wheel stiffness and inertial parameters of the body allows to improve the smoothness of the combine and comfort at the operator's workplace only for a given mode or speed. An effective way to stabilize the FH is probably to use some semi-active and active actuators in the hanger or attachments.

References

[1] OplandLindaSE. Size classification of passenger cars // Master's Thesis in the Master Degree Programme Civil Engineering. Chalmers university of technology Göteborg, Sweden 2007. p. 150.
[2] Peek N. Coleman J. Design machines // SIGGRAPH 2015: Studio 2015 1-1. doi: 10.1145/2785585.2792578.
[3] Królczyk, Jolanta&Legutko, Stanislaw &Krolczyk, Grzegorz. (2014). Dynamic Balancing of the Threshing Drum in Combine Harvesters – The Process, Sources of Imbalance and Negative Impact of Mechanical Vibrations. Applied Mechanics and Materials. 693. 424 - 429.
[4] Špokas L., Adamčuk V., Bulgakov V., Nozdrovický L. (2016): The experimental research of combine harvesters. Res. Agr. Eng., 62: p. 106-112.
[5] Petre M, Heinz-Dieter K.Modeling and simulation of grain threshing and separation in threshing units—Part I. Computers and Electronics in Agriculture. 60. 96-104. 10.1016/j.compag.2007.07.003.
[6] Okayasu Takashi, Ken Mori, Yuko Ueka, Masami Matsui, Muneshi Mitsuoka. Eiji Inoue Turbulent Flow Characteristics of the Cleaning Wind in Combine Harvester. Engineering in Agriculture, Environment and Food, 2012. p. 102-106.
[7] Sirotin P.V., Lebedinsky I.Yu., Kravchenko V.V. Analysis of vibro-acoustic loading of the workplace of operator of combine harvesters // Modern high technology. Regional application. 2018 No. 1 (53). pp. 113-121.
[8] Sirotin P.V., Sapegin A.G., Zlenko S.V. Experimental assessment of the smooth running of a self-propelled forage harvester // Transactions of NAMI - 2017 4 (271) p. 67-74.
[9] Sirotin P.V., Sapegin A.G., Zlenko S.V. Experimental studies of ride quality of self-propelled combine harvester // XIV International Scientific-Technical Conference "Dynamic of Technical Systems" (DTS-2018). September 12-14, 2018. - Rostov-on-Don, Russian Federation,: MATEC Web of Conferences Volume 226 (2018)., 2018.
[10] Belousov, B., Ksenevich, T.I., Vantsevich, V., Komissarov, D. 8×8 platform for studing terrain mobility and traction performance of unmanned articulated ground vehicles with steered wheels (2013) SAE Technical Papers, 9.
[11] Belousov, B.N., Shelomkov, S.A., Ksenevich, T.I., Kupreyanov, A.A. Experimental verification of a mathematical model of the wheel-supporting surface interaction during nonstationary rolling motion(2009) Journal of Machinery Manufacture and Reliability, 38 (5), pp. 501-505.
[12] Gorelov, V.A., Komissarov, A.I. Mathematical model of the straight-line rolling tire - Rigid terrain irregularities interaction (2016) Procedia Engineering, 150, pp. 1322-1328.
[13] Vol'skaya, N.S., Zhileykin, M.M., Zakharov, A.Y. Mathematical model of rolling an elastic wheel over deformable support base (2018) IOP Conference Series: Materials Science and Engineering, 315 (1), article № 012028.
[14] Gorelov, V.A., Komissarov, A.I., Miroshnichenko, A.V. 8×8 wheeled vehicle modeling in a multibody dynamics simulation software (2015) Procedia Engineering, 129, pp. 300-307.
[15] Keller, A.V., Gorelov, V.A., Anchukov, V.V. Modeling truck driveline dynamic loads at differential locking unit engagement (2015) Procedia Engineering, 129, pp. 280-287.
[16] Keller, A.V., Gorelov, V.A., Vdovin, D.S., Taranenko, P.A., Anchukov, V.V. Mathematical model of all-terrain truck (2015) Proceedings of the ECCOMAS Thematic Conference on Multibody Dynamics 2015, Multibody Dynamics 2015, pp. 1285-1296.

[17] Kupreyanov, A.A., Morozov, M.V., Belousov, B.N., Ksenevich, T.I., Vantsevich, V.V. Experimental research of tire elastomer-surface tribologicalproperties(2014) Proceedings of the ASME Design Engineering Technical Conference, 3.

[18] Klubnichkin, V.E., Klubnichkin, E.E., Kotiev, G.O., Beketov, S.A., Makarov, V.S. Interaction between elements of the track ground contacting area with the soil at curvilinear motion of the timber harvesting machine (2018) IOP Conference Series: Materials Science and Engineering, 386 (1), article № 012016.

[19] Wong, J.Y. Theory of Ground Vehicles / J.Y. Wong. – New York: Wiley IEEE, 2001. – 560 p.

[20] Sarach, E., Kotiev, G., Beketov, S. Methods for road microprofile statistical data transformation (2018) MATEC Web of Conferences, 224, article № 04009.

[21] Rybansky, M. The impact of terrain on cross-country mobility geographic factors and their characteristics / M. Rybansky [et. al] // 18th International Conference of the International Society for TerrainVehicle Systems, ISTVS 2014. – Seoul, 2014.

[22] Rocard Y. L'instabiliteenmecanique. Automobiles. Avions. Pontssuspendus. MASSON ET Cie, EDITEURS, Paris, 1954.

Analysis of the cornering stiffness uncertainty impact on the steering sensitivity of a two-axle automobile

V Gorelov[1], A Komissarov[1,2] and I Vozmishcheva[1]

[1]Bauman Moscow State Technical University

[2]E-mail: komisarov@bmstu.ru

Abstract. In order to provide a reliable estimate of the steering sensitivity at the early design stage it is necessary to know the impact of the ucertainty of the main parameters affecting this characteristic. These parameters include the tire cornering characteristics featuring a large degree of uncertainty. The objective of the research was to find the impact of the tire cornering characteristics variation on the steering sensitivity. The analysis was performed in a multi-body simulation program with the use of a spatial non-linear model of the automobile motion. The model takes into account the detailed kinematics of the suspension and steering linkages. Interaction of the tires with the road is described by an original semi-empirical model providing indepentent calibration of the tire longitudinal and lateral force characteristics in the contact patch. The reseach has demonstrated the degree of importance of considering the tire cornering stiffness variation during the steering sensitivity analysis.

Introduction

Steering sensitivity is an important characteristic of the automobile handling.Reliability of this characteristic estimation at early design stages depends on the reliability of the analysis input data.Steering sensitivity depends mostly on the tire cornering stiffness, automobile moment of inertia about its vertical axis, and longitudinal coordinate of the vehicle center of mass [1].The tire cornering characteristics feature a high degree of uncertainty.As it has been shown in [2], the variation of the cornering stiffness obtained for the same tire on different test rigs can reach up to 46%.In order to estimate reliability of the steering sensitivity analysis results it is necessery to estimate the impact of the tire cornering stiffness variation on this characteristic.

At present, estimation of the performance of the vehicle at early design stages is often performed by its dynamics simulation with the use of spatial non-linear multi-body models with suspension kinematics [3–9]. Models of this type provide good accuracy of the automobile handling characteristics estimation [1]. For complex multi-body mechanical systems, to which category automobiles with independent suspensions can be referred as well, there exists a problem not only to solve the equations of thier motion but also to derive these equtions.A solution to this problem is to use multi-body dynamics simulation (MBS) software [10–12]. In the MBS software, the user describes the mechanical system as a set of rigid bodies, joints, and force interactions from the library of standard elements, and the software generates the equations of motion automatically and provides built-in means for their numerical solution. MBS programs usually include built-in tire – road interaction models [13–17]. However, these tire models not always satisfy the analysts since their force calculation algorithms are either complicated and not fully disclosed in the manuals or need large amount of experimental data which are not available.As a result, an analyst doesn't fully trust the built-in model of the tire – road interaction. In this case, the development of relatively simple custom models of the tire – road interaction for specific problems by means of the MBS program or by means of an external programing language is an acceptable alternative [18].

The objective of this research is to analyse the impact of the tire cornering characteristics variation on the automobile steering sensetivity. The analysis will be performed in the Universal Mechanism MBS program with the use of a spatial non-linear multi-body model.

1. Automobile model

The research was done for a two-axle M1 category passenger vehicle with gross mass of 1262 kg, rear drive axle, and independent McPherson strut suspensions of all wheels. The automobile multi-body model consists of the subsystems shown in figure 1.

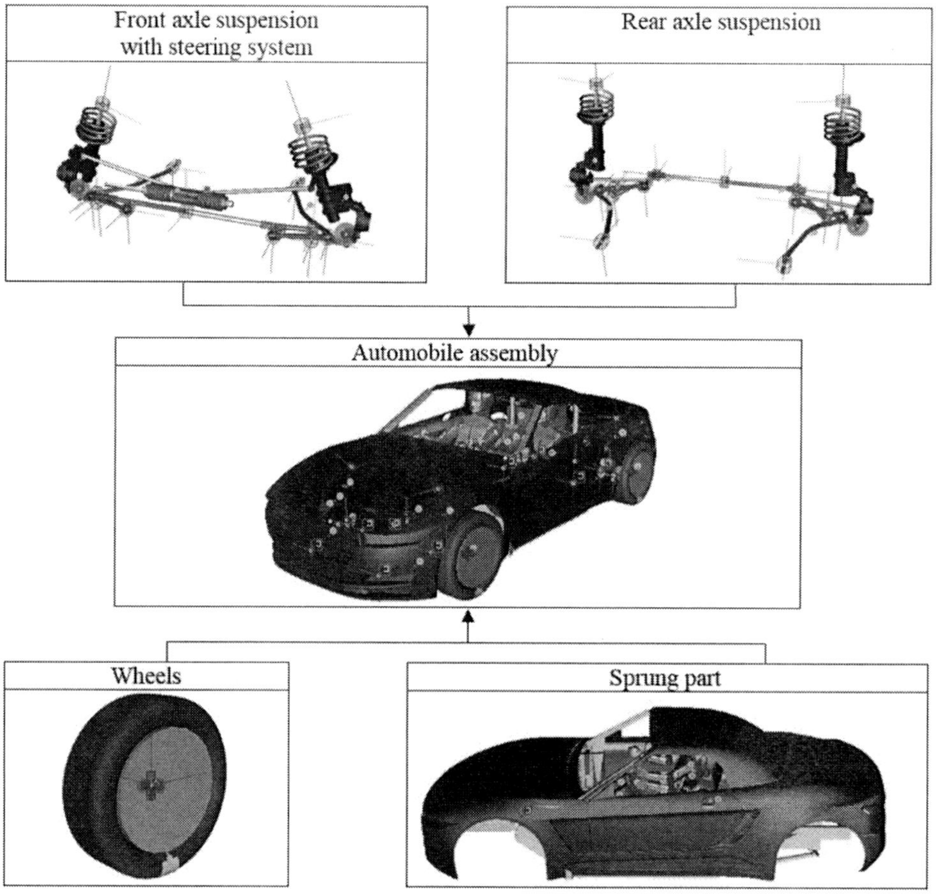

Figure 1. Structure of the 4x2 vehicle multi-body model

The sprung part model consists of the mass-inertial model of the body with spoilers and moldings and mass-inertial model of the frame with the main unsprung elements of the vehicle.

Suspension linkages are modeled as a set of rigid bodies connected by means of ideal joints and force elements.The masses and moments of inertia of the suspension bodies are calculated by their geometric models.The force elements are modeled as a pair of parallel spring and damper.

The steering system model couples the steering wheel rotation with the translation of the steering rack.

IOP Conf. Series: Materials Science and Engineering **820** (2020) 012024 doi:10.1088/1757-899X/820/1/012024

The vehicle model also includes the submodel of the powerplant with constant gear ratio transmission distributing the engine torque generated according to a control signal equally between the rear drive wheels, which corresponds to the transmittion with all open differentials.

2. Tire – road interaction model

The forces and moments acting on the wheel from the road are calculated by a custom tire model developed in the Universal Mechanism MBS program.The tire – road interaction model is shown in figure 2.The model uses the following coordinate systems (see figure 2a):

– wheel coordinate system (WCS) *OXYZ* – a movable coordinate system with the origin at the wheel center, *Z* axis is perpendicular to the road plane, *X* axis is perpendicular to the wheel rotation axis;

– road fixed coordinate system (FCS) O_r $X_rY_rZ_r$– an orthogonal coordinate system fixed on the road.

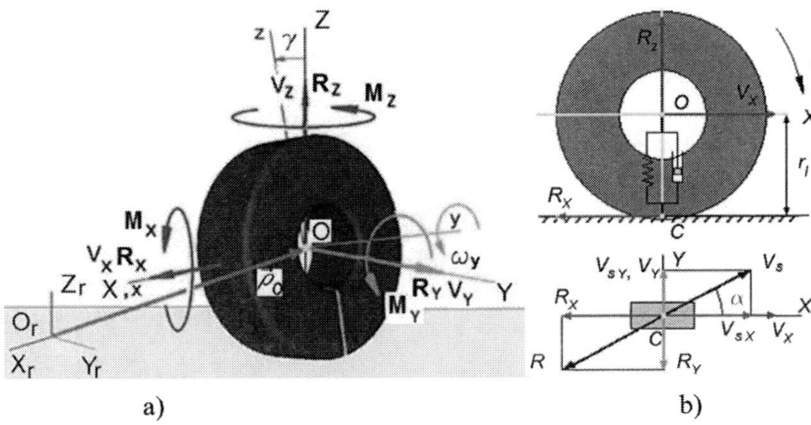

Figure 2. Tire – road interaction model: forces, moments and coordinate systems of the model – b) – forces and velocities in the longitudinal and lateral planes

The vertical reaction is calculated by the visco-elastic model:

$$R_Z = P_{z\_st} \cdot \left(\frac{h_z}{h_{z\_st}} \right)^{1,5} - b_z \cdot V_Z, (1)$$

where P_{z_st} is the wheel static load, h_{z_st} is the tire static deflection; b_z is vertical visco-elastic resistance force coefficient; h_z is the tire deflection:

$$h_z = \max\left(0, r_0 \cdot \cos(\gamma) - r_l\right), (2)$$

where r_0 is the unloaded tire radius; r_l is the loaded tire radius.

The reaction in the road plane:

$$R = \mu_{sxy}(S) \cdot R_Z, (3)$$

here S is the tire slip ratio; $\mu_{sxy}(S)$ is the tire – road friction coefficient.

$$\mu_{sxy}\left(S_k\right)=\begin{cases}0,\ \mu_{sx}\mu_{sy}<0;\\ \mu_{sx},\ \mu_{sy}=0,\mu_{sx}>0;\\ \mu_{sy},\ \mu_{sx}=0,\mu_{sy}>0;\\ \dfrac{\mu_{sx}\mu_{sy}}{\sqrt{\mu_{sx}{}^2 sin^2\alpha+\mu_{sy}{}^2 cos^2\alpha}},\ \mu_{sx}\mu_{sy}>0.\end{cases} \quad (4)$$

where μ_{sx},μ_{sy} are the tire – road friction coefficients in longitudinal and lateral directions.

$$\mu_{sx}=\mu_{sxmax}\left(1-e^{-\frac{S_x}{S_{0x}}}\right)\left(1+e^{-\frac{S_x}{S_{1x}}}\right), (5)$$

$$\mu_{sy}=\mu_{symax}\left(1-e^{-\frac{S_y}{S_{0y}}}\right)\left(1+e^{-\frac{S_y}{S_{1y}}}\right), (6)$$

where S_x, S_y are the tire longitudinal and lateral slip ratios; S_{0x}, S_{1x}, S_{0y}, S_{1y} are parameters defining the shape of the μ_{sx} and μ_{sy} characteristics; μ_{sxmax}, μ_{symax} are the friction ellipse parameters.

The slip ratios S_x,S_y are calculated by the following formulae:

$$S_x=\frac{V_{sX}}{\omega_y r_{e0}}, (7)$$

$$S_y=\frac{V_{sY}}{\omega_y r_{e0}}, (8)$$

where r_{e0} is the effective radius of a free rolling tire:

$$r_{e0}=\frac{3\cdot r_l}{1+\dfrac{2\cdot r_l}{r_0}}, (9)$$

V_{sX},V_{sY} are longitudinal and lateral slip velocities:

$$V_{cкX}=V_X-\omega_y\cdot r_{e0}; (10)$$

$$V_{sY}=V_Y-\omega_x\cdot r_l. (11)$$

The tire – road friction force $\vec{R}$ is directed opposite to the slip velocity $\overline{V_s}$ defined as:

$$V_s=\sqrt{V_{sX}^2+V_{sY}^2}; (12)$$

The angle α between the slide velocity vector and the wheel WCS X axis can be found from the following equations:

$$\sin\alpha=\frac{V_{sY}}{V_s}, (13)$$

$$\cos\alpha = \frac{V_{sX}}{V_s}. (14)$$

The projections of the tire – road friction force in the road plane are calculated in the following way:

$$R_X = -R \cdot \cos\alpha; (15)$$

$$R_Y = -R \cdot \sin\alpha. (16)$$

The moments acting on the wheel are defined as follows:

$$M_X = R_Y \cdot r_l - R_Z \cdot r_l \cdot \frac{\sin(\gamma)}{\cos(\gamma)}; (17)$$

$$M_Y = -R_X \cdot r_l + M_f; (18)$$

$$M_Z = 0, (19)$$

where M_f is the tire rolling resistance moment:

$$M_f = -R_Z \cdot r_{e0} \cdot f \cdot sign(\omega_y), (20)$$

where f is the tire rolling resistance coefficient:

$$f = f_0 + k_f \cdot (V_X)^2, (21)$$

where f_0 is the tire rolling resistance coefficient at a low speed (about 5 km/h); k_f is the tire rolling resistance growth factor describing increase in the rolling resistance with the growth of the forward velocity.

3. Step steer test model

According to Russian State Standard 31507-2012 [19], steering sensitivity analysis is a part of the step steer test. During this test the steering wheel of a vehicle moving with a constant velocity (80 km/h for M1 category) is rotated to a specified angle and held in that position till the begining of the steady cornering. The angle of the steering wheel rotation is increased incrementally at each new test until the lateral acceleration reaches the the tire – road adhession limit value or the maximum value determined by the safety conditions.

The steering sensitivity analysis results in the steering sensitivity characteristic which is a dependancy between the steering wheel rotation angle φ_{steer} and steady state lateral acceleration a_y of the vehicle.

During the simulation all the bodies of the vehicle model are given initial speed 80 km/h which is then kept constant by a PI speed controller model. The steering wheel is rotated according to the step transient function shown in figure 3.

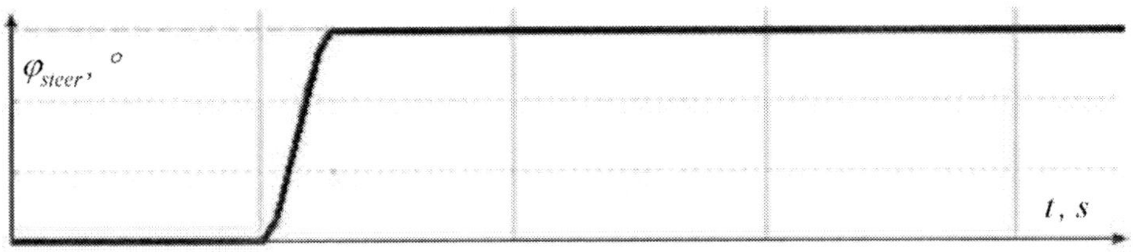

Figure 3. Steering wheel angle

IOP Conf. Series: Materials Science and Engineering **820** (2020) 012024 doi:10.1088/1757-899X/820/1/012024

Figure 4 shows animation frames of the step steer simulation. The vector *a* shows current direction of the vehicle acceleration.

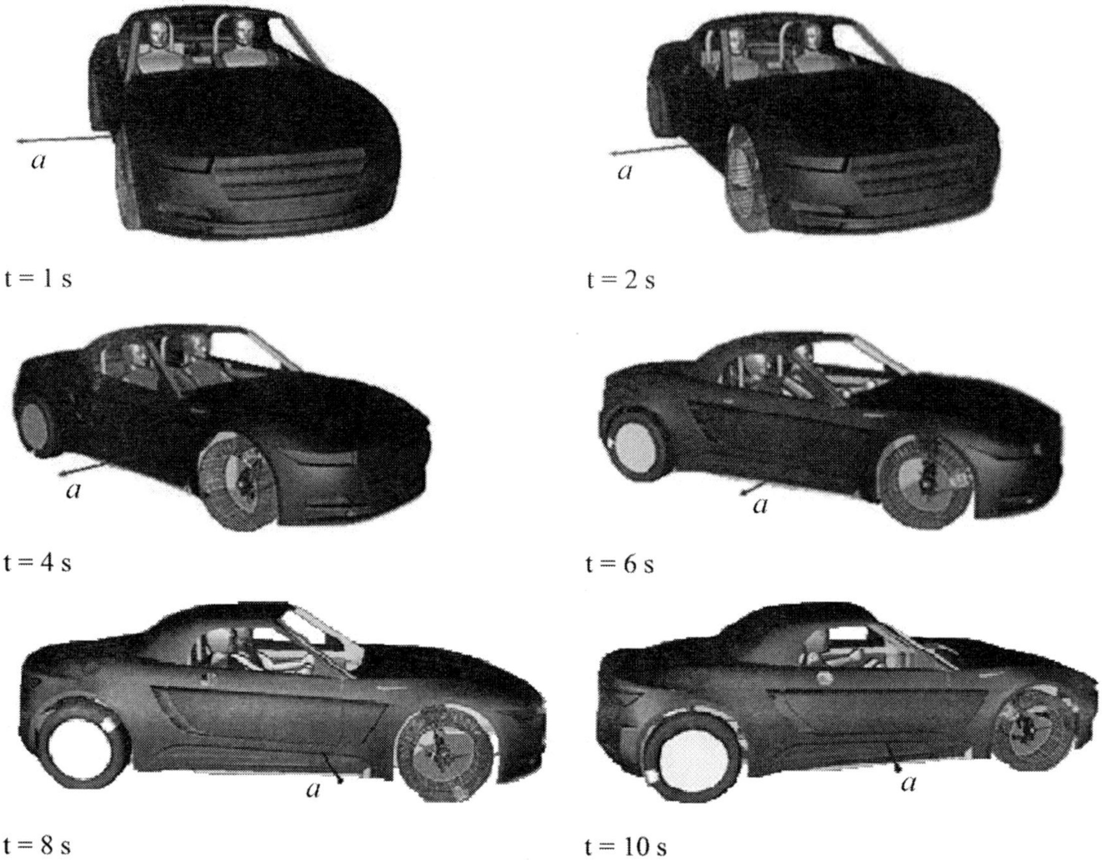

Figure 4. Animation frames of the step steer simulation

4. Analysis of the cornering stiffness variation impact

The analysis consisted in the repeated simulations of the step steer within the range of attainable lateral accelerations for the tires with the basic lateral friction coefficient – slip ratio characteristics and for less and more laterally stiff tires.Lateral friction coefficient – slip ratio characteristics for the three tire variants are shown in figure 5.The parameters of the model of the tire with basic characteristics have been identified from the available experimental data for the Continental 205/55 R16 tire [20].

Lateral accelerations obtained at different tire characteristics are shown in figure 6.

Table 1 shows results of the analysis of the model sensitivity to the changes in the stiffness of the lateral friction coefficient – slip ratio characteristics.

The analysis results show that the change in lateral accelerations due to the change of the tire lateral stiffness increases with the increase in the steer angle.At the same time, the mean accelerations for the stiff tire are higher than the ones for the basic tire, and the accelerations for the soft tire are lower than the accelerations for the basic tire.In the analysed range of the steer angle from 20° to 40°, when the tire lateral friction coefficient – slip ratio characteristics stiffness changes by 30 %, the change in the lateral accelerations is not higher than 3.1%.It means that the known ucertainty of the tire cornering stiffness [2] will not have a considerable impact on the vehicle steering sensitivity: its change will not exceed a few percent.

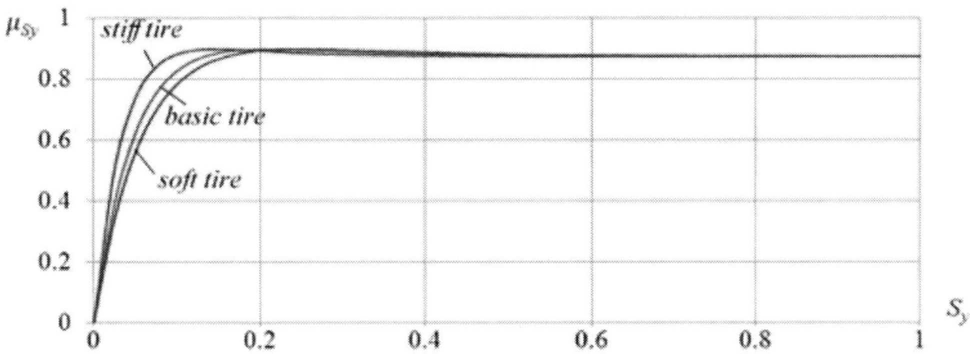

Figure 5. Lateral friction coefficient – slip ratio characteristics.

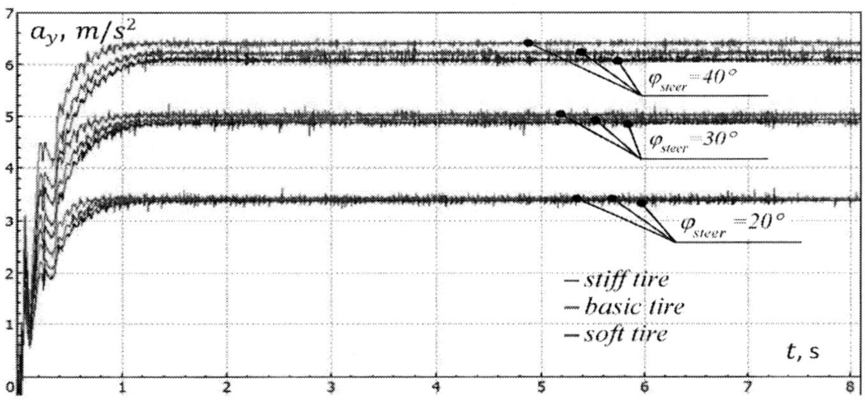

Figure 6. Time histories of the lateral accelerations at different steering angles and tire characteristics

Table 1. Impact of the tire lateral characteristics stiffness variation on the lateral acceleration

Steer angle φ_{steer}, °	Tire characteristics	Stiffness of the tire characteristics, %	Mean stationary lateral acceleration a_y, m/s^2	Change of a_y, %
20	Basic	100	3.40	–
	Stiff	70	3.44	1.2
	Soft	120	3.38	0.6
30	Basic	100	4.95	–
	Stiff	70	5.04	1.8
	Soft	120	4.89	1.2
40	Basic	100	6.22	–
	Stiff	70	6.41	3.1
	Soft	120	6.09	2.1

5. Conclusion

The paper has demonstrated the procedure of the analysis of the tire cornering characteristics uncertainty impact on the passenger vehicle steering sensitivity in an MBS program. The analysis was performed with the use of a spatial non-linear multi-body model of the vehicle with an original semi-empirical tire – road interaction model providing independent calibration of the longitudinal and lateral characteristics of the tire. The analysis has shown that simultaneous change in cornering

IOP Conf. Series: Materials Science and Engineering **820** (2020) 012024 doi:10.1088/1757-899X/820/1/012024

stiffness of all the tires by 30% has little impact on the steering sensitivity. This result means that during the steering sensitivity analysis the 30% accuracy of the tire cornering stiffnes is quite tolerable.

References

[1] Vilela Daniel & Barbosa Roberto 2011 Analytical Models Correlation for Vehicle Dynamic Handling Properties *Journal of the Brazilian Society of Mechanical Sciences and Engineering* 33 437 – 444.

[2] Van Oosten J J M, Unrau H J, Riedel A, & Bakker E 1999 Standardization in tire modeling and tire testing– Tydex workgroup, Time project *Tire Science and Technology* 27(3) pp 188– 202.

[3] Blundell M & Harty D 2015 *Multibody systems approach to vehicle dynamics* Elsevier.

[4] Gorelov V A, Komissarov A I, &Miroshnichenko, A V 2015 8×8 wheeled vehicle modeling in a multibody dynamics simulation software *Procedia engineering*, 129, 300– 307.

[5] Gorelov V A, Komissarov A I, Kositsyn B B 2016 Passenger car motion modeling in a multibody dynamics simulation software *Zurnal AAI*, (1), 18– 23.

[6] Fischer E 2007 Standard multi-body system software in the vehicle development process *Proc. Inst. Mech. Eng., part K: journal of multi-body dynamics*, 221(1), 13-20.

[7] Prado M, Cunha R H, Neto Á C, Costa A, Mancosu F, Savi C &D'Elboux J E 2001 Bus handling analysis and experimental validation using the multibody system technique (No. 2001-01-3966) *SAE Technical Paper*.

[8] Hegazy S, RahnejatH, Hussain K 2000 Multi-body dynamics in full-vehicle handling analysis under transient manoeuvre. *Vehicle System Dynamics*, 34(1), 1-24.

[9] Mastinu G &Plöchl M (Eds.) 2014 *Road and off-road vehicle system dynamics handbook*. CRC Press.

[10] Information on http://www.euler.ru.

[11] Information on http://www.umlab.ru.

[12] Information on http://www.mscsoftware.com/product/adams.

[13] Gipser M 2007 FTire – the tire simulation model for all applications related to vehicle dynamics. *Vehicle System Dynamics*, 45(S1), 139-151.

[14] Gallrein A, BäckerM 2007CDTire: a tire model for comfort and durability applications. *Vehicle System Dynamics*, 45(S1), 69-77.

[15] Schmeitz A J C ,Besselink I J M, & Jansen, S T H 2007 TNO MF-SWIFT *Vehicle System Dynamics*, 45(S1), 121-137.

[16] Pacejka H B, Bakker E 1992 The magic formula tyre model *Vehicle system dynamics*, 21(S1), 1 – 18.

[17] Dessort R, Chucholowski C & Rill G 2016. Parametrical approach for modeling of tire forces and torques in TMeasy 5. In 16. *InternationalesStuttgarter Symposium* pp. 435-449

[18] Gorelov V A &Komissarov A I 2016 Mathematical Model of the Straight-line Rolling Tire – Rigid Terrain Irregularities Interaction. *Procedia Engineering*, 150, 1322– 1328.

[19] Russian State Standard 31507-2012. Road vehicles. Handling and stability. Technical requirements. Testmethods.

[20] Schouten H E 2005 Evaluation of the MF-Tyre/MF-Swift 6.0 tyre model.

IOP Conf. Series: Materials Science and Engineering **820** (2020) 012025 doi:10.1088/1757-899X/820/1/012025

Mathematical modelling of torque vectoring differentials

G Ankinovich[1], A Verzhbitski[1] and A Antonyan[1,2]

[1]Bauman Moscow State Technical University

[1]E-mail: antonyan.akop@yandex.ru

Abstract. Active safety systems which distribute forces between the wheels are an integral part of a modern vehicle. The vast majority of systems are based on the brake systems, and reducing engine power, which certainly reduces the vehicle speed. However, the needs of car owners are constantly changing, there is a need for devices capable of ensuring the distribution of traction between the wheels without having to reduce speed. With this task perfectly cope with the main transmission containing the mechanisms of power distribution (MPD). Power distribution mechanisms have a greater impact on vehicle handling. More often meet term, Torque Vectoring Differential (TVD). They began to be used for the first time since the mid-1990s, and found their application in sports cars, and then in executive cars. The described main gears are an integral part of dynamic stabilization systems. During the vehicle development, these systems are developed and investigated using mathematical modeling methods. Therefore, the paper describes the design, operation principle and kinematic design of the most common. The equations systems describing the dynamics of the links and the mechanism as a whole are also composed. The mathematical modeling of vehicles equipped with TVD is modelled.

1. Introduction

Modern cars are equipped with various dynamic stabilization systems based on the operation of the brake system or reducing engine power [1]. However, since the mid-1990s, Mitsubishi has been using main drives with power distribution mechanisms known as Torque Vectoring Differential (TVD). Their distinctive feature compared to conventional main gears, is that they are able to transmit more torque to the high-speed wheel, thus creating an increased turning torque of the vehicle. Power distribution mechanisms, in turn, contain additional links and controls that ensure the distribution of torques [1–10].

PDM containing friction elements which activated by an electronic control system. However, to ensure effective operation of the control system, first, at the design stage, conducting the mathematical simulation of the vehicle motion [11–15], equipped with such a mechanism. In the development of mathematical models of multilink final drives made the following assumptions: the links of the final drive perfectly rigid, the losses in the gearing can be neglected, the ratio auxiliary links taken greater than one.

Thus, the paper analyzes the existing mechanisms, their comparison and writing of equations systems for mathematical modeling in the program MATLAB & Simulink.

2. Final drive Mitsubishi AYC

The final drive of Mitsubishi AYC was first used in 1996 on the Mitsubishi Lancer Evolution IV. [2] This final drive is based on using bevel differential. PDM is located on the right and is connected by one link to the differential housing and the right axle through friction clutches.

2.1. Final drive design

Mitsubishi AYC final drive design shown on the figure 1.

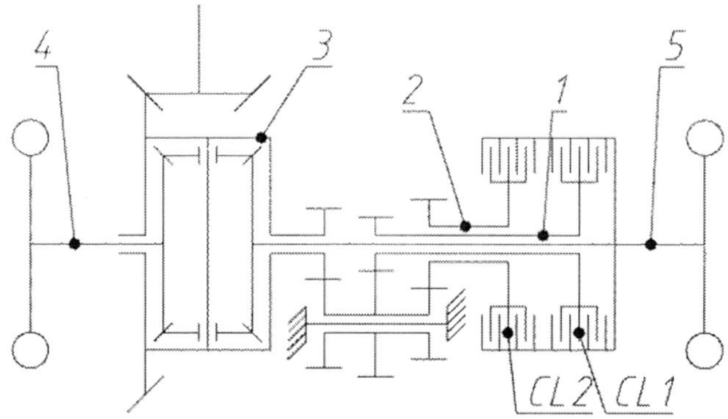

Figure 1. Mitsubishi AYC final drive design:
1 – fist auxiliary link;
2 – second auxiliary link;
3 – differential housing;
4 – left half-axis;
5 – right half-axis;
CL1 – right clutch;
CL2 – left clutch.

In the final drive, depending on right friction clutch CL1 engaged, part of the torque from the differential housing 3 is transmitted to the first auxiliary link 1 through a multiplier that accelerates the rotation of the right half-axis 5. The increase in torque is due to the summation of torques from the first auxiliary link 1 and the half-axial differential gear. Depending on left friction clutch CL2 engaged, part of the torque from the differential housing 3 will be transferred to the second auxiliary link 2 through the gearbox, slowing down the angular speed of the right half-axis 5. The resistance torque, which occurs on the left clutch CL2, creates a braking torque relative to the differential housing 3.

2.2. Final drive equations system

To compile an equations system revealed that the mechanism of five moving links and three gearing. Thus, it is necessary to make five dynamic equations and three kinematic equations. Let the first two equations (1, 2) describe the dynamics of the auxiliary links:

$$J_1 \cdot \dot{\omega}_1 = M_{link\,1} - M_{CL1} \tag{1}$$

$$J_2 \cdot \dot{\omega}_2 = M_{link\,2} - M_{CL2} \tag{2}$$

J_1, J_2 first and second auxiliary links inertias, $kg \cdot m^2$; $\dot{\omega}_1$, $\dot{\omega}_2$ first and second auxiliary links angular accelerations, rad / s^2; $M_{link\,1}$, $M_{link\,2}$, first and second auxiliary links torques $N \cdot m$; M_{CL1}, M_{CL2}, first and second clutches friction torques, $N \cdot m$.

In equation (3) torques $M_{link\,1}$ и $M_{link\,2}$ written below:

$$J_w \cdot \dot{\omega}_{w\,right} = \frac{M_{diff}}{2} - M_{link\,1} + M_{link\,2} - M^{dr}_{right} \tag{3}$$

$$J_w \cdot \dot{\omega}_{left} = \frac{M_{diff}}{2} - M^{dr}_{left} \tag{4}$$

J_w, wheel inertia, $kg \cdot m^2$; $\dot{\omega}_{w\,right}$, $\dot{\omega}_{left}$, right and left wheel angular accelerations rad / s^2; M_{diff}, differential housing torque, $N \cdot m$; M^{dr}_{right}, M^{dr}_{left}, right and left half-axis drag torques, $N \cdot m$.

Equations (5), (6), (7) describe links interaction which included in the final drive:

$$\dot{\omega}_{input} = i_{FD} \cdot \frac{\dot{\omega}_{w\,right} + \dot{\omega}_{left}}{2} \quad (5)$$

$$\dot{\omega}_{input} = i_{FD} \cdot \frac{\dot{\omega}_1}{u_1} \quad (6)$$

$$\dot{\omega}_{input} = i_{FD} \cdot \dot{\omega}_2 \cdot u_2 \quad (7)$$

$\dot{\omega}_{input}$, input shaft angular acceleration, rad/s^2; i_{FD}, final drive gear ratio; u_1, u_2, ratio between differential housing and first and second auxiliary links.

The equation of dynamics of the differential housing (8), reduced to the input shaft, is written below:

$$J_{input} \cdot \dot{\omega}_{input} = M_{input} - \frac{M_{diff}}{i_{FD}} - \frac{M_{link\,1}}{i_{FD} \cdot u_1} - \frac{M_{link\,2} \cdot u_2}{i_{FD}} \quad (8)$$

The considered final drive equations system written bellow:

$$\begin{cases} J_1 \cdot \dot{\omega}_1 = M_{link\,1} - M_{CL1}; \\ J_2 \cdot \dot{\omega}_2 = M_{link\,2} - M_{CL2}; \\ J_w \cdot \dot{\omega}_{w\,right} = \frac{M_{diff}}{2} - M_{link\,1} + M_{link\,2} - M^{dr}_{right}; \\ J_w \cdot \dot{\omega}_{left} = \frac{M_{diff}}{2} - M^{dr}_{left}; \\ \dot{\omega}_{input} = i_{FD} \cdot \frac{\dot{\omega}_{w\,right} + \dot{\omega}_{left}}{2}; \\ \dot{\omega}_{input} = i_{FD} \cdot \frac{\dot{\omega}_1}{u_1}; \\ \dot{\omega}_{input} = i_{FD} \cdot \dot{\omega}_2 \cdot u_2; \\ J_{input} \cdot \dot{\omega}_{input} = M_{input} - \frac{M_{diff}}{i_{FD}} - \frac{M_{link\,1}}{i_{FD} \cdot u_1} - \frac{M_{link\,2} \cdot u_2}{i_{FD}} \end{cases} \quad (9)$$

3. Final drive Magna

The Magna final drive was first used in 2009 on the Audi S4 [2]. It contains a symmetrical conical differential and multipliers located on the right and left.

3.1. Final drive design

Magna final drive design shown on the figure 2.

In the Magna final drive, when the right clutch CL1 is activated, part of the torque from the differential housing 3 is transmitted to the first auxiliary link 1 through the first stage of the multiplier, then through the friction right clutch CL1 to the second stage of the multiplier. This accelerates the rotation of the right half-axis 5. The increase in torque is due to the summation of moments from the first auxiliary link 1 and the right half-axis 5. In the case of activation of the left clutch CL2 part of the torque from the differential housing 3 will be similarly transmitted to the left axle 4.

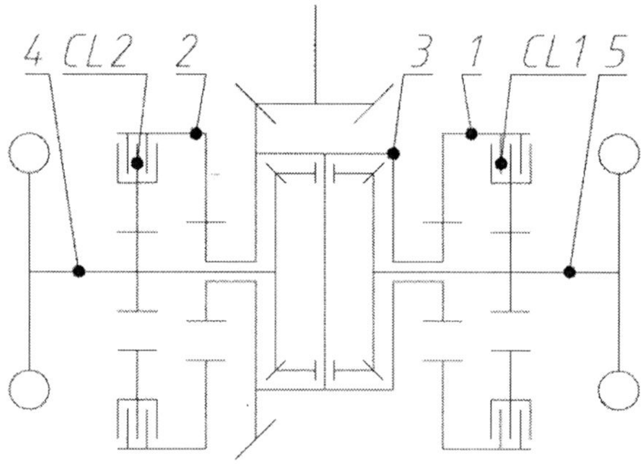

Figure 2. Magna final drive design:
1 – fist auxiliary link;
2 – second auxiliary link;
3 – differential housing;
4 – left half-axis;
5 – right half-axis;
CL1 – right clutch;
CL2 – left clutch.

3.2. Final drive equations system

The equations describing the auxiliary units dynamics do not differ from the equations (1, 2) written above. Next, dynamics equations of the half-axis are written bellow:

$$J_w \cdot \dot{\omega}_{w\,right} = \frac{M_{diff}}{2} - \frac{M_{link\,1}}{u_1} - M^{dr}_{right}; \tag{10}$$

$$J_w \cdot \dot{\omega}_{w\,left} = \frac{M_{diff}}{2} - \frac{M_{link\,2}}{u_2} - M^{dr}_{left}; \tag{11}$$

u_1, u_2, accordingly ratio between right and left half-axis and first and second auxiliary links.

Next, kinematic equations for auxiliary links are written bellow:

$$\dot{\omega}_{input} = i_{FD} \cdot \dot{\omega}_1 \cdot u_{diff}; \tag{12}$$

$$\dot{\omega}_{input} = i_{FD} \cdot \dot{\omega}_2 \cdot u_{diff}; \tag{13}$$

u_{diff}, the gear ratio between the differential housing and supporting links. The dynamics equation of the differential housing, reduced to the input shaft, is written, as well as (8).

$$\begin{cases} J_1 \cdot \dot{\omega}_1 = M_{link\,1} - M_{CL1}; \\ J_2 \cdot \dot{\omega}_2 = M_{link\,2} - M_{CL2}; \\ J_w \cdot \dot{\omega}_{w\,right} = \dfrac{M_{diff}}{2} - \dfrac{M_{link\,1}}{u_1} - M^{dr}_{right}; \\ J_w \cdot \dot{\omega}_{w\,left} = \dfrac{M_{diff}}{2} - \dfrac{M_{link\,2}}{u_2} - M^{dr}_{left}; \\ \dot{\omega}_{input} = i_{FD} \cdot \dfrac{\dot{\omega}_{w\,right} + \dot{\omega}_{left}}{2}; \\ \dot{\omega}_{input} = i_{FD} \cdot \dot{\omega}_1 \cdot u_{diff}; \\ \dot{\omega}_{input} = i_{FD} \cdot \dot{\omega}_2 \cdot u_{diff}; \\ J_{input} \cdot \dot{\omega}_{input} = M_{input} - \dfrac{M_{diff}}{i_{FD}} - \dfrac{M_{link\,1}}{i_{FD} \cdot u_{diff}} - \dfrac{M_{link\,2}}{i_{FD} \cdot u_{diff}} \end{cases} \tag{14}$$

4. Final drive ZF Torque Vectoring

The final drive ZF Torque Vectoring was first introduced in 2008 [2] and was installed on BMW cars. Unlike the above-mentioned mechanisms, the power distribution is due to the planetary multipliers located on the right and left.

4.1. Final drive design

ZF Torque Vectoring final drive design shown on the figure 3.

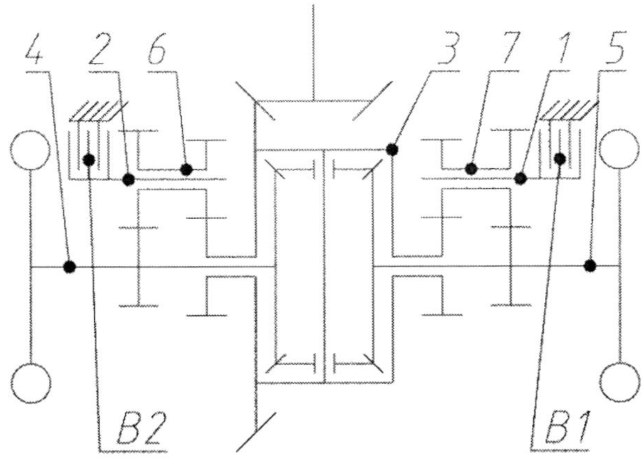

Figure 3. ZF Torque Vectoring final drive design:
1 – right planetary multiplier carrier;
2 – left planetary multiplier carrier;
3 – differential housing;
4 – left half-axis;
5 – right half-axis;
6 – satellite unit;
7 – satellite unit;
B1 – right friction break;
B2 – left friction break.

In ZF Torque Vectoring through the activation of the right friction break B1, the right planetary multiplier carrier. Due to this, the gear right half-axis 5 is accelerated with respect to the differential housing 3. The increase in moments occurs by summing the moments from the half-axis gear and the multiplier. However, in this mechanism, the closing of the brake link occurs on the final drive housing, so part of the moment is transferred to the final drive housing. That is, the torque on the input shaft, reduced to the drive axle, is the sum of the moments on the half-axis 4 and 5 minus the friction moment in the brake friction. Similarly, there is an increase in speed and torque on the left axle 4.

4.2. Final drive equations system

The dynamics equations for links 15 and 16 are written below:

$$J_{carr} \cdot \dot{\omega}_{carr1} = M_{carr1} - M_{B1} \tag{15}$$

$$J_{carr} \cdot \dot{\omega}_{carr2} = M_{carr2} - M_{B2} \tag{16}$$

J_{carr}, planetary multiplier carrier inertia, $kg \cdot m^2$; $\dot{\omega}_{carr1}$, $\dot{\omega}_{carr2}$, first and second planetary multiplier carrier angular accelerations, rad / s^2; M_{carr1}, M_{carr2}, first and second planetary multiplier carrier torques, $N \cdot m$; M_{B1}, M_{B2}, first and second planetary multiplier carrier friction breaks torques, $N \cdot m$.

Next, dynamics equations of the half-axes are written bellow:

$$J_w \cdot \dot{\omega}_{w\,right} = \frac{M_{diff}}{2} - M^{halfaxis}_{sun\,right} - M^{dr}_{right} \tag{17}$$

$$J_w \cdot \dot{\omega}_{w\,lfet} = \frac{M_{diff}}{2} - M^{halfaxis}_{sun\,left} - M^{dr}_{left} \tag{18}$$

$M^{halfaxis}_{sun\,right}$, right half-axle shafts sun gear torque 5: $M^{halfaxis}_{sun\,right} = k \cdot M^{diff}_{sun\,right}$, $N \cdot m$; $M^{halfaxis}_{sun\,left}$, left half-axle shafts sun gear torque 4: $M^{halfaxis}_{sun\,left} = k \cdot M^{diff}_{sun\,left}$, $N \cdot m$.

Below are the kinematics equations for planetary mechanisms:

$$\dot{\omega}_{input} = i_{FD} \cdot \frac{\dot{\omega}_{right} + k \cdot \dot{\omega}_{carr1}}{1+k} \tag{19}$$

$$\dot{\omega}_{input} = i_{FD} \cdot \frac{\dot{\omega}_{left} + k \cdot \dot{\omega}_{carr2}}{1+k} \tag{20}$$

The equations system describing ZF Torque Vectoring is written below:

$$\begin{cases} J_{carr} \cdot \dot{\omega}_{carr1} = M_{carr1} - M_{B1}; \\ J_{carr} \cdot \dot{\omega}_{carr2} = M_{carr2} - M_{B2}; \\ J_w \cdot \dot{\omega}_{w\,right} = \frac{M_{diff}}{2} - M^{halfaxis}_{sun\,right} - M^{dr}_{right}; \\ J_w \cdot \dot{\omega}_{w\,lfet} = \frac{M_{diff}}{2} - M^{halfaxis}_{sun\,left} - M^{dr}_{left}; \\ \dot{\omega}_{input} = i_{FD} \cdot \frac{\dot{\omega}_{w\,right} + \dot{\omega}_{left}}{2}; \\ \dot{\omega}_{input} = i_{FD} \cdot \frac{\dot{\omega}_{right} + k \cdot \dot{\omega}_{carr1}}{1+k}; \\ \dot{\omega}_{input} = i_{FD} \cdot \frac{\dot{\omega}_{left} + k \cdot \dot{\omega}_{carr2}}{1+k}; \\ J_{input} \cdot \dot{\omega}_{input} = M_{input} - \frac{M_{diff}}{i_{FD}} - \frac{M_{carr1}}{i_{FD}} - \frac{M_{carr2}}{i_{FD}} \end{cases} \tag{21}$$

5. Final drive mathematical modelling in MATLAB&Simulink

In this section, the simulation of the vehicle motion [16–20] with AWD system, on the drive axis of which is the final drive with PDM. The final drive Mitsubishi AYC is used as an example. The total vehicle weight is 1700 kg.

5.1. Study of the movement of vehicles with and without PDM

The vehicle motion on the road "ice with snow" (with the coefficient of friction of the wheel with the support $\mu_{s\,max} = 0{,}35$). The front wheels are steering. Start speed $v = 25\,км/ч$, and limited $v = 35\,км/ч$. Figure 4 shows graphs of vehicle yaw rate.

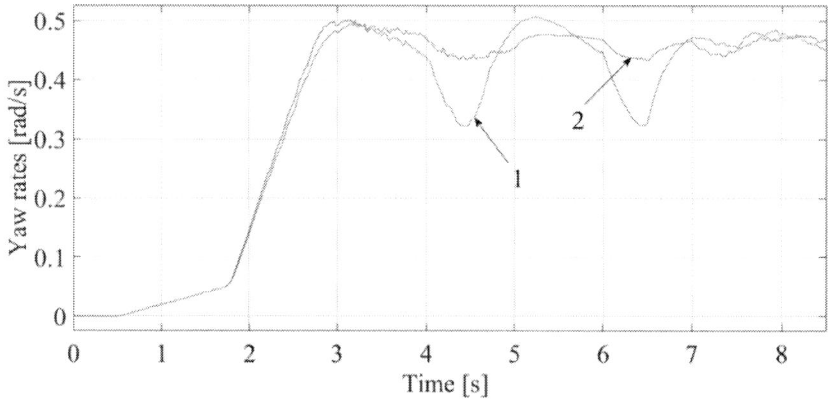

Figure 4. Vehicle yaw rate [rad/s]:
1 – equipped with conventional final drive,
2 – equipped with conventional final drive Mitsubishi AYC

IOP Conf. Series: Materials Science and Engineering **820** (2020) 012025 doi:10.1088/1757-899X/820/1/012025

6. Conclusion

In the paper presents the analysis of kinematic designs of power distribution mechanisms, their purpose, design and principle of operation. Also derived equations systems that can be used for mathematical modeling and study of the vehicle motion. The vehicles yaw rates are presented.

References

[1] Sawase K, Ushiroda Y 2009 S-AWC (Super All Wheel Control), Integrated Vehicle Dynamics control for High Performance All Wheel Drive Vehicle. (9-th European All-Wheel Drive Congress Grass)

[2] Granzow C, Arzner M 2014 Torque Vectoring for Drivetrain Systems (Encyclopedia of Automotive Engineering, Online John Wiley & Sons, Ltd) pp 285 - 296

[3] Mehdi Jaafari S M, Shirazi K H 2016 A comparison on optimal toque vectoring strategies in overall performance enhancement of a passenger car, Proc IMech Part K: J Multi-body Dynamics 230 (4).

[4] Koehler S, Droll S, Bringmann O, Brunner O 2013 Torque vectoring systems to improve energy-efficiency and vehicle dynamics (Research Controlling), pp 46 – 53

[5] Seyed Mohammad Mehdi Jaafari and Kourosh Heidari Shirazi 2016 A comparison on optimal torque vectoring strategies in overall performance enhancement of a passenger car (Journal of Multi-body Dynamics), pp 469 – 488

[6] Tomo Kato and Kaoru Sawase 2012 Classification and analysis of electric powered lateral torque-vectoring differentials (Journal of Multi-body Dynamics), pp 713 – 724

[7] Hancock M J, Williams R A, Gordon T J and Best M C 2003 A comparison of braking and differential control of road vehicle yaw-sideslip dynamics (Loughborough University Institutional Repository), pp 309 – 327

[8] Sawase K and Inoue K 2008 Classification and analysis of lateral torque-vectoring differentials using velocity diagrams (Automobile Engineering), pp 1527 – 1541

[9] Yu-Fan Chen, I-Ming Chen, Joshua Chang and Tyng Liu 2017 Design and Analysis of a New Torque Vectoring System with a Raveneaux Gearset for Vehicle Applications (Energies)

[10] Wheals J, Barnbrook R, Parkinson R, Dean M, Donin R 2007 Torque Vectoring (interdisciplinarity in engineering scientific international conference), pp I-33-1 – I-33-6

[11] Rajesh Rajamani 2012 *Vehicle Dynamics and Control Second edition*, 496 p

[12] Reza N Jasar *Vehicle Dynamics, Theory and Application*, 1015 p, 2008

[13] Guzzella L and Sciaretta A 2013 *Vehicle Propulsion Systems. Introduction to Modeling and Optimization Third edition*, 409p

[14] Waschl H, Kolmanovsky I, Steinbuch M and Luigi del Re 2014 *Optimization and Optimal Control in Automotive Systems*, 326p

[15] Popp K, Schnielen W 2010 *Ground Vehicle Dynamics*, 348p

[16] Vol'skaya N S, Zhileykin, M M , Zakharov, A Y 2018 Mathematical model of rolling an elastic wheel over deformable support base (IOP Conference Series: Materials Science and Engineering), 315 (1), article № 012028

[17] Gorelov V A, Komissarov A I, Miroshnichenko A V 2015 8×8 wheeled vehicle modeling in multibody dynamics simulation software (Procedia Engineering), 129, pp. 300-307.

[18] Zhileykin M M, Kotiev G O, Nagatsev M V 2018 Comparative analysis of the operation efficiency of the continuous and relay control systems of a multi-axle wheeled vehicle suspension (IOP Conference Series: Materials Science and Engineering, 315 (1)), article № 012030

[19] Zhileykin M M, Kotiev G O, Nagatsev M V 2018 Synthesis of the adaptive continuous system for the multi-axle wheeled vehicle body oscillation damping (IOP Conference Series: Materials Science and Engineering, 315 (1)), article № 012031

[20] Beloousov B, Ksenevich T, Vantsevich, V Komissarov D 2013 8×8 platform for studying terrain mobility and traction performance of unmanned articulated ground vehicles with steered wheels (SAE Technical Papers), 9

IOP Conf. Series: Materials Science and Engineering **820** (2020) 012026 doi:10.1088/1757-899X/820/1/012026

Research of dynamics of movement of grain and forage harvesters by methods of mathematical and imitating modeling

M Zhileykin[1] and P Sirotin[2,3]

[1]Bauman Moscow State Technical University
[2]Platov South Russian State Polytechnic University (Novocherkassk, Rostov region, Russia)

[3]E-mail:spv_61@mail.ru

Annotation. The mathematical model of movement of the grain and forage harvester is considered, the possibility of its algorithmization in the software complex is shown, as well as the data confirming its adequacy. The mathematical model includes dynamic and kinematic equations of relation of angular and linear velocities with angular and spatial coordinates. The peculiarity of the model is in the use of a set of coordinate systems, which allowed to take into account not only the weight and size and layout features of the type of machines under study, but also the nature of interaction of the elastic wheel with the damped base. In the given model the speed of movement of the combine is set not by the forced change of the coordinate of the center of mass of the skeleton, but is formed by the modeling of the process of interaction of driving wheels with the support base. This approach provides a more adequate description of the process of curvilinear movement of the combine, allows you to simulate the movement of the machine, acceleration, braking, overcoming obstacles, slipping and slipping processes, taking into account the characteristics of the tire and traction properties of the soil. The approach adopted ensures sufficient accuracy of the model with a minimum set of factors and requirements for calculation. Experimental and computational data obtained as a result of simulation are presented. Comparison of active forces and acceleration spectral densities in the main parts of the combine showed sufficient convergence of results. The revealed inaccuracy is caused by the discrepancy between the numerical description of the supporting surface and the test road, the presentation of the combine's frame as an absolutely rigid body, and differences in the elastic-viscous properties of tires. Conclusions are made and directions of the further researches providing improvement of operational properties of self-propelled combines at the expense of systems of suspension of wheels and working elements are defined.

Introduction

Today the main means of harvesting crops are self-propelled grain and forage harvesters, the efficiency of which largely determines the profitability of the agricultural sector as one of the priority areas of the economy [1, 2]. The harvesters are equipped with more efficient working and functional organs, powerful power plants and increase their working and transport speeds to increase productivity [3-6]. Such changes in design and operational properties, as a rule, lead to an increase in the curb weight of machines, which is unacceptable by the requirements of agroecology and is limited by the norms of maximum pressure on the ground [7-9]. In this case, the development and implementation of systems to reduce dynamic loads on combines and improve their smooth running is an actual practical task of modern agricultural engineering [10, 11].

Purpose of the study
Nowadays the creation of machines and the study of their motion dynamics should be based on the methods of mathematical and simulation modeling, the basis of which is a mathematical description of the processes under study [12-18]. The development of mathematical models is a complex and knowledge-intensive process that determines the reliability of the results, the duration of research, the cost of research and development, as well as the properties of finished products. Thus, the main purpose of this work was to develop and verify the mathematical model of movement of self-propelled wheeled grain and forage harvesters.

Materials and methods
The requirements for a mathematical model of the dynamics of combines are determined by a set of tasks, in the solution of which it is necessary to obtain information for the evaluation of performance, in particular:

- the model should describe the joint dynamics of the frame, engine and running gear elements of combines with the accuracy necessary for stability when driving on non-deformable soils;
- the design features of wheeled propellers should be taken into account, as well as the unstoppable nature of the links imposed on the combine;
- the movement of the combine should be modeled taking into account the characteristics of resistance and grip of the ground, as the traction and coupling characteristics affect the speed of the machine.

When calculating the differential equations of the harvester's movement, a number of assumptions were made, which, on the one hand, should ensure the fulfillment of the requirements for the mathematical model, and, on the other hand, should limit the number of simulated parameters of the system to the most necessary ones. In accordance with the requirements to the mathematical model, the following assumptions are made:

- the weights of the harvester's sprung elements are reduced to the load-bearing system;
- the support base is considered non-deformable (the necessary pliability to the ground can be taken into account in the corresponding characteristics of the tyres of the wheels, and the tangential pliability of the soil is taken into account in the characteristics of its traction properties).

When modeling the dynamics of a combine harvester its body was considered as a spatial absolutely firm structure. The bond between kinematic parameters and external perturbations was described by means of differential equations that make up the mathematical model of the combine's motion.

The system of equations of motion of a self-propelled combine contains:

- the dynamic equations describing the movement of the combine, obtained on the basis of the law of preservation of the amount of movement and the moment of the amount of movement;
- the kinematic equations of the bond between angular and linear velocities and angular and spatial coordinates, obtained on the basis of equations of the bond between different coordinate systems.

Results and discussion
The developed model uses three different coordinate systems (Fig. 1), which is due to the structure and form of the equations of motion of the object. The first, stationary coordinate system $O_2X_2Y_2Z_2$, is used to simulate the given ground and road conditions of movement. The origin of the system coordinates (point O_2) coincides with the beginning of the simulated trace (Fig. 1 a). The second, semi-located coordinate system $O_1X_1Y_1Z_1$, is characterized by the fact that its beginning point O_1 always coincides

with the center of mass of the combine and moves with it in space. The axes O_1X_1, O_1Y_1, O_1Z_1 are parallel to the corresponding axes of the unrelated coordinate system (Fig. 1 a). The third coordinate system *OXYZ*, used for mathematical description of the combine's motion, is a mobile coordinate system (MCS), its center of gravity *O* always coincides with the center of gravity *C*, and the axes coincide with the main axes of inertia of the combine (Fig. 1 a). The equations of the combine harvester dynamics were recorded in the linked coordinate system, therefore the projections of linear (V_X, V_Y, V_Z) and angular (*X, Y, Z*) velocities on the linked axes were taken as the motion parameters. The use of a linked coordinate system to record the self-propelled combine dynamics equations is determined by the following considerations:

- Let's consider that mobile axes with the beginning of coordinates in the center of masses of the wheeled car are the main axes of inertia of a body and moments of inertia concerning them do not depend on change of kinematic parameters;
- The main external forces acting on the coordinate system are oriented relative to the body and the most simply expressed in the coordinate axes rigidly connected to it.

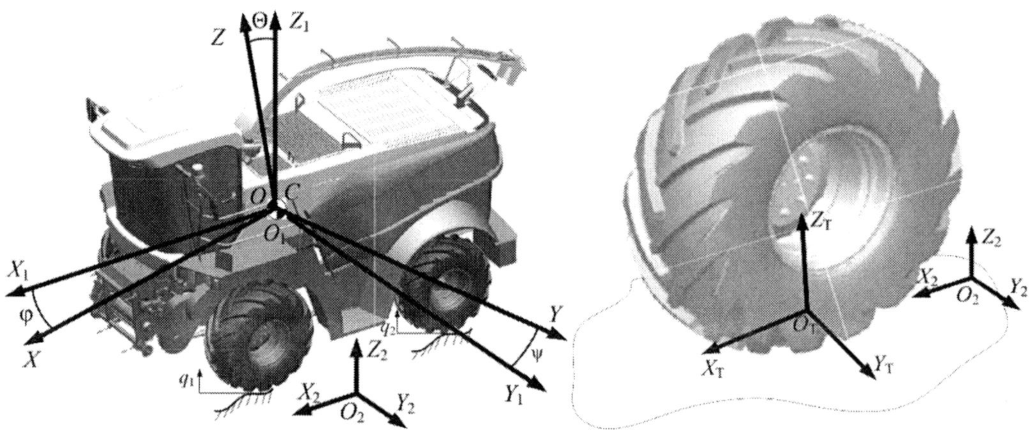

Fig. 1. Position of the self-propelled harvester in space (a) and micro-mobile coordinate system (b): C - center of mass of the harvester; φ, ψ, Θ - angles of trim, roll, course, respectively; qi - vertical coordinate of the road surface

In this case, the self-propelled combine dynamics equations recorded in MCS are the simplest and most convenient for the subsequent solution with a sufficiently complete reflection of the processes of interaction between the moving body and the environment. We introduce a micromovable coordinate system (MMCS) to determine the forces acting on the ground side of the combine and understand it as the $O_TX_TY_TZ_T$ system, the center of which $O_TX_TY_TZ_T$ coincides with the geometric center of the wheel contact spot, the O_TX_T axis coincides with the projection of the longitudinal axis of symmetry of the wheel on the bearing surface, and the O_TY_T axis respectively with the projection of the wheel axis (Fig. 1 b).

We obtain the general equations of motion of a self-propelled combine using the theorems of changing the amount of body movement and the amount of motion moment in projections on the MCS axis:

$$\begin{cases} m_\kappa \dfrac{dV_{CX}}{dt} + m_\kappa\left(\omega_Y V_{CZ} - \omega_Z V_{CY}\right) = G_X + F_X + \sum\limits_{i=1}^{N} R_{Xi}; \\ m_\kappa \dfrac{dV_{CY}}{dt} + m_\kappa\left(\omega_Z V_{CX} - \omega_X V_{CZ}\right) = G_Y + F_Y + \sum\limits_{i=1}^{N} R_{Yi}; \\ m_\kappa \dfrac{dV_{CZ}}{dt} + m_\kappa\left(\omega_X V_{CY} - \omega_Y V_{CX}\right) = G_Z + F_Z + \sum\limits_{i=1}^{N} R_{Zi}; \\ I_X \dfrac{d\omega_X}{dt} + \omega_X\omega_Z\left(I_Z - I_Y\right) = M_X(F) + \sum\limits_{i=1}^{N} M_X\left[R_{Zi}\right] - \sum\limits_{i=1}^{N} M_Z\left[R_{Yi}\right]; \\ I_Y \dfrac{dy}{dt} + \omega_Z\omega_X\left(I_X - I_Z\right) = M_Y(F) - \sum\limits_{i=1}^{N} M_Y\left[R_{Xi}\right] + \sum\limits_{i=1}^{N} M_Y\left[R_{Zi}\right]; \\ I_Z \dfrac{d\omega_Z}{dt} + \omega_X\omega_X\left(I_Y - I_X\right) = M_Z(F) + \sum\limits_{i=1}^{N} M_Z\left[R_{Yi}\right] - \sum\limits_{i=1}^{N} M_Z\left[R_{Xi}\right] + \sum\limits_{i=1}^{N} M_{nki} \end{cases}$$

where G_X, G_Y, G_Z – projections of gravity vector on the MCS *OXYZ* axis; F_X, F_Y, F_Z - projections of external force vector on the axis MCS *OXYZ*; R_{Xi}, R_{Yi}, R_{Zi}– projections of the interaction force between the wheel and the reference surface on the MCS *OXYZ* axis; $M_X(F)$, $M_Y(F)$, $M_Z(F)$ – projections of torque from the force of external influence on the MCS *OXYZ* axis; $M_X\left[R_{Yi}\right]$, $M_Y\left[R_{Xi}\right]$, $M_Z\left[R_{Yi}\right]$, $M_Z\left[R_{Xi}\right]$– projections of moments on the MCS axis from projections of the interaction forces between the ground and the wheel on the MCS axis; M_{nki} – projections of the moment of resistance to the rotation of the i-th wheel on the axis *Z* MCS; I_X, I_Y, I_Z– moments of inertia of the combine relative to the axes MCS *OXYZ*; m_κ– weight of the combine; N – number of the combine wheels.

The position of the self-propelled combine in space at any time is determined by the relative position of the semi-linked and mobile coordinate systems, which are characterized by three angular coordinates - Euler-Krylov's angles: Θ yaw angle; φ "trim" angle; ψ"roll" angle. The connection of Euler-Krylov's angles with other kinematic parameters of rotational motion - projections of angular velocity on the connected axes - is established on the basis of kinematic relations which are called equations of rotational motion connection:

$$\begin{cases} \dfrac{d\psi}{dt} = \omega_X \cos\varphi + \omega_Z \sin\varphi; \\ \dfrac{d\Theta}{dt} = \dfrac{\omega_Z \cos\varphi - \omega_X \sin\varphi}{\cos\psi}; \\ \dfrac{d\varphi}{dt} = \omega_Y - tg\psi\left(\omega_Z \cos\varphi - \omega_X \sin\varphi\right). \end{cases} \quad (1)$$

Since the case of a machine overturn is not considered, i.e. $\psi<\pi/2$, the system (1) does not degenerate.

When examining the dynamics of the combine harvester's movement, it is necessary to take into account the existing moving parts as a complex dynamic system. The axle of driving wheels is the only moving part of the existing assemblies of self-propelled combines, the calculated scheme of angular oscillations of which is presented in Fig. 2.

The differential equation of the angular oscillations of the bridge relative to the longitudinal axis passing through the point *D* in MCS has the form

IOP Conf. Series: Materials Science and Engineering **820** (2020) 012026 doi:10.1088/1757-899X/820/1/012026

$$J_{MX}\frac{d\omega_M}{dt}=\frac{\left(R_{z2}-R_{z4}\right)Bcos\left(\varphi_M\right)}{2},$$

where J_{MX} – moment of inertia of the bridge to the axis X.

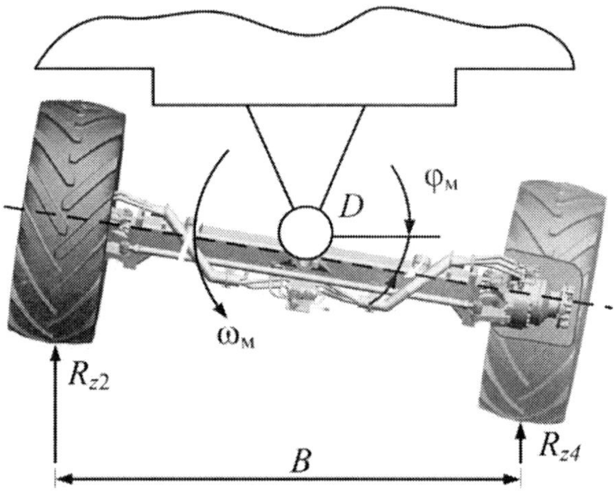

Fig. 2. Calculated diagram of angular oscillations
the rear axle of the self-propelled machine relative to the body:
ω_M,φ_M - angular speed and rotation angle of the axle relative to the longitudinal axis passing through point D; B - wheel track

The specifics of interaction of the wheel with the support base are presented in accordance with the known and approved methods [9-15] as a rolling elastic wheel on the roughness of the non-deformable base. In this case, in the mathematical model, the speed of the combine is set not by changing the coordinates of the center of mass of the body, but by modeling the process of interaction of driving wheels with the support base. This makes it possible not only to more adequately represent the motion of MSC along the unevenness, but also to simulate machine movement, acceleration, braking, overcoming obstacles, slipping and slipping processes, taking into account the characteristics of the tire and ground traction properties.

a)

b)

c)

Fig. 3. RSM 2650 forage harvester (a), its schematic (b) and 3D animated model in «Simulink»

The developed model was implemented in the «Simulink» of the MATLAB software package, which allowed to evaluate its performance on the example of the RSM 2650 forage harvester (Fig. 3 a). As the initial data we used mass and dimensional characteristics of the main elements of the combine in the most common and commercially available modification. The performance characteristics of tyres are based on data provided by the tyre manufacturer. Fig. 3 shows the generated schematic animation of modeling results in "Simulink" (Fig. 3 b), as well as 3D-animation obtained by superimposing on the model of the combine's appearance and visualizing the dynamics of its movement (Fig. 3 c).

The experimental researches and modeling of the dynamics of the combine's movement RSM2650 showed the spectrograms of the operating forces (reactions) in the vertical plane, transmitted from the wheels to the beam of the driving axle (T_Z) (Fig. 4), as well as the acceleration spectra on the driving axle (G_Z) (Fig. 5) when the combine moves along the unpaved road at a speed of 8 km / h.

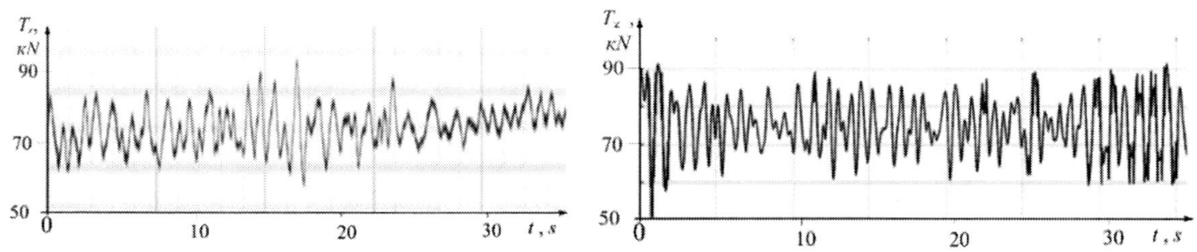

a) b)

Fig. 4. Oscillograms of force on the beam of driving wheels in the vertical plane, obtained experimentally (a) and by results of simulation modeling (b)

The data obtained indicate sufficient convergence of results. Thus, the average values of loads differ from the experiment by 4÷7%, and peak values by 8÷17%. The differences are caused, first of all, by the discrepancy between the numerical description of the bearing surface and the road on which the full-scale tests were carried out. The accuracy of the calculation was also affected by the presentation of the combine model as an absolutely rigid body, although in reality the load-bearing structure has a certain pliability, which is difficult to take into account at this stage of research.

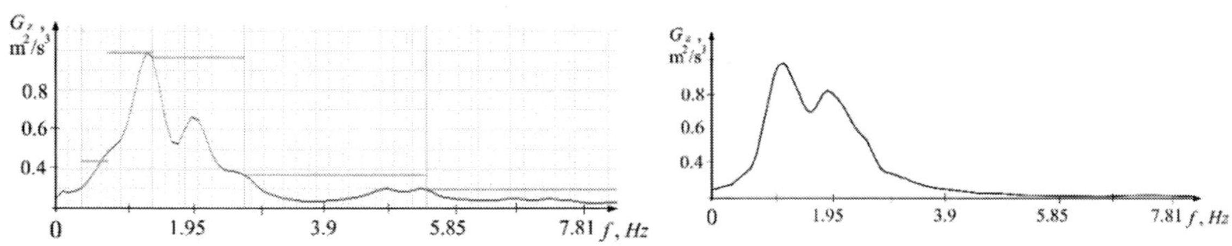

a) b)

Fig. 5. Acceleration spectral density on the axle of driving wheels obtained experimentally (a) and by results of simulation modeling (b)

The high convergence is also noted in the diagrams reflecting the spectral density of signals on the axle of the front wheels (Fig. 5), where two modes of vibration are distinguished, which, as shown by previous studies [19, 20], reflect the vertical and longitudinal angular vibrations of the combine on pneumatic tires. The graphs show some shift in the maximum signal frequency value: the curve obtained as a result of modeling has a maximum at a frequency of 1.1 Hz, and according to the results of experimental studies - 1.2 Hz. The difference can be explained by the difference between the parameters of the elastic-dissipative properties of the tires in the model and the actual values.

Conclusion

1. The developed mathematical model based on fundamental laws of dynamics possesses sufficient accuracy for research of dynamics of movement of grain and forage harvesters and can be used for the purpose of estimation of level of operating loads in various parts and elements of harvesters that allows to apply it for researches, designing of new machines or modernization of existing ones.

2. It is impossible to reduce significantly the loads from kinematic disturbances for the adopted design and layout of combines because of the lack of suspension systems and rigid links between the main elements, and therefore it is necessary to consider non-traditional methods of stabilization for self-propelled machines, for example, due to dynamic damping of vibrations in the adapter suspension.

References

[1] Olt J., Küüt K., Ilves R., Küüt A. (2019): Assessment of the harvesting costs of different combine harvester fleets. Res. Agr. Eng., 65: p. 25-32.

[2] Miroslav M., Miroslav K., Frantisek K. (2017): Risk analysis of the business profitability in agricultural companies using combine harvesters. Res. Agr. Eng., 63: p. 99-105.

[3] Królczyk, Jolanta &Legutko, Stanislaw &Krolczyk, Grzegorz. (2014). Dynamic Balancing of the Threshing Drum in Combine Harvesters – The Process, Sources of Imbalance and Negative Impact of Mechanical Vibrations. Applied Mechanics and Materials. 693. p. 424 - 429.

[4] Špokas L., Adamčuk V., Bulgakov V., Nozdrovický L. The experimental research of combine harvesters. Res. Agr. Eng., 62: (2016): p. 106-112.

[5] Okayasu Takashi, Ken Mori, Yuko Ueka, Masami Matsui, Muneshi Mitsuoka. Eiji Inoue Turbulent Flow Characteristics of the Cleaning Wind in Combine Harvester. Engineering in Agriculture, Environment and Food, 2012. p. 102-106.

[6] Belousov, B., Ksenevich, T.I., Vantsevich, V., Komissarov, D. 8×8 platform for studing terrain mobility and traction performance of unmanned articulated ground vehicles with steered wheels (2013) SAE Technical Papers, 9.

[7] Benes L., Hermanek P., Novak P. Tensile resistance of wheeled combine harvester. Engineering for Rural Development // MM Science Journal, Volume. - 2018, pp. 2481-2483; doi:10.17973/mmsj.2018_10_201848.

[8] Shaheb M.R., Grift T., Godwin R., Dickin E., White D., Misiewicz P. Effect of tire inflation pressure on soil properties and yield in a corn - soybean rotation for three tillage systems in the Midwestern United States. -2018. doi: 10.13031/aim.201801834.

[9] Gorelov, V.A., Komissarov, A.I., Miroshnichenko, A.V. 8×8 wheeled vehicle modeling in a multibody dynamics simulation software (2015) Procedia Engineering, 129, pp. 300-307.

[10] Hassani H.S., Jafari A., Mohtasebi S.S., Setayesh A.M. Investigation on Grain Losses of the JD 1165 Combine Harvester Equipped with Variable Pulley and Belt for Forward Travel. American Journal of Food Technology, -2011, 6: p. 314-321.
doi: 10.3923/ajft.2011.314.321.

[11] Xu L., Chai X., Gao Z., Li Y., Wang Y. Experimental study on driver seat vibration characteristics of crawler-type combine harvester. International Journal of Agricultural and Biological Engineering. (2019). 12. DOI: 90-97. 10.25165/j.ijabe.20191202.3657.

[12] Keller, A.V., Gorelov, V.A., Anchukov, V.V. Modeling truck driveline dynamic loads at differential locking unit engagement (2015) Procedia Engineering, 129, pp. 280-287.

[13] Keller, A.V., Gorelov, V.A., Vdovin, D.S., Taranenko, P.A., Anchukov, V.V. Mathematical model of all-terrain truck (2015) Proceedings of the ECCOMAS Thematic Conference on Multibody Dynamics 2015, Multibody Dynamics 2015, pp. 1285-1296.

[14] Gorelov, V.A., Komissarov, A.I. Mathematical model of the straight-line rolling tire - Rigid terrain irregularities interaction (2016) Procedia Engineering, 150, pp. 1322-1328.

[15] Vol'skaya, N.S., Zhileykin, M.M., Zakharov, A.Y. Mathematical model of rolling an elastic wheel over deformable support base (2018) IOP Conference Series: Materials Science and Engineering, 315 (1), article № 012028.

IOP Conf. Series: Materials Science and Engineering **820** (2020) 012026 doi:10.1088/1757-899X/820/1/012026

[16] Wong, J.Y. Theory of Ground Vehicles / J.Y. Wong. – New York: Wiley IEEE, 2001. – 560 p.
[17] Turakhia T., Modi M. Mathematical Modeling and Simulation of a Simple Quarter Car Vibration Model // International Journal for Scientific Research & Development. (2016). DOI:3. 2321-613.
[18] Kupreyanov, A.A., Morozov, M.V., Belousov, B.N., Ksenevich, T.I., Vantsevich, V.V. Experimental research of tire elastomer-surface tribological properties(2014) Proceedings of the ASME Design Engineering Technical Conference, 3.
[19] Sirotin P.V., Sapegin A.G., Zlenko S.V. Experimental studies of ride quality of self-propelled combine harvester // XIV International Scientific-Technical Conference "Dynamic of Technical Systems" (DTS-2018). September 12-14, 2018. - Rostov-on-Don, Russian Federation,: MATEC Web of Conferences Volume 226 (2018)., 2018.
[20] Sirotin P.V., LebedinskyI.Yu., Kravchenko V.V. Analysis of vibroacoustic loading of the workplace of combine harvester operators // Modern high technologies. Regional application. 2018. №1 (53). p. 113-121.

IOP Conf. Series: Materials Science and Engineering **820** (2020) 012027 doi:10.1088/1757-899X/820/1/012027

Improvement of the active safety of the vehicle for driving on the irregularities by the method of control of vertical reactions on wheels

V Ryazantsev[1,3], I Balabin[1] and A Balsam[2]

[1] Bauman Moscow State Technical University
[2]Higher Institute for Applied Sciences and Technology (HIAST)
Syria DamasP.O. BOX 31983

[3]E-mail: ryazantsev1@yandex.ru

Abstract. Countries with large territories have numerous roads of relatively nobody's quality, characterized by the presence of large irregularities. There are improved roads, but they are badly damaged by inappropriate exploitation or by geophysical and climatic hazards. Such roads are characterized by unevenness, periodic or randomness. The speedy movement of medical transport, rescue vehicles and special services can be difficult because of the possible loss of vehicle stability, which leads to accidents. Often, the loss of stability in such situations is associated with the appearance of resonance vibrations of the vehicle's wheels, which leads to a deterioration of the wheel to the road, reducing the transverse stability of the axle and thus allowing it to slide transversely. There are many active safety systems, such as ABS, directional stability system and others, which lose their effectiveness in these circumstances. This inconvenience can be avoided by applying the stabilization system for vertical reactions to the vehicle's wheels (SVRDK).

Introduction

The control of the suspension has been one of the priority tasks in the field of car construction for a long time. The widely used uncontrolled suspension vehicle does not provide sufficient running quality in the required driving conditions. This defect has been corrected by using a controlled suspension. Another quality that can be related to the vehicle's body vibrations is the stability of the vehicle. Many works are devoted to the study of the vehicle's stability and the active safety systems. The problem of vehicle stability can be reflected in different aspects. Thus, for special vehicle designers, the question of the stability of the vehicle in the event of a tire breakage can be asked [1]. Traffic safety, including driving stability, is ensured by the numerous active safety systems. System analysis of the active vehicle safety problem is presented in the article [2]. The active suspension is used among the active safety features in cars. The active suspension systems can minimize or completely reduce the vehicle's roll to ensure dynamic ride while improving safety and comfort [3]. The characteristics of modern active suspension designs allow their adjustment directly when the vehicle is in motion, depending on the road conditions, both in automatic and manual mode [4]. In particular, adaptive shock absorbers, elastic elements with non-linear characteristics and improved running quality have been developed. Often the road surface is seriously damaged for many reasons, such as natural disasters, landslides, landslides, improper operation of the road, etc. [5]. The traffic with high speeds of emergency vehicles is difficult. Traffic with high speeds of emergency vehicles is difficult. This problem requires the use of active safety systems that can help to accelerate the progress of special service vehicles. This paper [6] discusses the strategy for using a particular turning trajectory, the strategy for emergency braking in order to ensure that the vehicle moves sustainably. The article [7] discusses the issue of transporting seriously sick patients, which is desirable to perform at higher speeds. The problem of active suspension control for a vehicle experiencing impact perturbations is appropriately formulated as a mixed control problem FTS/H∞ [8]. This paper [9] studies the control for the model of an active vehicle suspension system with input time delay in the presence of an external disturbance. The control under consideration allows trade-offs between energy consumption, processing performance, and smooth running. The estimation of the tire's slip angle to control the vehicle's stability is presented in [10]. In work [11], the magnetorheological damper of the rear semi-active suspension is controlled on the basis of sensor signals located in the front of the vehicle. The paper [12] presents the method of adaptive control of the active suspension of the vehicle. This method allows the active suspension of the vehicle to be automatically adjusted so that external disturbances do not affect the dynamic characteristics of the system. The article [13] is devoted to the design of the antilock braking system (ABS) regulator to adjust the slip of the wheel to its preferred value by controlling the slip mode of the wheel. ABS should be resistant to external influences, such as changes in friction between the tire and the road caused by changes in road conditions, loads, etc. The paper [14] considers methods of converting statistical data of road microprofile. The interaction of wheel elements with the ground at curvilinear motion is investigated in the paper [15]. The operation of the antilock brake control system, the structure of the system is shown in the work [16]. The mathematical model of a straight-line rolling tire, which allows studying its interaction with the roughness of the relief surface, is presented in [17]. The paper [18] presents the research within the framework of synthesis of the adaptive suspension system used for damping vibrations in multiaxle wheeled vehicles. A review of existing methods of friction assessment with the development of a new method, which is used to develop an effective ABS control system, is presented in [19].The main attention in the article [20] is given to the control of braking torque of electric vehicles. The recommendations concerning the topology of hierarchical system with the distribution of torque between electric and hydraulic brakes and accurate accounting of hybrid energy storage are given.

When the vehicle is moving on uneven terrain as a result of vertical wheel oscillations, especially in wheel resonance and close to it frequencies, at certain moments the vertical reactions of the road to the wheels can be quite small. Sometimes, when the wheel is taken off the road, these values can become zero. In this case, as practice and calculations show, the lateral stability of the vehicle during driving is dramatically reduced.The stability of the vehicle can be improved by using a system to stabilize the vertical road to wheel response. Such a system can be considered as a new active safety system that increases the stability of the vehicle on an uneven road.

Theoretical Background

Figure 1 schematically shows the system for stabilizing the vertical wheel response. The dynamic model of the stabilization system under consideration includes the suspended mass simulating the car body, unsprung mass simulating the mass of the car wheel with moving parts of the suspension, bearing unit and brake mechanism. In the suspension structure, the actuator (actuator) is switched on in parallel with the elastic element and shock absorber. As an actuator, almost any linear motor can be used: hydraulic or electromagnetic. Thepresentedmodelisdescribedbytheequations:

$$m_2 \cdot \ddot{z_2} = c_c \cdot (z_1 - z_2 + A_n) + k_a \cdot (\dot{z_1} - \dot{z_2}) + c_c \cdot (z_1 - z_d - r_d) - m_2 g$$

$$m_1 \cdot \ddot{z_1} = -c_c \cdot (z_1 - z_2 + A_n) - k_a \cdot (\dot{z_1} - \dot{z_2}) - c_c \cdot (z_1 - z_d - r_d) - m_1 g +$$
$$+c_t \cdot (z_d - z_1 + A_t) + k_t \cdot (\dot{z_d} - \dot{z_1})$$

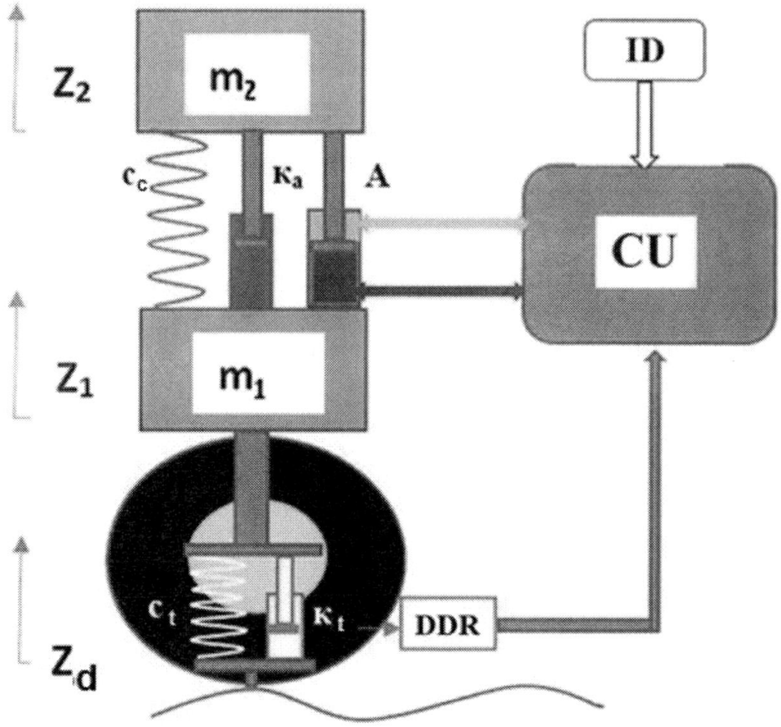

Figure 1. Control diagram of the vertical reaction of the road to the car's wheel when driving on a periodic profile.

It's right here: m_2-sprung mass; m_1-unsprung mass; c_c -rigidity of elastic element; k_a -rigidity of shock absorber; c -transfer coefficient, which determines the value of force created in the actor; c_t - the tire radial stiffness; k_t -coefficient of inelastic resistance in the tire; z_2, z_1-vertical coordinates of the suspended and unsprung masses; z_d -vertical coordinate of the road profile; $A_{n,}$, A_a -constant coefficients; A-actuator (e.g., hydraulic cylinder);DDR-sensor of dynamic radius; ID-source of pressure.

Analysis of the vertical response of the road to the wheel

The vibrational properties of the system, in particular, the resonance zone of the wheel, can be seen on the diagram of changes in the vertical response from the road in the function of time when simulating the movement of the car on the sinusoidal profile with a variable frequency of exposure of the road to the wheel (Figure 2).

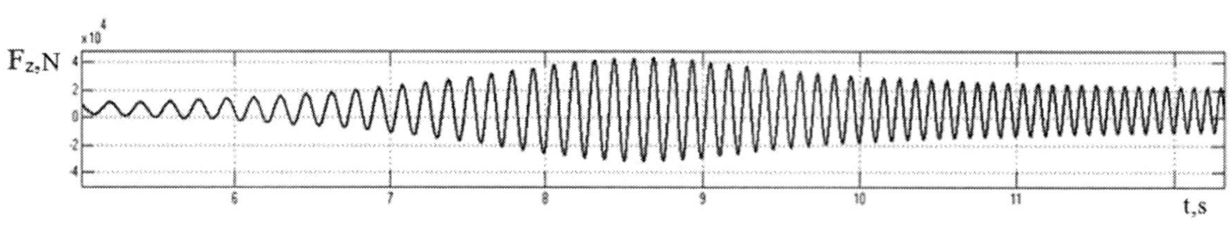

Figure 2. The vertical response of the road to the wheel in the passive linear suspension when the vehicle is moving along a periodic profile at an increasing speed.

The analysis of this graph shows that the natural frequency of the unsprung mass imitating the mass of the wheeled-steam node is approximately 8 hertz. In the real movement of the car in the resonance mode, the wheels may be detached from the road. In this case, the mathematical description of the relationship between the wheel and the road becomes non-linear.For the decision of a nonlinear problem of such plan the method of the imitation modeling realised in package Matlab Simulink is applied.

When driving with uncontrolled suspension in the wheel resonance mode, the character of the change in the value of the vertical reaction of the wheel for the given parameters of the system of suspension of the car and the road profile is represented by the diagram in Figure 3a. In these conditions of movement the value of vertical reaction of the road to one wheel varies from 0 to 15000 N. The peculiarity of the graph in Figure 3a is the presence of zones, in which the vertical reaction of the road is equal to zero. Such a change in the vertical response of the road increases the likelihood of loss of stability of the car. Figure 3b shows the same process with the vertical response stabilization system (SVRDK).

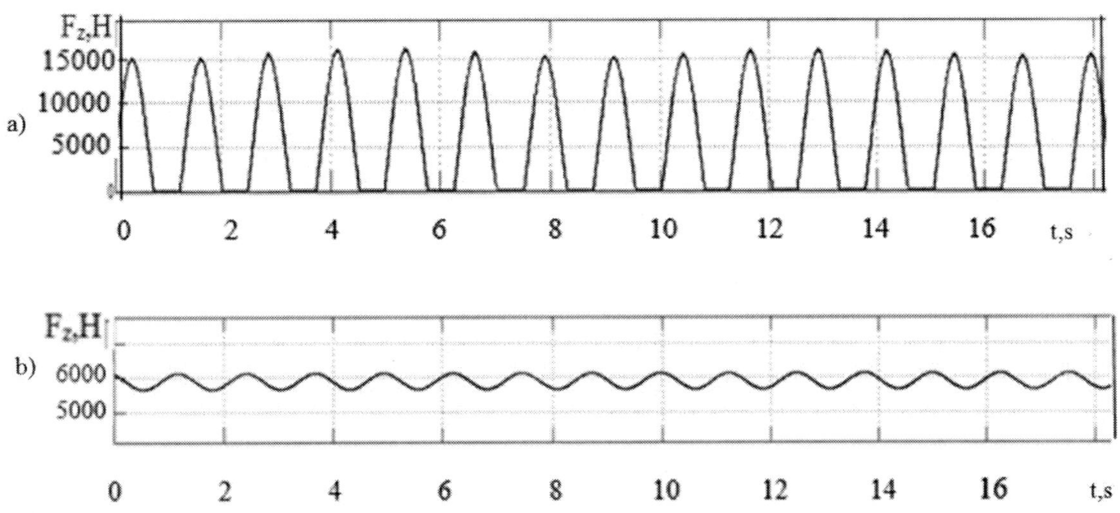

Figure 3. The vertical response of the road to the car's wheel when driving in resonance mode on a sinusoidal profile with an amplitude of 0.02 m: a) without suspension control, b) with suspension control.

The use of this control system reduces the vibration amplitude of the vertical road response as shown in Figure 3b. The vertical reactions in the system of suspension with the control of these reactions sharply decreases at the influencing of the road on the resonance frequency of unsprung masses of the passive system of suspension amplitude of, in the given case, approximately in 30 times.

The calculations show that the applied control requires a power consumption of less than 2 kW per wheel for the system under consideration. Its maximum values do not exceed 3.5 kW. Maximum values of forces created by it, acting on the wheel and the body of the car, are 4500 N. The result is the stabilization of the vertical reaction of the road to the wheel with a deviation of its value from the static of less than 10%.

Analysis of the system of stabilization of vertical road reactions

The system of stabilization of vertical road reactions under consideration increases the effectiveness of a number of active safety systems when driving on uneven terrain. Let's consider the influence of the system of stabilization of vertical reactions of the road on the efficiency of such active safety system as the system of control of the car wheels convergence (SARS). Figure 4 shows the results of calculating the traffic on a circular car on a road with a profile amplitude of 0.01 m using various combinations of SVRDK and SARS. The presented trajectories show the effectiveness of the SVRDK in the vehicle's circular motion with a periodic profile that causes oscillations at the resonance frequency of the vehicle's wheels. The efficiency of SARS in combination with SVRDK is also shown.

Similarly, the system for stabilizing vertical reactions has a positive effect on the efficiency of the anti-lock braking system. This can be judged by the results of calculations given in Table 1. The table shows the values of the braking distance of the vehicle *D* (*m*) in the function of amplitude *A* road profile and application of the SVRDK.

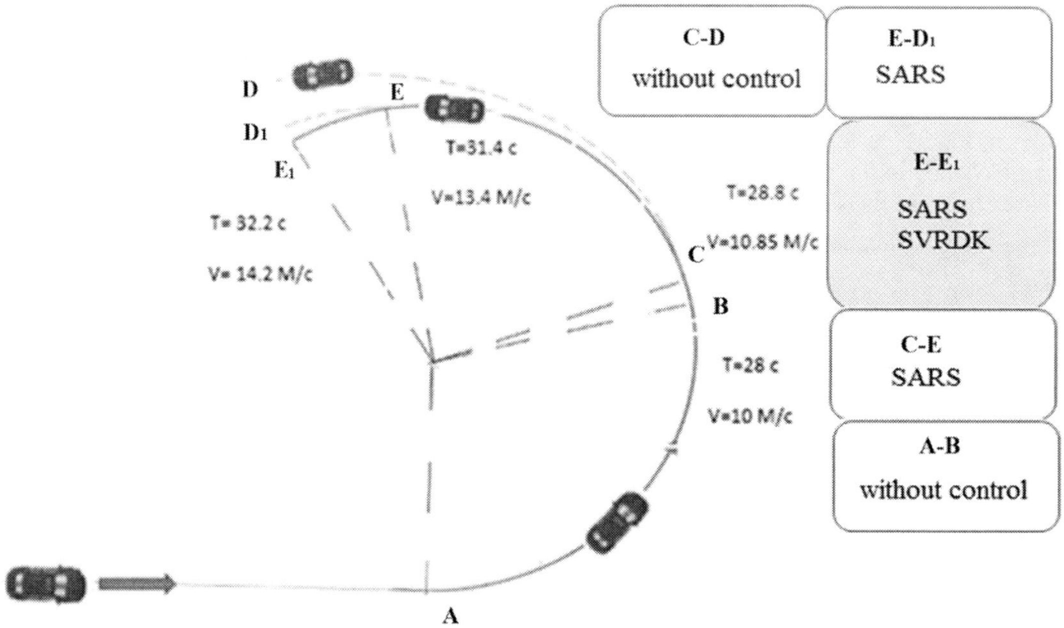

Figure 4. The trajectory sections of the car's roll-on trajectory along the sinusoidal profile with an amplitude of 0.01 m with different combinations of vertical response stabilization system (SVRDK) and car wheel convergence control system (SARS). The trajectory sections and the corresponding switched on control systems are shown within the limits.

Similarly, the system for stabilizing vertical reactions has a positive effect on the efficiency of the anti-lock braking system. This can be judged by the results of calculations given in Table 1. The table shows the values of the braking distance of the vehicle *D* (*m*) in the function of the amplitude *A* of the road profile and the application of the SVRDK.

Table 1. The braking distance of the vehicle *D* (*m*).

Amplitude A of the road profile *A (cm)*	0	1	2	3
Braking distance of the vehicle *D* (*m*) without SVRDK	26.2	27.3	34	45
Braking distance of the vehicle *D* (*m*) with SVRDK		26.2	26.2	26.2
Increasing value *D* (*m*), %		4.5	24	42

The simulation also confirms that SVRDK under consideration stabilizes the vertical response even when driving on a random road profile. This fact extends the range of modes in which it is reasonable to use the system.

As a result of the executed researches it is received that the considered system solves problems of stabilization of vertical reaction of road on a wheel at movement of the car on roughness. In addition, the system under consideration additionally solves the problem of reducing the amplitude of vibrations of the vehicle body at its resonance frequency.

In the process of modeling the vibrations experienced by the vehicle, it has been found that at the frequency of the vehicle body's own vibrations from the action of the sinusoidal profile road with an amplitude of 0.05 m there is an impact on the suspension system of the vehicle. Figure 5 shows the calculation of the vertical position of the vehicle's centre of gravity without control (first 10 seconds) and the following 10 seconds of driving with control of vertical reactions. According to the diagram, the amplitude of vertical center of gravity vibrations of the car during the motion without using the system of vertical reaction stabilization is 0.26 m, and when using this system 0.069 m.

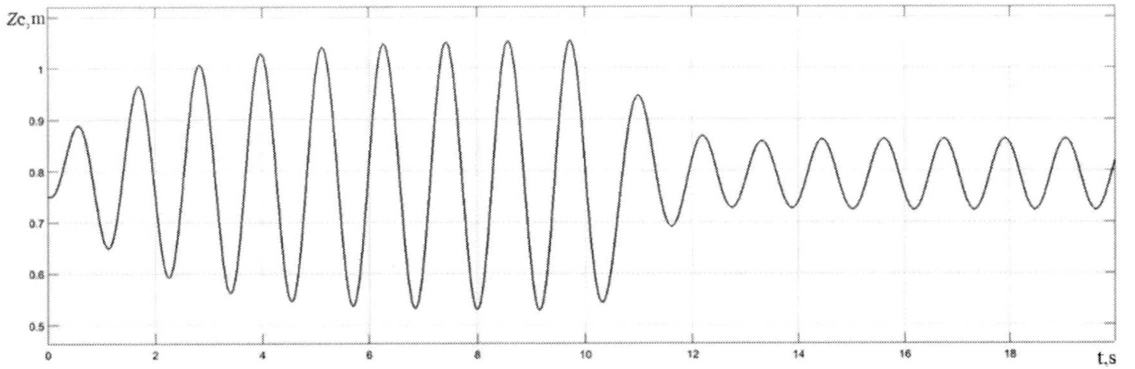

Figure 5. Changes in the vertical position of the vehicle's centre of gravity when the body vibrates at the natural frequency of the vehicle's own vibrations when driving on a sinusoidal profile with an amplitude of 0.05 m without control (0-10 s) and with control of vertical reactions (10-20 s)

Thus, the amplitude of oscillations of the center of gravity of the vehicle body when using the system of stabilization of vertical reactions to the wheels is 26% of the amplitude of oscillations of this point in the absence of control. Similar results were obtained in the calculation of angular, longitudinal and transverse body vibrations.

Conclusion

Our study has shown a positive effect of the work of the system of stabilization of vertical reactions of the road on the transverse stability of the car when driving on the unevenness. It is effective when the car is driven by the side wind and on bends. In the process of vehicle, movement at transition from a mode of movement on a highway in a mode of movement on rough road SVRDK allows to keep efficiency of work of some systems of active safety.

References

[1] H Jing and Z Liu.Gain-scheduling robust control for a tire-blow-out road vehicle. 2019. Proceedings of the Institution of Mechanical Engineers, Part D: Journal of Automobile Engineering 233(2), c. 344–362.

[2] S E Buznikov. System analysis of vehicle active safety problem.2018. IOP Conference Series: Materials Science and Engineering315(1),012005.

[3] D Song , K Schwaiger, M Jaensch andSGoetz.Highly dynamic multiphase wide-bandgap DC-DC converter for automotive active suspension systems. IEEE 3rd International Future Energy Electronics Conference and ECCE Asia, IFEEC - ECCE Asia 2017. 7992094, c. 528–533.

[4] A Dubrovskiy,SAliukov, S Dubrovskiy and A Alyukov.Basic Characteristics of Adaptive Suspensions of Vehicles with New Principle of Operation. 2017. SAE International Journal of Commercial Vehicles. 10(1), c. 193–203.

[5] J R Keaton. Envision rating system applied to vehicle access restoration on a low-volume road in USA. 2016. Proceedings of the Institution of Civil Engineers: Engineering Sustainability 171(2), c. 102–111.

[6] P Treffner, R Barrett and A Petersen.Stability and skill in driving 2002 Human Movement Scie.21(5–6), c. 749–784

[7] M Gebremichael, U Borg,N M Habashi, (...), M McCunnandH N Reynolds.Interhospital transport of the extremely ill patient: The mobile intensive care unit. 2000. Critical Care Medicine 28(1), c. 7.

[8] X Wenping, L Kangji, C Qian and L Guohai. Mixed FTS/H∞ $_{\infty}$ control of vehicle active suspensions with shock road disturbance. 2018. VEHICLE SYST DIN

[9] K Karim Afshar, A Javadi.Constrained H $_{\infty}$ control for a half-car model of an active suspension system with actuator time delay by predictor feedback. 2019. JVC/Journal of Vibration and Control. 25(10), c. 16

[10] E Joa, K Yi and Y Hyun.Estimation of the tire slip angle under various road conditions without tire–road information for vehicle stability control. 2019. Control Engineering Practice 86, c. 129–143

[11] Kamalakkannan. Input based control of full car model (ATV) semi active suspension system-simulation. 2013. SAE Technical Papers4

[12] E Esmailzadeh and H Bateni.Optimal active vehicle suspensions with full state feedback control. SAE Technical Papers

[13] S SMoosapour, S.B FazeliAsl and M Azizi.Adaptive fractional order fast terminal dynamic sliding mode controller design for antilock braking system (ABS). 2019. International Journal of Dynamics and Control 7(1), c. 368–378

[14] E Sarach, G Kotiev and S Beketov. Methods for road microprofile statistical data transformation (2018) MATEC Web of Conferences, 224, article № 04009.

[15] V E Klubnichkin, E.E Klubnichkin, G O Kotiev, S A Beketov and V S Makarov. Interaction between elements of the track ground contacting area with the soil at curvilinear motion of the timber harvesting machine (2018) IOP Conference Series: Materials Science and Engineering, 386 (1), article № 012016

[16] A AErgin, M B Kolomejtseva and G O Kotiev. Antiblocking control system of the brake drive of automobile wheel(2004) PriboryiSistemyUpravleniya, (9), pp. 11–13

[17] V A Gorelov and A I Komissarov. Mathematical model of the straight-line rolling tire - Rigid terrain irregularities interaction (2016) Procedia Engineering, 150, pp. 1322–1328

[18] M MZhileykin, G O Kotiev and M V Nagatsev. Synthesis of the adaptive continuous system for the multi-axle wheeled vehicle body oscillation damping (2018) IOP Conference Series: Materials Science and Engineering, 315 (1), article № 012031

[19] S Rajendran, S K Spurgeon, G TsampardoukasandRHampson.Estimation of road frictional force and wheel slip for effective antilock braking system (ABS) control. International Journal of Robust and Nonlinear Control . 29(3), c. 736–765

[20] V Vodovozov, E Petlenkov, Z Raud and A Aksjonov.Torque control in blended antilock braking systems of electric vehicles. 2019 Proceedings of the Biennial Baltic Electronics Conference, BEC

IOP Conf. Series: Materials Science and Engineering **820** (2020) 012028 doi:10.1088/1757-899X/820/1/012028

Vehicle vibration safety estimation area

L Zheglov[1] and A Fominykh[1,2]

[1]Bauman Moscow State Technical University

[2]E-mail: fab@bmstu.ru

Annotation. It is known that the problem of mathematical modeling of the car vibration isolation system can be solved in the frequency and time domain. As the primary vehicle vibration isolation system has non-linear elements, the question arises: how does the linearized dynamic system solution in the frequency domain corresponding to the calculations of the accepted parameters in the time domain? The problem is solved at random kinematic perturbation from the road surface. Therefore, when working in the time domain to estimate the adequacy of solutions, it is necessary to make a choice of the method of statistical linearization from the known in practice design of automatic control systems. Four methods of statistical linearization are considered, using which calculations in the frequency domain have been carried out. For the chosen dynamic system with initial and statistically linearized nonlinear elements similar actions in the time domain were carried out. It is shown that the first method of statistical linearization is the most flexible, according to the amplitude-frequency response of the system. Such calculations were carried out for two surfaces corresponding to the cobblestone and subsoil roads at different speeds of the vehicle movement. The results of the research are separate-frequency and integral parameters. The last ones do not give any priority in the choice of the calculation field, under the condition of vehicle movement safety, i.e. there is no tire contact loss with the supporting surface.

Introduction

The solution of the problem of determining the parameters and characteristics of nonlinear dynamic systems uses two approaches to its realization depending on the chosen area of solution. There are two such areas - frequency [1-4] and time domain [5-10]. The choice of the area of solution in the assessment of the quality of the analyzed system of the object is an important task, which is important in terms of the calculation accuracy and the possibility of using its equivalent characteristics for the research and choice of its optimal parameters. It should be noted that in the analysis of dynamic systems both estimation areas of determined values are used. The methods of statistical linearization are used to solve problems in the frequency domain. Simulation of the analyzed system can also be carried out in the time domain [11-17] with the obtained data. The deviation of the calculated parameters may be incorrect when solving problems in the frequency domain. This is proved by the formulation of problems of modernization of methods of statistical linearization [18-24].

This research focuses on a dynamic system that is equivalent to a vehicle vibration isolation system with non-linear characteristics of the primary suspension system and tires. The perturbation is a random process determined by the microprofile of the road surface [25] and the vehicle speed. The problem of estimating the convergence of the obtained results is set when the research is carried out in the frequency and time domain. For this purpose, the efficiency of four methods of statistical linearization was compared with the results of calculations of the initial dynamic system in the time domain. The choice of the method of statistical linearization is carried out in order to estimate the adequacy of the determined spectral characteristics of the system and the parameters of vibration safety in the frequency and time domain. To confirm the conformity of the obtained data, the dynamic system is tested according to the calculations of the analyzed characteristics in the frequency and time domain. The test parameters are the own and the resonant frequencies of this system. The unequivacity of the carried out actions is provided by estimation of calculation accuracy of considered spectral characteristics of the analyzed object.

Mathematical model of non-linear vibration isolation system

To solve this problem, a non-linear dynamic system equivalent to the primary system of suspension, vibration isolation of vehicle components and constructions is considered (Fig. 1). This dynamic system was considered when calculating the required spectral characteristics as a result of modeling in the frequency and time domain.

The generalized coordinates are the center of mass displacements and angular displacements in the longitudinal and transverse planes relatively to the axes passing through the center of mass of systems I, II, III and V, respectively. Each system IV and VI has one degree of freedom - the vertical displacement of their mass centers.

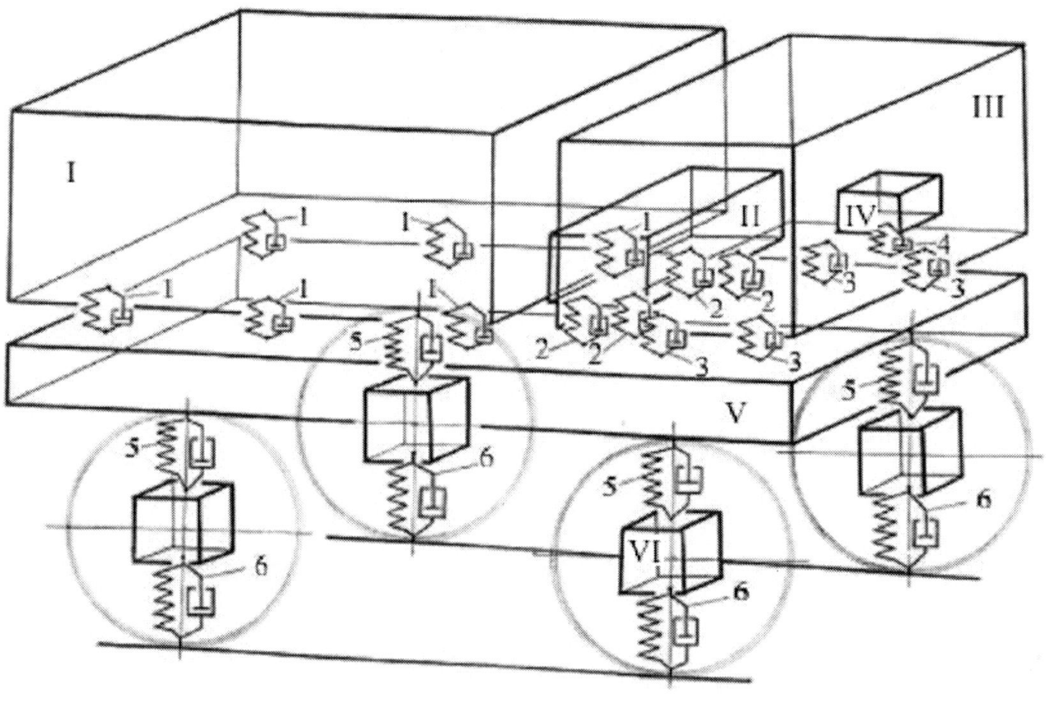

Fig. 1. Dynamic system scheme equivalent to the vehicle's vibration isolation system:
I - vehicle body, II - engine, III - driver's cabin, IV - driver's seat, V - undercarriage, VI - unsprung mass, 1, 2, 3, 4, 5, 6 - elements (elastic and damping) of the vibration isolation system of the body, engine, driver's cab, seat, undercarriage and tire, respectively.

The solution to this problem is to determine the spectral density of the analyzed vibration signal power, for example, vibration acceleration on the driver's seat. In the frequency domain, we solve this problem by transforming the initial linear system of differential equations into a system of algebraic equations using the Laplace transformation. Using the generalized Kramer rule, we determine the vector of the frequency characteristics of the dynamic system using generalized coordinates. This allows for the spectral characteristics of the perturbation to estimate the characteristics of the analyzed vibration signals to determine the parameters of vibration safety and coefficients of statistical linearization. In this case, four known methods of statistical linearization are considered. They allow calculating constant coefficients in three variants and frequency response of a nonlinear element. In the time domain, the considered problem was solved by the numerical method using the recurrent difference equations. The spectral densities of the vibration signals were calculated using the Fourier transform algorithm at the final range.

Analysis of the mathematical modeling results

During the first modeling stage, a dynamic system with seventeen degrees of freedom equivalent to the vehicle's vibration isolation system was tested with a linear problem set. The solution of this problem was carried out on the spectrum of natural frequencies and amplitude-frequency response of the system (resonance frequencies). It showed full matching of the obtained results in frequency and time areas. In addition, the accuracy of the analyzed characteristics was estimated. It is shown that for the spectral density of the process power at linear and nonlinear systems the normalized root-mean-square error is 9% . The amplitude-frequency response of the analyzed system is determined according to the expression

$$H_{qyc}(\omega)=\frac{\left|G_{qyc}(\omega)\right|}{G_q(\omega)},$$

where $G_{qyc}(\omega)$ – cross-spectral density of vibration signals$q(t)$and $y_c(t)$; $q(t)$- input vibration signal - disturbance from road surface microprofile; $y_c(t)$- output vibration signal - vibration acceleration in the driver's seat; $G_q(\omega)$ - spectral density of power perturbation $q(t)$.

In this case, the normalized random error of the frequency response of the system is calculated as follows

$$\varepsilon_H(\omega)=\frac{\left[1-\gamma^2{}_{qyc}(\omega)\right]^{\frac{1}{2}}}{\left|\gamma_{qyc}(\omega)\right|\cdot\sqrt{2n_d}}$$

where $\gamma_{qyc}(\omega)$ - vibration coherence function $q(t)$ и $y_c(t)$.

We obtain that in the considered frequency range, i.e. in the range of intensive vibration output signal 0-50 (60) sec^{-1}, the calculated error of amplitude-frequency response is 8-14%. It should be noted that the initial spectral characteristics for the analysis were determined taking into account the features of the digital spectral analysis. In this case, the function of coherence in the analyzed frequency range changes for the vibration signal output in the range 0,6-1,0 depending on the conditions of the vehicle movement. The calculations are carried out the variants of perturbation from the microprofile of the cobblestone and subsoil roads with the mean-square deviations of the road heights of 20 mm and 30 mm in the speed range of 20-80 km/h.

Therefore, according to the amplitude-frequency response of the system, the choice of the method of statistical linearization was made, with the use of which we have the smallest error of linearized and nonlinear system equivalence. This system analysis was carried out when the vehicle moved along the cobblestone road at a speed of v_a=40 km / h. At this "resonance" speed, the low-frequency resonance areas of the dynamic system and the maximum perturbation power are combined. This results in a significantly non-linear system. According to the calculations (Fig. 2), it is possible to conclude that the first method of statistical linearization is significant.

The system was therefore tested over the full range of vehicle movements in order to avoid errors in analysis and to ensure the validity of the data obtained. These results show that a significant difference in the amplitude-frequency response of the nonlinear and linearized dynamic system is detected only in the mode of "resonance speed" when the vehicle is moving on a dirt road. Thus, there is a problem of calculations accuracy in the frequency domain. For this case, we turn to the parameter which describes the tire's contact with the supporting surface. The possibility of this effect is defined as following

$$P_{hш} = 0,5 \cdot (1 - \Phi(\mu)),\ \Phi(\mu) = \frac{2}{\sqrt{\pi}} \int_0^{\mu} e^{-v^2} dv,\ \mu = \frac{h_ш}{\sqrt{2 \cdot D_{hш}}},$$

where $\Phi(\mu)$ - probability integral, $h_ш$ - static tire deflection, $D_{hш}$ - tire deflection variance.

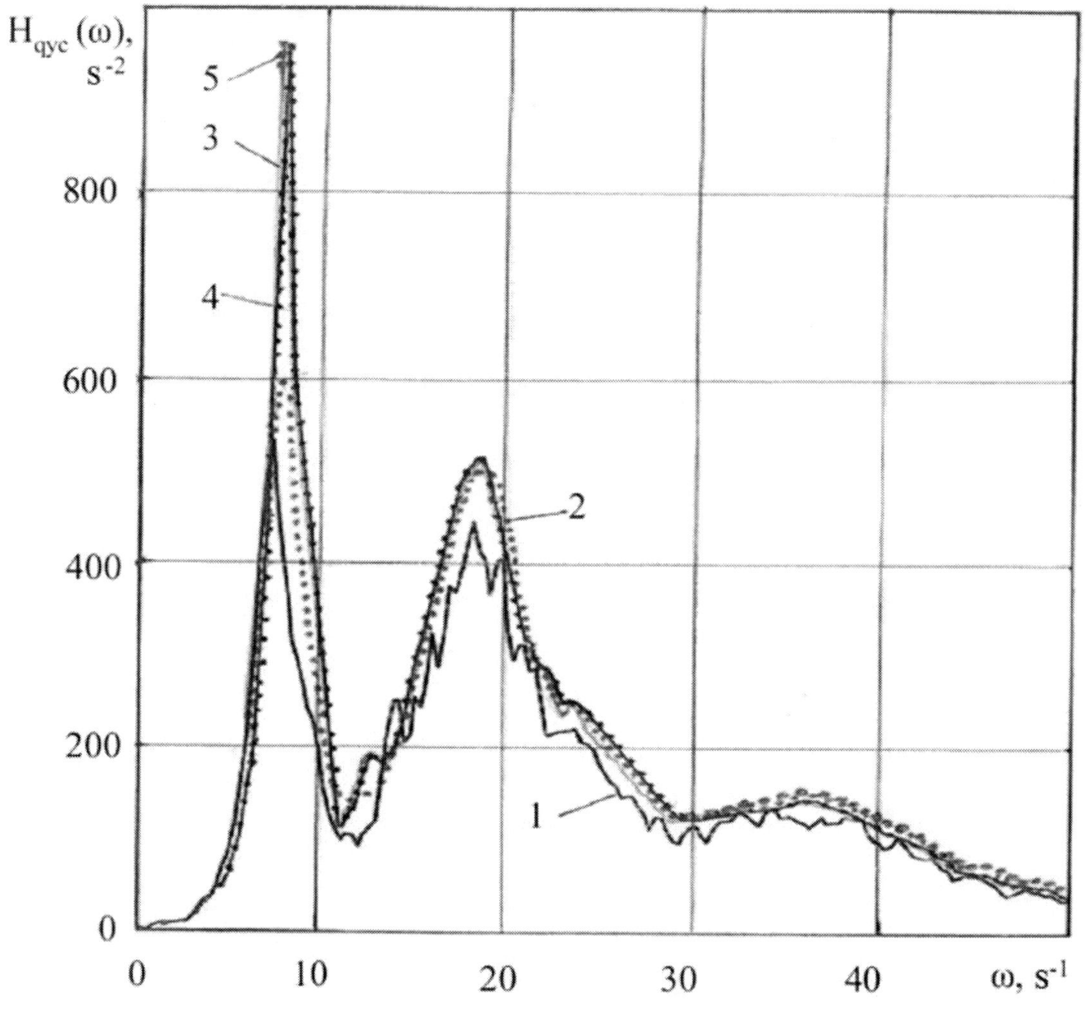

Fig.2. Amplitude-frequency response of the dynamic vehicle vibration isolation system on the driver's seat. Cobblestone road, v_a =40 km/h:
1 - non-linear vibration isolation system; 2, 3, 4, 5 - statistically linearized system of the 1st, 2nd, 3rd, 4th method of statistical linearization

In this case there is a high probability of up to 35% loss of tire contact with the ground surface when the vehicle moves along the dirt road (Fig. 3). Thus, there is a safety problem with the vehicle, and the estimation of vibration safety is incorrect. The following diagrams show that when the vehicle moves along the cobblestone road at a speed of 40 km/h and the probability of contact loss of 20% is rather high, the results of calculations are comparable. Confirmation of the obtained results is provided by separate-frequency and integral estimation of vibration safety parameters (Fig. 4). In this case, as well as in the analysis of amplitude-frequency response, the greatest difference between the estimated parameters occurs when the vehicle moves along the dirt road at a speed of 40 km/h.

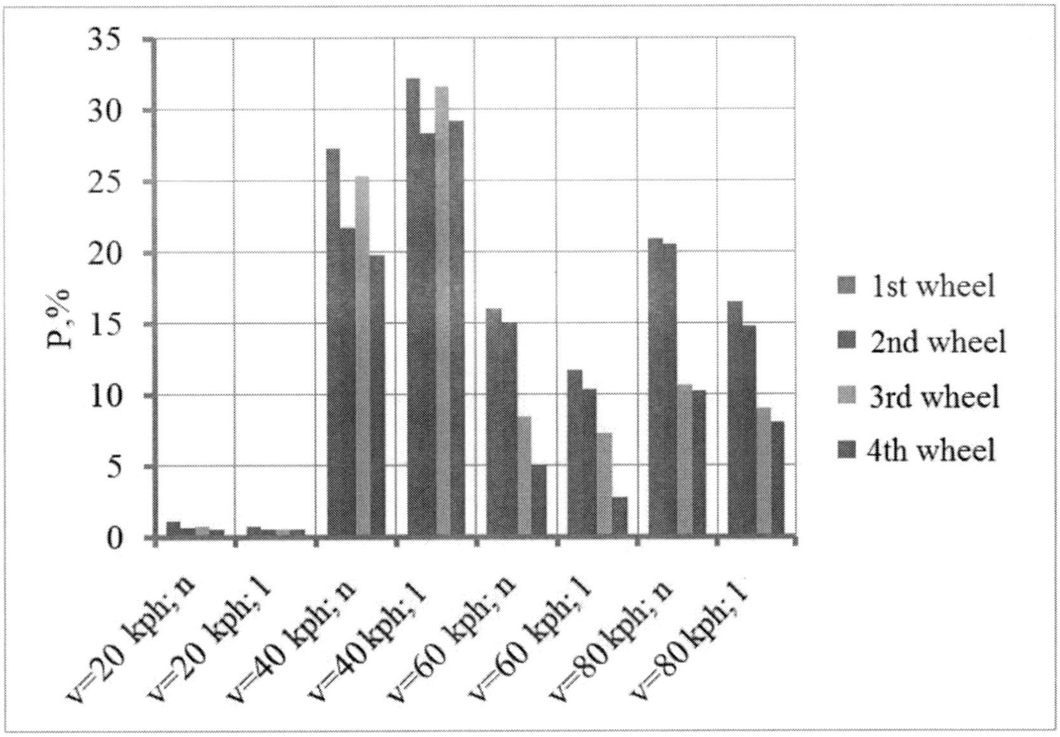

Fig.3. Possibility of loss of contact between the tyre and the ground surface. Dirt road: n, non-linear, l- linearized dynamic system

Conclusions

The results of the calculations of spectral characteristics for the nonlinear system of vehicle vibration isolation in the frequency and time domain, as well as the parameters of vibration safety allow to make the following conclusion:

- created methods of analysis of dynamic systems in the frequency and time domain make it possible to carry out an estimation of the object vibration state;
- the best approximation of the results of calculations of nonlinear and linearized systems is possible using the first method of statistical linearization;
- the deviation from the adaptability of calculations in the frequency and time domain takes place at a significant probability of loss of contact of the tire with the ground surface;
- the research shows that vibration safety can be estimated in the frequency and time domain.

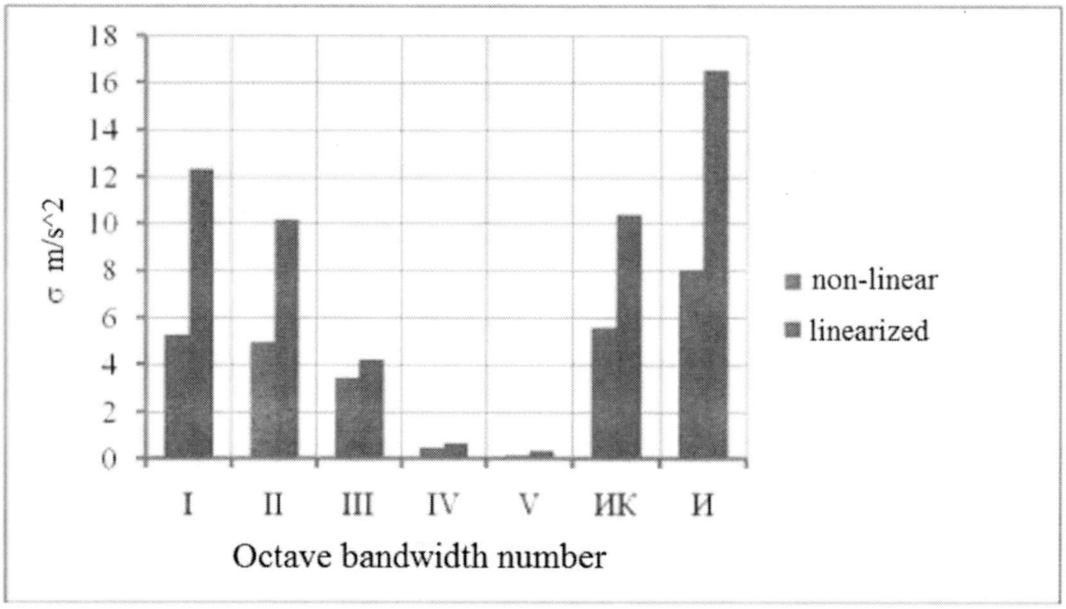

a

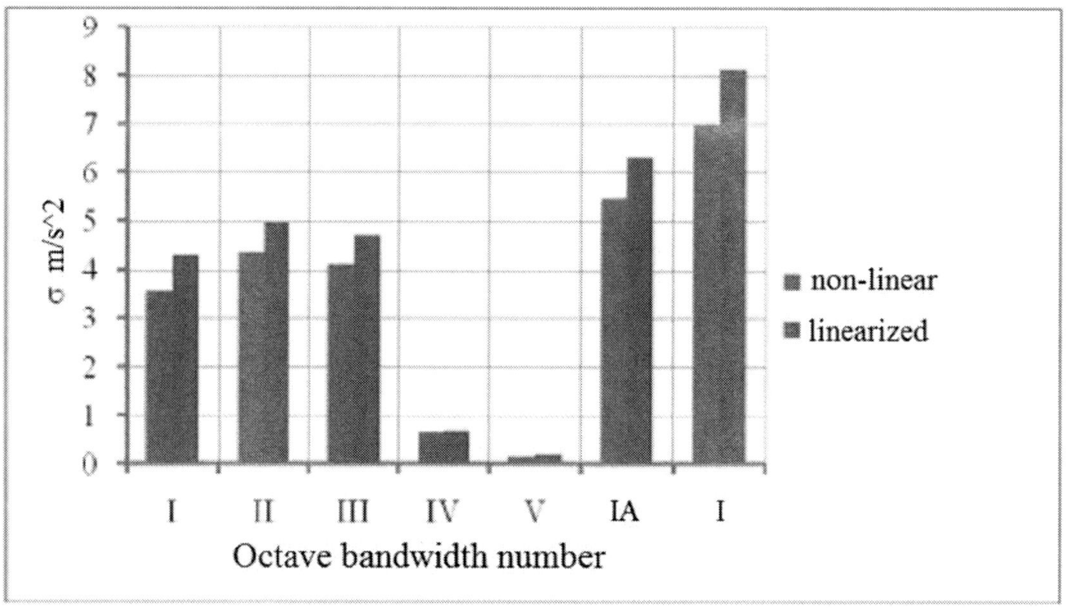

b

Fig. 4. Octave spectra of vibration acceleration in the driver's seat. Dirt road:
a, b, - v_a =40 kph, v_a =60 kph, IA-integrated adjusted estimation; I - integrated estimation

References

[1] Ovchinnikov, I.N. Its own spectrum band - A property of mechanical systems (2018)IOP Conference Series: Materials Science and Engineering468(1),012045

[2] Zhao, N., Huang, G. Efficient Nonstationary Stochastic Response Analysis for Linear and Nonlinear Structures by FFT (2019)Journal of Engineering Mechanics145(5),04019023

[3] Cheng, Z., Li, Z., Kong, F. Statistical linearization of nonlinear stiffness in hydropneumatic suspension (2018)MATEC Web of Conferences153, 04006

[4] Kougioumtzoglou, I.A., Fragkoulis, V.C., Pantelous, A.A., Pirrotta, A.Random vibration of linear and nonlinear structural system (2017)Journal of Soundand Vibration404, c. 84-101

[5] Pashchenko, A.F., Pashchenko, F.F.Application of the method of statistical linearization in problems of identification of nonlinear systems (2012) Proceedings of the 2012 7th IEEE Conference on Industrial Electronics and Applications, ICIEA 2012 6360966, c/1529-1532

[6] Gopala Rao, L.V.V., Narayanan, S. Preview control of random response of a half-car vehicle model traversing rough road (2008) Journal of Sound and Vibration310(1-2), c. 352-365

[7] Gorelov, V.A., Komissarov, A.I., Miroshnichenko, A.V.8×8 wheeled vehicle modeling in a multibody dynamics simulation software(2015) Procedia Engineering, 129, pp. 300-307.

[8] Zhileykin, M.M., Kotiev, G.O., Nagatsev, M.V.Comparative analysis of the operation efficiency of the continuous and relay control systems of amulti-axle wheeled vehicle suspension(2018) IOP Conference Series: Materials Science and Engineering, 315 (1), article № 012030.

[9] Zhileykin, M.M., Kotiev, G.O., Nagatsev, M.V.Synthesis of the adaptive continuous system for the multi-axle wheeled vehicle body oscillationdamping(2018) IOP Conference Series: Materials Science and Engineering, 315 (1), article № 012031.

[10] Novikov, V.V., Pozdeev, A.V., Diakov, A.S.Research and testing complex for analysis of vehicle suspension units(2015) Procedia Engineering, 129, pp. 465-470.

[11] Tlibekov, A.K., Yakhutlov, M.M., Batyrov, U.D., Dosko, S.I. Identification of mechanical systems in the frequency area (2017) Proceedings of the 2017 International Conference"Quality Management, Transport and Information Security, Information Technologies", T and QM and IS 2017 8085812, c. 284-286

[12] Ma, G., Tan, R., Wu, R.Non-linear mathematic model of hydro-pneumatic suspension in crane vehicles and its simulation(2002)JixieGongchengXuebao/Chinese Journalof Mechanical Engineering38(5), c. 95-99

[13] Dong, M.-M., Luo, Z.-X., Liu, B.-G.Optimum damping match for the single wheel vehicle nonlinear suspension (2015) Beijing LigongDaxueXuebao/Transaction of Beijing Institute of Technology35(7), c. 697-700

[14] Zhang, Z., Cao, S., Ruan, C. Statistical Linearization Analysis of a Hydropneumatic Suspension System With Nonlinearity (2018)IEEE Access6, 8554267, c. 73760-73773

[15] Jin, Z., Li, X., Zhu, Y., Qiang, S.Random Vibration Analysisof Nonlinear Vehicle-bridgeDynamic Interactions (2017)TiedaoXuebao/Journal of the China Railway Society39(9),
c. 109-116

[16] Dong, M., Luo, Z. Statistical linearization on 2 DOFs hydropneumatic suspension with asymmetric non-linear stiffness (2015) 12th International Conference on Applications of Statistics and Probability in Civil Engineering, ICASP 2015

[17] Zhang, L., Chen, X. Statistical linearization study on nonlinear random response of vehicle (2004) ZhendongyuChongji/Journal of Vibration and Shock 23(3), c. 33-36

[18] Asada, H.H., Sotiropoulos, F.E. Dual faceted linearization of nonlinear dynamical systems based on physical modeling theory (2019)Journal of Dynamic Systems, Measurement and Control, Transactions of the ASME141(2), 021002
[19] De Domenico, D., Ricciardi, G. Improved stochastic linearization technique for structures with nonlinear viscous dampers (2018) Soil Dynamics and Earthquake Engineering113, c. 415-419
[20] Socha, L.Statistical linearization of stochastic dynamic systems under external excitations (2008)Lecture Notesin Physics730, c. 103-146
[21] Cai, G.Q., Suzuki, Y. On statistical quasi-linearization (2005) International Journal of Non-Linear Mechanics 40(8), c. 1139-1147
[22] Chernyshov, K.R. Statistical linearization based on the maximal correlation (2007) IEEE International Siberian Conference on Control and Communications, SIBCON-2007; Proceedings 4233274, c. 29-36
[23] De Domenico, D., Ricciardi, G.Improved stochastic linearization technique for structures with nonlinear viscous dampers (2018) Soil Dynamics and Earthquake Engineering113, c. 415-419
[24] Atalik, T.S., Utku, S. Stochastic linearization of multi-degree-of-freedom non-linear systems (1976)Earthquake Engineering & Structural Dynamics4(4), c. 411-420
[25] Sarach, E., Kotiev, G., Beketov, S. Methods for road microprofile statistical data transformation (2018) MATEC Web of Conferences, 224, article № 04009

IOP Conf. Series: Materials Science and Engineering **820** (2020) 012029 doi:10.1088/1757-899X/820/1/012029

Investigation of the influence of the dynamic system parameters on its frequency and modal spectrums

M Lahtyukhov[1] and A Fominykh[1,2]

[1]Bauman Moscow State Technical University

[2]E-mail: fab@bmstu.ru

Abstract. The influence of parameters of dynamic system on its frequency and modal spectra is investigated by the example of chain unbranched dynamic system with finite number of degrees of freedom. The general regularities between the parameters of the chain dynamic system and its frequency and modal spectra are determined. It is established that according to the own spectrum of the chain dynamic system the inflated estimation of the interval width of eigenfrequency change at variation of any parameter may be given. The criterion, allowing to determine the structural parameters of the system, which significantly affect the given natural frequencies, is obtained.

Introduction

The efficiency of power transmissions of wheeled and tracked vehicles is largely determined by their dynamic processes and, in particular, resonance effects. Therefore, it is not a wonder that a large number of fundamental research [1-3] and articles [4-22] are focused on the research of dynamic processes in the transmission and reduction of loads in them. One of the easiest ways to reduce the dynamic loads in the transmission is the change of its elastic-inertial parameters, which ensures the removal of dangerous resonant operating modes outside the operating range. There is a practice of solving problems of removing resonant operating modes of transmission outside its operating range. The influence of dynamic system parameters on its natural frequencies and oscillation modes is different. Therefore, the possibility of targeted change of frequency and modal spectra is determined by the choice of changing parameters and their number. It is necessary to determine the parameters that significantly affect the frequency spectrum. They are assigned as variable parameters. It is necessary to determine the parameters that significantly affect the frequency spectrum. They are assigned as variable parameters. Such an approach allows, on the one hand, to increase variation limits of the frequency range, and, on the other hand, to reduce problem dimensions (the number of variable parameters). Thus, the problem solving efficiency is increased.

The influence of dynamic system parameters on its frequency and modal spectra

Investigation of the influence of dynamic system parameters on its frequency and modal spectra was carried out as an example of a 7-mass chain unbranched system. Each of the system's parameters changed over a rather wide range. Calculation of frequency and modal spectra was carried out by the Givens method together with the inverse iteration method. The research results are shown on Fig. 1. The following diagrams describe a well-known feature: the natural frequencies of unbranched chain systems are different.

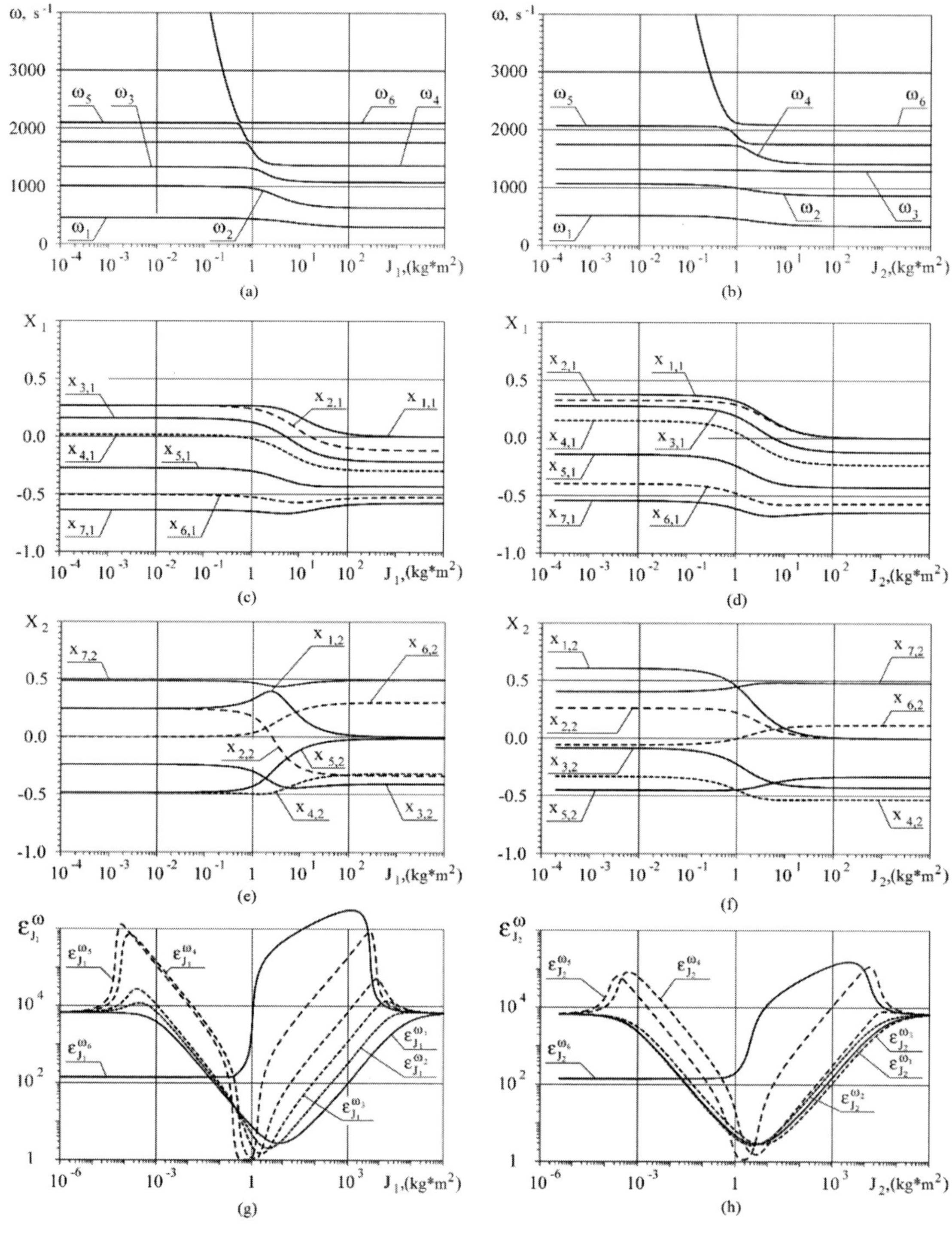

Fig. 1. Frequency spectrum, normalized amplitudes of free oscillations of $x_{j,i}$ j- th mass at i-th natural frequency and sensitivity coefficients $\varepsilon_{J_k}^{\omega_i}$ natural frequency ω_i to the change of parameter J_k, $(k=1,2)$ of the investigated dynamic system at variation of moments of inertia J_1 and J_2

The unlimited increase or decrease of any taken by itself k –th parameter of system p_k (moment of inertia or compliance coefficient) occurs monotone decrease or increase of all natural frequencies of system respectively. For each natural frequency of the system, except the supreme frequency, there is a finite upper and lower limit of its change. The upper limit of change of the i –th natural frequency does not exceed the lower limit of change of the $i+1$ –th natural frequency.

A significant change of the random i –th natural frequency, except supreme one, is carried out at a limited interval of $\left[p_k^{\text{л}\omega_i}, p_k^{\text{п}\omega_i}\right]$ values of the k –th parameter. The width of this interval depends on the order number of the natural frequency, the difference between the upper and lower limits of this frequency change, the position of the variable parameter in the system, the specific numerical values of the system parameters and its structure.

As the natural frequency ordinal number increases (when frequencies are numbered in ascending order of frequency) and decreases the difference between the upper and lower limits of the frequency change, the width of the interval $\left[p_k^{\text{л}\omega_i}, p_k^{\text{п}\omega_i}\right]$, in which the frequency changes significantly decreases.

In cases where the lower limit of change of the natural frequency differs from zero and there is a finite upper limit, the ratio $p_k^{\text{п}\omega_i} / p_k^{\text{л}\omega_i}$ does not exceed 10^2. If the lower limit of change of the first natural frequency is equal to zero, (it takes place in systems with fixed support), the interval $\left[p_k^{\text{л}\omega_i}, p_k^{\text{п}\omega_i}\right]$ is extended. And then $p_k^{\text{п}\omega_i} / p_k^{\text{л}\omega_i} < 10^4$.

The supreme natural frequency does not have a finite upper limit.

For any chain unbranched n mass system without restraint, the following ratios are correct

$$\begin{aligned} 0 = \omega_0^{(O)} &< \lim_{p_k \to \infty} \omega_1 < \omega_1^{(O)} < \lim_{p_k \to 0} \omega_1 \le \lim_{p_k \to \infty} \omega_2 < \omega_2^{(O)} < \lim_{p_k \to 0} \omega_2 \le \ldots \\ &\le \lim_{p_k \to \infty} \omega_{i-1} < \omega_{i-1}^{(O)} < \lim_{p_k \to 0} \omega_{i-1} \le \lim_{p_k \to \infty} \omega_i < \omega_i^{(O)} < \lim_{p_k \to 0} \omega_i \le \ldots \\ &\le \lim_{p_k \to \infty} \omega_{n-1} < \omega_{n-1}^{(O)} < \lim_{p_k \to 0} \omega_{n-1} = +\infty, \end{aligned} \tag{1}$$

where $\omega_0^{(O)}, \omega_1^{(O)}, \ldots, \omega_{n-1}^{(O)}$ – natural frequencies of the system at initial values of its parameters.

Arbitrary i –th natural oscillation mode (when it's normalized $\mathbf{x}_i^T \mathbf{x}_i = 1$, where $\mathbf{x}_i^T$ – transposed vector $\mathbf{x}_i$) change significantly at a limited parameter variation range $\left[p_k^{\text{л}\mathbf{x}_i}, p_k^{\text{п}\mathbf{x}_i}\right]$. The width of the $\left[p_k^{\text{л}\mathbf{x}_i}, p_k^{\text{п}\mathbf{x}_i}\right]$ range is slightly greater than the width of the $\left[p_k^{\text{л}\omega_i}, p_k^{\text{п}\omega_i}\right]$ range for the corresponding natural frequency. And the $\left[p_k^{\text{л}\omega_i}, p_k^{\text{п}\omega_i}\right]$ range is inside the $\left[p_k^{\text{л}\mathbf{x}_i}, p_k^{\text{п}\mathbf{x}_i}\right]$ range.

The dependence of the amplitudes of natural oscillations on the parameter is not always monotonous. In case of unlimited increase or decrease of any parameter p_k the normalized amplitudes of natural oscillations tend to some limiting values.

The properties of the frequency spectrum considered allow express estimation of the influence of any parameter of the system on its natural frequencies.

The method is based on the following idea. When analyzing the influence of any k - th parameter p_k on the eigenst spectrum of the dynamic system, four auxiliary dynamic systems are created. Each system has the same structure and dimension as the original system. All parameters of the auxiliary systems, except for k parameters calculated by the following equations:

$$p_k^{(1)} = 10^{-4} p_k^{(O)}, \quad p_k^{(2)} = 10^4 p_k^{(O)}, \quad p_k^{(3)} = 0.5 p_k^{(O)}, \quad p_k^{(4)} = 2 p_k^{(O)}, \tag{2}$$

where $p_k^{(O)}$ – value of the k - th parameter in the initial system,

$p_k^{(l)}$ – value of the k - th parameter in the l - th auxiliary system,
are equal to the corresponding parameters of the initial system

$$p_m^{(l)} = p_m^{(O)}, \quad m = \overline{1, n_p} (m \neq k), l = \overline{1,4}, \tag{3}$$

where n_p – system parameter number.

Using the natural spectra of the first two auxiliary systems, approximate estimates of the upper and lower limits of change in the natural frequencies of the system can be calculated by changing each k - th parameter p_k on the $(0, +\infty)$ range:

$$\lim_{pk \to 0} \omega_i \approx \omega_i^{(1)},$$

$$\lim_{pk \to +\infty} \omega_i \approx \omega_i^{(2)},$$

where $\omega_i^{(1)}$, $\omega_i^{(2)}$ – i -th natural frequency 1-st and 2-nd auxiliary systems ($i = \overline{1, n-1}$ – for systems without fixed support and $i = \overline{1, n}$ – for systems with fixed support).

1.1. By the difference

$$\Delta\omega_i^{(1-2)} = \left(\omega_i^{(1)} - \omega_i^{(2)}\right) \tag{4}$$

and quotient

$$\varepsilon_{\omega i}^{(1-2)} = \omega_i^{(1)} / \omega_i^{(2)} \tag{5}$$

can be inferred by the possible effect of the k -th parameter on the i -th natural frequency. The larger these values, the more the natural frequency ω_i changes when the parameter p_k varies over a rather wide range $\left[10^{-4} \cdot p_k^{(O)}, 10^4 \cdot p_k^{(O)}\right]$.

The parameter $p_k^{(O)}$ belongs to the range $\left[p_k^{л\omega_i}, p_k^{n\omega_i}\right]$. At this range there is a significant change of frequency ω_i, can be set using the sensitivity coefficient $\varepsilon_{pk}^{\omega_i}$ of natural frequency ω_i to the change of parameter p_k, calculated by the equation

$$\varepsilon_{pk}^{\omega_i} = \frac{\omega_i^{(1)} - \omega_i^{(2)}}{\omega_i^{(3)} - \omega_i^{(4)}}.$$

The analysis of the change in the coefficient $\varepsilon_{pk}^{\omega_i}$ from the parameter p_k (Figures 1 (d), 1 (f)) shows that when the parameter is inside the range $\left[p_k^{л\omega_i}, p_k^{n\omega_i}\right]$ (Fig. 2, (a)), where a change in the frequency ω_i is significant, independent of $\Delta\omega_i^{(1-2)}$ for all frequencies, except the supreme one, the sensitivity coefficient $\varepsilon_{pk}^{\omega_i}$ does not exceed 4–5. When the point $p_k^{(O)}$ on the numerical axis is outside the range $\left[p_k^{л\omega_i}, p_k^{n\omega_i}\right]$, but is close to its boundaries (Fig. 2, (b)), the inequality:

$$\varepsilon_{pk}^{\omega_i} < 10 - 20.$$

If the distance of the point $p_k^{(O)}$ from the range $\left[p_k^{\text{л}\omega_i}, p_k^{\text{п}\omega_i}\right]$ boundaries (Fig. 2, (c)) is significant, the coefficient $\varepsilon_{pk}^{\omega_i}$ increases to $6{,}8\cdot10^3$ and more. Therefore, to find the k - th parameter $p_k^{(O)}$ inside the range $\left[p_k^{\text{л}\omega_i}, p_k^{\text{п}\omega_i}\right]$, where a significant change in the frequency ω_i, occurs, or close to the boundaries of this range, it is necessary to the fulfillment of the condition:

$$\varepsilon_{pk}^{\omega_i} < \sigma = 5-20. \tag{6}$$

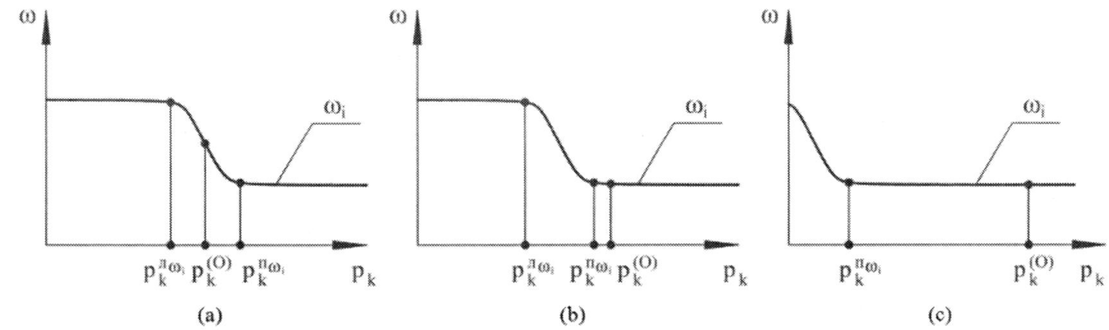

Fig. 2. Possible positions of a point $p_k^{(O)}$ on a numerical axis relative to the range $\left[p_k^{\text{л}\omega_i}, p_k^{\text{п}\omega_i}\right]$

Estimation of sensitivity of natural frequencies and coupled modes of dynamic system oscillations to changes in its elastic-inertial parameters is possible by analyzing the values of the corresponding partial derivatives of the parameter [23-25]. However, if a similar approach is used, the point $p_k^{(O)}$ on the numerical axis is outside the range $\left[p_k^{\text{л}\omega_i}, p_k^{\text{п}\omega_i}\right]$ and in close to its left or right boundary (Fig. 2, (b)). The value of derivative $\dfrac{\partial\omega_i}{\partial p_k}$ at this point will be small. When the problem is solved due to the change of the frequency spectrum, the k - th parameter can be wrongly removed from the list of variable parameters and thus the limits of the change of natural frequencies. As a result, the range of natural frequencies may be narrowed.

Conclusions

1. The general conformations between the parameters of the chain dynamic system and its frequency and modal spectrum are established.
2. It is determined that the eigen spectrum of a chain dynamic system can be used to give an inflated estimate of the range width of change in the natural frequencies when a randomly chosen parameter. p_k (moment of inertia or compliance coefficient) is varied. Thus for any random i -th natural frequency $\omega_i^{(O)}$ of the investigated system the following ratio is correct:

$$\omega_{i-1}^{(O)} < \lim_{pk\to0}\omega_{i-1} \le \lim_{pk\to\infty}\omega_i < \omega_i^{(O)} < \lim_{pk\to0}\omega_i \le \lim_{pk\to\infty}\omega_{i+1} < \omega_{i+1}^{(O)}$$

3. The criterion, allowing to find out the design parameters of the system, which significantly influence on the given natural frequencies, is obtained.

References

[1] Wilkinson J.H. *The Algebraic Eigenvalue Problem*. London, Oxford University Press, 1965.

[2] Wilkinson J.H., Reisch C. *Handbook for Automatic Computation. Linear Algebra.* Heidelberg New York Springer-Verlag Berlin, 1974.

[3] Belousov B.N. and Popov S.D., *Heavy-duty wheeled vehicles: Design, Theory, Calculations* SAE International ISBN 978-0-7680-7723-0, 2014, 800 p.

[4] Moler C.B. and Stewart G.W. An *Algoritm for Generalized Matrix Eigenvalue Problems*, SIAM J. Nume. Anal. – 1973. – №10. – P. 241-256.

[5] Nour-Omid B., Parlett B.N., Taylor R.L. *Lanczos versus subspace iteration for solution of eigenvalue problem* // Int. J. Numer. Meth. Eng. – 1983. – 19, № 6. – P. 859 – 871.

[6] Taratorkin, A., Derzhanskii, V., Taratorkin, I. *Decrease in dynamic loading of transmission elements of the vehicle.* 2013, Lecture Notes in Electrical Engineering 198 LNEE (VOL. 10), c. 495-504

[7] Zubov, N.E., Mikrin, E.A., Misrikhanov, M.S., Ryabchenko, V.N. *Output control of the spectrum of a descriptor dynamical system.* 2016. Doklady Mathematics 93(3), c. 259-261.

[8] Zubov, N.E., Mikrin, E.A., Misrikhanov, M.S., Proletarskii, A.V., Ryabchenko, V.N. *Reduction of large-scale dynamical systems by the Krylov subspaces method: Analysis of approaches.* 2015 Journal of Computer and Systems Sciences International 54(2), c. 165-183.

[9] Ovchinnikov, I.N. *Its own spectrum band - A property of mechanical systems.* 2018 IOP Conference Series: Materials Science and Engineering 468(1),012045.

[10] Chernikov, S.A. *Expansion of the suppression band of a vibroprotective system by a feedback dynamic damper.* 2015 Journal of Machinery Manufacture and Reliability 44(5), c. 439-444.

[11] Ovchinnikov, I.N. *Its own spectrum band - A property of mechanical systems.* 2018 IOP Conference Series: Materials Science and Engineering 468(1),012045.

[12] Taratorkin, I., Taratorkin, A., Derzhanskii, V. *Reduction in Dynamicl of the Hydromechanical Transmission on the Wheeled Chassis.* 2015 SAE Technical Papers 2015-September.

[13] Taratorkin, I., Derzhanskii, V., Taratorkin, A. *Oscillation damping in the power unit when starting the engine equipped with the Common Rail system*/ 2018 MATEC Web of Conferences 224, 02037.

[14] Taratorkin, A., Derzhanskii, V., Taratorkin, I. *Stability of the dynamic shifting process in the vehicle transmission with the input shaper.* 2018 MATEC Web of Conferences 211,02007.

[15] Taratorkin, A., Taratorkin, I., Derzhanskii, V., Alexander, V. *Oscillation elimination method in transient processes of controlling the transport vehicle transmission.* 2018 FISITA World Automotive Congress 2018.

[16] Taratorkin, I., Derzhansky, V., Taratorkin, A. *Prediction of the parametrical resonance oscillations of transmission friction discs on the basis of high precision determination of their modal characteristics.* 2016 AIP Conference Proceedings 1785,040086.

[17] Taratorkin, I., Derzhanskii, V., Taratorkin, A. *Reduction in dynamic loading of hydromechanical transmission of multiple wheel chassis.* 2014 FISITA 2014 World Automotive Congress – Proceedings.

[18] Novikov, V.V., Pozdeev, A.V., Diakov, A.S., *Research and testing complex for analysis of vehicle suspension units*, Procedia Engineering, 129, pp. 465-470.

[19] Kotiev G.O., Miroshnichenko A.V., Stadukhin A.A., Kositsyn B.B. *Estimating operation modes for the individual wheel electric drive of the all-wheel drive vehicle with the use of the driving simulator.* Volume 534, Issue 1, 12 June 2019, International Automobile Scientific Forum: Intelligent Transport System Technologies and Components, IASF 2018.

[20] Evseev, K.B., Kartashov, A.B., Dashtiev, I.Z., Pozdeev, A.V. *Analysis viscoelastic properties of fiber-reinforced composite spring for the all-terrain vehicle*, MATEC Web of Conferences 224, 02039 (2018).

[21] Kotiev G.O., Miroshnichenko A.V., Stadukhin A.A., Kositsyn B.B. *Determination of mechanical characteristics of high-speed tracked vehicles traction motor with individual drive wheels.* Journal of Physics: Conference Series. Volume 1177, Issue 1, 10 April 2019,

IOP Conf. Series: Materials Science and Engineering **820** (2020) 012029 doi:10.1088/1757-899X/820/1/012029

№012058, 3rd Science and Technology Seminar on Mobility of Transport and Technological Machines, MTTM 2018; Nizhny Novgorod; Russian Federation; 28 August 2018.

[22] V.V. Novikov, A.V. Pozdeev, A.S. Diakov, *Research and testing complex for analysis of vehicle suspension units*, Procedia Engineering, Vol. 129, pp. 465-470 (2015) DOI: 10.1016/j.proeng.2015.12.153

[23] Carl S.Radissil and Yee-Yeen-Chu. *Numerical Methods for Evaluating the Derivatives of Eigenvalues and Eigenvectors.* //AIAA Journal. – 1975. – v.13, N6. – P. 834 – 837.

[24] Augustaitis V.V. Optimization synthesis method of the oscillating system parameters with the help of the sensitivity coefficients (in Russian) // Machine science. - – 1985. №3. - – C. 3 – 9.

[25] *Silovye peredachi transportnyh mashin: Dinamika i raschet*/ S.V. Alekseeva, V.L. Vejc, F.R. Gekker, E.A. Kochura. L.: Mashinostroenie, Leningr. otd-nie, 1982. 256 s.

IOP Conf. Series: Materials Science and Engineering **820** (2020) 012030 doi:10.1088/1757-899X/820/1/012030

The study of parametric optimization algorithms on example of vehicle bumper crashworthiness

R Goncharov[1,2] and V Zuzov[1]

[1]Bauman Moscow State Technical University

[2]E-mail: goncharov.roman@bmstu.ru

Abstract. Different algorithms of parametric optimization (implemented in the program LS-OPT) to solve high-speed and high-nonlinear problems of impact character on the example of vehicle bumper optimization are considered and compared. Different algorithms of response surface optimization such as linear polynomial model, quadratic polynomial model, Feedforward, Radial Basis Function, Kriging and Support Vector Regression were studied and analyzed; algorithms of choosing point selection scheme were considered. The analysis of optimization results allowed us to determine which of these algorithms are most effective in terms of accuracy and computational time. The problem of impact on the vehicle's bumper is solved in the paper, the optimization is based on the application of metamodel (RBF). The method provided a reduction of its mass by 16% while maintaining the initial parameters of crashworthiness.

1. Introduction

The evolution of the automotive industry has reached such a level that it is impossible to imagine further improvement of structures without using optimization, especially in the field of crashworthiness [1-6].

Parametric optimization methods are divided into direct optimization and metamodel-based optimization. The direct optimization is used to solve single-criteria problems with a relatively small number of variables, the metamodel-based optimization is used to solve complex multi-criteria problems with many variable parameters.

Structural scheme of the optimization algorithm is shown in Figure 1, where you can see that the algorithm of creating a metamodel includes three main stages: the choice point selection scheme, the construction of response surfaces for the criteria and limitations and the search for the optimal solution (minimum target function).

Let's consider and compare different optimization algorithms based on the construction of a metamodel (implemented in the program LS-OPT) to solve high-speed and high-nonlinear problems of crashworthiness on the example of a vehicle bumper.

The purpose of the work is to select the preferred optimization algorithm based on the construction of the response surface (metamodel) in relation to the processes of impact on the supporting structures of vehicle to reduce their mass and meet the requirements of crashworthiness at acceptable values of accuracy and time of solving.

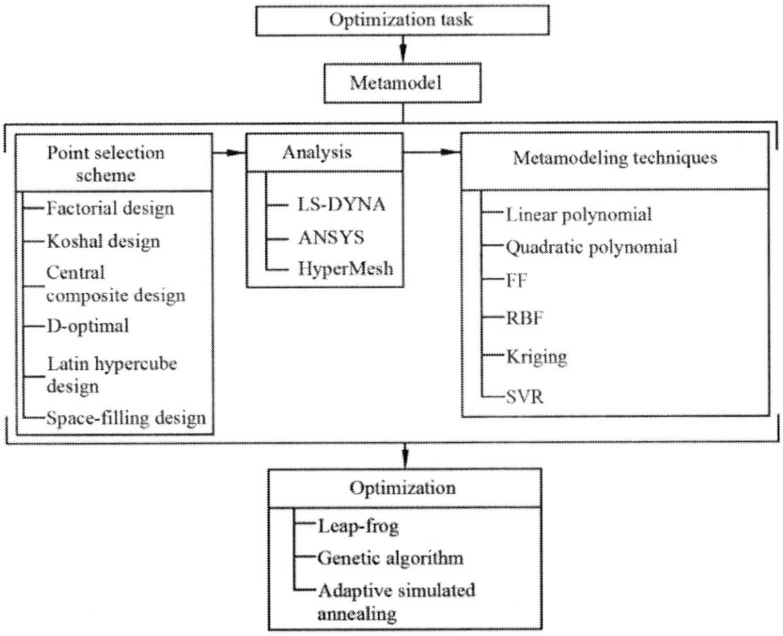

Figure 1. Optimization algorithm block diagram

2. Optimization methods

2.1. The first stage of the optimization algorithm

At the first stage of the optimization it is necessary to choose the model of the response surface, which can be obtained in different ways. Let's consider the following types of approximations: linear polynomial model, quadratic polynomial model, neural networks: Feedforward (FF) and Radial Basis Function (RBF), Kriging and Support Vector Regression (SVR).

Linear polynomial model is the simplest and least costly in terms of computing time. It is built by approximating the n-th number of individual test results. Calculations show that the accuracy of such a model is satisfactory only for preliminary evaluations.

The FF and RBF models are based on the principles of the neural networks. The direct-spread neural networks like FF has a clear multilevel topology [7]. Each unit performs a biased weighted sum of their inputs and passes this value through a transfer (activation) function to produce the output. The outputs of each layer of neurons are inputs to the next level. RBF uses radial basic functions as activation functions [8]. The output of the neural network is a linear combination of radial basic functions of inputs and unit parameters.

The Kriging model is based on the interpolation method, for which interpolated values are simulated by the Gaussian process defined by previous covariance.

The SVR model, which is a variation of the reference vector method, has good generalization properties [9]. Instead of minimizing the empirical incoherence, the model is based on the structural minimization of incoherencies.

2.2. The second stage of the optimization algorithm

At the second stage of the optimization problem it is necessary to choose the point selection scheme, which depend on the type of response model under consideration. The set of points and types of response surface determines the metamodel for further optimization. There are many ways to select the point selection scheme. The following methods are known from the theory of experiment planning: factorial design (factor experiment), Koshal design, Central composite design, D-optimal, Latin hypercube design, Space-filling design.

The method of the complete factor experiment involves calculating the number of experiments by $N = m^n$, where m — the number of levels of each factor, n — number of factors. The disadvantages of this method include its redundancy in terms of the number of experiments, which leads to irrational waste of time.

The D-optimal method, most often used for polynomial models, uses a subset of all possible points of a complete factor experiment [10]. For linear polynomial model, the minimum number of points is determined by $1{,}5(n+1)+1$, for the quadratic polynomial model, by $0{,}75(n+1)(n+2)+1$. This method is a compromise of achieving high accuracy of forecasting and low calculation time.

The space-filling method optimizes the minimum distance between experimental design points for a given number of points [11]. This method is convenient to combine with FF, RBF and Kriging models. The disadvantage of this method is that it does not "recommend" the number of necessary experiments.

2.3. The third stage of the optimization algorithm

At the third stage the algorithm of searching for the minimum of the target should be chosen. The program provides three variants of the minimum search: Leap-Frog Optimizer [12] (the algorithm used when the minimum of the target function is searched for only one criterion), genetic algorithm (multi-criteria and multi-parameter algorithm) [13, 14] and Adaptive Simulated Annealing [15, 16] with the ability to switch to Leap-Frog Optimizer to search for the local minimum..

In this paper, the D-optimal method is used for polynomial metamodels, and space-filling method is used for the rest of them [17]. The number of points to obtain an exact solution is generally unknown. The sequential RSM method is used to determine this parameter. As a result, the number of points depended on the convergence of three criteria: the accuracy of the metamodel, the accuracy of the optimization criterion, and the accuracy of the restrictions (the minimum number of points for one iteration cannot be less than $1{,}5(n+1)+1$). All criteria are defined by the engineer and depend on the task.

3. Mathematical model and optimization model

In order to determine the most effective algorithm of parametric optimization, a study of the vehicle bumper on impact was conducted. The main purpose of the bumper is to protect against impact by maximizing energy absorption. In the process of impact in the construction of the bumper there are numerous zones with plastic deformations, there comes a loss of stability and buckling. This type of problem is well suited for the analysis of optimization algorithms, because it describes the main nuances of deformation arising in the elements of vehicle construction in road accidents. The test pattern and 3D model of the car bumper are shown in Figure 2.

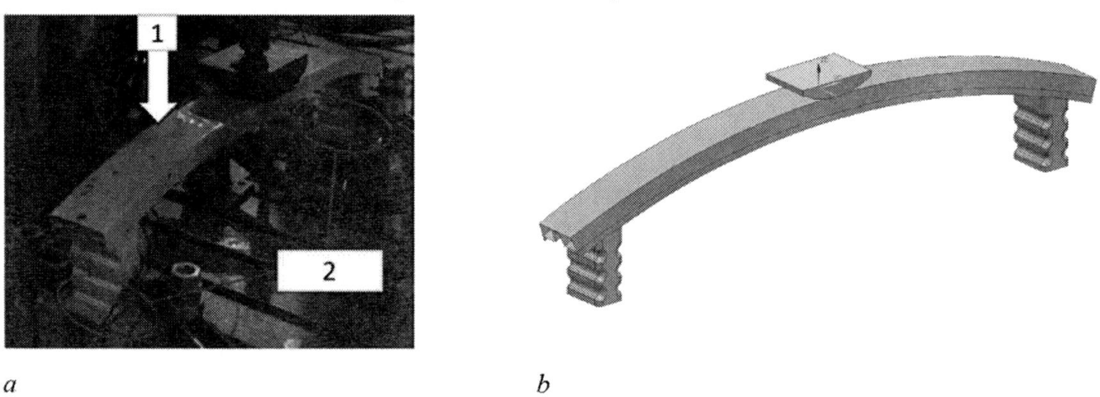

Figure 2. Test pattern (a) and 3D model (b) of the car bumper:
1 - Test bench; 2 - Constraint

Geometric dimensions, properties of materials, conditions of interaction between bumper and drummer are considered in detail in the paper [18].

The loads and simulation results obtained using the LS-DYNA program are shown in Figure 3. The drummer's mass is 25 kg, the speed is 10 m/s, and the calculation time is limited to 0.025 s. The maximum movement of the drummer is 100 mm.

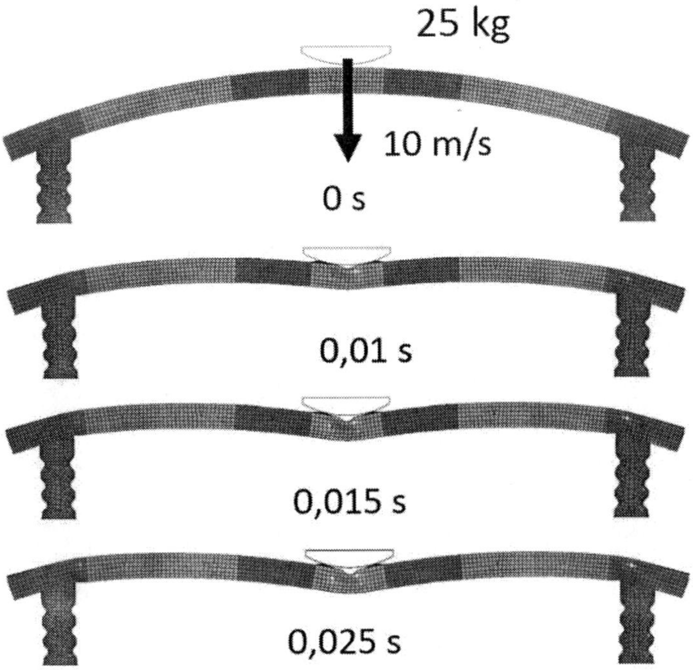

Figure 3. Simulation results

The goal of the optimization is to create a structure similar in crashworthiness to the basic model with a minimum mass. The target function (criterion) is the mass of the bumper. The restriction was the maximum movement of the drummer, which should correspond to the experimental data.

The bumper structure, consisting of external and internal power elements, is broken into parts. Each of them had an independent variable - the thickness of the internal and external t_{nj} parts respectively (i = 1...5, j = 1...5). The range of change of the variable is 0,5...1,5 mm. Figure 4 shows finite element models of bumper power elements broken down into parts.

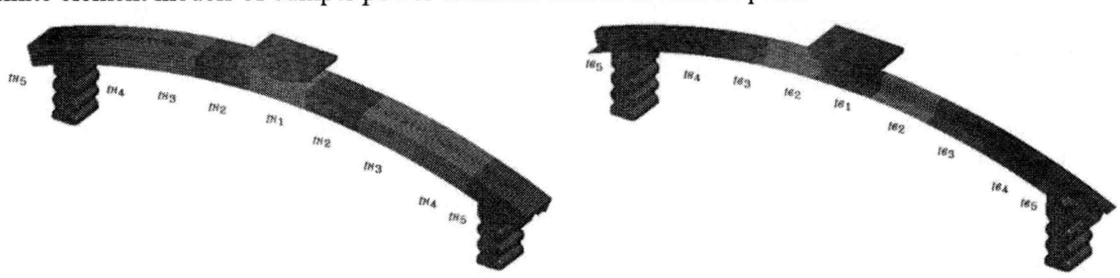

a *b*

Figure 4. Finite element models of external (a) and internal (b) bumper power elements

IOP Conf. Series: Materials Science and Engineering **820** (2020) 012030 doi:10.1088/1757-899X/820/1/012030

The Adaptive Simulated Annealing algorithm is used to optimize the bumper as the most effective algorithm among those related to the crashworthiness [19].

It is important to note that all algorithms are very sensitive in terms of accuracy and duration of the solution to the assignment of criteria of convergence of the optimization task. In this case, the sequential RSM method is used, i.e. the initially unknown number of experiments was determined in the process of iteration solution. The result was obtained when the convergence criteria did not exceed 0.01 values. Setting too high values may lead to inappropriate results due to errors in the solution.

4. Result and discussion

Analysis of the results of application of different metamodels for vehicle bumper optimization is given in the Table.

Table. Results of the bumper optimization for the metamodels

Type of metamodel	Bumper's mass		Moving the drummer		Solution convergence (number of experiments)
	Value, kg	Change, %	Value, mm	Accuracy, %	
Base model	3,14	–	100	–	–
linear polynomial	2,68	14,6	115	15	No (391)
quadratic polynomial	2,25	28	162	62	No (391)
FF	2,46	22	158	58	No (391)
RBF	2,64	16	101	1	Yes (221)
Kriging	2,39	24	148	48	No (391)
SVR	2,46	22	128	28	No (391)

The results showed that the convergence of the optimization problem solution has been achieved only for the RBF metamodel (221 experience, 13 iterations). For other cases, the Table presents the results of the best intermediate iterations.

Figure 5 shows the response surfaces method [20] obtained with the help of the considered meta-models, where L is the movement of the drummer; t_{v1} is the thickness of the inner part; t_{n1} is the thickness of the outer part.

For all types of metamodels, the impactor's movement error (for the best intermediate iterations) varied from 15 % (linear polynomial model) to 62 % (FF), except for the RBF, for which all convergence conditions were met, and the impactor's movement was 101 mm. Consequently, only the RBF model is applicable to optimization tasks related to impacts specific to bumpers, bodies and cabins of vehicles.

For the problem of crashworthiness of the vehicle bumper the use of the optimization allowed to reduce the mass by 16% (from 3.14 to 2.64 kg) while maintaining the crashworthiness of the basic model.

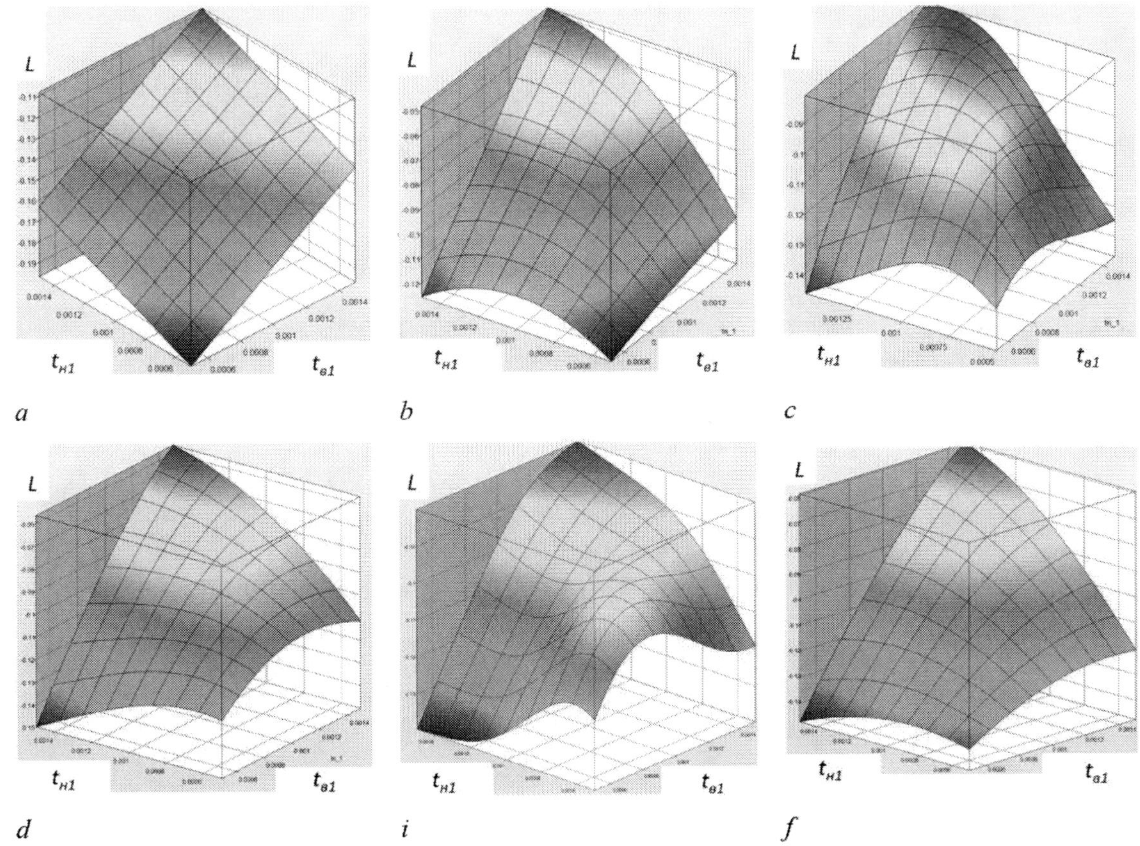

Figure 5. Response surfaces obtained with different metamodels
a — linear polynomial; b — quadratic polynomial; c — FF; d — RBF; i — Kriging; f — SVR

5. Conclusion

The use of control, FF, Kriging or SVR metamodels does not allow to get the result for the solution of the optimization task. The accuracy of the drummer movement for intermediate iterations exceeds 28%, it is too high.

The linear polynomial model with D-optimal is advisable to use only when performing preliminary multi-variant calculations for impact problems, as it is rational in terms of accuracy and machine time.

The use of software based on RBF metamodel allowed to reduce the mass of the vehicle bumper by 16% (from 3.14 to 2.64 kg) while maintaining the initial parameters of crashworthiness.

References

[1] Kotiev G O, Padalkin B V, Kartashov A B and Diakov A S 2017 *ARPN J. of Eng. and Appl. Sci.* **12** 1064-71
[2] Vdovin D and Chichekin I 2016 *Procedia Engineering* **150** 1276-79
[3] Skotnikov G I, Jileykin M M and Komissarov A I 2018 *IOP Conf. Series: Materials Science and Engineering* **315** 012027
[4] Diakov A S and Kotiev G O 2018 *MATEC Web of Conf.* **224** 02096
[5] Kotiev G O, Butarovich D O and Kositsyn B B 2018 *IOP Conf. Series: Materials Science and Engineering* **315** 012014
[6] Evseev K B, Kartashov A B, Dashtiev I Z and Pozdeev A V 2018 *MATEC Web of Conf.* **224** 02039
[7] Hornik K, Stinchcombe M and White H 1990 *Neural networks* **3** 551–560

[8] Buhmann M D 2003 *Radial Basis Functions: Theory and Implementations* (Cambridge: Cambridge University Press) p 271
[9] Jin R, Chen W and Sudjianto A 2002 *Proc. Of ASME 2002 Design Engineering Technical Conferences and Computers and Information in Engineering Conference DETC2002/DAC-34092 (Montreal)*
[10] Roux W J, Stander N and Haftka R 1998 *International Journal for Numerical Methods in Engineering* **42** 517-534
[11] Wilson B, Cappelleri D J, Frecker M I and Simpson T W 2001*Optimization and Engineering* **2** 31–50
[12] Snyman J A 2000 *Computers and Mathematics with Applications* **40** 1085–1096
[13] Holland J H 1992 *Adaptation in Natural and Artificial Systems* (Cambridge: MIT Press) p 225
[14] Goldberg D E 1989 *Genetic Algorithms in Search. Optimization and Machine Learning* (Boston: Addison-Wesley Publishing Company, Inc.) p 432
[15] Kirkpatrick S, Gelatt C D and Vecchi M P 1983 *Science, New Series* **220** 671–680
[16] Pincus M A 1970 *Operations Research* **18** 1225–28
[17] Oman M and Nilsson L 2010 *Structural and Multidisciplinary Optimization* **41** 797–815
[18] Zhi Xiao, Jianguang Fang, Guangyong Sun and Qing Li 2015 *Int. J. Advances in engineering software* **85** 81-95
[19] Schuur P C 1997 *Mathematics of Operations Research* **22** 266–275
[20] Myers R H, Montgomery D C and Anderson-Cook C M 1995 *Response Surface Methodology, ed D J Balding et al* (New Jersey: Wiley) p 856

IOP Conf. Series: Materials Science and Engineering **820** (2020) 012031 doi:10.1088/1757-899X/820/1/012031

Automatization of calculation of dynamic characteristics of the car suspension system

M Lahtyukhov[1,2] and L Zheglov[1]

[1]Bauman Moscow State Technical University

[2]E-mail: motor@bmstu.ru

Abstract. The article deals with the questions of finding the parameters of generalized coordinates for plane linear dynamic systems of any structure with degrees of freedom, used in the study of suspension systems. The appropriateness of separation of a number of subsystems for each of which potential energy and dissipative function are calculated by the same type equations is shown. As an example, the differential equations describing in the general case the linear movements of the masses of the dynamic system are given. It is shown that in order to find the parameters of generalized coordinates of a flat linear dynamic system with degrees of freedom at the kinematic harmonic disturbance determined by the structure of the road surface, a solution of the system of linear algebraic equations is required.

Introduction

The solution of problems of vibration safety and loading of vehicle components (systems) is carried out on condition of choosing a rational scheme of dynamic system of the analyzed object. The design of the main suspension system can be carried out using multi-mass dynamic systems [1-6]. In case of research of automatic suspension systems, the structure of the dynamic system is determined taking into account the control algorithm being applied [7-12]. The linear [13-16] and nonlinear [17-20] definition of the problem for plane and three-dimensional schemes of dynamic systems can be used. Execution of the algorithm of structural synthesis of a dynamic system should be implemented using the modular approach of its construction [21-25].

Plane linear dynamic systems consisting of a finite number of masses connected to each other and in contact with the road surface through inertial elastic and dissipative bonds are considered. Two types of masses with one or two degrees of freedom, respectively, are usually used to construct a dynamic system scheme. Afterwards, for the convenience of presentation, depending on the number of degrees of freedom, the corresponding masses will be called masses of the first or second type. The masses of both types can be moved vertically, and the masses of the second type can also be rotated around the transverse axis of the vehicle passing through the center of mass. The arbitrary i -th mass of the first type is only the mass m_i, and i -th mass of the second type is the mass m_i and the moment of inertia J_i.

If the dynamic system has n_1 masses of the first type and n_2 masses of the second type, the total number of degrees of freedom is $n = n_1 + 2n_2$.

IOP Conf. Series: Materials Science and Engineering **820** (2020) 012031 doi:10.1088/1757-899X/820/1/012031

Calculation of amplitude-frequency response of plane linear dynamic systems

Differential equations of the holonomic dynamic system with lumped constants and stationary constraints can be obtained using the Lagrange second order differential equation

$$\frac{d}{dt}\frac{\partial T}{\partial \dot{q}_i} - \frac{\partial T}{\partial q_i} + \frac{\partial V}{\partial \dot{q}_i} + \frac{\partial U}{\partial q_i} = Q_i \qquad i = \overline{1, n}, \tag{1}$$

where T, U – the kinetic and potential energies of the system, respectively; V – dissipative function of the system; Q_i – i –th generalized disturbing force; q_i, $\dot{q}_i$ – i –th generalized coordinate and speed; t – time.

For the generalized coordinates for all masses of the system, we will select the vertical displacements, calculated upwards from the position of static equilibrium of the centers of masses, and for masses of the second type, in addition, the angular displacements, calculated from the position of equilibrium of these masses counterclockwise.

At small oscillations near the static equilibrium position, the kinetic and potential energies of the system, as well as its dissipative function, can be expressed by quadratic forms from generalized coordinates:

$$T = \frac{1}{2}\sum_{i=1}^{n}\sum_{j=1}^{n} a_{ij}\dot{q}_i\dot{q}_j , \quad U = \frac{1}{2}\sum_{i=1}^{n}\sum_{j=1}^{n} c_{ij}q_i q_j , \quad V = \frac{1}{2}\sum_{i=1}^{n}\sum_{j=1}^{n} b_{ij}\dot{q}_i\dot{q}_j , \tag{2}$$

where a_{ij} – inertial coefficients, c_{ij}, b_{ij} – stiffness and dissipation coefficients, respectively.

In general, the numbering of masses can be done in random order. We will assign a number from 1 to n_1 to the masses of the first type for simplicity. To the masses of the second type from $n_1 + 1$ up to n_2. Then the first $n_1 + n_2$ vector $\mathbf{q}$ components are vertical displacements of the first and second type masses:

$$q_i = x_i, \ i = \overline{1, n_1 + n_2},$$

and the last ones n_2 vector $\mathbf{q}$ components is the angular displacement of type two masses

$$q_{n_2+i} = \alpha_i, \ i = \overline{n_1 + 1, n_1 + n_2}.$$

The kinetic energy of the system is determined by the equation:

$$T = \frac{1}{2}\left[\sum_{i=1}^{n_1+n_2} m_i\dot{q}_i^2 + \sum_{j=n_1+n_2+1}^{n} J_{j-n_2}\dot{q}_j^2\right]. \tag{3}$$

In order to simplify the following calculations in the initial system depending on the type of mass to be connected, it is useful to identify a number of subsystems, for each of which both the potential energy and the dissipative function are calculated by the same type of equations. There are usually five subsystems:

- two masses of the first type with elastic linkage between them (Fig. 1, a);
- mass of the first type and mass of the second type with elastic linkage between them (Fig. 1, b);
- mass of the first type, supported on the road surface by means of elastic linkage (Fig. 1, c);
- two masses of the second type with several elastic linkages between them (Fig. 1, d);
- mass of the second type, supported on the road surface by several elastic linkages (Fig. 1, e),if necessary, other subsystems can be added.

Then both potential energy and dissipative function of the dynamic system can be represented in the sum of potential energies and dissipative functions of separate subsystems (Table 1)

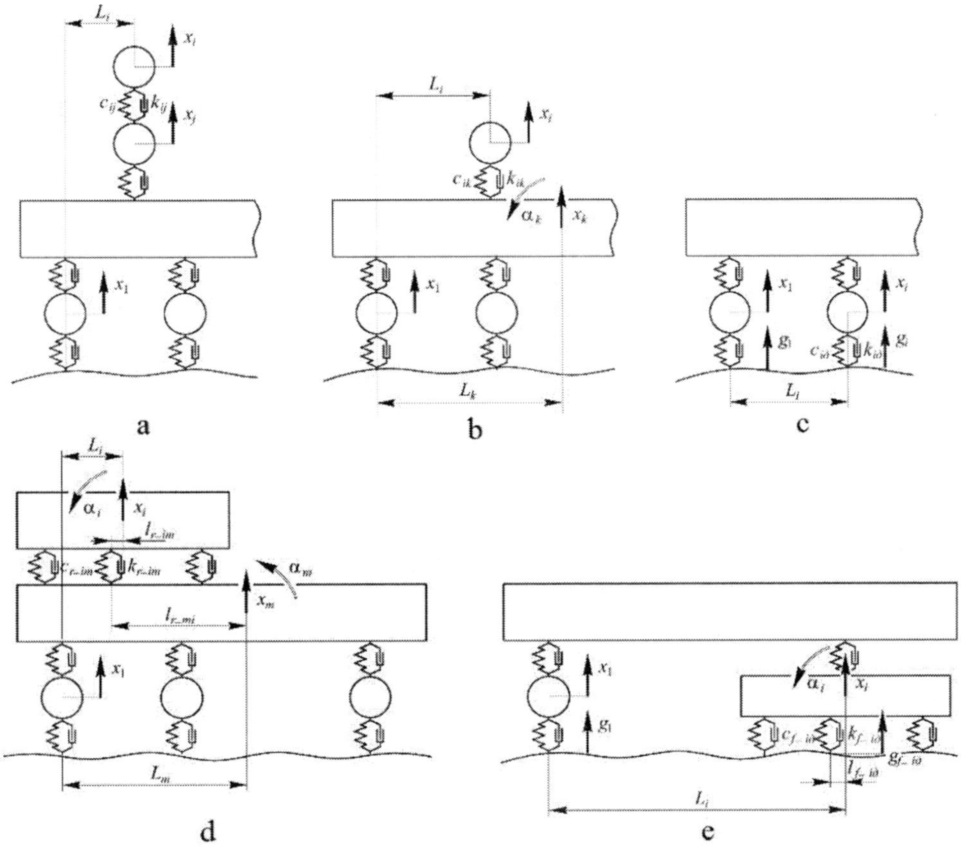

Fig. 1. Typical dynamic subsystems

$$U = \sum_{l=1}^{nl} U_l^{(1-1)} + \sum_{m=1}^{nm} U_m^{(1-2)} + \sum_{p=1}^{np} U_p^{(1-\partial)} + \sum_{s=1}^{ns} U_s^{(2-2)} + \sum_{t=1}^{nt} U_t^{(2-\partial)} \qquad (4)$$

and

$$V = \sum_{l=1}^{nl} V_l^{(1-1)} + \sum_{m=1}^{nm} V_m^{(1-2)} + \sum_{p=1}^{np} V_p^{(1-\partial)} + \sum_{s=1}^{ns} V_s^{(2-2)} + \sum_{t=1}^{nt} V_t^{(2-\partial)} \qquad (5)$$

where nl – the number of mass pairs of the first type connected to each other; nm – the number of pairs of masses of the first and second type connected to each other; np and nt – the number of masses of the first and second type supported by the road surface, respectively; ns – the number of pairs of masses of the second type connected to each other.

As the dynamic system is exposed only to kinematic disturbance from the support surface when calculating the amplitude-frequency response, all generalized disturbing forces in (1) are equal to zero:

$$Q_i = 0\,, \qquad i = \overline{1,n}. \qquad (6)$$

Using equations (1), (3), (4), (5) and (6), it is not difficult to obtain an equation of motion for any of the masses of the dynamic system. For example, if we assume that the random mass of the first type has elastic linkages with all masses and with the road surface, then the differential equation of its motion after the rearrangement of the terms will take the form:

$$
\begin{gathered}
m_i\ddot{x}_i + \overbrace{\left(\sum_{j=1}^{n_1} k_{ij}\right)\dot{x}_i}^{1-1} + \overbrace{\left(\sum_{k=n_1+1}^{n_1+n_2} k_{ik}\right)\dot{x}_i}^{1-2} + \overbrace{k_{i\partial}\dot{x}_i}^{1-\partial} - \overbrace{\sum_{j=1}^{n_1}\left(k_{ij}\dot{x}_j\right)}^{1-1} - \overbrace{\sum_{k=n_1+1}^{n_1+n_2}\left(k_{ik}\dot{x}_k\right)}^{1-2} - \overbrace{\sum_{k=n_1+1}^{n_1+n_2}\left(k_{ik}\left(L_i - L_k\right)\dot{\alpha}_k\right)}^{1-2} + \\
+ \overbrace{\left(\sum_{j=1}^{n_1} c_{ij}\right)x_i}^{1-1} + \overbrace{\left(\sum_{k=n_1+1}^{n_1+n_2} c_{ik}\right)x_i}^{1-2} + \overbrace{c_{i\partial}x_i}^{1-\partial} - \overbrace{\sum_{j=1}^{n_1}\left(c_{ij}x_j\right)}^{1-1} - \overbrace{\sum_{k=n_1+1}^{n_1+n_2}\left(c_{ik}x_k\right)}^{1-2} - \\
- \overbrace{\sum_{k=n_1+1}^{n_1+n_2}\left(c_{ik}\left(L_i - L_k\right)\alpha_k\right)}^{1-2} = \overbrace{k_{i\partial}\dot{g}_i}^{1-\partial} + \overbrace{c_{i\partial}g_i}^{1-\partial} \; .
\end{gathered} \tag{7}
$$

The differential equation of motion of the i-th mass of the second type, corresponding to the linear displacements of its center of mass, has the following form:

$$
\begin{gathered}
m_i\ddot{x}_i + \overbrace{\left(\sum_{k=1}^{n} k_{ik}\right)\dot{x}_i}^{1-2} + \overbrace{\left(\sum_{m=1}^{n2}\sum_{r=1}^{nr_m} k_{r\_im}\right)\dot{x}_i}^{2-2} + \overbrace{\left(\sum_{f=1}^{nf_i} k_{f\_i\partial}\right)\dot{x}_i}^{2-\partial} - \overbrace{\sum_{k=1}^{n}\left(k_{ik}\dot{x}_k\right)}^{1-2} - \overbrace{\sum_{m=1}^{n2}\left[\left(\sum_{r=1}^{nr_m} k_{r\_im}\right)\dot{x}_m\right]}^{2-2} + \\
+ \overbrace{\left(\sum_{k=1}^{n} k_{ik}\left(L_k - L_i\right)\right)\dot{\alpha}_i}^{1-2} + \overbrace{\left(\sum_{m=1}^{n2}\sum_{r=1}^{nr_m} k_{r\_im}l_{r\_im}\right)\dot{\alpha}_i}^{2-2} + \overbrace{\left(\sum_{f=1}^{nf_i} k_{f\_i\partial}l_{f\_i\partial}\right)\dot{\alpha}_i}^{2-\partial} - \overbrace{\sum_{m=1}^{n2}\left[\left(\sum_{r=1}^{nr_m} k_{r\_im}l_{r\_im}\right)\dot{\alpha}_m\right]}^{2-2} + \\
+ \overbrace{\left(\sum_{k=1}^{n} c_{ik}\right)x_i}^{1-2} + \overbrace{\left(\sum_{m=1}^{n2}\sum_{r=1}^{nr_m} c_{r\_im}\right)x_i}^{2-2} + \overbrace{\left(\sum_{f=1}^{nf_i} c_{f\_i\partial}\right)x_i}^{2-\partial} - \overbrace{\sum_{k=1}^{n}\left(c_{ik}x_k\right)}^{1-2} - \overbrace{\sum_{m=1}^{n2}\left[\left(\sum_{r=1}^{nr_m} c_{r\_im}\right)x_m\right]}^{2-2} + \\
+ \overbrace{\left(\sum_{k=1}^{n} c_{ik}\left(L_k - L_i\right)\right)\alpha_i}^{1-2} + \overbrace{\left(\sum_{m=1}^{n2}\sum_{r=1}^{nr_m} c_{r\_im}l_{r\_im}\right)\alpha_i}^{2-2} + \overbrace{\left(\sum_{f=1}^{nf} c_{f\_i\partial}l_{f\_i\partial}\right)\alpha_i}^{2-\partial} - \overbrace{\sum_{m=1}^{n2}\left[\left(\sum_{r=1}^{nr_m} c_{r\_im}l_{r\_im}\right)\alpha_m\right]}^{2-2} = \\
= \overbrace{\sum_{f=1}^{nf_i}\left(k_{f\_i\partial}\dot{g}_{f\_i}\right)}^{2-\partial} + \overbrace{\sum_{f=1}^{nf_i}\left(c_{f\_i\partial}g_{f\_i}\right)}^{2-\partial} \; .
\end{gathered} \tag{8}
$$

In both equations above the curly brackets there are the types of subsystems mentioned above, and under the brackets there are the corresponding terms.

The matrix equation describing the motion of the linear dynamic model of the suspension system with degrees of freedom looks like:

$$A\ddot{\mathbf{q}} + B\dot{\mathbf{q}} + S\mathbf{q} = \mathbf{f}, \tag{9}$$

where A - matrix of inertial coefficients (dimension $n \times n$), S, B - Matrix of stiffness and dissipation (damping) coefficients $(n \times n)$, $\mathbf{q}$ - n- dimensional vector of generalized coordinates, $\mathbf{f}$ - n- dimensional vector of external forces, caused by the microprofile of the road surface.

It is easy to make sure that the right side i–th of the differential equation f_i can be represented by

$$f_i = f_{1\_i}\cos\omega t + f_{2\_i}\sin\omega t. \tag{10}$$

IOP Conf. Series: Materials Science and Engineering **820** (2020) 012031 doi:10.1088/1757-899X/820/1/012031

Table 1: Typical dynamic subsystems, their potential energy and dissipative function

Sub-system type	Sub system number	Number and type of connected mass	Potential energy	Dissipative function
$U_p^{(1-1)}$	p	i -th mass of the first type and j - th mass of the first type	$\frac{1}{2}c_{ij}\left(x_i - x_j\right)^2$	$\frac{1}{2}k_{ij}\left(\dot{x}_i - \dot{x}_j\right)^2$
$U_s^{(1-2)}$	s	i - th mass of the first type and k - th mass of the second type	$\frac{1}{2}c_{ik}\left[x_i - \left(x_k + \left(L_i - L_k\right)\alpha_k\right)\right]^2$	$\frac{1}{2}k_{ik}\left[\dot{x}_i - \left(\dot{x}_k + \left(L_i - L_k\right)\dot{\alpha}_k\right)\right]^2$
$U_t^{(1-\partial)}$	t	i - th mass of the first type and road surface	$\frac{1}{2}c_{i\partial}\left(x_i - g_i\right)^2$	$\frac{1}{2}k_{i\partial}\left(\dot{x}_i - \dot{g}_i\right)^2$
$U_u^{(2-2)}$	u	i - th mass of the second type and m - th mass of the second type	$\frac{1}{2}\sum_{r=1}^{nr_u} c_{r\_im}\left[\left(x_i + l_{r\_im}\alpha_i\right) - \left(x_m + l_{r\_mi}\alpha_m\right)\right]^2$	$\frac{1}{2}\sum_{r=1}^{nr_u} k_{r\_im}\left[\left(\dot{x}_i + l_{r\_im}\dot{\alpha}_i\right) - \left(\dot{x}_m + l_{r\_mi}\dot{\alpha}_m\right)\right]^2$
$U_v^{(2-\partial)}$	v	i - th mass of the second type and road surface	$\frac{1}{2}\sum_{f=1}^{nf_v} c_{f\_i\partial}\left[\left(x_i + l_{f\_i\partial}\alpha_i\right) - g_{f\_i}\right]^2$	$\frac{1}{2}\sum_{f=1}^{nf_v} k_{f\_i\partial}\left[\left(\dot{x}_i + l_{f\_i\partial}\dot{\alpha}_i\right) - \dot{g}_{f\_i}\right]^2$

Note: Numeric and alphabetic indexes in upper brackets for U and V, separated by a hyphen, represent the type of mass to be connected or the road surface; c_{ij} and k_{ij} – stiffness and damping coefficients of the elastic linkage between the i – thand j – th masses; $c_{i\partial}$ and $k_{i\partial}$ – stiffness and damping coefficients of the elastic linkage of the i - th mass of the first type with the road surface; c_{r_im} and k_{r_im} – stiffness and damping coefficients r - th elastic linkage between i - th and m - th masses of the second type; nr_u – the number of elastic linkages between i - th and m - th masses of the second type in u - th subsystem; $c_{f\_i\partial}$ and $k_{f\_i\partial}$ – stiffness and damping coefficients f - th elastic linkage between i – th mass of the second type with the road surface; $nf_v - nr_u$ – the number of elastic linkages between i – i – th mass of the second type with the road surface in v - th subsystem; L_i – distance from the first axle of the vehicle to the centre of mass (CM); i – th mass (distance is taken with sign "–", if the CM is located to the left of the front axle of the vehicle, and with a sign "+", if the CM is located to the right of the front axle of the vehicle); l_{r_im} and l_{r_mi} – the distance between r – th elastic linkage and CM i – th and m – th mass of the second type respectively (distance is taken with sign "–", if the r – th elastic linkage is located to the left of the CM, and with a sign "+", if the r – th elastic linkage is located to the right of the CM), and $l_{r\_mi} = L_i + l_{r\_im} - L_m$; $l_{f\_i\partial}$ – distance between f – th elastic linkage and CM i – th mass of the second type (distance is taken with sign "–", if the f – th elastic linkage is located to the left of the CM i – th mass, and with a sign"+", if the f – th elastic linkage is located to the right of the CM i – th mass); g_i – road surface shape under i - th mass of the first type; g_{f_i} – road surface shape under f – th elastic linkage, connecting the i – th mass of the second type with the road surface.

IOP Conf. Series: Materials Science and Engineering **820** (2020) 012031 doi:10.1088/1757-899X/820/1/012031

Due to the energy dissipation, free vibrations with frequencies different from the disturbance frequency ω are out over time and stationary vibrations with the frequency ω, independent of the initial conditions, are installed. The solution of equation (9) corresponding to such stationary oscillations is presented in the form

$$q_i = q_{Mi}\sin(\omega t + \alpha_{Mi}) = q_{1i}\cos\omega t + q_{2i}\sin\omega t, \qquad (11)$$

where q_{Mi} and α_{Mi} – amplitude and phase angle i- th generalized coordinate, $q_{1i} = q_{Mi}\sin\alpha_{Mi}$, $q_{2i} = q_{Mi}\cos\alpha_{Mi}$.

By substitution (10) and (11) in (9) and separating the variables containing sinuses and cosines, the initial system from n differential equations is transformed to a system of linear algebraic equations:

$$D\mathbf{Q} = \mathbf{h}, \qquad (12)$$

where $D = \begin{bmatrix} S-\omega^2 A & \omega B \\ -\omega B & S-\omega^2 A \end{bmatrix}$ – matrix of $2n \times 2n$ dimension; $\mathbf{Q} = \begin{bmatrix} \mathbf{q}_1 \\ \mathbf{q}_2 \end{bmatrix}$ – vector $2n$ length; $\mathbf{h} = \begin{bmatrix} \mathbf{f}_1 \\ \mathbf{f}_2 \end{bmatrix}$ – vector $2n$ length; $\mathbf{q}_1$, $\mathbf{q}_2$ – vectors n length containing q_{1i} and q_{2i}, respectively; $i = \overline{1, n}$, $\mathbf{f}_1$, $\mathbf{f}_2$ – vectors n length containing f_{1_i} and f_{2_i}, $i = \overline{1, n}$, respectively

Obtained by solving the system of linear equations (12) elements of vector $\mathbf{Q}$ can be used to find the amplitudes and phases of all generalized coordinates. Thus, to find the amplitudes and phases of the generalized coordinates of the dynamic system with n degrees of freedom at the kinematic harmonic perturbation the solution of the system of $2n$ linear algebraic equations is required.

At a given speed of v_a vehicle movement, the formation of A, B, S matrices and $\mathbf{h}$ vector (12) can be automated. The system of linear algebraic equations (12) is solved using numerically stable matrix methods of linear algebra.

Conclusions

1. The method of automated calculation of amplitude-frequency response of a plane linear dynamic system with degrees of freedom and random connections is developed.
2. The expediency of separation of a number of subsystems is presented, for each of which both the potential energy and the dissipative function are calculated by the same type of equations.
3. Examples of differential equations are given that generally describe linear mass displacements.
4. It is shown that the solution of the system of linear algebraic equations is required to find the amplitudes and phases of the generalized coordinates of the plane linear dynamic model with degrees of freedom at the kinematic harmonic perturbation from the road surface.

References

[1] Belousov B.N. and Popov S.D., Heavy-duty wheeled vehicles: Design, Theory, Calculations SAE International ISBN 978-0-7680-7723-0, 2014, 800 p.
[2] Chumakov, D.A.,Chernyshov, K.V., Novikov, V.V., Diakov, A.S., Suchenina, A.S.Mathematical model of motor vehicle air suspension with a combined damping system – Journal of Physics: Conference Series, Volume 1177, Issue 1, 10 April 2019, article №012049, 3rd Science and Technology Seminar on Mobility of Transport and Technological Machines, MTTM 2018; Nizhny Novgorod; Russian Federation; 28 August 2018; 147340
[3] Vdovin D., Chichekin I. Loads and Stress Analysis Cycle Automation in the Automotive Suspension Development Process– Procedia Engineering, Volume 150, 2016, Pages 1276-1279, 2nd International Conference on Industrial Engineering, ICIE 2016; Chelyabinsk; Russian Federation; 19 May 2016 to 20 May 2016; 123270

[4] Gorelov, V.A., Komissarov, A.I., Miroshnichenko, A.V.8×8 wheeled vehicle modeling in a multibody dynamics simulation software(2015) Procedia Engineering, 129, pp. 300-307.

[5] Keller, A.V., Gorelov, V.A., Vdovin, D.S., Taranenko, P.A., Anchukov, V.V.Mathematical model of all-terrain truck(2015) Proceedings of the ECCOMAS Thematic Conference on Multibody Dynamics 2015, MultibodyDynamics 2015, pp. 1285-1296.

[6] Evseev, K.B., Kartashov, A.B., Dashtiev, I.Z., Pozdeev, A.V. Analysis viscoelastic properties of fiber-reinforced composite spring for the all-terrain vehicle, MATEC Web of Conferences 224, 02039 (2018).

[7] Velupillai S., Guvenc L., Oncu S., Ozcan D. Vehicle chassis control using adaptive semi-active suspension– IFAC Proceedings Volumes (IFAC-PapersOnline), Volume 17, Issue 1, PART 1, 2008, 17th World Congress, International Federation of Automatic Control, IFAC; Seoul; South Korea; 6 July 2008 to 11 July 2008; 79403

[8] FüsunAlışverişçi G., Bayıroğlu H., Balthazar J.M., Felix J.L.P., da Fonseca Brasil R.M.L.R. On dynamic behavior of a nonideal torsional machine suspension structure(Conference Paper) – Springer Proceedings in Mathematics and Statistics, Volume 181, 2016, Pages 1-10, 13th International Conference on Dynamical Systems, 2015; Łódź; Poland; 7 December 2015 to 10 December 2015; 181529

[9] Song X., Cao D. Design of semiactive vehicle suspension controls in frequency domain– Proceedings of the ASME International Design Engineering Technical Conferences and Computers and Information in Engineering Conference 2009, DETC2009, Volume 6, 2010, Pages 1047-1052, 2009 ASME International Design Engineering Technical Conferences and Computers and Information in Engineering Conference, DETC2009; San Diego, CA; United States; 30 August 2009 to 2 September 2009; 80638

[10] Connolly T.J., Longoria R.G. An approach for actuation specification and synthesis of dynamic systems– Journal of Dynamic Systems, Measurement and Control, Transactions of the ASME, Volume 131, Issue 3, May 2009, Pages 1-15

[11] Margolis, D.L.Bond graphs for automated simulation and control of nonlinear vehicle systems– SAE Technical Pa pers1987Future Transportation Technology Conference and Exposition; Seattle, WA; United States; 10 August 1987 to 13 August 1987; 90059

[12] Zhileykin, M.M., Kotiev, G.O., Nagatsev, M.V.Synthesis of the adaptive continuous system for the multi-axle wheeled vehicle body oscillationdamping(2018) IOP Conference Series: Materials Science and Engineering, 315 (1), article № 012031.

[13] Chernikov, S.A.Expansion of the suppression band of a vibroprotective system by a feedback dynamic damper (2015) Journal of Machinery Manufacture and Reliability44(5), c. 439-444

[14] Jabeen, S.D. Vehicle vibration and passengers comfort– Advances in Intelligent Systems and Computing, Volume 509, 2017, Pages 357-372, 1st International Conference on Computational Intelligence, ICCI 2015; Ranchi; India; 10 December 2015 to 11 December 2015; 186569

[15] Chen K.-C., Shih H.-Y.Research and Simulation of the Electrical Vehicle Based Dynamica l System – Ma thematical Problems in Engineering, Volume 2015, 2015, article № 929701

[16] Tlibekov, A.K., Yakhutlov, M.M., Batyrov, U.D., Dosko, S.I.(2017) Identification of mechanical systems in the frequency area. Proceedings of the 2017 International Conference "Quality Management, Transport and Information Security, Information Technologies", IT and QM and IS, 2017 8085812,c. 284-286

[17] Le V.Q., Zhang J., Liu X., Wang Y. Nonlinear dynamic analysis of interaction between vehicle and road surfaces for 5-axle heavy truck– Journal of Southeast University (English Edition), Volume 27, Issue 4, December 2011, Pages 405-409

[18] Liang Y.-J., Li N., Gao D.-X., Wang Z.-S.Optimal vibration control for nonlinear systems of tracked vehicle half-car sus pensions–International Journal of Control, Automation and Systems, Volume 15, Issue 4, 1 August 2017, Pages 1675-1683

[19] Milanese M., Novara C., Pivano L. Structured SM identification of vehicle vertical dynamics– Mathematical and Computer Modelling of Dynamical Systems, Volume 11, Issue 2, June 2005, Pages 195-207

[20] Zhileykin, M.M., Kotiev, G.O., Nagatsev, M.V.Comparative analysis of the operation efficiency of the continuous and relay control systems of amulti-axle wheeled vehicle suspension(2018) IOP Conference Series: Materials Science and Engineering, 315 (1), article № 012030.

[21] Belousov B., Ksenevich T.I., NaumovS.Automated system to control steering and wheel springing parameters in vehicle locomotion module– SAE Technical Papers, 14 January 2015, 14th Symposium on International Automotive Technology, SIAT 2015; ARAI CampusPune; India; 21 January 2015 to 23 January 2015; 110846

[22] Metallidis P., Stavrakis I., Natsiavas S. Parametric identification and health monitoring of large scale strongly nonlinear vehicle models– 2007 Proceedings of the ASME International Design Engineering Technical Conferences and Computers and Information in Engineering Conference, DETC2007, Volume 1, PART B, 2008, Pages 1397-1406, 21st Biennial Conference on Mechanical Vibration and Noise, presented at - 2007 ASME International Design Engineering Technical Conferences and Computers and Information in Engineering Conference, IDETC/CIE2007; Las Vegas, NV; United States; 4 September 2007 to 7 September 2007; 72165

[23] Novikov, V.V., Pozdeev, A.V., Diakov, A.S.Research and testing complex for analysis of vehicle suspension units(2015) Procedia Engineering, 129, pp. 465-470.

[24] Ovchinnikov, I.N. Its own spectrum band - A property of mechanical systems (2018) IOP Conference Series: Materials Science and Engineering 468(1), 012045

[25] Tlibekov, A.K., Yakhutlov, M.M., Batyrov, U.D., Dosko, S.I.Identification of mechanical systems in the frequency area (2017) Proceedings of the 2017 International Conference "Quality Management, Transport and Information Security, Information Technologies", IT and QM and IS 20178085812, c. 284-286

IOP Conf. Series: Materials Science and Engineering **820** (2020) 012032 doi:10.1088/1757-899X/820/1/012032

A method for selecting parameters of the electromechanical transmission of an industrial tractor

B Padalkin[1,2], A Stadukhin[1], B Kositsyn[1] and K Balkovskiy[1]

[1]Bauman Moscow State Technical University

[2]E-mail: padalkin@bmstu.ru

Abstract. The article describes a method for selecting the torque — speed characteristics of the drive motors and parameters of the electromechanical transmission of an industrial tractor.In order to provide the maximum performance of the tractor the transmission developed in the research has two gears:transportation range and technological range.The distinctive feature of the transmission is the use ofthe ZK tank steering systemdeveloped by the researchers of BMSTU (G.I. Zaychik, M.A. Kreynes,M.K. Kristi).The authors have studied the impact ofthe design parameter of the planetary gear sets of the ZK mechanism on the tractor performance.The torques and drive motor rotor speeds providing required performance in a turn as a functions of the planetary gear design parameter have been calculated. Also the time of the tractor full turn around its center of mass and around one braked track when the transmission operates in the transport and technological ranges has been found.An finaly, the impact of the tractor transmission parameters on the maximum tractor speed at manoeuvring has been analysed. The results of the analysis were used for selection of the electric motor output power providing a comfortable work for the tractor operator.

1. Introduction

An industrial tractor is used for performing working operations involving high tractive efforts [1]. However the tractor operation efficiency depends not only on the speed of the working maneuvers execution but also on the speed of its transfer between the sites in the transport mode.

This requires the tractor transmission to have at least two gears: a working gear and a transport gear.Besides, when the transmition is operating in each gear, the tractor should be able to move without shifting gears and to turn smoothly in order to provide its operation by drivers with lower professional qualification and to increase the working efficiency.

One of the ways to provide such qualities of the industrial tractor is to use the electromechanical transmission.In the simplest case, the smooth turn of the tracked vehicle can be provided by the use of an individual electric drive of its driving wheels.But in this case during turnning in the transport gear, the torque of the drive sprocket of the outside track is not sufficient, which forces the driver to lower the tractor speed and shift into the working gear.In the course of this problem solution, different transmission layouts for tracked vehicles have been analyzed [2], [3], [4], [5], [6].

The solution that meets the specified requirements is the use of the ZK steering system(developed by the researchers G.I. Zaychik, M.A. Kreynes andM.K. Kristi from the BMSTU) providing a smooth turnig and increased tractive performance of the tracked vehicle [7].

The electromechanical transmission to be installed inside the frame of the commercially available tractors should be extremely compact.Figure 1 shows a kinematic diagram of the tractor transmission intended to solve this problem.The kinematic diagram contains identical gear sets which provide a high reduction in the working gear and the functionality of the ZK steering system during transportation operations.Switching between the modes is carried out by means of four control elements (since it is assumed that there will be no switching between the ranges while driving). Cam clutches can be used as the control mechanisms.

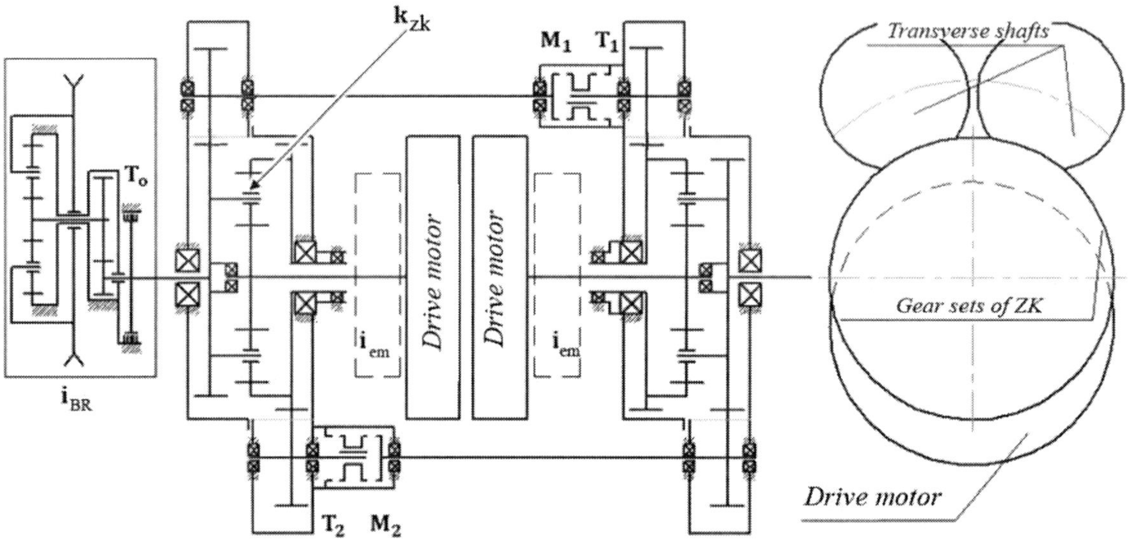

Figure 1. Kinematic diagram of the tractor transmission.

When the transmission is in the working gear, the brakes T1 and T2 are locked (see figure 1). The motors individually drive the wheels.The gear sets of the ZK steering system are disengaged and operate as the gear sets of the planetary gear train with the stopped central gear (the gear ratio of such a planetary gearis $k_{zk}+1$).

Turning of the tracked vehicle is provided by the high reduction on each track and with the help of the stopping brakes Ts.

When the transmission is in the transport gear, the clutches M_1 and M_2 are locked (see figure 1) and provide the reduction required for the straight-line motion.When the tractor is turning, the tractive effort on the tracks is provided by the ZK steering system (with the respective decrease in the tractor speed).

2. Selection of the main parameters of the transmission

In order to reduce the requirements for the accuracy of the gear manufacturing and accordingly reduce the cost of the transmission, the speed of the drive motor shaft should be limitted by the value of 4000 rpm.In this case, the required gear ratio of the transmission top gear (the transport gear) is determined by the maximum speed of the tractor:

$$i_{tr}^{v}=\frac{n_{max}\cdot r_k}{V_{max}}, \tag{1}$$

here n_{max} is the maximum speed of the motor shaft; r_k is the radius of the drive wheel; V_{max} is the maximum speed of the tractor.

When the transmission operates in the transport gear in case of a straight-line motion, all the elements of the ZK mechanism rotate together without torque reduction. Hence, the transmission gear ratio in the transport gear is defined by the gear ratio of the final reduction gear i_{BR} and motor reductor ratio i_{em} (see figure 1).

As it is known, in the average operation conditions, theneed for a tractive effort to wheight ratio to be higher than 0.4 is very unlikely [8].Then, the required maximum torque of the motor is defined in the following way:

$$M_{max} = \frac{mg \cdot D_v \cdot r_k}{i_{tr}^v \cdot 2\eta_{tr}\eta_{gus}}, \tag{2}$$

where M_{max} is the motor maximum torque; m is the gross mass of the tractor; g is the acceleration of gravity; η_{tr} is the mechanical efficiency of the transmission ; η_{gus} is the mechanical efficiency of the track. D_v is the tractive effort to weight ratio of the tractor.

For any planetary gear set with negative gear ratio an important parameter isthevalue $k_{zk} = Z_{BCK}/Z_{MCK}$,where$Z_{BCK}$is the number of teeth of the ring gear, Z_{MCK} is the number of teeth of the sun gear).Recomended values of the ratio k_{zk} for such a planetary gear set are selected in the range from 4/3 to 4 [9].The tractive effort to weight ratio and maximum speed of the tractor in the first gear as functions of the ratio k_{zk} of the ZK mechanism planetary gear sets can be calculated in the following way:

$$D_n = \frac{M_{max} \cdot i_{tr}^v \cdot 2 \cdot \eta_{tr} \cdot \eta_{gus} \cdot (k_{zk}+1)}{m \cdot g \cdot r_k} = D_v \cdot (k_{zk}+1), \tag{3}$$

$$V_{max}^n = \frac{\pi \cdot n_{max} \cdot r_K}{30 \cdot i_{tr}^v \cdot (k_{zk}+1)} = \frac{V_{max}}{(k_{zk}+1)}, \tag{4}$$

where D_n is the tractive effort to weight ratio in the first gear; V_{max}^n is the maximum speed of the tractor in the first gear.

Let us consider an industrial tractor as an example. The tractor specifications are given in table 1.

Table 1. Industrial tractor specifications.

Parameter	Value
Mass, kg	41500
Maximum tractive effort to weight ratio in the transport gear, minimum	0.4
Maximum speed in the transport gear, km/h	20
Drive sprocket radius, m	0.434
Wheel base, m	3.3
Track width, m	2.2
Height of the center of mass, m	1.8
Maximum tractive effort to weight ratio in the working gear, minimum	0.9
Maximum speed in the working gear, km/h, minimum	3
Maximum mechanical efficiency of the transmission	0.922

The tractor characteristics calculated by formulae 1 and 2 are shown in table 2.

Thus (according to formulas 3 and 4), for the considered tractor, the range of permissible values of the ratiok_{zk}provides a tractive effort to weight ratio from 0.933 to 2, and maximum speeds from 8.6 km/h to 4.0 km/h in the transport mode.

IOP Conf. Series: Materials Science and Engineering **820** (2020) 012032 doi:10.1088/1757-899X/820/1/012032

Table 2. Calculated characteristics of the industrial tractor.

Parameter	Value
Maximum required torque of the drive motor, Nm	1250
Maximum rotor speed of the drive motor, rpm	4000
Gear ratio of the trasport gear	32.76

In order to calculate performance characteristics of the tractor we should use the drive sprocket required torque/tractor speed — tractor turning radius curves.In a general case this kind of analysis should be performed taking into account the characteristics of the road irregularities [10] and with the use of the appropriate model of the track shoe — road interaction [11, 12], for instance the model based on the friction ellipse concept [13]. In this study, to reduce the amount of input data, it is advisable to use a simpler solution of this problem [14, 15, 16].

The most suitable method is the one described in [16]. The method consists of the following steps (for the case when the tractor moves in the most severe road conditions — on a turfy soil).

1. Calculation of the theoretical turning radius R_t:

$$R_t = R_f \cdot \frac{B}{L \cdot k}, \tag{5}$$

where B is the track width; L is the wheel base; k ratio of the actual turning radius to the theoretical turning radius for the given tracked vehicle; $R_f$is the actual turning radius.

2. Calculation of the coefficient of lateral resistance μby the A. O. Nikitin formula [17]:

$$\mu = \frac{\mu_{max}}{0.925 + 0.15\frac{R_t}{B}}, \tag{6}$$

where $\mu_{max}$is the maximum coefficient of lateral resistance on the given terrain(in the case of the turfy soil μ_{max}= 0.85 [18]).

3. Calculation of the required track speeds V_2 and V_1 of the outside and inside tracks:

$$V_2 = V\left(1 + \frac{B}{2R_t}\right);\ V_1 = V\left(1 - \frac{B}{2R_t}\right), \tag{7}$$

where V is the speed of the tractor center of mass.

4. Calculation of the required tractive efforts P_2 andP_1 on the outside and inside tracks [8]:

$$P_2 = \frac{m \cdot g}{2}\left(f_{gr}\left(1 + \frac{2V^2 \cdot H_z}{B \cdot R_f \cdot g}\right) + \frac{\mu \cdot L}{2B}\left(1 - \left(\frac{V}{V_{kr}}\right)^4\right)\right) + \frac{m \cdot V^4 \cdot L}{4 \cdot R_f^3 \cdot \mu \cdot g}; \tag{8}$$

$$P_1 = \frac{m \cdot g}{2}\left(f_{gr}\left(1 - \frac{2V^2 \cdot H_z}{B \cdot R_f \cdot g}\right) - \frac{\mu \cdot L}{2B}\left(1 - \left(\frac{V}{V_{kr}}\right)^4\right)\right) + \frac{m \cdot V^4 \cdot L}{4 \cdot R_f^3 \cdot \mu \cdot g}, \tag{9}$$

where f_{gr} is the tractors traight –line motion resistance coefficient for the given soil (in the case of a turfysoil $f_{gr} = 0.07$ [18]); H_z is the height of the tractor center of mass; V_{kr} is the skidding critical speed ($V_{kr} = \sqrt{\mu_{max} \cdot g \cdot R_f}$).

5. Calculationof the mechanical efficiency η_2 and η_1 of the outside and inside tracks [19]:

$$\eta_2 = 0.95 - 0.005 \cdot 3.6V_2;\ \eta_1 = 0.95 - 0.005 \cdot 3.6V_1. \tag{10}$$

6. Calculation of the required power N_2 and N_1 on the sprockets of the outside and inside tracks:

$$N_2 = \frac{P_2 \cdot V_2}{\eta_2}; \begin{cases} N_1 = P_1 \cdot V_1 \cdot \eta_1, \ if \ P_1 < 0 \\ N_1 = P_1 \cdot V_1 / \eta_1, \ if \ P_1 > 0 \end{cases}. \tag{11}$$

7. Calculation of the required drive torques M_2 and M_1 on the sprockets of the outside and inside tracks:

$$M_2 = \frac{P_2 \cdot r_k}{\eta_2}; \begin{cases} M_1 = P_1 \cdot r_k \cdot \eta_1, \ if \ P_1 < 0 \\ M_1 = P_1 \cdot r_k / \eta_1, \ if \ P_1 > 0 \end{cases}. \tag{12}$$

Then, based on the data taken from the given characteristics the required torque — speed curve in the transport gear is calculated.The calculation is performed according to the following method:

1. Calculation of the rotation speeds ω_2 and ω_1 for the outside and inside tracks:

$$\omega_2 = V_2 / r_k; \quad \omega_1 = V_1 / r_k. \tag{13}$$

2. Calculation of the rotation speeds ω_{e2} and ω_{e1} for the drive motors of the outside and inside tracks:

$$\omega_{e2} = \left(\left(1 + k_{zk}\right) \cdot \omega_2 - k_{zk} \cdot \omega_1 \right) \cdot i_{tr}^v; \ \omega_{e1} = \left(\left(1 + k_{zk}\right) \cdot \omega_1 - k_{zk} \cdot \omega_2 \right) \cdot i_{tr}^v. \tag{14}$$

3. Calculation of the required torques M_{e2} and M_{e1} of the drive motors of the the outside and inside tracks:

$$M_{e2} = \left(\frac{1 + k_{zk}}{1 + 2k_{zk}} M_2 + \frac{k_{zk}}{1 + 2k_{zk}} M_1 \right) \frac{1}{i_{tr}^v}; M_{e1} = \left(\frac{1 + k_{zk}}{1 + 2k_{zk}} M_1 + \frac{k_{zk}}{1 + 2k_{zk}} M_2 \right) \frac{1}{i_{tr}^v}. \tag{15}$$

4. Calculation of the required torques N_{e2} and N_{e1} of the drive motors of the the outside and inside tracks:

$$N_{e2} = M_{e2} \cdot \omega_{e2}; \ N_{e1} = M_{e1} \cdot \omega_{e1}. \tag{16}$$

In addition, using the required speeds of the tracks the time spent on the full turn of the tractor around its vertical axis T can be calculated:

$$T = \frac{2\pi \cdot B}{V_2 - V_1}. \tag{17}$$

Figures 2–4 show the torque — speed and output power — speed curves of the drive motor at different turning radii and tractor speed 10 km/h in transport gear on a turfy soil.

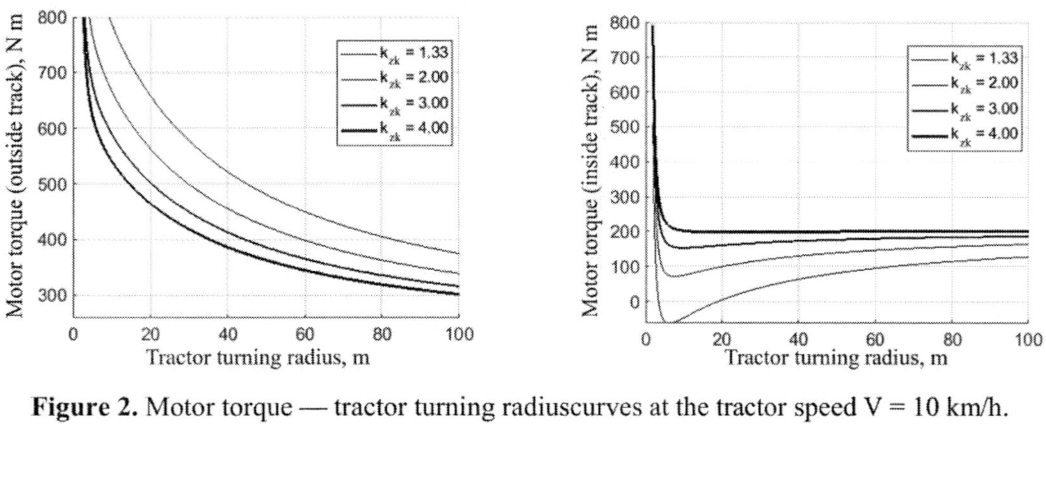

Figure 2. Motor torque — tractor turning radiuscurves at the tractor speed V = 10 km/h.

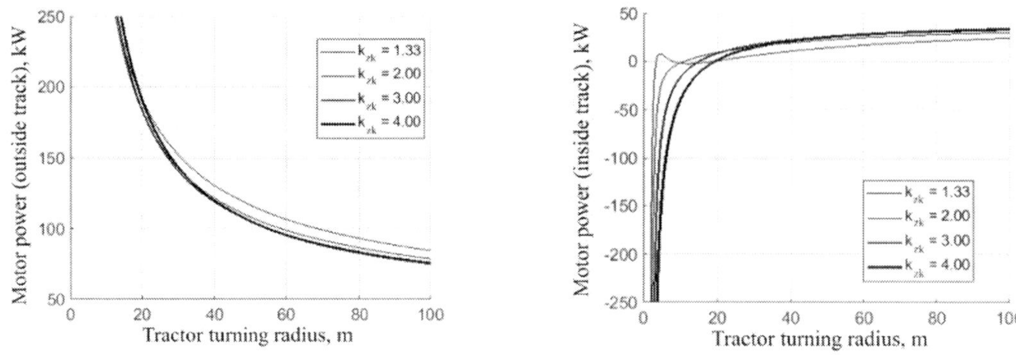

Figure 3. Motor output power– tractor turning radius curves at the tractor speed V = 10 km/h.

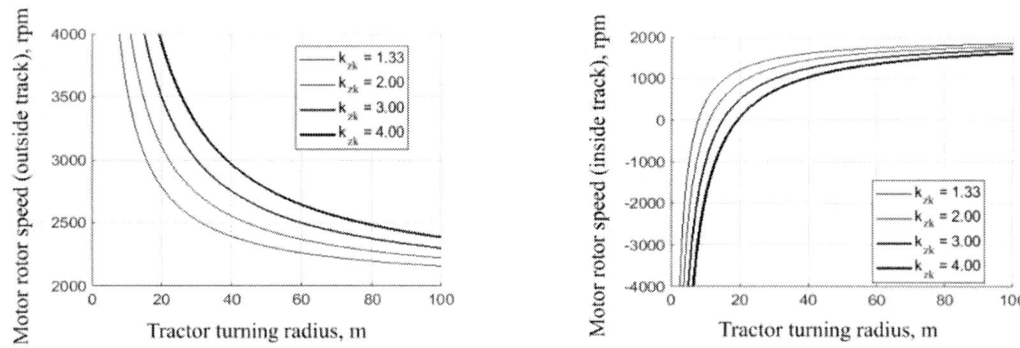

Figure 4. Motor rotor speed — turning radius curves at the tractor speed V = 10 km/h.

These results show that with the decrease in the parameter The power required for the tractor maneuver under the assumptions made is weakly dependent on the parameter k_{zk} of the ZK planetary gear sets.

During its operation the tractor makes turnswith small radii including full turn about a stopped track and full turn around the tractor center of mass (when this maneuver is possible).In this regard, it is advisable to consider the impact of the parameter on the operation of the transmission when performing such maneuvers in the transport and working gears.The required parameters of the drive motor when the tractor pivots about its stopped track are calculated by the formulae (5–17). For the case of the tractor pivoting around its center of mass the speeds V_1 and V_2 of the tracks and the required trative efforts P_1 and P_2 on the tracks are calculated according to the following method.The speed of the drive motors was set to its maximum value ($|n_{1,2}| = 4000\,rpm$).

1. Calculation of the required track speeds Vand V_2 of the outside and inside tracks during pivoting around the tractor center of mass:

$$\omega_2 = \left(\frac{1+k_{zk}}{1+2k_{zk}} \cdot \omega_{e2} + \frac{k_{zk}}{1+2k_{zk}} \cdot \omega_{e1} \right) \cdot \frac{1}{i_{tr}^{v}}; \tag{18}$$

$$\omega_1 = \left(\frac{1+k_{zk}}{1+2k_{zk}} \cdot \omega_{e1} + \frac{k_{zk}}{1+2k_{zk}} \cdot \omega_{e2} \right) \cdot \frac{1}{i_{tr}^{v}}; \tag{19}$$

$$V_2 = \omega_2 \cdot r_k; \; V_1 = \omega_1 \cdot r_k. \tag{20}$$

2. Calculation of the required tractive efforts Pand P_2 on the outside and inside tracks during pivoting around the tractor center of mass:

$$P_2 = mg \cdot \left(-0.5 f_{gr} + \frac{\mu_{max} \cdot L}{4 \cdot B} \right); \; P_1 = mg \cdot \left(0.5 f_{gr} - \frac{\mu_{max} \cdot L}{4 \cdot B} \right). \tag{21}$$

The calculation results are shown in tables 3–6.

Table 3. Parameters of the tractor pivoting about its stopped track in the ransport gear

	parameter k_{zk}			
	1.33	2.00	3.00	4.00
Required torque $M_{e1}$of the drive motor of the inner track,Nm	-178.6	-43.7	42.0	95.0
Required torque $M_{e2}$of the drive motor of the outer track, Nm	785.2	636.8	542.1	483.2
Required output power $N_{e1}$of the drive motor of the inner track,kW	22.3	6.5	-6.9	-16.7
Required output power $N_{e2}$of the drive motor of the outer track,kW	172.0	139.5	118.9	105.9
Duration t of the tractor full turn, s	11.1	14.6	19.0	23.8

Table 4. Parameters of the tractor pivoting about its stopped track in the working gear

	parameter k_{zk}			
	1.33	2.00	3.00	4.00
Required torque $M_{e1}$of the drive motor of the inner track, Nm	-626.7	-473.7	-364.5	-291.5
Required torque $M_{e2}$of the drive motor of the outer track, Nm	886.0	665.7	510.2	407.0
Required output power $N_{e1}$of the drive motor of the inner track,kW	0	0	0	0
Required output power $N_{e2}$of the drive motor of the outer track,kW	185.7	139.7	107.0	85.2
Duration t of the tractor full turn, s	11.6	15.3	19.9	24.9

Table 5. Parameters of the tractor pivoting about its center of mass in the transport gear.

	parameter k_{zk}			
	1.33	2.00	3.00	4.00
Required torque $M_{e2}(-M_{e1})$ of the drive motor of the outer (inner) track, Nm	454.0	319.3	234.1	181.5
Required output power$N_{e2}(N_{e1})$ of the drive motor of the outer (inner) track,kW	190.2	133.7	98.1	76.0
Duration t of the tractor full turn, s	4.55	6.41	8.7	11.2

Table 6. Parameters of the tractor pivoting about its center of mass in the working gear.

	parameter k_{zk}			
	1.33	2.00	3.00	4.00
Required torque $M_{e2}(-M_{e1})$ of the drive motor of the outer (inner) track, Nm	725.5	542.6	414.4	329.7
Required output power $N_{e2}(N_{e1})$ of the drive motor of the outer (inner) track,kW	303.9	227.3	173.6	138.1
Duration t of the tractor full turn, s	2.89	3.88	5.0	6.2

3. In contrast to the rotation with a large turning radius, in the case when the tractor turns around its center of mass and around its stopped track, the parameterhas an impact on the required power of the drive motor.The results given in tables 2–5 show that with the increase in the parameter required power and torque of the drive motor decrease and the duration of the tractor full turn rises.When the parameter k_{zk}risies from 1.33 to 4, the duration of the tractor full turn increases by the factor of 2.22 at the average.

3. Calculation of the required output power of the drive motors

The maximum power of the drive motor should provide turning maneuvers at the tractor maximum speedlimitedby comfort conditions of the driver.It is known that the comfort limit of the lateral acceleration ay is 0.05g [20].Then, the tractor maximum speed in a turn (limited by the driver comfort) can be calculated as:

$$V_{pr} = \sqrt{a_y \cdot R_f} = \sqrt{0.05g \cdot R_f}\,. \qquad (22)$$

Next, by the method of successive approximations the maximum power of the drive motor providing maneuvering at a given tractor speed can be calculated (see figure 5).

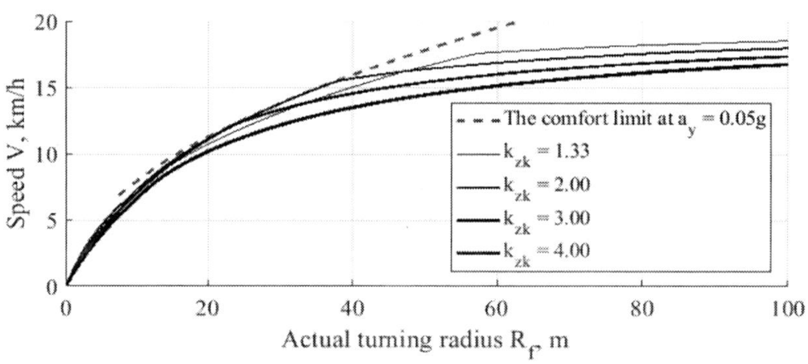

Figure 5. Relationship between the tractor maximum speed and its turning radius

Thus, the required power of the drive motor of the tractor in question is 415 kW (the power to weight ratio is 10 kW / t).Besides, in order to minimize the mass and overall dimensions of the drive motor and to provide the maximum efficiency of the steering system (mechanical energy recuperation between the tracks at low turning radii) based on the data in figures 2–5 the value of the parameter k_{zk} should be 2.

Tables 2 and 7 as well as figure 6 show the calculated performace characteristics of the tractor for the selected parameters of the electromechanical transmission.

Table 7. Calculated performance characteristics of the tractor.

Parameter	Value
Maximum tractive effort to weight ratio in the working gear	1.232
Maximum speed in the the working gear, km/h	6.5
Working gear ratio	98.27
Duration t of the tractor full turn around its center of mass in the transport gear, s	6.4
Duration t of the tractor full turn around its center of mass in the working gear, s	3.9
Duration *t* of the tractor full turn around its stopped track in the transport gear, s	14.6
Duration *t* of the tractor full turn around its stopped track in the working gear, s	15.3

IOP Conf. Series: Materials Science and Engineering **820** (2020) 012032 doi:10.1088/1757-899X/820/1/012032

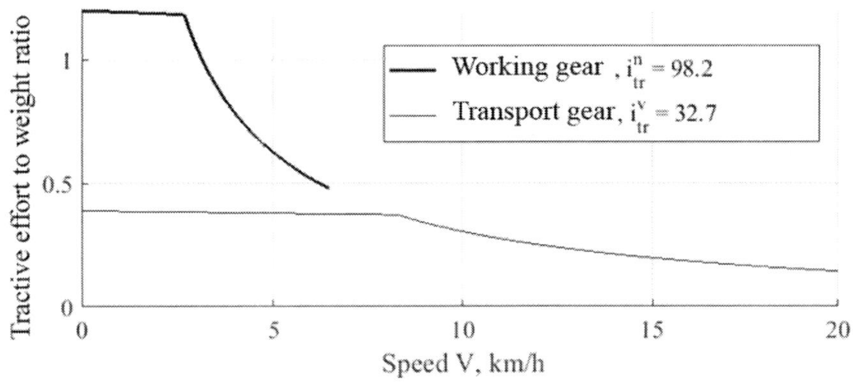

Figure 6. Tractive effort to weight ratio — speed characteristics of the tractor.

4. Conclusion

The proposed layout of the electromechanical transmission is mostly suitable for the industrial tactors.The transmission layout features planetary gear sets as reduction units when operating in the working gear and the use of the same planetary gear sets as the elements of the ZK steering system in the transport gear.This approach is advatageous in terms of the tractor systems packaging and can provide better performance characteristics for the existing tractors.

The method described in the article has allowed giving recommendations the main parameters of the proposed transmission: parameter kzk of the ZK planetary gear sets is 2, maximum tractive effort to weight ratio in the working gear is 1.2, maximum tractive effort to weight ratio in the trasport gear is 0.4.

The research results has shown that in order to provide the maximum operation efficiency of the tractor with the limitations imposed by the driver comfort conditions there is no need to increase the power to weight ratio of the tractor drive motors higher than the value of 10 kW/t.

References

[1] Ginzburg Yu V, Shved A I, Parfenov A P 1986 Industrial tractors. Mashinostroenie.

[2] Research on a Novel Type of Hybrid Propulsion System and Control Strategy for Tracked Vehicle / Jiangtao Gai, Tian Ma, Shoudao Huang, Xin Zhang, Xiaofeng Ma, Gongding Bai // International Conference on Electrical and Control Engineering — 2010. pp 4066–69.

[3] Dynamic Modeling and Simulation on Automatic Transmission of Tracked Vehicle Using Fuzzy Control / Wu Shijing, Li Qunli, Zhu Enyong, Zhang Dawei, Xie Jing // 27th Chinese Control Conference, Kunming, Yunnan, China — 2008. pp 387–390.

[4] The Shift Modeling and Simulation of Hydro-Mechanical Continuously Variable Transmission of Tracked Vehicle in Turning Conditions / Cao Fuyi, Guo Guanglin, Lei Yanlong // IEEE Advanced Information Management, Communicates, Electronic and Automation Control Conference (IMCEC) — 2016. pp 1375–78.

[5] Simulation of Steady Turning Movement Trace of Tracked Vehicle of DoublePower Flow Transmission / Cao Fuyi, Zhou Zhili, Xu Liyou, Zhang Mingzhu // International Conference on Intelligent Computation Technology and Automation — 2010. pp 333–336.

[6] Control and Parameters Matching of Straight Running for High speed Tracked Vehicle with Hydrostatic Drive / HaiLing Zhang, Heyan Li, Biao Ma, Zhifeng Gao // IEEE International Conference on Mechatronics and Automation, Chengdu, China — 2012. pp 887–891.

[7] Padalkin B V, Kharitonov S A, Kotiev GO2018.Analysis of the methods for development of electric transmissions of tracked vehicles *Actual problems of development of rocket and space technology and weapons systems* pp 228–244.

[8] Chobitok V.A. 1984Locomotion Theory of Tanks and IFVs. Voenizdat

[9] Krasnen'kov V I, Vashets A.D. 1986 Development of planetary mechanisms of transport vehicles. Mashinostroenie.
[10] Methods for road microprofile statistical data transformation / Evgeniy Sarach, George Kotiev and Sergey Beketov // MATEC Web of Conferences — 2018. V224. article № 04009.
[11] Designs and development of Russian scientific schools in the field of cross-country ground vehicles building / George O. Kotiev, Boris V. Padalkin, Alexander B. Kartashov and Alex S. Diakov // ARPN Journal of Engineering and Applied Sciences — 2017. V 12.I4. pp 1064–1071.
[12] Automatic determination of soil parameters by roboticvehicles / Naumov, V.N., Mashkov, K.Y., Byakov, K.E //Journal of Physics: Conference Series — 2019. V 1177. I 1. article № 012016.
[13] Interaction between elements of the track ground contacting area with the soil at curvilinear motion of the timber harvesting machine/ Klubnichkin, V.E., Klubnichkin, E.E.,Kotiev, G.O.,Beketov, S.A.,Makarov, V.S. // IOP Conference Series: Materials Science and Engineering — 2018. V 386. I 1.article №012016.
[14] Control strategy based on BP neutral network plus PID algorithm for dual electric tracked vehicle steering / Chen Ze-yu, Zhang Cheng-ning // 2nd International Conference on Advanced Computer Control — 2010. pp 584–587.
[15] Study on Steering Power Balance Simulation for Hydrostatic Drive Tracked Vehicle / Lei Yang, Biao Ma, He Y. Li, and Bao R. Chen //IEEE Intelligent Vehicles Symposium — 2009. pp 1204–08.
[16] Kotiev, G.O., Miroshnichenko, A.V., Stadukhin, A.A., Kositsyn, B.B. Determination of mechanical characteristics of high-speed tracked vehicles traction motor with individual drive wheels(2019) Journal of Physics: Conference Series, 1177 (1), article № 012058
[17] Nikitin A O1962 Tank theory. MoscowAcademy of Armored Forces.
[18] Platonov V F, Leiashvili G R1986 Tracked and wheeled transport vehicles and tractors. Mashinostroenie
[19] Zabavnikov N A 1975 Fundamental theory of the Fundamentals of the theory of transport tracked vehicles Mashinostroenie.
[20] Rotenberg R V Automobile suspension. 1972 Mashinostroenie.

IOP Conf. Series: Materials Science and Engineering **820** (2020) 012033 doi:10.1088/1757-899X/820/1/012033

Crashworthiness design of truck's cabin using topological and parametric optimization

R Goncharov[1,2] and V Zuzov[1]

[1]Bauman Moscow State Technical University

[2]E-mail: goncharov.roman@bmstu.ru

Abstract. In the given work the basic problems of search of optimum parameters of a cabin of the truck based on parametrical and topological optimization for the purpose of maintenance of requirements on crashworthiness under the international rules and reception of its minimum mass were considered. The article presents the developed finite element models of the cabin and drummer in relation to the optimization tasks, allowing to obtain results with acceptable accuracy and minimum time of solution when using LS-OPT and LS-TaSC programs with LS-DYNA solver. Steel linings and a foam filler are used to reinforce the cabin. In order to solve the problem and more fully assess the impact of parameters, several options were considered to refine the cabin. Topological optimization was carried out in order to obtain a picture of the best distribution of filler on the frame of the cabin. Parametric optimization was carried out by selection of filler properties (aluminium foam) and thickness of structural elements of the cabin. As a result, the passive safety requirements were achieved with optimal mass distribution as a result of combined use of linings and filler. The cabin's mass increase was 20%.

1. Introduction

Increasing the requirements of ensuring the passive safety of people in the car is the most important task for designers working in this field [1–5]. Increase in the number of cars, speed of traffic on city streets and highways, growth of load-carrying capacity of newly developed cars [6–9] leads to necessity of creation the truck cabins that meet the requirements of passive safety with minimum mass and enough crashworthiness.

Passive safety testing is a complex and expensive task, so it is advisable to use the capabilities of modern computational analysis based on computer modelling to reduce the time and cost of design development. This kind of problems are solved by the finite element method (FEM) using an explicit method of solving differential equations.

Modern software packages allow solving highly non-linear problems with contact interaction between structural elements. They implement various optimization algorithms, including those based on hybrid cellular automata method (HCA) for topological optimization and various variation algorithms for parametric optimization

In this paper, topological and parametric optimization was carried out to select the parameters of aluminium foam and the properties of the elements of the frame type cabins to achieve the requirements for crashworthiness under international rules and at the optimal value of mass. The main problem in solving optimization problems associated with the impact is the machine calculation time (from a few days to several weeks). The LS-TaSC and LS-OPT programs with the explicit LS-DYNA solver were used for the solution, which gives good results in terms of the ratio of accuracy to computational time. Similar problems were solved in the works [10–13], but they do not give the methods of searching for the optimal parameters for the truck cabs from the standpoint of crashworthiness.

The purpose of the work is to meet the requirements of the passive safety rules of the truck cabin and minimize mass due to the optimal distribution and selection of aluminium foam parameters and thicknesses of frame type elements in the cabin.

2. Mathematical model

2.1. Analysis of the stress-strain states of the cabin during crashworthiness tests

Crashworthiness test are designed to simulate real-world emergency scenarios. In the case of trucks, these situations include frontal impact, rear-end impact of the braked load, and tipping over of the truck cabin. In accordance with VVVFS 2003:29, the crashworthiness of the cabin is verified by the following tests, as shown in Figure 1: static load on the cabin roof; front impact by a cylindrical drummer on the top front corner of the cabin; right-angle impact by a rectangular drummer on the rear wall of the cabin.

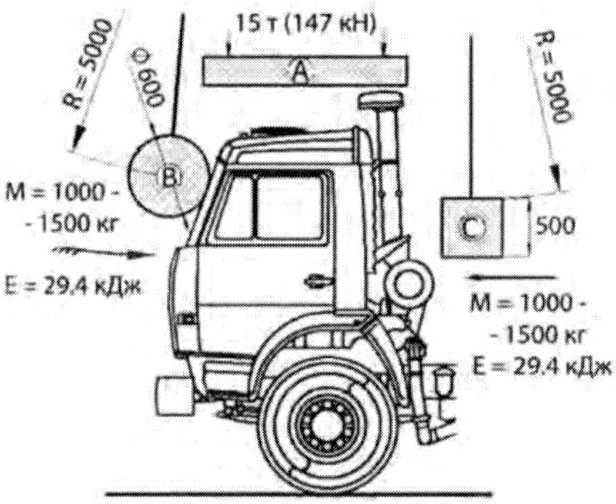

Figure 1. Crashworthiness tests VVFS 2003:29

The truck cabin is considered to have passed all types of tests under the following conditions:

– The supporting structure of the cabin or parts, as well as the fastening units, have not been damaged, and no significant cracks or deformations have occurred;

– There is still living space in the cabin for the driver and passengers;

– When dynamically loaded (impacted), the entire impact energy is consumed (absorbed) by the cabin without the drummer slipping over the roof.

The object of research in this paper is cabin of a truck. The cabin is shown in Figure 2, a.

The FEM of the cabin is shown in Figure 2, b. The size of the finite element is 35 mm, the number of elements is 28521. Cabin elements were modelled mainly by 4-knot shell elements such as Belytschko-Tsay. The thickness of internal and external panels is 1 mm, the cabin spars are 4 mm. Cabin material is steel 08kp with kinematic hardening, cabin mass 250 kg.

To check the performance of the cabin, a calculation was made to assess compliance with the requirements of passive safety rules. In this article the most complicated mode for the design is considered: drummer impact test with the mass of 1000 kg with the initial speed of 7,6 m/s on the front post of the cabin at an angle of 15°.

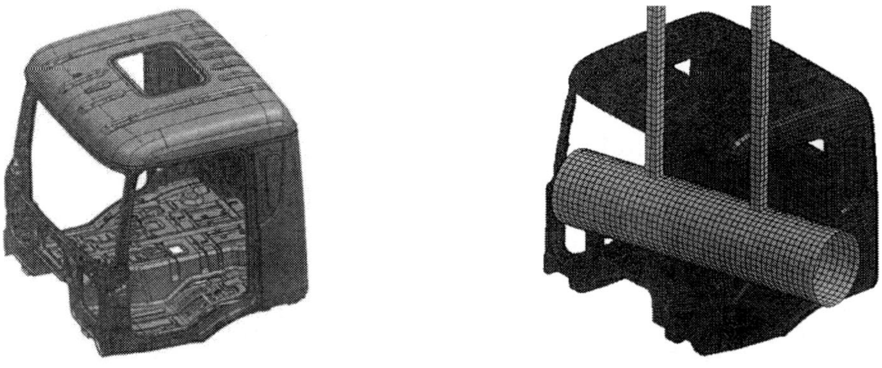

a b

Figure 2. Object of research:
a — 3D model; b — FEM at the moment of pendulum impact

The time of calculation in the LS-DYNA program is 41 minutes, which in case of solving the optimization problem is acceptable from the point of view of time, provided that the deformation character is identical to the experimental one. With a larger finite element size, the resulting calculation error does not produce adequate results [14]. If more accurate FEMs are used, the calculation time is increased many times, resulting in excessive time consumption. The calculation results are shown in Figure 3.

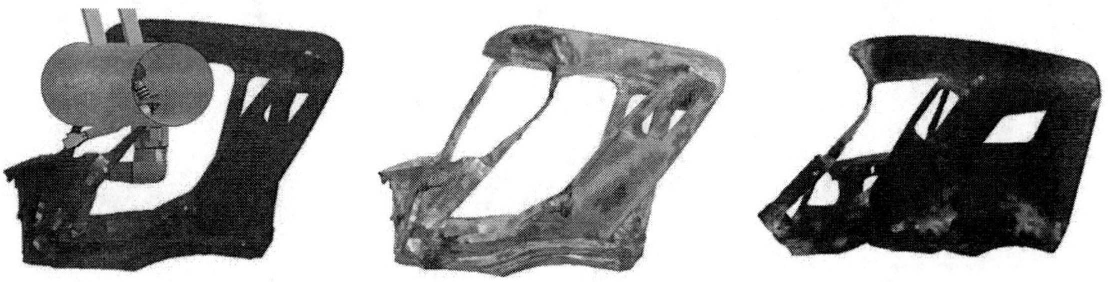

a b c

Figure 3. Cabin results:
a — FEM in the final phase of impact; b — stress Mises; c — deformed state of the cabin

Results show that the residual life space is significantly less than the required values. Extensive plastic deformations have arisen in the design, the doorway has been severely deformed, and collapse zones are observed at the points where the thresholds are connected to the front of the cabin.

Based on the analysis carried out, the aim was to find optimal solutions for the cabin in order to meet the requirements for crashworthiness and minimize mass.

In order to achieve this objective and to better assess the effect of the parameters, the approach of sharing an aluminium foam filler in frame type elements and linings has been considered.

2.2. Topological optimization

In order to optimally distribute the filler (aluminium foam) over the cabin frame, a topological optimization was carried out for the three loading modes (VVFS 2003:29) shown in Figure 4.

To solve the problems of crashworthiness was created FEM, consisting of shell elements (cabin) and solid elements (such as tetra 10 — aluminium foam). The whole set of solid elements constitutes the design space, which is involved in topological optimization. The size of the element is 35 mm (at the place of impact and stress zones, the size of the element is 10–15 mm), and the number of

IOP Conf. Series: Materials Science and Engineering **820** (2020) 012033 doi:10.1088/1757-899X/820/1/012033

elements is 67890. The aim of the optimization is to provide living space in the structure and to minimize mass. The main objective of the calculation is to obtain a picture of the optimal distribution of the material in the cabin frame in the event of impact loading with a drummer.

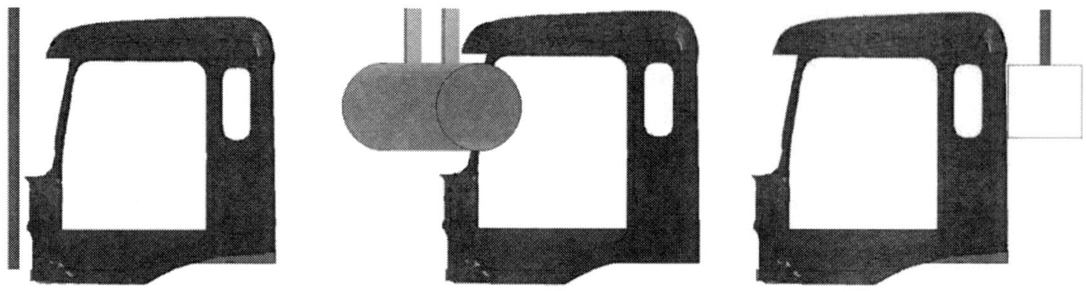

a b c

Figure 4. Test types for topological optimization:
a — frontal strike; b — rack strike; c — rear strike

Optimization was carried out by the hybrid cellular automata method (HCA) in the program LS-TaSC, as one of the most effective for such problems [15–18]. The results of topological optimization are presented in Figure 5.

Figure 5. Topological optimization results

Results show that the main elements that transmit the load on the brackets of the cabin to the frame, when hitting the front drummer are the floor and side members. Aluminium foam must be in the cavities along the contour of the doorway, tightens the structure, preventing the emergence of plastic hinges when hitting the post. Frame-type elements of the front panel reinforced with filler close the door apertures and spars into a single contour, thus uniting the structure into a single whole.

The following conclusions can be drawn from the analysis of the results:

– It is necessary to fill the cavities around the doorway and the front panel with aluminium foam;

– The cabin floor and spars are the main elements that transmit the impact load on the brackets of the cabin to the frame;

The use of medium level FEM for the topological optimization task at the preliminary stage of development allows getting the best picture of the filler distribution by the volume of the cabin with acceptable computational time consumption.

2.3. Parametric optimization

Based on the results of topological optimization and comprehensive analysis of the stress-strain state of the cabin, the variant of distribution of filler (aluminum foam) and linings on the design of the cabin was considered.

To obtain a solution to the optimization problem, the cabin was divided into several parts (subzone) and filled with aluminum foam. The division into subzones allows to limit the number of variable parameters in such a way that their number was enough to obtain appropriate results at the minimum cost of estimated time. Based on preliminary studies, the subzones around the doorway, windscreen, spars and the rear wall of the cabin were identified. In addition to the filler, the doorways and the front part were reinforced with linings. The thickness of the spars also varied to increase the crashworthiness of the cabin. The scheme of division of the structure into subzones is shown in Figure 6.

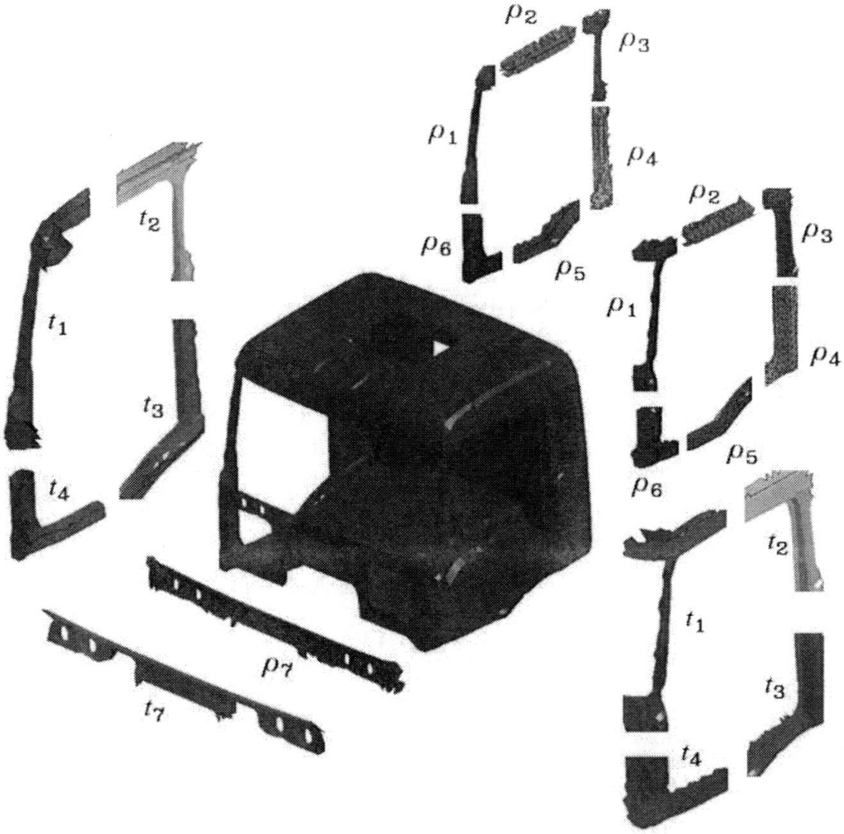

Figure 6. Cabin redesign scheme for optimization

The aluminum foam in the program LS-DYNA is described by the model proposed by Deshpande and Fleck [19]. In this model, the properties of the filler depend on the density, which allows it to be used to solve optimization problems, choosing density as a variable parameter. The use of this model in relation to structural optimization problems is presented in [20–22].

The variable parameters are the density of foam aluminum ρ_i, which varies from 100 kg/m^3 to 800 kg/m^3, and elements of frame type with variants of thickness t_j : 0,1 mm; 1 mm; 2 mm; 3 mm; 4 mm. They are analogous to the linings, increasing the crashworthiness of the cabin, all parameters vary independently of each other. The aluminum foam fills the area around the doorway and the subzone under the windscreen. The use of aluminum foam increases the crashworthiness of thin-walled closed profiles [23], such as truck cabin racks.

Based on previous analysis, the maximum displacement of the rack at the checkpoint should not exceed 350 mm (the checkpoint is the node on the rack that has the maximum movement). The optimization was solved by using metamodel (Radial basis function) and adaptive simulated annealing algorithm (ASA). The optimization results are in Table 1.

Table 1. Results of combined optimization of foam density and lining's thickness

Aluminum foam	ρ_1	ρ_2	ρ_3	ρ_4	ρ_5	ρ_6	ρ_7
Density, kg/m^3	800	100	680	100	120	435	640
Lining	t_1	t_2	t_3	t_4	t_7		
Thickness, mm	4	1	0,1	0,1	0,1		
Mass, kg				306			
Displacement of the checkpoint, mm				358			

It is necessary to avoid extensive plastic deformation of the subzone ρ_1 and ρ_3 to fill with aluminum foam with a density of 800 kg/m^3, 680 kg/m^3, respectively, and to strengthen the front post with a 4 mm lining, in order to meet the requirements for the crashworthiness of the cabin when hitting the front post by a drummer. As a result, displacement of the checkpoint does not exceed the specified value of 350 mm, and the cabin mass is 306 kg

3. Result and discussion
Table 2 shows the results for basic and advanced cabin variants.

Table 2. Comparison of basic and advanced cabin variants

	Displacement of the checkpoint, mm	Cabin's mass, kg	Mass increase, %	Satisfying safety standards
Basic cabin	850	250	-	No
Cabin with fillers and linings	349	306	22	Yes

After optimization, the cabin was modified for practical use. Filler zones ρ_4, ρ_5 and frame type elements with a thickness of 0,1 mm were removed from the structure, and the t_1 element was replaced by a 4 mm equivalent stiffness overlay.

As a result of the final modification, the cabin has become compliant with the requirements for crashworthiness, but the mass of the cabin has increased by 20% to 300 kg compared to the original design.

4. Conclusion
The use of the proposed combined approach (topological optimization for the distribution of filler and parametric optimization of the thicknesses of linings and aluminum foam parameters) allowed to achieve the specified properties of crashworthiness of the truck cabin design with a minimum increase in mass (20%).

Application of the approach based on the division of the task into subtasks allowed us to increase the efficiency of the solution of the optimization task.

References
[1] Gorelov V A and Komissarov A I 2016 *Procedia Engineering* **150** 1322-28
[2] Gorelov V A, Komissarov A I and Miroshnichenko A V 2015 *Procedia Engineering* **129** 300-7
[3] Keller A V, Gorelov V A and Anchukov V V 2015 *Procedia Engineering* **129** 280-7
[4] Kotiev G O, Padalkin B V, Kartashov A B and Diakov A S 2017 *ARPN J. of Eng. and Appl. Sci.* **12** 1064-71
[5] Volskaya N S, Zhileykin M M and Zakharov A Y 2018 *IOP Conf. Series: Materials Science and Engineering* **315** 012028

[6] Vdovin D and Chichekin I 2016 *Procedia Engineering* **150** 1276-79
[7] Klubnichkin E E, Klubnichkin V E and Kotiev G O 2018 *IOP Conf. Series: Materials Science and Engineering* **386** 012025
[8] Zhileykin M M, Kotiev G O and Nagatsev M V 2018 *IOP Conf. Series: Materials Science and Engineering* **315** 012031
[9] Sarach E B, Kotiev G O and Beketov S A 2018 *MATEC Web of Conf.* **224** 04009
[10] Skotnikov G I, Jileykin M M and Komissarov A I 2018 *IOP Conf. Series: Materials Science and Engineering* **315** 012027
[11] Diakov A S and Kotiev G O 2018 *MATEC Web of Conf.* **224** 02096
[12] Kotiev G O, Butarovich D O and Kositsyn B B 2018 *IOP Conf. Series: Materials Science and Engineering* **315** 012014
[13] Evseev K B, Kartashov A B, Dashtiev I Z and Pozdeev A V 2018 *MATEC Web of Conf.* **224** 02039
[14] Goncharov R B and Zuzov V N 2018 *Transactions of NNSTU n.a. R.E. Alekseev* **3** 130-6
[15] Tovar A, Patel N M, Kang B S and Renaud J E 2009 *Journal of mechanical design* **131** 1-12
[16] Patel N M, Tillotson D, Rehaud J E, Tovar A and Izui K 2008 *AIAA Journal* **46** 1963-75
[17] Penninger C L, Patel N M, Niebur G L, Tovar A and Rehaud J E 2008 *Mechanics Research Communications* **35** 32-42
[18] Gan N, Yao S, Xiong Y and Hong X 2018 *Engineering Optimization* **50** 2054-70
[19] Deshpande V S and Fleck N A 2000 *J. Mech. Phys. Solids* **48** 1253–83.
[20] Zhaokai Li, Qiang Yu, Xuan Zhao, Man Yu, Peilong Shi and Cilei Yan 2017 *Adv. Mech. Eng.* **9** 1-21
[21] Liuwei Guo and Jilin Yu 2011 *International Journal of Impact Engineering* **38** 85-94
[22] Zhi Xiao, Jianguang Fang, Guangyong Sun and Qing Li 2015 *Int. J. Advances in engineering software* **85** 81-95
[23] Yi Z, Si-yuan H, Jia-gui L, Wei Z, Xiao-lu G and Jin Y 2019 *Composite Structures* **220** 451-459

IOP Conf. Series: Materials Science and Engineering **820** (2020) 012034 doi:10.1088/1757-899X/820/1/012034

One of the approaches for the analytical description of the interaction of an elastic wheel with a solid surface

S Popov[1]

[1]Bauman Moscow State Technical University

[2]E-mail: unesco-tvet-01@yandex.ru

Abstract. The design of branched transmissions of multi-axle vehicles, multistage steering systems and automated workflow control systems requires their preliminary dynamic calculation. These calculations are based on mathematical models of structural elements corresponding to the modern level of development of system theory, i.e. taking into account dynamic interrelations between the system links. The aim of the work is to build a mathematical model of the automobile wheeled propulsor intended for use in the dynamic model of the "car-road" system or models of separate automobile systems. The approach to mathematical modeling of the automobile wheeled propulsor is offered, consisting in its description by methods of the control theory as an element of dynamic system. The method of construction and analysis of the complete system of equations of kinematic bonds superimposed in the contact zone on the movement of points of the external surface of the tire of linear elastic wheel at its perturbed rectilinear rolling on a solid uneven support surface is developed. The method of construction of a transfer matrix of automobile linear elastic wheel at which its structure practically does not depend on features of the settlement scheme of the tyres is developed too in this study. The method of obtaining the tire characteristics necessary for the calculation of the transfer matrix elements by the generalized method of successive approximations, which gives the qualitatively correct results in zero approximation, is proposed. The method of qualitative theoretical analysis of the features of rectilinear stationary rolling of the automobile wheel is given in this paper. The results allow to carry out researches and calculations of dynamic characteristics of automobile systems, units and assemblies directly interacting with the wheel; they can serve as a basis for further improvement of mathematical models of automobile ruts by methods of system theory.

1. Introduction

The scientific literature devoted to the study of the elastic wheel in the general dynamic model of a car, the processes occurring during the rolling of elastic bodies and the development of their mathematical models is very extensive [1–27], therefore, when analyzing it is advisable to describe the whole scientific directions on the example of the most complete and complete works. In the context of this work, the scope of modeling tasks is limited to the case of rolling the wheel on a hard road, as it is the most widespread, although the mathematical model of the wheel of high cross-country vehicles should allow for the possibility of generalization in the event of interaction with pliable soil without disrupting its structure or principal processing.

The following principal approaches to the modeling of wheeled propulsion properties can be distinguished: 1) the use of "conceptual models" based on hypotheses about the processes occurring during rolling; 2) the use of simplified tire models; 3) adaptation of already known solutions of similar problems, most often - the classical contact problem of the elasticity theory; 4) the creation of special mathematical models.

The main limitation of the first direction of work (the hypothesis of "apparent sliding" and lateral withdrawal) is the stationarity of the problem. The peculiarity of conceptual models is that their equations contain unknown in advance parameters of interaction of tyres about the road, and for their calculation various simplified models of tyres are offered. The current high level of solution of internal tire mechanics problems allows to develop methods of calculation of parameters of conceptual models with full account of the peculiarities of tire construction. The more general results are obtained on the basis of the classical contact problem of elasticity theory, developing it into the theory of rolling elastic bodies. Such models differ in their distraction from the design of real tyres, stationary mode of rolling and difficulty in spreading the results to the spatial motion of the wheel. In general, in the study of stationary rolling of the wheel on a solid road a high level of theoretical generalization and accumulated significant experimental material. Otherwise, it is the case with the study of unsteady rolling of the wheel - here the course of research was determined by the works based on a priori hypotheses about the interaction of the tyre with the road and requiring experimental determination of its characteristics. The need for experiments with real tires is a serious inconvenience that makes it difficult to use the results at the design stage, especially in cases where the experimental determination of tire parameters is difficult or impossible, - for example, for large-size tires of heavy vehicles.

2. Mathematical model of the automobile wheel

2.1. Requirements for the mathematical model

In general, there is a tendency to create a structure of mathematical model of the automobile wheel, which would allow to combine the calculations on the basis of the classical contact problem of the theory of elasticity with a detailed description of the design of the tire and would cover the non-stationary modes of motion. This avoids the use of a priori hypotheses about the rolling mechanism, and the results are sought in a form suitable for the use of the model in the dynamic system "vehicle-road". In addition, there is a gradual separation of two main directions: the study of internal processes (i.e., those occurring in the tire and depending on the peculiarities of its construction, materials, etc.) and the study of the interaction of the elastic wheel with the road and its features (i.e., affecting the operation of systems and units connected to the wheel). In other words, the tasks of external and internal mechanics of the automobile wheel are distinguished. It should be noted that the solutions of external problems are important in the design of automobiles. Hence the desire to create such a model of elastic wheel, the structure of which would depend as less as possible on the peculiarities of the design scheme of the tyres, as the design of tyres are diverse and the internal processes in the tyres are very complex.

Thus, the level of development of the car theory, as well as the logic and trend of development of the car wheel theory require the construction of a mathematical model of the wheel that meets the following conditions:

- model should be based on a limited number of general hypotheses that allow for direct experimental testing (axiomaticity),
- model should be describe the spatial non-stationary movement of the wheel on an uneven road,
- model should be universal in relation to different tyres design schemes and be consistent with their possible mathematical description,
- model should be provide for the possibility of generalization for more complex cases of movement (on the ground, etc.),
- structure of the model should be convenient for use in the calculation of control systems.

In our paper we consider an insulated wheel consisting of a rigid rim and an elastic tyre fixed to it. The wheel rolls on a smooth, hard, uneven support surface. The reactions at the points where the wheel rim is connected to the suspension, transmission and steering are led to the main vector and the main torque, which are set by six scalar time functions. The movement of the wheel rim is also described by six scalar time functions, which form the vectors of its linear speed and angular velocity.

2.2. Theoretical background

In the general case, the model of the automobile wheel that meets the above requirements is considered a nonlinear operator $A(\bullet;\bullet)$, who connected the space of vectors of the input parameters of the model $R[\bar{u}]$ and the space of vectors of the state of the model $R[\bar{\alpha}]$ with the space of vectors of its output parameters $R[\bar{y}]$:

$$\bar{y}(t)=A\left(\bar{\alpha}(t_0);\bar{u}(t,t_0)\right) \tag{1}$$

In this paper we use two equivalent definitions of the vector of input parameters - the main ("power input") $\bar{u}\left(f^{(n)};\bar{M};\bar{F}\right)'$ and auxiliary ("kinematic input") $\bar{u}\left(f^{(n)};\bar{V};\bar{\Omega}\right)'$, which are most often found in the applied tasks of the vehicle's theory. They correspond to the vectors of output parameters $\bar{y}=\left(\bar{V};\bar{\Omega}\right)'$ and $\bar{y}=\left(\bar{M};\bar{F}\right)'$. In these formulas $f^{(n)}$ — derivatives from the function describing the shape of the road; $\bar{F}$ — the main vector of external forces acting on the wheel rim, $\bar{M}$ — the main moment of these forces; $\bar{V}$ — the linear speed of the wheel rim; $\bar{\Omega}$ — its angular speed.

Because of the extreme complexity of the nonlinear problem in the general formulation, it is advisable to limit oneself to the case of perturbed rectilinear motion of the wheel, defined as a set of independent basic motion (evolution) and oscillation near the evolutionary trajectory. For simplification purposes, only rectilinear stationary evolutionary motion is considered, and the solution is performed in the space of images of the two-way Laplace transformation, therefore (1) disintegrates into two matrix equations - evolutionary

$$\bar{Y}=A\left(\bar{\alpha}_0;\bar{u}\right) \tag{2}$$

and outraged movement

$$\tilde{y}(p_t)=\tilde{A}\left(\bar{\alpha}_0;\bar{Y}\right)\circ\tilde{u}(p_t) \tag{3}$$

The solution of the problem can be considered achieved if the matrix structure $\tilde{A}$ ("transfer matrix of linear elastic wheel") corresponding to the above mentioned conditions is found and described in terms of system theory, the possibility of using for construction of operators A and $\tilde{A}$ any of the existing tire models is proved, and the identification of mathematical model with the real prototype of the automobile wheel is performed.

The Lagrangian variation principle is used to derive the equations of motion, the labour intensity of which is justified by the possibility of successively opening up all the necessary simplifications and obtaining the conditions for determining the boundary of the contact zone; the reactions of all the bonds are explicitly introduced, as it allows the simplest way to obtain a closed system of equations of the elastic wheel

$$M\frac{d\bar{V}}{dt}-\bar{F}-\int_{S_K}\bar{Q}dS_\kappa=0 \tag{4}$$

$$\left\{J_{\chi\kappa}\frac{d\Omega^{\kappa}_{\chi o\delta}}{dt}\right\}-\left[\bar{\Omega}_{o\delta};\left\{J_{\chi\kappa}\Omega^{\kappa}_{\chi o\delta}\right\}\right]-\bar{M}-\int_{S_K}\left[\bar{Q};\bar{r}^{\circ}+\bar{\xi}\right]dS_\kappa=0 \tag{5}$$

$$L\left[\bar{\xi}\right]=L_{s\beta}\left[\bar{\xi}\right]+L_t\left[\bar{\xi}\right]-\chi(S_\kappa)\bar{Q}=0 \tag{6}$$

$$\left(\bar{Q};\xi\right)\Big|_{\Gamma_\kappa} = 0 \tag{7}$$

In formulas (4...7) M — total weight of a wheel; $\bar{V}_{об}$ — linear speed of a rim; $\bar{\Omega}_{об} = \left\{\Omega^{\kappa}_{x\,об}\right\}$ — its angular speed; J^k_x — components of a tensor of inertia of the deformed wheel; $\bar{F}$ — main vector of external forces acting on the rim; $\bar{M}$ — main moment of these forces; $\bar{Q}$ — distributed reaction in the area of contact of the tire with the road; $\bar{r}^{\circ}$ — radius-vector, drawn from the center of the wheel and describing the position of the free external surface of the tire relative to the rim of the wheel; ξ — vector of moving the points of the outer surface of the tire from the initial position to the actual one; S_κ — display of the area of contact of the tire with the road on the initial outer surface of the tire; Γ_κ — boundary of the area S_κ; $\chi(S_\kappa)$ — characteristic function of the area; $L[\;]$ — linear differential operator; s; β — coordinates on the outer surface of the free tire; t — time.

The equations (4...7) are derived under the following assumptions:

- the rim and tire are symmetrical with respect to the rolling plane to the values of the same order of magnitude with the square of motion of the points of the outer surface of the tire,
- the starting point of the rim and tire center of gravity and inertia coincide with the center of the wheel,
- the material characteristics and design of the tire are independent of changes in the temperature in the vicinity of the tire of a steady state,
- the tire material is incompressible,
- the initial state of the bus does not depend on time, it is physically realizable and unambiguously determined by the finite number of constructive parameters;
- the motion of the tire elements can be studied within the framework of a geometrically linear (infinitesimal) model;
- the properties of the tire are described by differential equations, and the internal connections in the tire are known and allow for the calculation of deformations of the inner areas of the tire at known displacements and deformations of its outer surface;
- The tire temperature does not change significantly in the course of perturbation.

Instead of equation (6), it is proposed that it be treated in a mutually unambiguous manner, which, if it is not possible to travel on a ledge, to leave the road, etc., will take the form of

$$\lim_{t_0 \to -\infty} \xi(s;\beta;t;t_0) = \int_{-\infty}^{+\infty} \int_{S_\kappa} G(s;\beta;t;s_0;\beta_0;t_0;r)\chi(S_{\kappa 0})\bar{Q}(s_0;\beta_0;\tau)\,dS_{\kappa 0}\,d\tau \tag{8}$$

where $G(\;)$ — operator's Green function $L[\;]$. In this case, the initial conditions are related to the point $t_0 \to -\infty$ and are excluded from consideration.

If the distribution of movements in the contact zone and its boundary are known, the reaction of the road in the space of images of the two-way Laplace transformation is determined by the formula (9), which allows to solve the problem, if the boundary Γ^*_κ of the contact zone S_κ^* and the distribution of movements of points of the outer surface of the tire in this zone will be determined.

$$\tilde{\bar{Q}}(s;\beta^*;t)\chi(S_\kappa^*) = \int_{S_\kappa^*} \tilde{G}^{-1}(s;\beta^*;\,s_0;\beta_0;p_t)\tilde{\bar{\xi}}(s_0;\beta_0;\,p_t)\,dS_\kappa^* - \chi(S_\kappa^*)\tilde{\bar{Q}}_1(s;\beta^*;p_t) \tag{9}$$

The points on the external surface of the tyre lying inside the contact area are in contact with the road, so their movements before leaving contact are dependent on changes in the position of the rim of the wheel relative to the stationary coordinate system associated with the road. The vector equation shall be accurate to the values of the second order of lowness, taking into account possible slippage in the contact zone

IOP Conf. Series: Materials Science and Engineering **820** (2020) 012034 doi:10.1088/1757-899X/820/1/012034

$$\bar{V}_0+\bar{V}+\left[\bar{\Omega}_{об};\hat{\bar{r}}\right]+\frac{\partial\bar{\xi}}{\partial x^{in}}V^i_{0x^*}+\frac{\partial\bar{\xi}}{\partial t}=\bar{V}_{ск};\ i=\overline{1;2}. \tag{10}$$

where $\bar{V}_0$ — velocity of evolutionary motion of the wheel rim center; $\bar{V}$ — velocity of perturbed motion of the wheel rim center; $\bar{\Omega}_{об}$ — angular velocity of the wheel rim; $\hat{\bar{r}}$ — radius-vector describing the outer surface of the deformed tire in the coordinate system connected with the wheel rim; $\bar{V}_{ск}$ — velocity of the tire points in the contact zone relative to the supporting surface (slip rate).

If we consider that the shape of the reference surface is known, we can proceed to the vector-matrix form of equation (10)

$$\begin{gathered}P_1\left\{V_0^i+V_z^i-V^i_{скz}\right\}+P_2\omega_0^1+P_3\left\{\omega^i\right\}+P_4\left\{\Phi^i\right\}+P_5\left\{l^i\right\}+P_6\left\{\xi^i\right\}+P_7\left\{\xi_{\beta^*}{}^{i'}\right\}+\\+P_8\left\{\xi_S{}^{i'}\right\}+P_9\left\{\xi_t{}^{i'}\right\}=0\end{gathered} \tag{11}$$

The coefficient P_i — nonlinear matrix-functions of evolutionary motion parameters, the shape of the external surface of the free tire and the parameters of the reference surface. If we consider the angles of the initial installation of the wheel to be small ($\leq 10°$) and assume that the height of unevenness of the hard roads has the same order of smallness with the movements of the tire, then within the contact zone, these coefficients become matrices-functions linear to the coordinates of points on the outer surface of the free tire. Then equation (11) is translated into the space of images of two-sided Laplace transformation.

By analyzing the dependence of the change in the distance between the two adjacent points of the outer surface of the tire in contact with the road on the sliding speeds of these points, it is possible to obtain, to the accuracy of the values of the second order of smallness, three more equations of connection in the images on the Laplace

$$\begin{cases}-\dfrac{\partial\tilde{\xi}^1}{\partial\beta^*}-\tilde{\xi}^2\cos\phi+\tilde{\xi}^3\sin\phi=\dfrac{1}{p_t}\left\{-\dfrac{\partial\tilde{V}^3_{скz}}{\partial\beta^*}\cos\beta^*-\dfrac{\partial\tilde{V}^1_{скz}}{\partial\beta^*}\sin\beta^*\right\};\\ -\dfrac{\partial\tilde{\xi}^2}{\partial\beta^*}-r\dfrac{\partial\tilde{\xi}^1}{\partial\beta^*}-\tilde{\xi}^1\sin\phi=-\dfrac{\sin\phi}{p_t}\dfrac{\partial\tilde{V}^2_{скz}}{\partial\beta^*}-\dfrac{r\sin\beta^*}{p_t}\dfrac{\partial\tilde{V}^1_{скz}}{\partial s}-\\ -\dfrac{r\cos\phi\cos\beta^*}{p_t}\dfrac{\partial\tilde{V}^1_{скz}}{\partial\beta^*}-\dfrac{\cos\phi\sin\beta^*}{p_t}\dfrac{\partial\tilde{V}^3_{скz}}{\partial\beta^*};\\ -\dfrac{\partial\tilde{\xi}^3}{\partial\beta^*}+\dfrac{1}{\rho}\tilde{\xi}^3=\dfrac{1}{p_t}\left\{-\dfrac{\partial\tilde{V}1_{скz}}{\partial\beta^*}\cos\phi\cos\beta^*-\dfrac{\partial\tilde{V}^2_{скz}}{\partial s}\sin\phi+\dfrac{\partial\tilde{V}^3_{скz}}{\partial s}\cos\phi\sin\beta^*\right\}\end{cases} \tag{12}$$

Finally, in the contact area, the outer surface of the tyre follows the shape of the support surface:

$$\forall m_0\in S_к:Пp_{z^3}\left(\bar{r}^\circ+\bar{\xi}\right)=f\left(y^{1*};\,y^{2*}\right) \tag{13}$$

The equations (11...13) form a complete system of equations of kinematic bonds that determine the distribution of displacements in the contact zone.

There are three possible cases: in the contact zone there is also a coupling zone and a sliding zone (general case), the coupling zone is degraded to a point and the coupling zone covers the entire contact area except for its rear edge (limiting cases). The first of the limiting cases is not of theoretical interest, and in the second case the equations of connection are reduced to a normal system, which is compatible and quite integrated with the accuracy of the initial hypothesis of geometric linearity of the model. It can be shown that to study the perturbed motion it is enough to consider only the system of equations

$$\begin{cases} \dfrac{\partial \tilde{\xi}^1}{\partial \beta^*} = l_1\left(\tilde{\xi}^1;\tilde{\xi}^2;\Delta\beta^*;\Delta\phi\right) - \gamma \dfrac{R_0}{\rho_0}\dfrac{\partial \tilde{\xi}^1}{\partial \phi} \\ \dfrac{\partial \tilde{\xi}^2}{\partial \beta^*} = l_2\left(\tilde{\xi}^1;\tilde{\xi}^2;\Delta\beta^*;\Delta\phi\right) - \gamma \dfrac{R_0}{\rho_0}\dfrac{\partial \tilde{\xi}^2}{\partial \phi} \end{cases} \qquad (14)$$

where $l_1; l_2$ — linear forms from the corresponding variables are designated, and the other equations are automatically satisfied with the accuracy of the initial hypothesis. The general integral of the system (14) is easily constructed in Laplace images and depends on four arbitrary functions C_κ and the transformation parameter p_t. The use of images has made it possible to bypass the serious difficulties associated with the solution of the boundary value problem for the equations of motion of bus elements together with the equations of communication at the previously unknown mobile boundary of the contact zone.

Further, the structure of the transfer matrix is formed, which describes the perturbed rectilinear motion of the elastic wheel on a horizontal uneven hard road. First, from condition (7) is determined in the first approximation the boundary of the contact area for evolutionary and perturbed cases by the Freshet method. The rim motion equations are simplified by taking into account the hypothesis of geometric linearity of the model and are reduced to a dimensionless form using the following scaling values: $G_{ном}$ — scale of force (nominal vertical wheel load; R_0 — linear, scale (equator radius of the outer surface of the free tyre); V_0 — scale of speed (speed of the main movement of the wheel rim centre).

After excluding from these equations the image of the reaction in the contact zone, a group of formulas

$$\begin{cases} \left\{A_{qo} + A_{q\kappa}C_\kappa + f_{эz}^{q}\right\} = 0; \quad q = 1,2,3;\ k = 1,2,..,4; \\ \left\{A_{qo} + A_{q\kappa}C_\kappa + m_{эz}^{q-3}\right\} = 0;\ q = 4,5,6;\ k = 1,2,..,4. \end{cases} \qquad (15)$$

for the evolutionary movement and

$$\begin{cases} \left\{\tilde{a}_{qi\delta}\tilde{K}_{вi\delta} - f_{вz}^{q}\right\} = 0; \quad q = 1,2,3;\ i = 1,2,..,8; \\ \left\{\tilde{a}_{qi\delta}\tilde{K}_{вi\delta} - f_{вz}^{q-3}\right\} = 0; \quad q = 4,5,6;\ i = 1,2,..,8. \end{cases} \qquad (16)$$

for an outraged movement.

The formulas (14...15) contain the following designations: $f_{эz}^{q}$ — components of the evolutionary part of the dimensionless main vector of the external forces acting on the rim; $m_{эz}^{q-3}$ — components of the evolutionary part of the dimensionless main vector of the main moment of the same forces; $f_{вz}^{q}$ — components of the perturbed part of the main vector; $f_{вz}^{q-3}$ — components of the perturbed part of the main moment vector, C_κ — evolutionary motion parameters obtained by solving the equations of coupling and subject to exclusion from the equations; $\tilde{K}_{вi\delta}$ — images of "dimensionless components of perturbed motion", which include dimensionless components of perturbed parts of the linear velocity vector of the rim center $\tilde{\bar{v}}$ angular speed of the rim $\tilde{\bar{\omega}}_\delta$ cog and the image of dimensionless functions that define the profile of the road and its derivatives; A_{qo}, $A_{q\kappa}$, and $\tilde{a}_{qi\delta}$ — equation coefficients that depend on the parameters of the initial state of the tire and the resonance of its function Green.

The formulas (16) allow constructing a transfer matrix of elastic wheel for two variants of determination of input parameters - kinematic and power inputs.

For the kinematic input, the transfer matrix looks like

$$\tilde{A}(p_t)=\{\tilde{a}_{qi\delta}\};q=1,2,3;\ i=1,2,..,8. \tag{17}$$

that is, it is composed of system coefficients (16).

In the case of the power input, the transfer matrix has a cellular structure

$$\tilde{A}_c(p_t)=\begin{pmatrix} -\{\tilde{D}_{q;i-2}\}\{\tilde{a}_{qi\delta}\} & 0 & \\ 0 & -\{\tilde{D}_{q;i-2}\}\{\tilde{a}_{qi\delta}\} & \varnothing \\ & \varnothing & \{\tilde{D}_{q;i-2}\} \end{pmatrix} \tag{18}$$

where $\tilde{\Delta}_{q;i-2}$ — algebraic adjunct of an element $\tilde{a}_{q;i-2}$, $\tilde{\Delta}=\det\tilde{A}$ — determinant, $\tilde{D}_{q;i-2}=\tilde{\Delta}_{q;i-2}\tilde{\Delta}^{-1}$ (the method of constructing the transfer matrix for other options for specifying the space of input parameters is also indicated).

The elements of the transfer matrix depend on the parameters of evolutionary rolling. Due to the independence of the main motion from oscillation, these parameters can be determined by any available in each case by experiments with the real tire, calculation by, simplified models, etc.

The analysis of the complete system of equations of kinematic bonds has allowed to establish that for an estimation of evolutionary process it is impossible to be limited to the decision of only the truncated system: (14). Taking into account additional kinematic equations: the equations made possible an analytical description of the relationship between the usual characteristics of evolutionary motion, such as rolling radius and torque, lateral force, lateral deflection angle and stabilizing moments; in particular, a nonlinear elastic characteristic of vertical wheel compression was constructed.

2.3. Results

Thus, a universal transfer matrix of linear elastic wheel is constructed; this one is universal because the structure of transfer matrix is identical in all cases and contains some function (Green's resonant), reflecting peculiarities of a concrete model of tyres.

Further, it is necessary to suggest such ways of construction of this function that they appeared to be compatible with the developed model of an automobile wheel. The first way is to generalize a special method of solving the ordinary differential equation developed by N.K. Kulikov and to construct the Green function taking into account the peculiarities of wheel rolling. So it is possible to investigate complex models of tyres, receiving qualitatively correct result already in zero approximation, however features of structure of function Green I. Its resonents remain latent. Another way is to use the method of eigenfunctions. Its main advantage is the explicit structure of the Green resonant, which allows to trace analytically the features of the transfer matrix of the elastic wheel.

3. Conclusions

In this work by methods of the theory of systems the indignant rectilinear movement of an automobile elastic wheel on nervous firm road is studied, the transfer matrix with universal structure in relation to settlement schemes of the tyre is constructed and ways of use of this matrix together with models of automobile systems for calculation of their general dynamic characteristics are offered; movement of a wheel are offered. The study is carried out from a single position without involving hypotheses, artificial methods or experimental determination of additional parameters.

References

[1] Popov S.D. About mathematical model of automobile wheel. Works / MVTU, 1973, № 166, p. 45–52.

[2] Popov S.D. Integral representation of the movements of an arbitrary point of the rolling elastic wheel. — M.:, 1975, —19 p. Dep. the Research Institute of Automobile Industry No. 137.
[3] Popov S.D. Equations of perturbed rectilinear rolling of a wheel with an elastic tyre on any solid surface. — M.:, 1975. — 18 p. Dep. in the Research Institute of Automotive Industry No. 138.
[4] Popov S.D. Problem setting on mathematical model of automobile wheeled propulsor. — Izv. Izv. Mechanical Engineering, 1976, No. 2, pp. 25–31.
[5] Popov S.D. About one of the ways of construction of mathematical model of automobile wheel. — Works / MSTU, 1979, No. 233, p. 58–67. (Studies of wheeled and tracked vehicles).
[6] Ignatushin A.P., Kupreyanov A.A., Smirnov G.A., Popov S.D. About the choice of rational schemes of the system "transmission-engineer" of the all-wheel drive cars", Automobile industry1984, No. 15.
[7] Guchkov D.K., Ignatushin A.P., Lovtsov A.N., Kupreyanov A.A., Smirnov G.A., Popov S.D. Experimental-calculation method of investigation of large-size tires. Automotive industry, 1985 № 12.
[8] Popov S.D. Side retraction theory for the case of unsteady movement of an elastic wheel. News from universities. Mechanical Engineering, 1988, No. 3
[9] Lovtsov A.N., Popov S.D. Amplitude-frequency characteristics of the elastic wheel at perturbed motion along the trajectory with insignificant curvature. News from universities. Mechanical Engineering. 1989, № 9.
[10] Popov S.D. Mathematical model of a multi-axle special vehicle. Automotive industry, 1999, № 11
[11] Komissarov, D.S., Popov S.D. Mathematical model of curvilinear motion of the wheeled propulsion module of the rack type on the deformed ground. Research and design of wheeled and tracked vehicles of high cross-country ability. Coll. of scientific works — M.: MADI (TU), 2001. 186 p. p. 56–63
[12] Belousov B.N., Lovtsov A.N., Popov S.D. All-wheel steering for the heavy-duty multi-axle automatic telephone exchanges. Automotive industry. No. 3, 2002, p. 11–14
[13] Belousov B.N., Popov S.D. Wheeled vehicles of especially high load-carrying capacity. Design, theory, calculation. -M.: Bauman Moscow State Technical University, 2006. — 728 p.: il. ISBN 5-7038-2713-2
[14] Popov S.D. External mechanics of car wheel with elastic tire. — Part 1: Static characteristics: textbook. — Moscow: Bauman Moscow State Technical University, 2011.
[15] Popov, S.D. External mechanics of the automobile wheel with the elastic tire. — P.2: Static characteristics: manual. — Moscow: Bauman Moscow State Technical University, 2011.
[16] Popov, S.D. Elastic wheel model for the case of its indignant motion along the trajectory of insignificant curvature. MSTU Bulletin after N.E. Bauman: electronic edition. 2012, Issue. 10
[17] Popov S., Sarkisov P. Active tire pressure control as a step to sustainable automotive mobility. Proceedings of Multidisciplinary Academic Conference on Transport, Logistics and Information Technologies (MAC - TLIT 2013). 1st ed., Prague. MAC Prague consulting Ltd. 2013/
[18] Boris N. Belousov, Sergey D. Popov. Heavy-Duty Wheeled Vehicles: Design, Theory, Calculations. Published by SAE International with a Product Code of R-419, ISBN of 978-0-7680-7723-0, and 800 pages in a hardbound binding.
[19] Popov S.D., Sarkisov P.I. Approach to simulation of rolling of a wheel with an elastic tire without sliding on the basis of kinematic bonds. Anniversary XXV International Innovation-oriented Conference of Young Scientists and Students (MIKMUS — 2013): Proceedings of the Conference (Moscow, November 13-15, 2013). / M: IMAASH Publishing House RAS, 2013 — 424 pp.
[20] Popov S.D., Sarkisov P.I. Overview of models of non-stationary wheel rolling with elastic tire on non-deformable support base. Engineering journal: science and innovation, issue 12. 2013.
[21] Popov S.D., Sarkisov P.I. Calculation and experimental method of modeling the non-stationary rolling of the automobile wheel at horizontal movement of its center. Journal of Automotive Engineers (Journal of Automobile Engineering), No. 5 (88), 2014

[22] Gladov G., Popov S, Prokop G, Sarkisov P. Physical tire model based on understanding of separate components of tire deformation. M. Rosenberger (Ed.) — Proceedings of the 24th International Symposium on Dynamics of Vehicles on Roads and Tracks (IAVSD 2015), 17–21.08.2015, Graz, Austria.

[23] Popov S, Prokop G, Sarkisov P. The non-steady-state tire model as a set of physical submodels for driver assistance systems analysis. P. Pfeffer (Ed.). Proceedings of 6th International Munich Chassis Symposium, S. 799–814.

[24] Popov S, Prokop G, Sarkisov P. Physical tire model based on understanding of separate components of tire deformation. M. Rosenberger (Ed.). Proceedings of the 24th International Symposium on Dynamics of Vehicles on Roads and Tracks (IAVSD 2015), 17–21.08.2015, Graz, Austria, Chapter 78. Physical tire model based on understanding of separate components of tire deformation, p. 749–758.

[25] Popov S, Prokop G, Sarkisov P. Opening the black box: decoupling the physical processes which determine tire handling behavior in a non-steady state/ Executive Track Report. 36th FISITA World Automotive Congress, Busan, South Korea, 26–30.09.2016.

[26] Popov S, Prokop G, Sarkisov P. Physical effects of tire transient handling behavior: carcass deflection and contact patch form. 12th Intelligent Tire Technology Conference, Dresden, 25. –27.10.2016.

[27] Kubenz J., Popov S, Prokop G., Sarkisov P., Optical measurement of tire deformation focused on transient handling properties/ Tire Technology Expo 2017, Hannover, 14. –16.02.2017

IOP Conf. Series: Materials Science and Engineering **820** (2020) 012035 doi:10.1088/1757-899X/820/1/012035

Analysis of the current state of research in the field of improving the smooth ride of vehicles equipped with suspension

V Novikov[1], A Pozdeev[1,4], K Chernyshov[1], I Ryabov[1], A Diakov[2] and K Chutkov[3]

[1] Volgograd State Technical University, Volgograd, Russian Federation
[2] Bauman Moscow State Technical University
[3] Combined Arms Academy of the Armed Forces of the Russian Federation

[4]E-mail: pozdeev.vstu@gmail.com

Abstract. In the article the analysis of a modern condition of research in the field of increase of smoothness of a course of various modern motor vehicles, including machines of special and military appointment is resulted. The reasons limiting the decision of a global problem of the further development of automatic telephone exchanges and growth of productivity and safety of work at performance of transport works are formulated. The basic ways of the decision of the given problem, consisting in application of regulation of characteristics of a suspension depending on driving conditions and character of fluctuations of a motor vehicle are analyzed. At the same time, it is proposed to develop fundamentally new innovative structures and algorithms to regulate the characteristics of the suspension, improving the smoothness of the ride and other operational properties of the motor vehicle.

1. Introduction

At present, the requirements to the operational properties of various modern motor vehicles are growing rapidly. This is especially true for special-purpose and military vehicles, which often have to move at high speeds on uneven roads and terrain and do not tire people with transport vibration to perform their tasks. The dump trucks have to limit their travel speed because of the intense body vibrations and the risk of overheating the tyres. «Difficult» road conditions cause up to 30% of engine power to be lost in the suspension of fast-tracked vehicles. The additional suspension of cabins and seats does not solve the problem of significant reduction of vibrations and increase of average speeds. For modern high-speed cars there is a problem of reducing the controllability, stability and braking properties on uneven roads due to the occurrence of large fluctuations and the separation of the wheels from the road, which reduces the active safety and leads to an increase in accidents.

These combined hold back the solution to the global problem of further development of motor vehicles and increased productivity and safety of transport operations, which leads to significant annual economic losses for the country, as well as affects the speed of special and military vehicles.

2. Potential solutions to the problem of smooth ride of suspension machines

One of the ways to solve this problem is to apply the regulation of suspension characteristics depending on traffic conditions and the nature of vehicle oscillations. However, the active suspension management systems used, for example, in some high-end luxury passenger vehicles do not always provide the required increase in average speed, but significantly increase their cost. Also, the regulation of shock absorbers on new models of special equipment, although it solves the problem of increasing the speed of movement, but leads to rapid overheating of the suspensions and reduce the efficiency of their operation. The latter is due to the fact that due to the phase shift of oscillations of the vehicle frame relative to the profile of the road in the cycle of oscillations, there are zones in which the shock absorber force coincides with the direction of movement of the suspended mass, i.e. on this course the shock absorber does not brake, but, on the contrary, pushes the frame, heating up itself uselessly [1, 2].

Therefore, new suspension structures with damping controls are needed to reduce these inefficient zones, or other types of shock absorbers that operate on new principles. In this case, in order to simplify the adjustable suspension, increase their reliability and reduce costs, it is extremely important to develop, first of all, self-regulating autonomous suspension systems that do not require external control and supply of energy and have small size and weight, as well as implementing new algorithms to regulate the characteristics of the minor modernization of serial pneumatic and pneumatic-hydraulic springs and hydraulic shock absorbers [3–6].

Therefore, the task of developing fundamentally new innovative structures and algorithms to regulate the characteristics of suspensions that improve the smoothness and stability of movement, fuel efficiency, controllability and braking properties of vehicles is relevant.

3. Analysis of the current state of research in the field of smooth ride of vehicles equipped with suspension

The feature of the kinematic disturbance of the vehicle's vibrations is the phase shifts between the vertical movements of the contact point of each wheel with the supporting surface and the unsprung mass, as well as between the vertical movements of unsprung and sprung mass. The last phase shift in the oscillation cycle results in the formation of two areas where the force of the shock absorber is directed towards the movement of the suspended mass, which increases its kinetic energy and contributes to increasing the amplitudes of its oscillations. Therefore, the operation of the shock absorber in these areas is ineffective in terms of reducing the amplitudes of vibrations of the suspended mass. The beginning of each inefficient shock absorber section coincides with the stop of the suspended mass in the upper or lower position, and the end of the section coincides with the stop of the rod in the shock absorber in the upper or lower position, i.e. at the moment of changing the direction of shock absorber deformation [1]. The shock absorber is switched off or the damping level of the shock absorber is reduced by means of regulation of the shock absorber when the zones of inefficient operation are passed in each cycle of vibration, which will lead to an increase in the efficiency of the shock absorber operation and a decrease in the amplitudes of vibrations of the absorbed mass [2].

3.1. Regulation of damping in the suspension

The first works on the possibility of regulating damping in the vibration cycle appeared in the 60s of the twentieth century. Thus, in his 1965 PhD thesis "Study of some issues of damping of car oscillations" [7] the employee of the Research Institute of Construction and Architecture of the BSSR State Construction and Architecture R.I. Furunzhiev, using the principle of maximum L.S. Pontryagin[8] has shown by the example of a two-mass model of a vehicle that the optimal regulation of damping is to switch it off immediately at those parts of the vibration cycle when the direction of motion of the suspended and unsprung masses coincide, and the absolute speed of the unsprung mass is greater than the absolute speed of the suspended mass in the vertical direction, with subsequent instantaneous inclusion when these conditions change to the opposite.

R.I. Furunzhiev also noted that the elastic damping characteristic of the suspension at such regulation (force dependence on its deformation) has the form of a "butterfly".

The first references in the European and American literature to the possibility of regulating damping in the vibration cycle date back to the first half of the 1970s. The algorithm of regulation similar to that of R.I. Furunzhiev was proposed as an alternative to expensive, very complex and energy-intensive active suspensions by American scientists D. Karnopp (D.S. Karnopp) and Crosby (M.J. Crosby) in "The Active Damper - a New Concept for Shock and Vibration Control" (1973) [9] and "Vibration Control Using Semi-Active Force Generations" (written in collaboration with R. Harwood in 1974) [10]. The suspension with this principle of damping regulation is called semi-active suspensions.

D. Karnapp also proposed another concept - a smooth change of damping in order to bring the action of the standard shock absorber closer to the action of the imaginary shock absorber, located between the suspended mass and a certain point stationary in the vertical direction (the concept of "skyhook").

Further studies of these damping control principles have revealed their significant disadvantages [11].

The first disadvantage related to the skyhook concept is the difficulty of enforcing the law of smooth damping, which is associated in English literature with the difficulty of determining the vertical velocity of the suspended mass. This disadvantage can be eliminated by introducing a constant value of the effective damping factor, thus moving to the algorithm "semi-active control suspension".

The second disadvantage is the discomfort felt by passengers when testing suspensions with "skyhook" and "semi-active" control algorithms, and significant noise when switching damping on and off. Experimental bench research of the semi-active suspension by the authors of this article has confirmed the calculated reduction of the amplitudes of vibrations of the suspended mass, accompanied by the presence of shocks and noise at a sharp change of damping in the suspension. Thus, the reduction of amplitudes of vibrations of the suspended mass at this method of regulation leads not to increase, but to decrease of smoothness of a course.

The third disadvantage is that the unsprung mass in the regulation of the algorithms of "skyhook" remains almost without damping, and when reaching its own frequency of unsprung mass of its oscillation on the mathematical model is infinitely increased, which in practice will correspond to the separation of the wheels from the road and the loss of stability and controllability of the vehicle.

In order to solve the problem of stability and controllability of the car, in 1998 the Czech scientist M. Valášek [12] proposed the concept of "groundhook".

The groundhook concept involves a smooth change in damping to bring the standard shock absorber closer to the action of an imaginary shock absorber that links the unsprung mass to the road surface. The main objective of this regulation is to reduce tyre deformation and wheel oscillation. As the theoretical studies of the authors have shown [11], the law of regulation "groundhook" is not workable, because checking the vibrating system, simulating the movement of the vehicle on the road, on the stability of this law of regulation gives a negative result.

Suspensions are used nowadays to simplify the damping control process, where inelastic resistance changes not in the vibration cycle, but depending on the steady-state vibration modes. For example, in the hydropneumatic suspension of a Citroen-C5, which is controlled with the help of an on-board computer, there are 3 modes of operation: comfortable, optimal and sporty. In this case, different from civilian vehicles on combat vehicles, additional adjustment conditions are required, for example, when driving through ski jumps, it is possible to "fly out" the machine and detach all the rollers from the ground, followed by landing, which in order to prevent suspension breakage during compression and rapid damping of vibrations of the hull it is necessary to sharply increase the hydraulic resistance of the suspension of all support rollers. However, the suspensions should have minimal suspension resistance in order to reduce the harmful effects of shaking on the crew when the machine is moving on a relatively flat surface with low unevenness. Solution of these problems with the help of a large number of sensors and on-board computer significantly complicates the suspension, reducing its reliability and significantly increasing the cost. For example, in comparison with the spring unregulated suspension, a Citroen car with a hydropneumatic adaptive suspension "Indractive-2" is more expensive by $5000.

3.2. Regulation of the stiffness of the elastic suspension element

Nowadays the experts in the field of vehicle suspension have shown interest in the possibility of increasing the smoothness of the car by changing the rigidity of the suspension in the cycle of vibration. The most obvious way to change the stiffness of the suspension is the use of additional volume with the possibility of its disconnection and inclusion in the air springs. Such a method of change of suspension stiffness in the cycle of vibrations is considered in many works, in which various algorithms of using the additional volume are proposed and considered [3, 13–15].

It is shown in [13] that high efficiency of damping of vibrations in the air suspension, comparable with the efficiency of hydraulic shock absorbers, can be achieved due to free air flow between the working and additional cavities of the air spring at certain moments of the cycle of vibrations.

It is shown in [13] that high efficiency of damping of vibrations in the air suspension, comparable with the efficiency of hydraulic shock absorbers, can be achieved due to free air flow between the working and additional cavities of the air spring at certain moments of the cycle of vibrations.

The paper [14] proposes the pneumatic springs with self-regulating air damper in amplitude and direction, as a result of which the damping force is almost always directed against the movement of the suspended mass, which provides effective damping of its vibrations and reduction of energy losses and heating of the suspension.

The algorithm of control of a suspension with S-shaped piecewise linear elastic characteristic of the air spring having two stages of rigidity, has been offered by Takagi A., Yoshimura T. in the Tokushima University (Japan) [15].

The large number of the proposed algorithms is connected with the absence of the theory of optimal regulation of suspension stiffness in the cycle of oscillations.

Studying the possibility of increasing the smoothness of the car's motion by regulating the stiffness of the elastic element, it is necessary to take into account the method of changing the stiffness, as well as the change in the static deformation of the suspension as a result of changes in stiffness. There are three ways to change the stiffness of the suspension:

1) blocking a part of the elastic element used, for example, cutting off a part of the cavity of the working chamber in the air suspension [13], blocking a part of the coils of the screw spring [13, 16, 17], blocking a part of the torsion;

2) use of a spring element that can be switched off in series, e.g. use of a series-connected additional volume in the air suspension [3, 14], installation of a series of additional springs;

3) use of a spring element which can be switched off in parallel, e.g. use of an additional torsion in the torsional suspension.

These methods provide two-stage stiffness, but in the same way you can get a suspension with a higher number of stiffness levels.

Although there are many different ways to change stiffness, there are two basic schemes of two-step change of stiffness: with constant stiffness of the steps and variable stiffness of the steps. Thus, the suspension with additional elastic element to be installed in series or in parallel with the switchable elastic element has a constant stiffness of steps and belongs to the first type of hangers, and the air suspension with additional volume has no fixed stiffness of steps and belongs to the second type of hangers.

The two-stage suspension with constant stiffness works as follows. The object of vibration protection (vehicle body) is sprung by two successively installed elastic elements — main and additional. The suspension is equipped with a mechanism to block the additional elastic element. When the additional elastic element is blocked, only the main elastic element remains in operation, and the rigidity of the suspension increases. When the elastic element is released, it is included in the work and the rigidity of the suspension is reduced. Thus, this suspension provides almost instantaneous two-stage switching of stiffness that allows to carry out adjustment of stiffness in a cycle of fluctuations of a suspension.

The two-stage suspension with variable stiffness disengages the part of the elastic element that is in the blocking zone when the blocking mechanism is activated. For example, the additional volume in the air suspension with disconnectable additional volume blocks part of the additional volume of gas that is in the additional volume at the moment of its disconnection.

The researches of the authors of this article have shown that the two-stage suspension with constant rigidity of stages has higher potential vibration-protective properties in comparison with the suspension having variable rigidity of stages which provides good vibration-protective properties only in the field of low-frequency vibrations. The optimum ratio of the stiffness of the main and additional elastic elements in the suspension with constant stiffness of the stages is about 3:1.

The specific feature of the suspension with instant change of stiffness is the discharge of potential energy of the blocked and free elastic elements (or, depending on the design, the sections of the elastic element) at the moment of unblocking the blocked elastic element (section). The instant release of the elastic element (region) in real designs creates a blow at the junction of the elastic element, causing high-frequency vibrations of the elastic element's body. For example, longitudinal spring oscillations in spring hangers. The presence of weights fixed on the springs (separator of the main and additional elastic elements) is the reason of excitation of own oscillations of loads, which does not exclude the own oscillations of springs, which are overtones in relation to the oscillations of the load. The energy of these vibrations is equal to the energy released when unblocking a part of the suspension.

It takes a long time to dampen high-frequency vibrations, so during the switching process the energy of the vibrations will accumulate, which will eventually lead to suspension failure. In order to avoid this, it is necessary to use special damping devices in the suspension with instant change of stiffness or to use a smooth, instantaneous change of stiffness of the suspension.

3.3. Use of dynamic vibration dampers

The dynamic vibration dampers are one of the means of reducing the vibration dynamic loading of various technical constructions and structures. The principle of their action consists in the use of an additional inertial element that affects the object of vibration protection directly or by means of elastic (elastic damping) element in such a way that this effect is directed against the movement of the object of vibration protection in most of the cycle of vibration.

One of the effective ways to reduce wheel oscillations and body accelerations in the field of high frequencies is the use of dynamic vibration wheel dampers (DVWD) in the suspension design [18–24]. Therefore, interest in DVWD has increased recently, but, nevertheless, dynamic dampers have not been widely used, as their introduction leads to complication and weighting of the car suspension due to the presence of additional inertial mass. In addition, due to the relatively large size of DVWD, their arrangement in the serial suspension of vehicles becomes more complicated.

Another use of DVWD, which not only does not increase the unsprung weight of vehicles, but even reduces it, is the integration of a dynamic damper with existing elements of the unsprung weight of the vehicle. As the inertial mass of DVWD it is proposed to use a disc brake caliper, which is installed on both front and rear wheels in modern vehicles. For this purpose, the caliper is mounted on a swivel arm with the possibility of angular movement relative to the wheel axle and is connected to the wheel stand by means of an elastic- damping bond, which is locked during braking and which includes a resilient element and a lockable hydraulic shock absorber. This reduces the unsprung mass by the weight of the caliper, and the mass of the caliper itself becomes a DVWD inertial mass, sufficient to effectively dampen the vibrations of the wheels while the vehicle is in motion. When the vehicle is braked by the pressure of the operating fluid of the brake system, the hydraulic shock absorber is blocked, so the braking properties of the vehicle do not change.

Of course, such a design DVWD makes certain adjustments to the overall design of the disc brake and suspension, but with the right calculation and choice of parameters, this idea is quite feasible and will have a significant effect.

In order to reduce truck body oscillations, which have high rigidity of elastic suspension elements, which is necessary for realization of the given load-carrying capacity of vehicles, it is expedient to use flywheel dynamic dampers (absorbers) of oscillations (inertial-friction or inertial-hydraulic shock absorbers). They due to alternating rotation of the flywheel provide an increase in the reduced mass and, as a consequence, a decrease in its own frequency of suspension oscillations and a significant increase in the smoothness of the ride. The studies of suspensions with flywheeled dynamic vibration dampers carried out by the authors of this article have shown their high efficiency [22, 23].

4. Conclusions

Thus, it is necessary to solve the following urgent problems in order to improve smooth running and further develop vehicles equipped with suspension:

- search for algorithms of smooth damping control and their design solutions, resulting in increased efficiency of the shock absorber [5, 6];
- development of adaptive suspensions with self-adjustable damping characteristics depending on vibration modes, which do not require external control and energy supply [3, 4, 24];
- development of the basis for the theory of optimal regulation of vehicle suspension stiffness [3];
- search for algorithms of suspension stiffness control and their design solutions, resulting in smoother ride of the vehicle [3, 13, 14];
- development of suspensions in which DVWD are integrated into serial suspensions without increasing their dimensions, for example, by installing air springs with rubber-cord casings in pistons, the variants of which are proposed by the authors [20, 21].
- production and testing of prototypes of inertial-friction or inertial-hydraulic shock absorbers with energy recovery [22, 24].

References

[1] Ryabov I M, Novikov V V, Pozdeev A V 2016 Efficiency of Shock Absorber in Vehicle Suspension *Procedia Engineering (International Conference on Industrial Engineering (ICIE-2016) vol 150)* ed A A Radionov (Elsevier publishing) pp 354–362

[2] Ryabov I M, Chernyshov K V, Pozdeev A V 2016 Energy Analysis of Vehicle Suspension Oscillation Cycle *Procedia Engineering (International Conference on Industrial Engineering (ICIE-2016) vol 150)* ed A A Radionov (Elsevier publishing) pp 384–392

[3] Pozdeev A V, Novikov V V, Diakov A S, Pokhlebin A V, Ryabov I M, Chernyshov K V 2013 Adjustable pneumatic and hydro-pneumatic springs of motor vehicles suspensions (Volgograd, Volgograd State Technical University) 244 p

[4] Pokhlebin A V, Pozdeev A V, Golyatkin I A 2016 Control algorithms of inelastic resistance forces of hydropneumatic springs *2nd International Conference on Industrial Engineering, Applications and Manufacturing (ICIEAM)* (Chelyabinsk, Russia) Publisher: IEEE 6 p

[5] Ryabov I M, Chernyshov K V, Pozdeev A V 2017 Vibroprotective and Energetic Properties of Oscillating System Equivalent to Vehicle Suspension with Pendular Regulator on Shock Absorber *Procedia Engineering (International Conference on Industrial Engineering (ICIE 2017) vol 206)* ed A A Radionov (Elsevier Ltd.) pp 392–400

[6] Ryabov I M, Chernyshov K V, Pozdeev A V 2017 Vibroprotective and Energetic Properties of Vehicle Suspension with Pendular Damping in a Single-Mass Oscillating System *Procedia Engineering (International Conference on Industrial Engineering (ICIE 2017) vol 206)* ed A A Radionov (Elsevier Ltd.) pp 519–526

[7] Furunzhiev R I 1965 Investigation of some issues of damping vehicle vibrations *Dissertation of Candidate of technical sciences (Minsk)*

[8] Pontryagin L S, Boltyansky V G, Gamkrelidze R V, Mishchenko E F 1976 Mathematical theory of optimal processes (Moscow, Nauka) 392 p

[9] Karnopp D C, Crosby M J 1973 The Active Damper — a New Concept for Shock and Vibration Control *43-rd Shock and Vibration Bulletin, Part H, June* pp 46–73

[10] Karnopp D C, Crosby M J, Harwood R A 1974 Vibration control using semi-active force generators *Transactions of the ASME (Journal of Engineering for Industry vol 96)* pp 619–626

[11] Chernyshov K V, Ryabov I M, Pozdeev A V, Pylinskaya T V 2018 Analysis of "skyhook" and "groundhook" principles of damping control in the vehicle's suspension *Truck: transport complex, special equipment* (Moscow, Russia) **10** pp 3–6

[12] Valášek M et. al. 1997 Extended ground-hook — new concept of semi-active control of truck's suspension *Vehicle system dynamics vol 27* **5-6** pp 289–303

[13] Kalashnikov B A 2008 Objects damping systems with discrete switching of elastic elements (Omsk, Omsk State Technical University) 344 p2008

[14] Pozdeev A V, Diakov A S, Novikov V V, Riabov I M 2013 Self-adjustable dual-chamber pneumatic springs with switching of chambers *Truck: transport complex, special equipment* (Moscow, Russia) **9** pp 2–5

[15] Yoshimura T, Takagi A 2004 Pneumatic active suspension system for a one-wheel car model using fuzzy reasoning and a disturbance observer *Journal of Zhejiang University SCIENCE* **5(9)** pp 1060–1068

[16] Chernyshov K V, Ryabov I M, Pozdeev A V 2019 Potential Vibration Isolation Qualities of Suspensions with Two-Step Stiffness Control in Oscillation Cycle *Proceedings of the 4th International Conference on Industrial Engineering (ICIE 2018)* (Moscow, Russia) ed A A Radionov (Lecture Notes in Mechanical Engineering (LNME) Springer) pp 437–448

[17] Chernyshov K V, Ryabov I M, Pozdeev A V 2019 Theoretical Foundations of Optimal Two-Step Control of Suspension Stiffness of Transport Vehicle in Oscillation Cycle *Proceedings of the 4th International Conference on Industrial Engineering (ICIE 2018)* (Moscow, Russia) ed A A Radionov (Lecture Notes in Mechanical Engineering (LNME) Springer) pp 421–436

[18] Novikov V V, Pozdeev A V, Chumakov D A, Kovalev A M 2017 The joint operation of the pneumatic suspension of the vehicle with a dynamic vibration damper and a hydraulic shock absorber *Mechanical Engineering Bulletin* **7** pp 34–39

[19] Ryabov I M, Chernyshov K V, Kovalev A M 2010 Potential vibroprotective properties of a vehicle suspension with a dynamic vibration damper of wheels *Automotive industry* **12** pp 13–16

[20] Novikov V V, Pozdeev A V, Ryabov I M, Chernyshov K V, Chumakov D A 2016 *Pneumatic suspension* (RU Utility model 167265)

[21] Novikov V V, Pozdeev A V, Ryabov I M, Chernyshov K V, Chumakov D A 2016 *Pneumatic suspension* (RU Utility model 169805)

[22] Ryabov I M, Novikov V V, Pozdeev A V, Chernyshov K V, Mitroshenko A S 2013 Types of designs of inertial-friction shock absorbers, their modeling and testing *Tractors and agricultural machinery* **4** pp 23–26

[23] Ryabov I M, Chernyshov K V, Pozdeev A V 2015 Comparative evaluation of the vibration isolation properties of a suspension with different flywheel dynamical absorbers of the car body oscillations *Procedia Engineering* (International Conference on Industrial Engineering (ICIE-2015) vol 129) ed A A Radionov (Elsevier publishing) pp 480–487

[24] Novikov V V, Ryabov I M, Chernyshov K V 2009 *The vibration isolation properties of the motor vehicles suspensions* (Volgograd, Volgograd State Technical University) 338 p

IOP Conf. Series: Materials Science and Engineering **820** (2020) 012036 doi:10.1088/1757-899X/820/1/012036

Perspectives of increasing smooth ride of wheeled vehicles without suspension with pneumatic damping system

A Pozdeev[1,3], I Ryabov[1], V Novikov[1], E Zhdamirova[1], A Diakov[2] and K Evseev[2]

[1]Volgograd State Technical University, Volgograd, Russian Federation
[2]Bauman Moscow State Technical University

[3]E-mail: pozdeev.vstu@gmail.com

Abstract. In the article the analysis of a modern research condition in the field of increase of smoothness of a course of wheeled vehicles without suspension with pneumatic damping system is resulted. The reasons limiting the solution of a global problem of the further development of wheeled vehicles without suspension and growth of productivity and safety of work at performance by them of transport, agricultural and road-building works are formulated. The basic ways of the decision of the given problem solution, consisting in increase of damping properties of pneumatic tyres for wheeled non-suspended vehicles without increase in resistance to rolling of wheels and decrease in a resource of tyres are analyzed. Thus development of essentially new innovative structures of wheels, improving smoothness of a course and other operation al properties of non-suspended vehicles is offered.

1. Introduction

The suspensionless wheeled vehicles (agricultural tractors and combines, self-propelled road-building and special-purpose machinery) operate at resonant speeds in the field and on broken roads, which leads to their intense fluctuations, reduced productivity and diseases of operators. The additional suspension of cabins and seats used in suspensionless vehicles does not solve the problem of significant reduction of vibrations and increase of average travel speeds. This constrains the solution to the global problem of further development of suspensionless vehicles and the growth of productivity and safety in the performance of their transport, agricultural and road-building works, which leads to significant annual economic losses for the country.

2. Possible solutions to the problem of smooth ride of wheeled vehicles without suspension

One of the directions of solution of the above mentioned problem is a significant increase of damping properties of pneumatic tyres for the suspended vehicles without increasing the rolling resistance of wheels and reducing the service life of tyres. This is possible by reducing the radial stiffness of the wheels while maintaining static deflection and increasing the inelastic resistance of the tyre when the vehicle body vibrates. To ensure these new qualities, it is necessary to search for new wheel structures that have no analogues in the world, and to conduct appropriate scientific research. The solution of this extremely difficult task will allow to reduce significantly the vibrations of tractors and road-building vehicles, increase their productivity and significantly improve the working conditions of operators.

In this regard, the challenge of developing innovative wheel structures that improve the smoothness and stability of the drive, fuel efficiency, controllability and braking properties of the suspensionless vehicles is urgent.

3. Analysis of the current situation in the field of smooth ride of wheeled vehicles without suspension

The tyres are continuously being improved and dozens of patents are granted each year for new tyre designs. However, the analysis of modern tyres shows that for the past 175 years since the day of invention in 1844, the design of the tyre has not changed fundamentally. As in R.W. Thomson's pioneering invention, a tyre is an elastic shell that forms elastic support surfaces around the rims, filled with compressed air. Nowadays it is impossible to create high-speed vehicles without tyres.

The development of tyre design has taken two directions: development of increasingly low-profile tyres for the «good roads» and tyres with high profile heights for the «bad roads» and off-roads. The introduction of tyres with a high profile height has made it possible to eliminate the use of spring suspension, which has made it possible to create suspensionless vehicles with a simpler design.

Nowadays, a variety of mobile wheeled vehicles with tyres with a high profile height have been designed, operated and continuously improved.

First of all, they are so-called suspensionless vehicles: wheeled tractors, combine harvesters, agricultural, road construction and special-purpose vehicles. These machines are widely used in transport, agricultural, construction and other works.

Secondly, they are vehicles with suspension: heavy dump trucks, various army vehicles, including mobile missile systems, for which the task of increasing the speed of movement on broken ground roads and terrain while ensuring the preservation of the transported product and the efficiency of the crew is very important.

It has been established by the conducted researches that the characteristics of tyres - their dynamic rigidity smoothing and absorbing abilities — have the greatest influence on smooth running of the vehicles having a suspension and tyres with the big height of a profile [1–4].

Many scientists, both in Russia and abroad, have studied the output characteristics and modeling of pneumatic tyres [1–25].

The analysis of domestic and foreign literary sources has shown that in order to solve this problem, first of all, it is necessary to develop a methodology for the synthesis of innovative pneumatic wheels with internal structural elements, providing for the solution of three contradictory scientific problems, each of which can be considered fundamental without exaggeration.

3.1. The first conflicting scientific task

The first conflicting task is to improve the vibration protection properties of pneumatic tyres. The problem situation, currently unresolved in domestic and foreign science, here is that it is necessary to reduce the radial rigidity of the tire without increasing its static deflection.

The studies of the smoothness of the 8x8 high cross-country vehicle movement conducted by foreign researchers under the guidance of B.D. van Deusen have shown that the greatest influence on the value of the mean square acceleration of the suspended parts in various road conditions is provided by the tire rigidity, which was regulated by the pressure change [2]. The reduction of suspension rigidity and the installation of shock absorbers had a much smaller effect. The greatest effect was obtained with simultaneous reduction of tire rigidity, suspension and installation of shock absorbers.

Professor N.N. Yatsenko notes that in order to increase the smoothness of the movement of wheeled suspensionless machines it is necessary to reduce the radial stiffness of tyres by 2–3 times while keeping the static deformation of the tyres [3, 4]. He also notes that the reduction of radial stiffness can be achieved in the design of ultra low pressure wide tyres. However, the static deformation of tyres is very high, which reduces their service life. In addition, these tyres have large dimensions, so they are not suitable for most common wheeled transport vehicles.

As the elastic characteristic of the tyre is almost linear, a reduction in stiffness is only possible if the static deformation of the tyre increases. However, the high static deformation of the tyres reduces their durability and increases the rolling resistance coefficient, which in turn leads to higher operating costs for the machine. Therefore, the only way to reduce the radial stiffness of a tyre without increasing its static deformation is to change the elastic characteristic from linear to non-linear — S-shaped. According to Professor Rothenberg R.V., this characteristic is considered to be ideal for the vehicle suspension, as it provides high vibration protection properties and reduces the probability of a breakdown. [5]. It is not possible to obtain such elastic characteristics of the tyre by changing its profile or the value of the filling pressure, because the elastic characteristics of tyres of all profiles and at any pressure are progressive. Their rigidity in the small initial part of the elastic characteristic increases with increasing deformation, and then becomes almost linear. The S-shaped elastic characteristic of a tyre must have a constant stiffness of 2 to 3 times less than that of a conventional tyre in its static deformation zone. The determination of this section from the mean square deformation of the tire, which occurs during the motion of the unevenness, is a different task. A wheel that implements the calculated S-shaped elastic characteristic of the tyre can only be created by introducing the internal structural elements into the wheel.

3.2. Second conflicting scientific task

The second conflicting task is to improve the damping properties of pneumatic tyres to effectively dampen the vertical vibrations of the wheels and the vehicle as a whole. The problem situation, which is currently also unresolved in domestic and foreign science, here is that it is necessary to increase the damping properties of tyres without increasing their coefficient of rolling resistance.

In order to evaluate the absorbing properties of the tire, an elliptical-steppe model of viscosity resistance in the tire material [6] has been spread, the use of which approximates the calculated statistical characteristics of the occurring processes to the similar characteristics obtained from observations and measurements in a field experiment. The absorbing or damping capacity of the tire material manifests itself in the vibrations of its deformation in the vertical, longitudinal and lateral directions and is determined by the amount of mechanical energy converted into thermal energy at its deformation. It can be estimated by the area of the hysteresis loop of the elastic characteristic, the dynamic coefficient of resonance oscillations of the load on the tire, the damping coefficient or the relative attenuation coefficient ψ.

Figure 1 shows the dependence of the damping factor in the tire on the deformation frequency and rolling speed in linear coordinates [7].

The figure shows that the damping in the tyres is very small. Up to a rolling speed of 30 km/h, the damping factor k_t decreases with the increase in the deformation frequency ω, and at speeds above 30 km/h, k_t is virtually independent of ω. Even a slight increase in travel speed from 0 to 10 km/h at a tyre deformation rate of 16 rads/s (typical for wheeled tractors in agricultural work) leads to a 4–5 times reduction in k_t. For trucks with speeds ranging from 0 to 60 km/h and frequencies of 45 rad/sec, and for cars moving at speeds ranging from 0 to 100 rad/sec at 75 rad/sec, the damping factor k_t is reduced by more than 10 times.

In the works of Professor N.N. Yatsenko and other scientists it was established that the relative attenuation coefficient of tyres is very small [3, 4]. For non-rotating tyres it is within the limits of ψ = 0.03 ... 0.05, and for rotating tyres it is 3 ... 5 times less than static tyres. Obviously, in [3] the rolling speeds of the tyres were low, because according to the data in Fig. 3 at typical truck and passenger car travel speeds, the damping factor for vertical vibration is more than 10 times lower than that for non-rotating tyres and is ψ = 0.003 ... 0.005.

That's why at high-frequency (resonance) influence from road roughness pneumatic tires weaken vibrations of the machine body, and at resonance frequencies - considerably strengthen (in 10 and more times). The experience of operation of wheeled unsuspended machines and the conducted research show that when performing transport work they have insufficient smoothness of movement and do not provide observance of norms of vibroload of the operator-driver and crew, even when driving on the improved roads. The reason for this is the low vibration protection properties of pneumatic tyres and their inability to absorb the energy of the body's vertical vibrations. In order to reduce the vibrations on the driver's body, as well as to ensure traffic safety, the driver has to reduce the speed of the machine, which impairs fuel efficiency, reduces productivity and efficiency of its use.

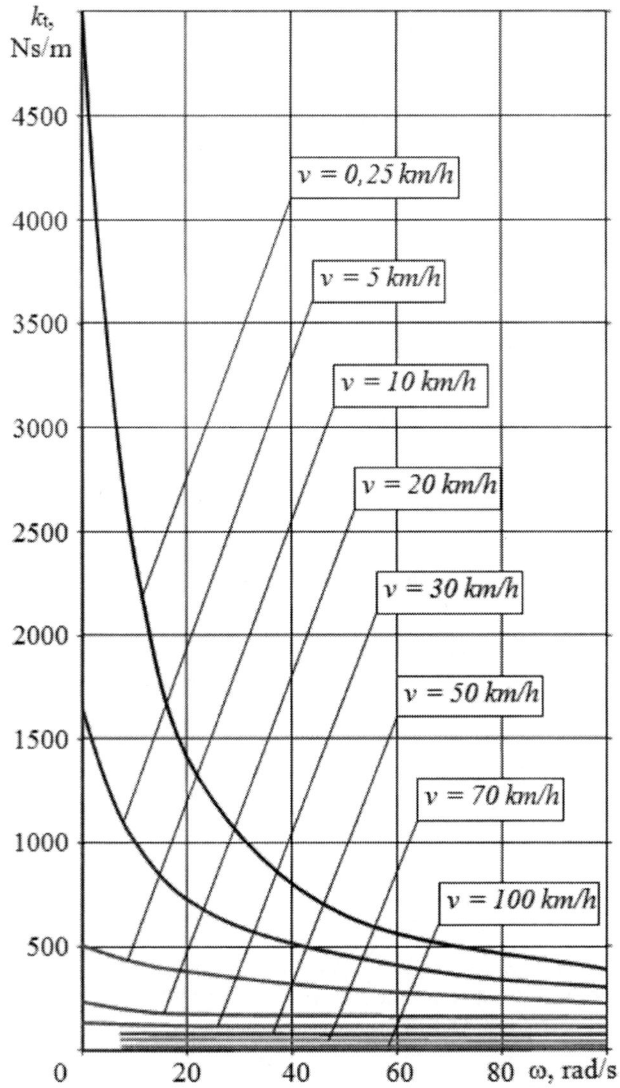

Figure 1. Dependence of the coefficient tyre damping from the frequency of its deformation at different speeds rolling stock

The analysis of the literature has shown that with increase in speed of movement of the machine influence of a hysteresis in a material of the tyres on damping of vertical fluctuations of the case of the machine decreases. When the tire rotates, its radial deformation is distributed around the tire circumference, as the vertical deformation rate is much lower than the linear speed of the tire. This results in a significant reduction in the effectiveness of the vertical vibration damping due to the loss of material in the tire material, and the higher the speed of rotation of the tire.

So experimentally it is established [8] that at rotation of the tyre of the wheeled car with linear speed of 10 km/h, characteristic for wheeled tractors on agricultural works, its absorbing properties decrease in 4–5 times in relation to the nonrotating tyre (figure 2) [9]. The optimum relative attenuation coefficient required for smooth running should be within the limits of $\psi = 0.15 \ldots 0.25$. Thus, the tyres have a very low damping capacity, which creates a problem of smooth running of wheeled suspensionless machines. Therefore, it is necessary to find an opportunity to increase the damping capacity of the tyres with vertical vibrations in 15–25 times to approach the required optimum value of $\psi = 0.25$ without increasing the coefficient of resistance to rolling. Moreover, the damping properties of the tyre should not be reduced when the rolling speed of the tyre increases.

Figure 2. Photos of experimental suspension systems for Tata Nano and VAZ vehicles with pneumatic wheels on the BISS ITW servohydraulic vibration test bench of the Automatic Installations Department of Volgograd State Technical University

In the works of Professor I.M. Ryabov it is shown that it is possible to improve the damping properties of tyres without increasing the coefficient of rolling resistance only due to internal structural elements [1, 9, 10]. Thus, in 1998 I.M. Ryabov proposed a way to solve this problem and received a patent of the Russian Federation for the invention № 2108240 for the wheel of a vehicle, the tire which contains an internal pneumatic damping system. It is made in the form of an inner shell mounted on the rim with the possibility of axial rotation of the inner shell with a truncated lower part, the cavity of which communicates with the tire cavity through valves and throttling holes [9]. However, the wheel proposed by I.M. Ryabov was not sufficiently studied on physical models and was not manufactured. Also, the theory and methods of optimization of damping system parameters were not developed. It has been established that this wheel can not provide the increase of damping of vertical vibrations of the suspensionless machine to the required level. In addition, the proposed wheel does not provide the necessary reduction of tire rigidity, which makes it necessary to search for new structures of internal elastic damping elements of the wheel and conduct research to optimize their parameters.

3.3. Third conflicting scientific task

The third conflicting task is in the complex area of increasing the smoothness of movement, controllability and stability of the wheeled vehicle. At the same time, it is necessary to reduce the stiffness of tyres and increase their resistance to side retraction, but it is very difficult to do so, because with a decrease in the rigidity of tyres (by reducing the pressure), their resistance to side retraction decreases.

The insufficient resistance to tyre sideways retraction impairs the stability and controllability of the wheeled vehicle, reduces the safety of the vehicle and the average speed of the vehicle, and leads to rapid fatigue for the driver and crew.

It is known that the resistance coefficient to the side retraction of the tyre is the maximum value at the nominal load on the tyre and decreases both with the increase of the normal load and with its reduction, and the underload of the tyre has a greater impact than its overload. The increase in the side drive resistance coefficient results in a reduction in the height of the tyre profile. However, its vibration protection properties decrease. To solve this controversial problem, that is to increase resistance to side retraction of pneumatic tyres without reducing the height of the profile, it is possible only through the introduction of internal structural elements into the wheel.

4. Conclusions

The above analysis of the current state of research in the field of influence of the structure and parameters of the internal elements of the innovative wheel with pneumatic tire on its elastic damping properties and frequency characteristics of oscillations in the process of rolling shows that these data are not enough to create innovative solutions that significantly improve the vibration-protective properties of pneumatic tires and the quality wheeled vehicles without suspension with pneumatic damping system.

Thus, the search for new wheel structures and the determination of the optimal parameters of the built-in elastic damping elements, which ensure a smooth ride without increasing the rolling resistance of the wheel, is an urgent task.

It should be noted that the this work was carried out at the Bauman Moscow State Technical University, with financial support from the government in the face of the Russian Ministry of Education under the project: №14.577.21.0272. (Identification number: RFMEFI57717X0272)

References

[1] Ryabov I M, Chernyshov K V, Sokolov A Yu 2007 Mathematical model of a wheel with pneumatic damping system for vehicles without suspension *News of VSTU. Series "Land Transport Systems" vol 2* (Volgograd, VSTU) **8** pp 51–53
[2] Van Deusen B D 1963 Ride Evaluation *Automobile Engineer* **13**
[3] Yacenko N N, Prutchikov O K 1969 Smoothness of trucks *(Moscow, Engineering)* 243 p
[4] Yacenko N N 1978 Absorbing and smoothing ability of tires *(Moscow, Engineering)* 133 p
[5] Rotenberg R V 1972 Car suspension *(Moscow, Engineering)* 392 p
[6] Rykov S P, Tarasyuk V N 2010 Studies of the output characteristics of pneumatic tires. Absorbing ability *Engineering problems. Systems Methods Technology* **2(6)** p 19–30
[7] Knoroz V I, Klennikov E V, Petrov I P 1967 *Transport (Moscow*) 237 p
[8] Novikov V V, Pozdeev A V, Diakov A S 2015 Research and testing complex for analysis of vehicle suspension units Procedia Engineering (International Conference on Industrial Engineering (ICIE-2015) vol 129) ed A A Radionov (Elsevier publishing) pp 465–470 Ryabov I M, Pozdeev A V, Sokolov A Yu, Malinin N N 2014 Wheel with internal pneumatic damping system and the study of its dynamic damper *News of VSTU. Series "Land Transport Systems" vol 8* (Volgograd, VSTU) **3(130)** pp 37–41
[9] Ryabov I M 1998 *Vehicle wheel* (RU Invention 2108240
[10] Ryabov I M, Pozdeev A V, Sokolov A Yu, Malinin N N 2014 Wheel with internal pneumatic damping system and the study of its dynamic damper *News of VSTU. Series "Land Transport Systems" vol 8* (Volgograd, VSTU) **3(130)** pp 37–41
[11] Pacejka H B 2012 Tyre and Vehicle Dynamics *Butterworth-Heinemann* 629 p
[12] Pacejka H B, Bakker E 1991 The Magic Formula Tyre Model *Proccedings of the 1st International Tyre Colloquium SWETS & ZEITLINGER BV vol 21* p 1–18
[13] Bakker E, Pacejka H B, Lidner L A 1989 New Tyre Model with Application in Vehicle Dynamics Studies *Proccedings of the 4th International Conference of Automotive Technologies* №890087 p 83–96
[14] Biris S.-S, Ungureanu N, Maican E, Murad E, Vladut V 2011 FEM model to study the in-fluense of tire pressure on agricultural tractor wheel deformations *Engineering for Rural Develop-ment* pp 223–228
[15] Burke A, Olatunbosun O A 1997 New techniques in tyre modal analysis using MSC/NASTRAN *Vehicle Design* **18(2)** pp 203–212
[16] Burhaumudin M S, Samin P M 2012 Modeling and Validation of Magic Formula Tire Model *International Conference on Automotive Mechanical and Materials Engineering (ICAMME'2012)* Penang (Malaysia) pp 113–117
[17] Gafvert M 2002 Construction of Novel Semi-Empirical Tire Models for Combined Braking and Cornering *Lund Institute of Technology* 66 p
[18] Gharibkhani M, Mardani A, Vesali F 2012 Determination of Wheel-Soil Rolling Resistance of Agricultural Tire *Australian Journal of Agricultural Engineering* pp 6–11

[19] Holscher H, Tewes M, Botkin N 2004 Modeling of Pneumatic Tires by a Finite Element Model for the Development a Tire Friction Remote Sensor pp 1–17
[20] Orengo F 2003 Modeling Tire Blow-Out in Roadside Hardware Simulations Using LS-DYNA *Proceedings of IMECE'03 (2003 ASME International Mechanical Engineering Congress&Exposition)* pp 1–10
[21] Sapragonas J, Darguzis A 2011 Model of Radial Deformations of Protector of Vehicle Tire *MECHANIKA* **17(1)** pp 21–29
[22] Shoop S A 2001 Finite Element Modelling of Tire-Terrain Interaction *Technical Report ERDC/CRREL TR-01-16* 69 p
[23] Matsuzaki R, Todoroki A 2008 Intelligent Tires Based on Measurement of Tire Deformation *Journal of Solid Mechanics and Materials Engineering* vol 2 **2** p 269–280
[24] Wang H, Al-Qadi I L, Stanciulescu I 2010 Effect of Friction on Rolling Tire — Pavement Interaction *Nextrans* 45 p
[25] Xia K 2010 Finite element modelling of tire/terrain interaction: Application to predicting soil compaction and tire mobility *Journal of Terramechanics* pp 1–11

IOP Conf. Series: Materials Science and Engineering **820** (2020) 012037 doi:10.1088/1757-899X/820/1/012037

Use of functional and cost design for the snowmobile and motorcycle all-terrain equipment

M Shustilova[1,2], A Diakov[1] and G Baev[1]

[1]Bauman Moscow State Technical University

[2]E-mail: shustilova-mariya@yandex.ru

Abstract. The article reveals the essence of functional-cost analysis describes the stages of functional-cost design for snowmobile and all-terrain vehicles, reviews the publications of other authors on this topic and a review of methods similar to functional-cost design.

1. Introduction

The development of the Arctic is an urgent task for the leadership of such Russian and other countries. For example, many types of natural resources can be found here. The Arctic deserts and ice cover account for up to 20 % of oil and more than half of domestic gas reserves, deposits of rare metals, gold, coal and other minerals.

One of the most important issues for the development of the Arctic is transport. What is the way to move around in the harsh Arctic conditions? The snowmobiles and all-terrain vehicles are able to cope with this task in the absence of roads with soil cover. Snowmobiles and motorcycles are already available.

But it is necessary to develop the models for particularly severe operating conditions, temperatures down to –60° C. At the same time, the equipment should be inexpensive, respecting the balance of price and quality. This is served by the functionally-cost design.

Functionally Cost Design (FCD) is a type of design technology that has a target functional and cost-oriented orientation and is focused on obtaining the optimal ratio between the quality of execution of functions and the cost of implementation in the sphere of production and operation [1]. In the course of FCD, the task of increasing the level of functionality of products as a prerequisite for improving quality is simultaneously solved. In the course of functional-value engineering the functional-value analysis is used.

Function-cost analysis (FCA) is a purposefully composed set of methods, the essence of which is to search and offer the best or even a fundamentally new solution to the functions of the analyzed object in order to increase the efficiency of its use [1].

FCA is characterized by the following features:

– the object, the increase of efficiency of which is the purpose of application of this analysis, is considered as a complex of functions;

– the functions, which reflect the behavior of the object, are evaluated in terms of value, costs and degree of their implementation. The functions that are too expensive, badly (not enough) performed or, on the contrary, performed above the required level are revealed by comparing such features. Thus, the directions and areas of further growth of the efficiency of using the object of analysis are determined. The decision of a problem — as differently, it is better, more effectively to reach performance of demanded function, is carried out further by means of the whole complex of methods of activation of creative thinking;

– the criterion of the effectiveness of the solution is the ratio between the level of satisfaction of public needs, expressed by the degree of performance of the function, and the costs of its provision. The purpose of each specific case of application of functional and cost analysis is to achieve the optimum benefit at minimum cost;

– when carrying out a functional cost analysis, a sequence of stages, steps and operations is introduced and reflected in the work plan;

– complexity of problem solving by means of functional and cost analysis requires organization of work of a group of specialists.

FCA can be implemented where it is possible to provide a function for people to do what they want and where the results of their work can be achieved in more than one way.

The object of FCA in the broad sense of the word is the activity that satisfies a public need. It can be an activity that meets the basic innovation premise, i.e. it can be carried out in at least two real ways. Depending on the degree of development of public knowledge, goals and selected scope of solutions, the analyzed activity as an object of functional and cost analysis can most often take the form:

– the technical system or part of it without regard to the degree of materialization. To such systems concern first of all products, the combined systems, subsystems as real objects or as objects at different stages of their design development. Further, this includes working tools and their parts, labour items such as materials, raw materials, semi-finished products, etc;

– process system of material and non-material character or its part. It, for example, such real processes, as technology, industrial process, operation, reception, movement, decision-making, the control etc." or processes of creation of the design documentation;

– combinations of both (technical and process) systems. They include material and energy or organizational and management systems, for example, organizational structures and processes occurring in them, etc.

FCA is applicable to various fields of activity.

Many authors in their publications [1]–[11] consider its practical application in this or that branch, for example, in construction, educational and many others. In the next chapter, we will look at several publications on the implementation, application and development of the FCD in different fields of activity.

2. Review of publications on the subject of functional-cost design.

How applicable is FCD to the task at hand and what is the practice of using FCD and FCA?

FCD and FCA were first developed in the 1940s, the beginning of the method was the work of the Soviet engineer Y.M. Sobolev and the American L.D. Miles. We are interested in how the FCD will be applied now.

Table 1. FCD applications by product and industry

Product, industry	Area of application	Sources
Construction room	Cost-cutting plans	Kulikova, Atanova, 2012 — [2]
Technological equipment	Analysis of compliance with the significance of cost functions for their implementation	Balashov, Cherepanov, 2014 — [3]
Electrical Engineering	Improvement of the manufacturing process of parts	Knutarev, 2014 — [4]
Educational process	Assessment of the quality of educational service in the field of "Commerce"	Larchenko, 2013 — [5]

End of table 1

Product, industry	Area of application	Sources
UNIVERSITY	Effective tool for cost management and increasing competitiveness in the educational services market	Sharashkina, Shilkina, 2013 — [6]
Regional development	Effective development strategy	Samostroenko, 2004 — [7]
Polypropylene pipes	New ways to improve the product	Devyataeva, Aniskina, Koles-nichenko, 2014 — [8]
Transport companies	Ways to improve the performance of the company	Savelieva, 2015 — [9]
Space rocket technology	Systematization of existing approaches to cost management and price forecasting at the stages of the life cycle of space rocket technology development	Falco, Tsisarsky, Baev, 2013 — [10]

According to the results of the authors' research review (Table 1), it was revealed that the FCD is used in different industries: construction, education, etc. However, according to the data [2] only 15 % of the surveyed builders really understand the meaning of FCD. Nevertheless, most of the researchers agree that the FCD confirms its effectiveness

3. Similar methods to functional-cost analysis (FCA)

Target Costing — is a method of managing the cost of production (production costs) of products. The essence is to reduce the cost of production throughout its entire production cycle, through the use of production, engineering, research and development.

Modern target costing (Japanese name — "genka kikaku") was born in Japan in the system of target costing management based on a very simple idea. First, the market price of this type of product is determined, then the desired profit margin is set, and then the maximum allowable cost of production is calculated.

Thus, the allowable size of the cost of production by the target costing method is determined as follows:

Price — Profit = Cost price

The market price in this method is called target price, the desired difference between the cost and sales price is called target profit, and the cost at which the product should be manufactured is called target cost.

The principles of the target costing system are as follows:

- primary and constant focus on the requirements of the market and customers;
- calculation of target costs for new products, as well as their components, to achieve the desired, pre-determined profit under existing market conditions;
- accounting for the impact on the cost of production of the wishes of consumers on the quality and timing of production;
- using the product life cycle concept.

For release of new production the enterprise carries out marketing research, positions a product in the market, then establishes the potential selling price for the given production. At the same time it is necessary to allocate such functional characteristics and properties of a product to which consumers give the greatest preference: so the component concept of the future manufacture is put and quality criteria are defined.

IOP Conf. Series: Materials Science and Engineering **820** (2020) 012037 doi:10.1088/1757-899X/820/1/012037

The so-called allowable costs are determined on the basis of the revenue forecast by the marketing department. Allowable costs are calculated as a deduction from the projected revenue, net of indirect taxes, of the planned profit attributable to the product. Target costs are calculated for all components and functions of the designed product. At the same time, it is necessary to determine the production costs taking into account the existing technologies and equipment of the enterprise. Such expenses in target costing name current expenses.

Monetization of innovations

The book (Madhavan Ramanujam, Georg Tuck, 2018) describes nine rules for monetizing innovation based on the experience of Simon-Kucher & Partners. For example, Monetizing Innovation argues that all failed products and services can be divided into four categories: overfunctioning, underpricing, hidden treasures and living dead — all of which are easily avoided. Real life examples show how some of the most innovative companies in the world (Uber, Porsche, LinkedIn, Dräger, Optimizely and Swarovski) apply the principles outlined in this book.

The authors of the book have a list of monetization errors, which are divided into four categories.

1) «Unnecessary functions»: trying to "stuff" a product with the maximum number of functions unnecessary — creates a product that does not meet the needs of customers and is often unreasonably expensive.

2) «Underpricing»: Even if the product is selected correctly and the market is looking forward to it, the price is too low to get all the potential income.

3) «Hidden treasure»: A potential bestseller that is not correctly brought to market because it goes beyond the scope of the core business.

4) «Living Dead»: Innovations that are not needed by customers but have been brought to market anyway, either because they are the wrong answer to the right question or because they are the answer to a question that no one has asked.

If we compare the target costing method and the monetization of innovations with functional cost design, we can draw the following conclusions.

The main criterion is the cost price and the cost/quality ratio. Methods complement each other, as FCD gives the technology, how to meet the set cost price.

The method of functionally cost design and target costing have a certain relationship at the stage of product design. FCD can be called the most important element of target costing, as they complement each other.

The method in the book "Innovation Monetization" works with the price of the product. Interrelation goes through the cost price (Price = Cost + Profit). The lower the cost of production, the higher the profit. The method provides ways to better understand the consumer.

All methods are good, but it is better to use a reasonable combination of them.

4. Use of FCD for snowmobile and cross-country vehicles

Here is an example of the use of FCD in the design of snowmobile and cross-country vehicles within the framework of the project on the development of the Arctic in the company "Russian Mechanics".

The essence of FCD at the stages of research and development is to prevent the emergence of functionally redundant costs and ensure the socially necessary level of quality of products.

The specifics of using FCD in the creation of new motorcycle products are as follows:

– design execution is carried out according to the scheme:

Needs → Goals → Functions → Elements;

– preliminary construction of a "tree of the purposes" of designing is provided;

– establishment of limits of expenses on functions, proceeding from their importance for consumers and complexity of realization is required;

– Technical and economic working out of decisions is carried out in parallel at observance of the established limits of expenses on functions and requirements to quality of their execution;

– Efforts are concentrated on prevention of occurrence of useless actions and the harmful phenomena in object, superfluous properties and unnecessary elements by check of level of the functional organization predetermining reliability and adaptability;

– orientation to multiple variants of execution of functions of a product is made, conditions for expansion of a field of search of decisions are created;

Let's consider a variant of application of FCD for snowmobile of "Russian mechanics" company.

"Russian Mechanics is a Russian company producing snowmobiles and motorcycles. Carries out designing, manufacture, marketing and service of made technics.

For example, let's take the snowmobile "Buran Leader" produced by the company "Russian Mechanics". The main function of the product is movement on snow. "Buran Leader" — the most affordable snowmobile, enduring, easy to operate and repair, confidently rides on deep snow, drags loaded sleds and performs the necessary economic tasks.

Schematically, let's depict the model of the product, based on the composition and number of required elements (parts) — material carriers of functions.

Example is a structural scheme of snowmobile (Fig. 1).

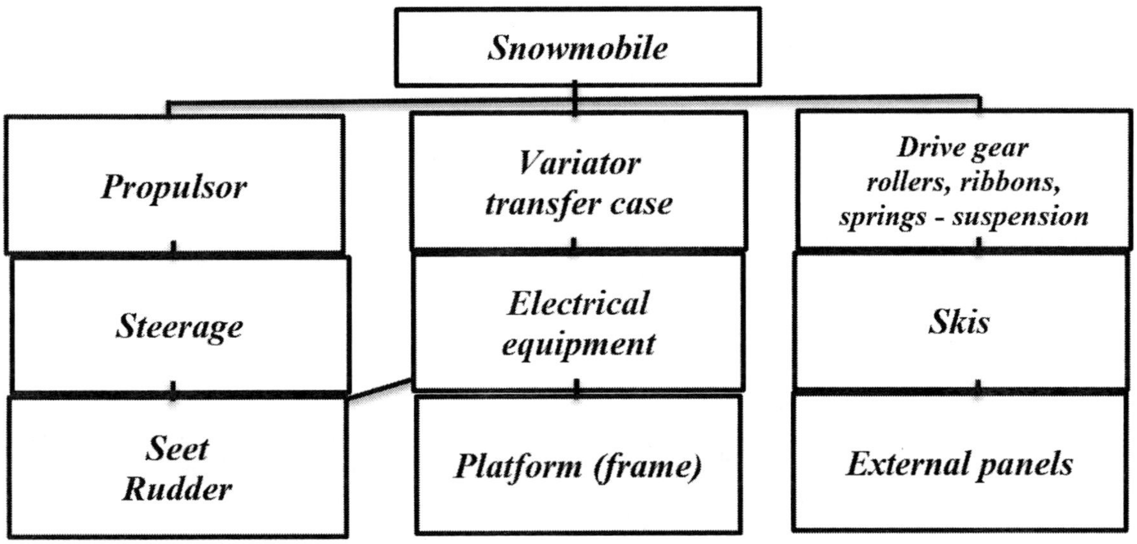

Fig. 1. Structure of the snowmobile

Further the matrix of interrelation of the basic functions and elements of a product (snowmobile) is made, taking into account the cost price of manufacturing of details, the analysis of importance and significance of functions.

The criteria of quality estimation are chosen on the basis of results of definition of requirements to a product and an estimation of their importance (for example, accuracy, reliability, maintainability etc.).

The degree of performance of functions, i.e. satisfaction of the set criteria of quality, is established by an expert method, in the simplest case by assigning scores (from 1 to 3 points) on variants. The unsatisfactory performance of the function corresponds to 0 points.

After assessing the quality of performance of functions by variants, the compliance of function costs with the established limits is checked. Operational costs of functions are determined by expert way or direct calculation using the results of assessment of reliability of elements in the performance of functions. Variants meeting the requirements of cost limits are presented for further development.

The last is a complex estimation and a final choice of a variant of construction of a product.

The aggregated algorithm of carrying out functional and cost design for snowmobile is shown in Figure 2.

IOP Conf. Series: Materials Science and Engineering **820** (2020) 012037 doi:10.1088/1757-899X/820/1/012037

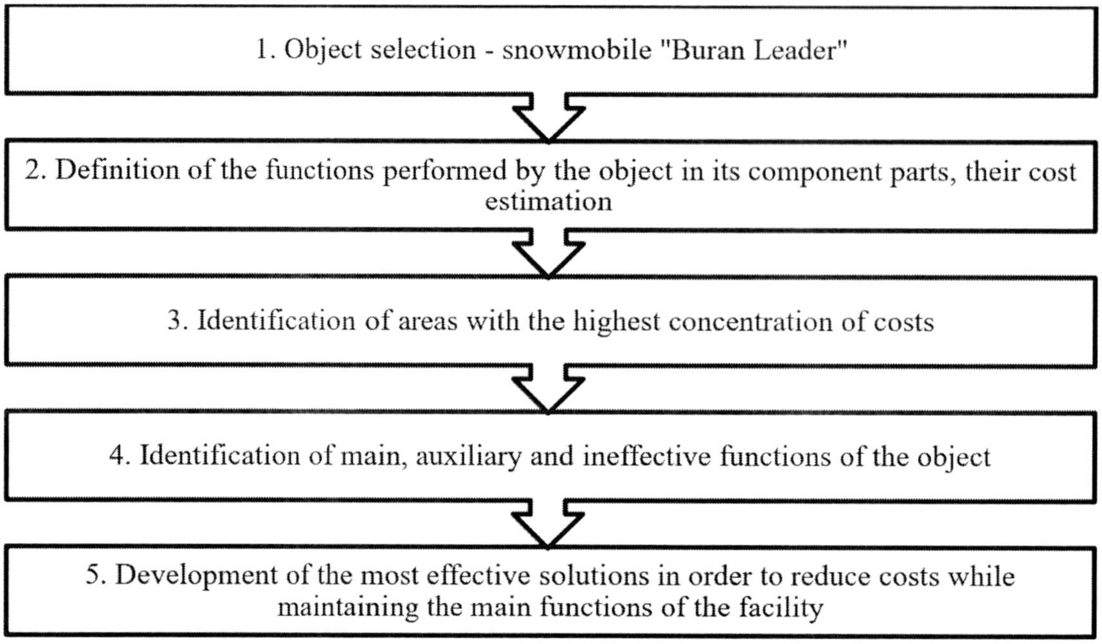

Fig. 2. Enhanced FCD algorithm

5. Conclusions

FCD as well as FCA develops and is used in different branches, if at first this method was applied only for perfection of let out products, increase of their technical and economic indicators, now it with success is used for perfection of designing, technology, the organisation of manufacture, improvement of management and planning, ordering of supply, etc.

FCD method is quite a popular technology, as it is used in many industries.

Advantages of the method:

– Better knowledge of product value enables you to make the right strategic decisions on the pricing of products, the right mix of products, the choice between the ability to produce or purchase independently, investment in research and development, process automation, promotion, etc.

– Greater clarity about the functions performed, which allows companies to focus more on management functions, such as increasing the efficiency of expensive operations; identify and reduce the volume of operations that do not add value to products.

However, the method also has disadvantages:

– The function description process can be overly detailed, and the model is sometimes too complex and difficult to maintain.

– Often the stage of collecting data on function data sources is underestimated

– The model is often outdated due to organizational changes.

– Implementation is often seen as an unnecessary "whimsy", not sufficiently supported by operational management.

Comparing FCD with other methods, it is possible to tell that functionally-cost designing perfectly combines with target costing and a method from the book "Monetization of innovations" as they well supplement each other at a stage of designing of a product.

FCD has shown itself to be a good example of snowmobile and cross-country vehicle design with the Russian Mechanics product. It helps to quickly identify the drawbacks of the constituent elements of the product and find the most effective solution to reduce costs while maintaining the main functions of the object.

References

[1] Moiseeva N.K. FCA in Mechanical Engineering. Moscow: Mechanical Engineering. 1987. 320 p.

[2] Kulikov, Ya.V., Atanova, V.S. Application of the functional-value engineering in the construction industry (in Russian) // Almanac of the modern science and education. 2012. № 6. pp. 87–91.

[3] Balashov, A.V.; Popova, T.S.; Cherepanov, D.A. Selection of the technological equipment and optimization of the cutting modes with the help of the apparatus of the functional-value engineering // Collection: prospects of the innovative development of the coal regions of Russia Proceedings of the IV international scientific-practical conference. Editorial board: Pudov E.Yu. (responsible editor), Klaus O.A. (Editor-in-Chief), Bershpolets S.I., Konoplya A.A. 2014. pp. 336–338.

[4] Knutarev A.S. FCA. Development, tendency, practical application // Collection: Week of Science of St. Petersburg State Pedagogical University materials of the scientific-practical conference with international participation. Editorial Board: M.S. Kokorin (responsible editor), etc. 2014. pp. 322–324.

[5] Larchenko Y.G., Litovchenko V. FCA quality of educational process // Modern problems of science and education. 2013. № 2. pp. 377.

[6] Sharashkina T.P., Shilkina A.T. FCA in the system of tools for improvement of processes in the university (in Russian) // Economic analysis: theory and practice. 2013. № 37 (340). pp. 32–37.

[7] Samostroenko G.M. FCA of the region development strategy // Security of Eurasia. 2004. № 1 (15). pp. 192–202.

[8] Devyataeva N.V., Aniskina E.V., Kolesnichenko M.S. FCA as the most important element of target costing on the example of polypropylene pipes production // Modern problems of science and education. 2014. № 3. pp. 312.

[9] Savelieva E.V. Management FCA and main business processes of transport enterprises // Collection: modern innovative society: dynamics of formation, development priorities, modernization: economic, social, philosophical, legal, scientific aspects, materials of the international scientific-practical conference. 2015. pp. 132–136.

[10] Falco, S.G., Tsisarskiy, A.D., Baev, G.O. Cost management and price forecasting on the stages of the rocket-space equipment creation life cycle (in Russian) // Rontrolling. 2013. № 47. pp. 70–74.

[11] Madhavan Ramanujam, Georg Taquet, "Monetization of innovations. How successful companies create a product around the price" // Kindle Edition. 2016. 256 pages

IOP Conf. Series: Materials Science and Engineering **820** (2020) 012038 doi:10.1088/1757-899X/820/1/012038

Formation of the law of steering angle control to maintain a trajectory of the vehicle movement

Y Levenkov[1,2], I Chichekin [1] and N Vol'skaya[1]

[1]Bauman Moscow State Technical University

[2]E-mail: levenkov_yy@bmstu.ru

Annotation. The object of the study is a control system for changing and maintaining the trajectory of the vehicle. Methods of maintaining a trajectory of the vehicle are the subject of research. Multi-body dynamics (MBD) methods are used to solve this problem. The purpose of the study is development of management model change the trajectory of the vehicle. On the example of a two-axle four-wheel drive truck, a mathematical model of the vehicle includes subsystems-transmission, springing system and steering. The study of the influence of the transmission and suspension to change the trajectory of the car is not considered. The paper deals with the formation of the steering angle control law to maintain a given trajectory. A detailed description of the developed control model and analytical dependencies for setting the parameters of the model. An example of selection of coefficients for its setting is shown, recommendations for other types of calculation modes are offered. In this paper, the calculation of one calculation case corresponding to the rearrangement of the car, followed by a return to the original lane. This example proves the adequacy of the model. The developed mathematical model can be used in the systems of calculation of solid body dynamos in the simulation of driver-driven cars, unmanned vehicles and adaptive driver assistance systems. The vehicle was described as a system of solid bodies connected by hinges and force interactions from the library of typical elements of the software package. According to this description of the system, the software package automatically formed a system of equations of motion and connections. The equations were solved using an implicit numerical method with variable pitch and automatic accuracy control.

1. Introduction

Currently, the evaluation of the projected design of the car, providing the specified performance properties, in the early stages of development is carried out using computer modeling and calculations. With the help of modern computer systems, it is possible not only to implement the known classical methods of calculating individual units or systems of the car, but also to create complex models of systems, units and the car as a whole [1–12]. The use of complex mathematical models based on the solution of the equation of dynamics of solids allows to take into account the mutual influence of aggregates and systems in the calculation, which is difficult in classical computational schemes.

The trajectory of the car, in turn, depends on many parameters-tires, suspension, transmission and especially steering. Also, there is a growing interest in the development of algorithms for controlling unmanned vehicles and adaptive driver assistance systems [13–15].

To maintain a given trajectory of the vehicle when modeling the dynamics of solids, it is necessary to develop a model for controlling the change in the trajectory of its movement. Setting the trajectory is beyond the scope of this work and is given as an example of the simplest required trajectory.

2. Description of the dynamic model of the car

The general view of the dynamic model of the car developed in the system of calculation of dynamics of bodies is presented in figure 1. The mathematical model of the car includes (figure 2) subsystems - a model of front and rear suspensions, all-wheel drive transmission with front and rear axle differentials, center differential, rack-and-pinion steering mechanism, tie rods, drive and steering wheel models [16–20]. The general technical characteristics of the simulated vehicle are presented in table 1.

Following assumptions were used:

– the links of the car are absolutely rigid, non-deformable bodies;

– the vehicle moves at a constant speed of the transfer case differential;

– there is no friction in the joints;

– the center of gravity of the car does not change when driving;

– simulation was carried out for the total weight of the car;

– wheel deformation is taken into account in the model of interaction of the wheel with the road surface, the forces on the wheel are applied in the center of the wheel;

– the road surface is non-deformable, absolutely flat.

The x-axis of the global coordinate system (GCS) coincides with the longitudinal axis of the car, the Z-axis of the GCS is directed vertically upwards, the y-axis is directed to the left. The HSC origin is at the intersection of the front axle and the plane of symmetry of the vehicle. The gravity vector is parallel to the Z axis of the HSC and directed in the opposite direction.

The MF-Tire model was used for the motion simulation, which is described in detail in the paper [21, 22].

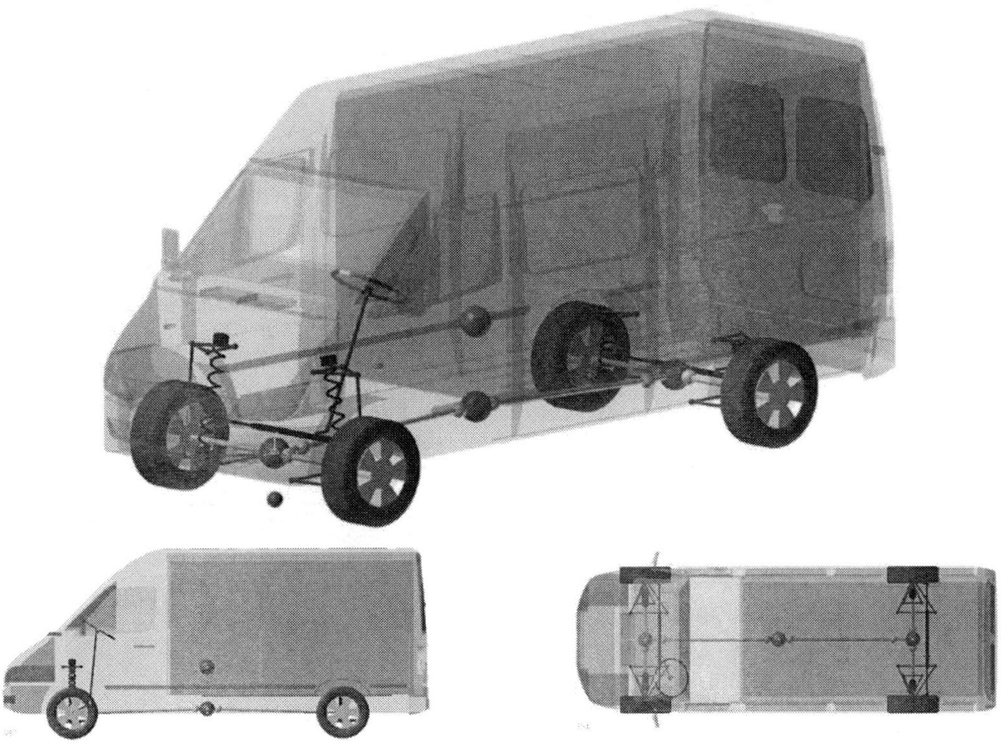

Figure 1. General view of the car model

IOP Conf. Series: Materials Science and Engineering **820** (2020) 012038 doi:10.1088/1757-899X/820/1/012038

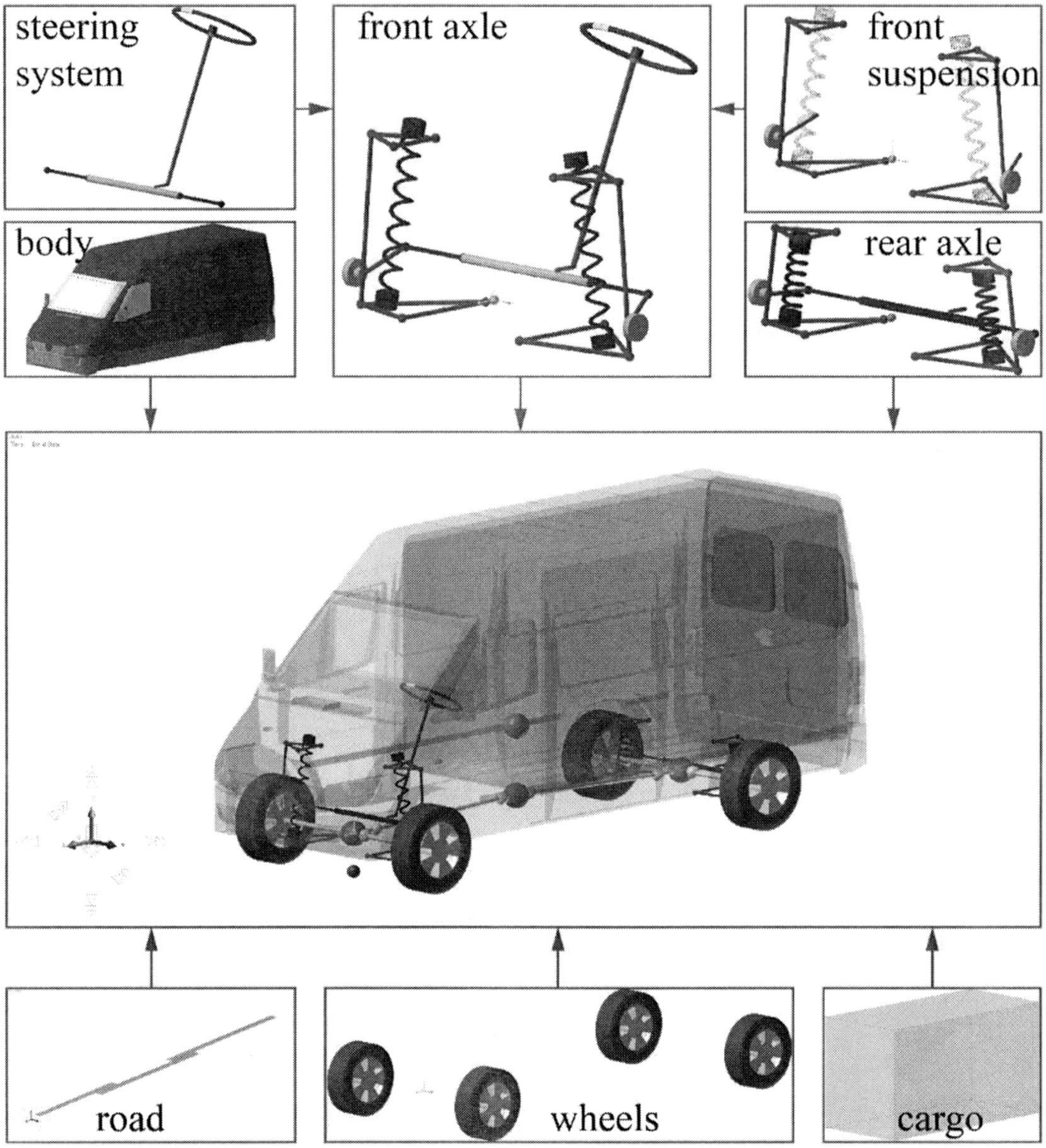

Figure 2. Components of the mathematical model of the car

Table 1. Technical characteristics of the car

The general technical characteristics of the car:	
Gross weight	3500 kg
Curb weight	1840 kg
The payload of	1660 kg
Base	3565 mm
Vehicle width	1700 mm
Vehicle height 2500 mm	2500 mm
Center of mass height	1200 mm

End of table 1

The general technical characteristics of the car:	
Distribution of the total mass on the axes of the car:	
Front axle	1460 kg, 41.7%
Rear axle	2040 kg, 58.3%
Vehicle suspension:	
Type	Independent, double-rod
Elastic element compression	Spring
Damping element	Telescopic shock absorber
Tyres, 235/55 R17	Tyres, 235/55 R17
Steering	Rack and Pinion steering mechanism Driven front wheels

For the virtual model of the driver in the mathematical model of the car added an auxiliary link 1 (figure 3) — the point of predicting the position of the body of the car, connected to the body of the car hinge allows only forward movement along the longitudinal axis. In the hinge connected to the steering mechanism, the law of movement of the steering link is formed. In the developed model, the control law is formed in the steering wheel rotation hinge 3 (figure 3).

Figure 3. Front axle of the car

Figure 4 shows the turning scheme of the vehicle. The change in the distance of the prediction point of the position of the car body 1, on the trajectory of the car 2, relative to the frame (figure 4) is given in proportion to the speed of the car (V_a):

$$L_{BT} = V_a \cdot k, \ m \tag{1}$$

k — delay the movement of the vehicle body along the trajectory of the prediction point of its position.

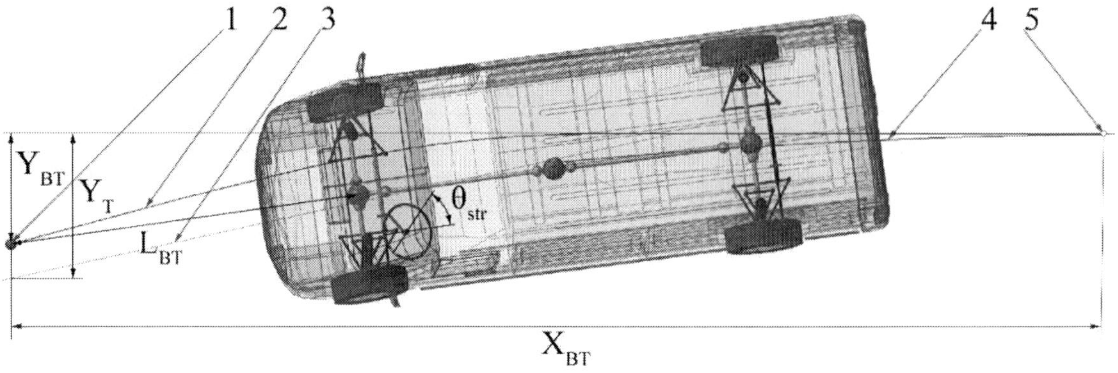

Figure 4. Scheme of the car in a turn

The desired trajectory of the car 3 (figure 4) can be given by an analytical equation $y = fun(x)$, x — longitudinal coordinate, y — lateral coordinate. You can also define a spline-a set of segments and interpolate the data using the built-in functions of the program for calculating the dynamics of solids.

The steering angle (figure 4) is proportional to the difference between the lateral displacement of the virtual point and the ordinate of the desired trajectory at the same distance in GSK:

$$\theta_{руя} = K \cdot \left(Y_{TP}(X_{BT}) - Y_{BT}\right) \tag{2}$$

where: θ_{str} — steering angle, rad;

K — steering ratio, rad / m;

Y_{TP} — ordinate the desired trajectory of the vehicle;

X_{BT} and Y_{BT} - the abscissa and ordinate of a virtual point in the XY plane of the GSC relative to the origin 5 (figure 4).

To analyze the simulation results on the car body, a point located in the center of gravity of the car is selected. The trajectory 4 is shown in figure 4. The paper shows a comparison of the deviation of the selected point on the car body from the transverse ordinate of the given trajectory.

3. Description of the results of modeling the curvilinear motion of the car

To analyze the operation of the virtual driver model, the evasive maneuver simulation was performed, followed by the return to the initial lane. Main plot sizes are selected according to GOST 31507-2012 (figure 5). The length of the plot on which the permutation is equal to 16 m. the program of calculation of dynamics of rigid bodies set to the desired trajectory of the virtual point.

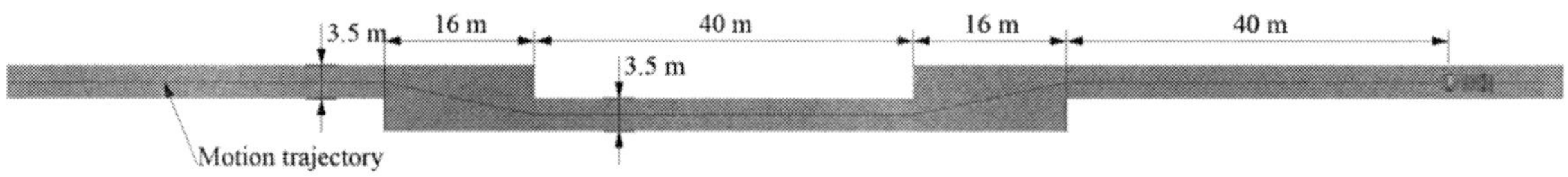

Figure 5. Main dimensions of the test area

On the initial section of the road the car was accelerated to a speed of 31.6 km/h. Acceleration was carried out kinematically-by setting the angular speed of rotation of the transfer case differential. All differentials used in the model are simple symmetric. There is no internal friction.

The simulation is performed for nine variants of the proportionality coefficient values (table 2) of the virtual point position k (1), as well as the coefficient of proportionality of the steering wheel K (2).

Table 2. The coefficients of proportionality

№ variant	k, s	K, rad/m
variant 1	0.36	-5
variant 2	0.36	-2.75
variant 3	0.36	-0.5
variant 4	0.72	-5
variant 5	0.72	-2.75
variant 6	0.72	-0.5
variant 7	1.08	-5
variant 8	1.08	-2.75
variant 9	1.08	-0.5

The simulation results for all variants are presented in figure 7. Since the selected point on the body is removed from the virtual point, the car will move along a predetermined path with a delay depending on the speed and distance L_{BT}.

According to the calculation results, it can be seen that with an increase in the coefficient of proportionality of the steering wheel K increased lateral displacement of the body (variants 3, 6, 9).

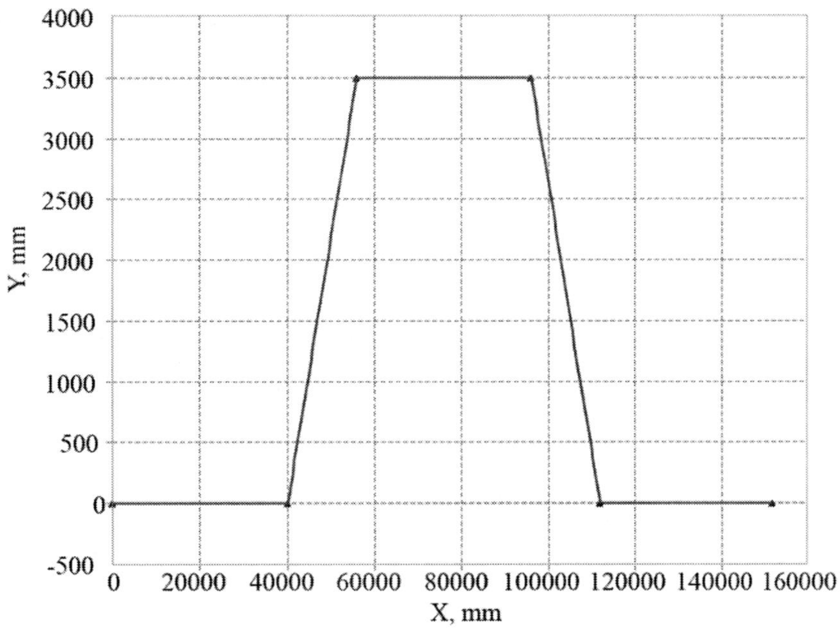

Figure 6. The specified trajectory of the vehicle

When the ratio of the virtual point position decreases, the trajectory of the selected point on the car body more precisely coincides with the specified trajectory, but the steering wheel turns frequently. It is recommended that you select lower values if you frequently change direction in short sections. When driving over long stretches, such as when modeling rough road traffic, it is recommended to increase the ratio of the virtual point position.

Based on the calculation results, the following values of the proportionality coefficients corresponding to option 5 are selected for the car in question, $k = 0.72$, s and $K = -2.75$, rad/m. Five screenshots of the motion simulation movie are shown in figure 8. The pictures show that the car didn't leave the specified corridor.

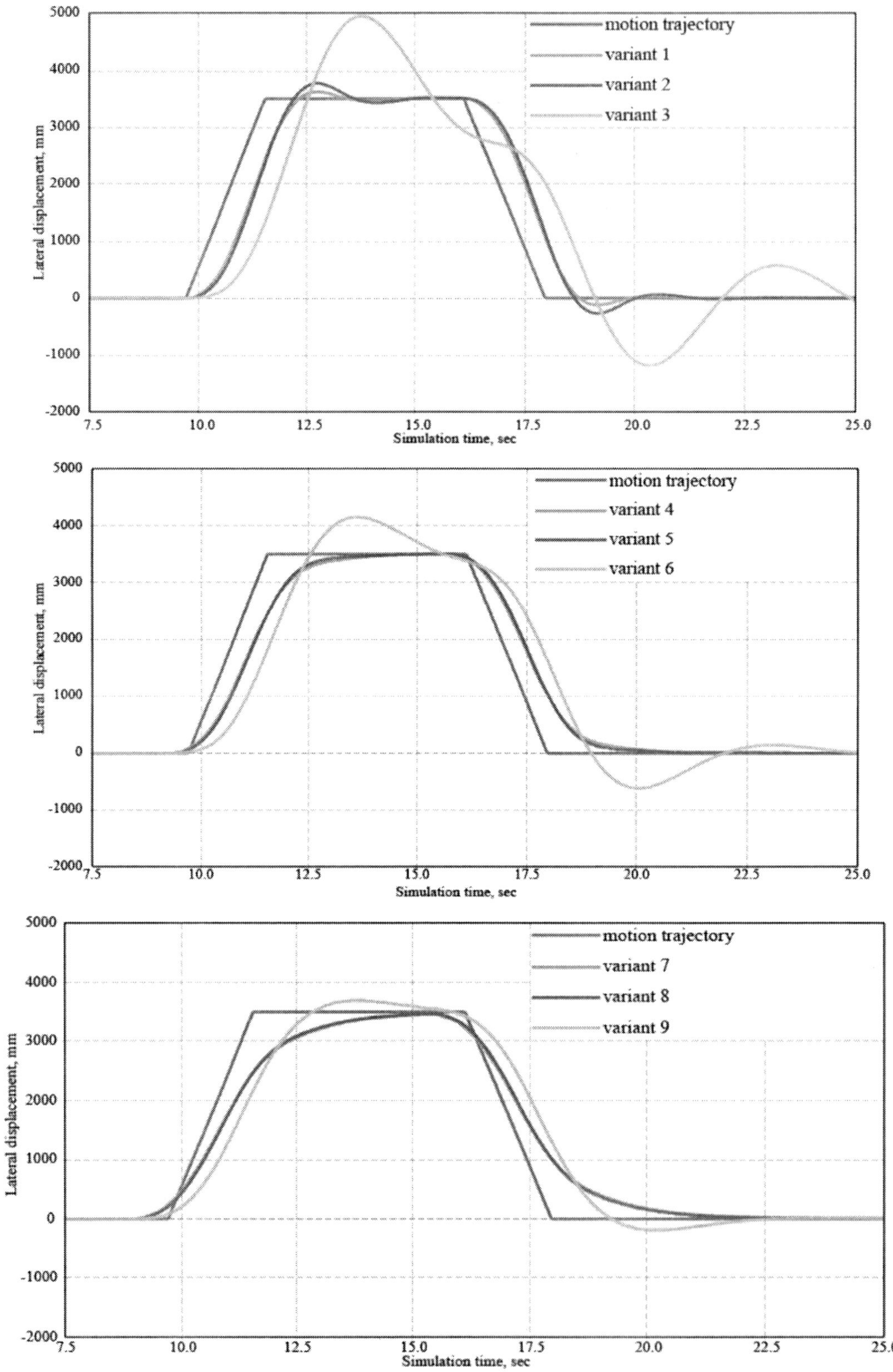

Figure 7. The trajectory of the selected point of the car body

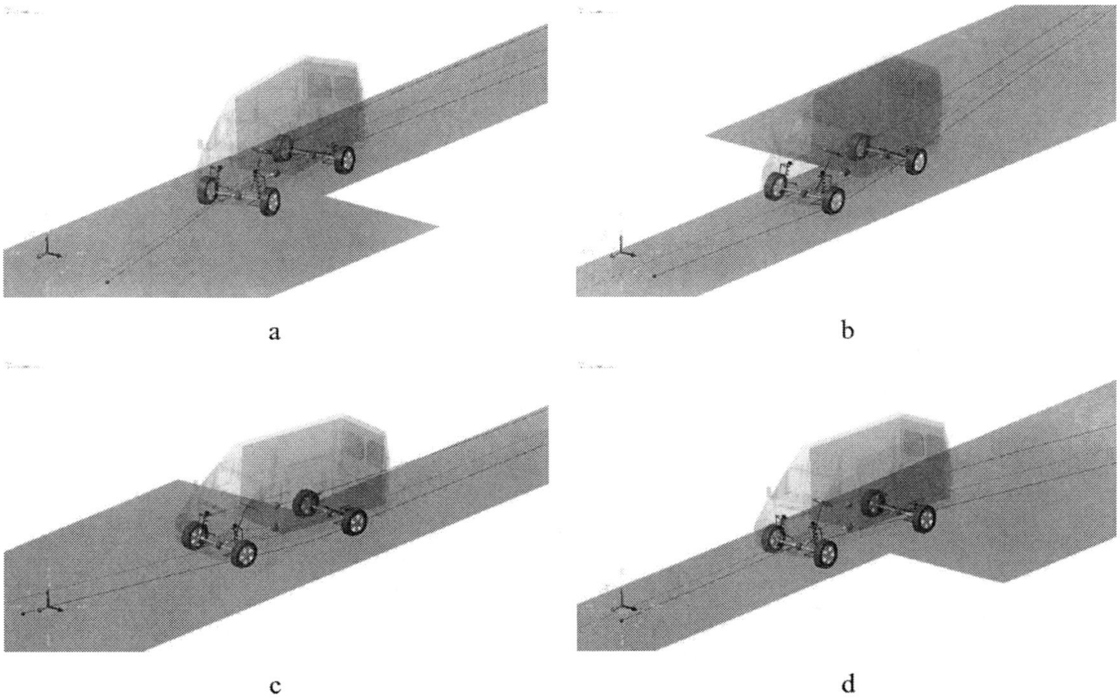

Figure 8. Screenshots of vehicle movement, with proportionality coefficients corresponding to option 5
where, the simulation time a-10.3 s, b-12.2 s, b-16.9 s, g-18.5 s

4. Conclusions

1. A model of driving allows you to simulate the movement of cars as a driver-driven and unmanned vehicles, or to simulate adaptive driver assistance systems.
2. The quantitative values of the proportionality coefficients for adjusting the regulators for the car in question are proposed. The values of the coefficients are as follows $k = 0.72$, s and $K = -2.75$, rad/m. For vehicles with other mass inertial parameters, geometric dimensions and kinematics of suspension and steering, the values of these coefficients should be obtained again.
3. To obtain numerical values of the proportionality coefficients, it is recommended to simulate additional computational cases, for example, a snake, movement on an uneven support surface, etc.
4. It is enough to set the desired trajectory approximately, the developed model of driving smoothes the trajectory itself. In this case, the trajectory of the selected point of the car body corresponds to the specified trajectory quite accurately
5. The developed model of driving is suitable for other types of vehicles, with different from the method of rotation considered in the article.

References

[1] Hashemi E, Pirani M, Khajepour A, Kasaiezadeh A 2016. *Vehicle System Dynamics* **54** 1736–61
[2] Wei Y, Liu Y, Li X, Oertel C 2016. *Vehicle System Dynamics* **54** 463–73
[3] Taheri M, Ahmadian M 2016. *Vehicle System Dynamics* **54** 653–66
[4] Rodriguez J, Freeman P T, Wagner J, Pidgeon P, Alexander K 2016. *Int. J. of Automotive Technol.* **17** 71–81
[5] Xia X, Xiong L, Sun K, Yu Z P 2016. *Int. J. of Automotive Technol.* **17** 991–1002
[6] Diakov A S, Kotiev G O 2018. *MATEC Web of Conf.* **224** 02096
[7] Evseev K B, Kartashov A B, Dashtiev I Z and Pozdeev A V 2018. *MATEC Web of Conf.* **224** 02039

IOP Conf. Series: Materials Science and Engineering **820** (2020) 012038 doi:10.1088/1757-899X/820/1/012038

[8] Kotiev G O, Padalkin B V, Kartashov A B, Diakov A S 2017. *ARPN J. of Eng. and Appl. Sci.* **12** 1064–71.
[9] Sarach E B, Kotiev G O, Beketov S A 2018. *MATEC Web of Conf.* **224** 04009
[10] Klubnichkin V E, Diakov A S, Klubnichkin E E, Zakharov A Y, Vakhidov U Sh, Suchenina A S and Basmanov I V 2019. J. of Phys.: Conf. Series **1177** 012048
[11] Volskaya N S, Zhileykin M M and Zakharov A Y 2018. *IOP Conf. Series: Materials Science and Engineering* **315** 012028
[12] Klubnichkin E E, Klubnichkin V E, Kotiev G O 2018. *IOP Conf. Series: Materials Science and Engineering* **386** 012025
[13] Zhileykin M M, Kotiev G O, Nagatsev M V 2018. *IOP Conf. Series: Materials Science and Engineering* **315** 012031
[14] Kotiev G O, Butarovich D O, Kositsyn B B 2018. *IOP Conf. Series: Materials Science and Engineering* **315** 012014
[15] Skotnikov G I, Jileykin M M and Komissarov A.I 2018. *IOP Conf. Series: Materials Science and Engineering* **315** 012027
[16] Gorelov V A, Komissarov A I 2016. *Procedia Engineering* **150** 1322–28.
[17] Gorelov V A, Komissarov A I, Miroshnichenko A V 2015. *Procedia Engineering* **129** 300–7
[18] Keller A V, Gorelov V A, Anchukov V V 2015. *Procedia Engineering* **129** 280-7.
[19] Vdovin D, Chichekin I 2016. *Procedia Engineering* **150** 1276–79
[20] Ejsmont J, Taryma S, Ronowski G, Swieczko-Zurek B 2016. *Int. J. of Automotive Technol.* **17** 237
[21] Pacejka, H B 2006. *Tyre and Vehicle Dynamics. Second Edition* (Oxford: Butterworth-Heinemann)
[22] Pacejka, H B, Besselink I J M 1997. *Supplement to Vehicle System Dynamics* **27** 234

Calculated evaluation of the efficiency of dynamic vibration isolators of the tractor cab suspension system

**V Shekhovtsov[1,3], M Lyashenk[1], N Sokolov-Dobrev[1]
A Diakov[2] and K Evseev[2]**

[1]Volgograd State Technical University, Volgograd, Russian Federation
[2]Bauman Moscow State Technical University

[3]E-mail: shehovtsov@vstu.ru

Abstract. The article gives a brief analysis of vibration-protective properties of standard rubber monoblock vibration isolators used in the systems of cabin suspension of domestic tracked tractors. The advantages of dynamic vibration dampers are described. The dynamic model of the tractor with standard and dynamic vibration isolators is described. The article presents the review of the methods of performed design researches and the analysis of their results.

Introduction
The quality of the cabin's suspension system has a significant impact on the vibration load of the traction vehicle operator's workplace. In these systems of domestic tracked agricultural tractors in most cases monoblock rubber vibration isolators are used (Fig. 1). They are installed between the frame and the floor of the cabin, have the most simple design and low cost [1, 2].

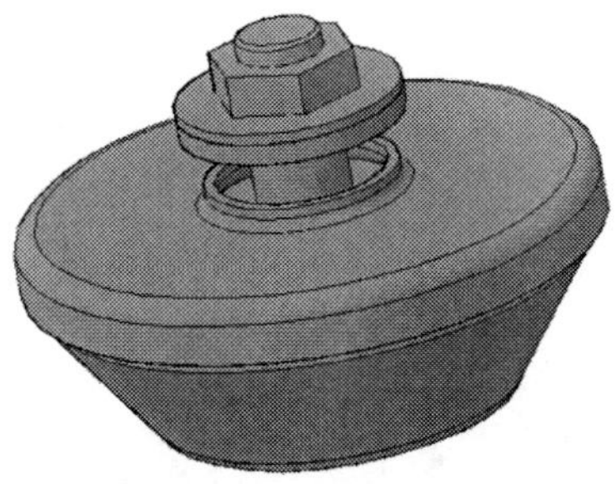

Fig. 1. The appearance of the rubber vibration isolator

During operation, a wide range of dynamic influences is transmitted to the tractor frame from the chassis system due to changes in traction resistance, tractor speed and direction of movement, from vertical, longitudinal and transverse angular oscillations of the scaffolding on the suspension and from rewinding of the tangential track. The frame also receives both high-frequency and low-frequency effects from the engine mounted on it, caused by changes in its speed and generated torque, as well as by the motor oscillations on its elastic supports. The task of the cab suspension system vibration isolators is to prevent this flow of disturbances and to provide the most complete vibration isolation of the cab floor as possible. Practice shows [3]–[18] that vibration isolators made of elastomer have not bad enough vibration-protective properties at the vibrations of cabins with high and medium frequencies and significantly worse protect the operator at low-frequency vibrations. In order to protect against such vibrations, a large, at least tens of millimeters, elastic course of the vibration isolator should be provided, while elastomeric vibration isolators provide an order of magnitude less elastic course. The best vibration protection properties are provided by vibration isolators with pneumatic elements with nonlinear elastic

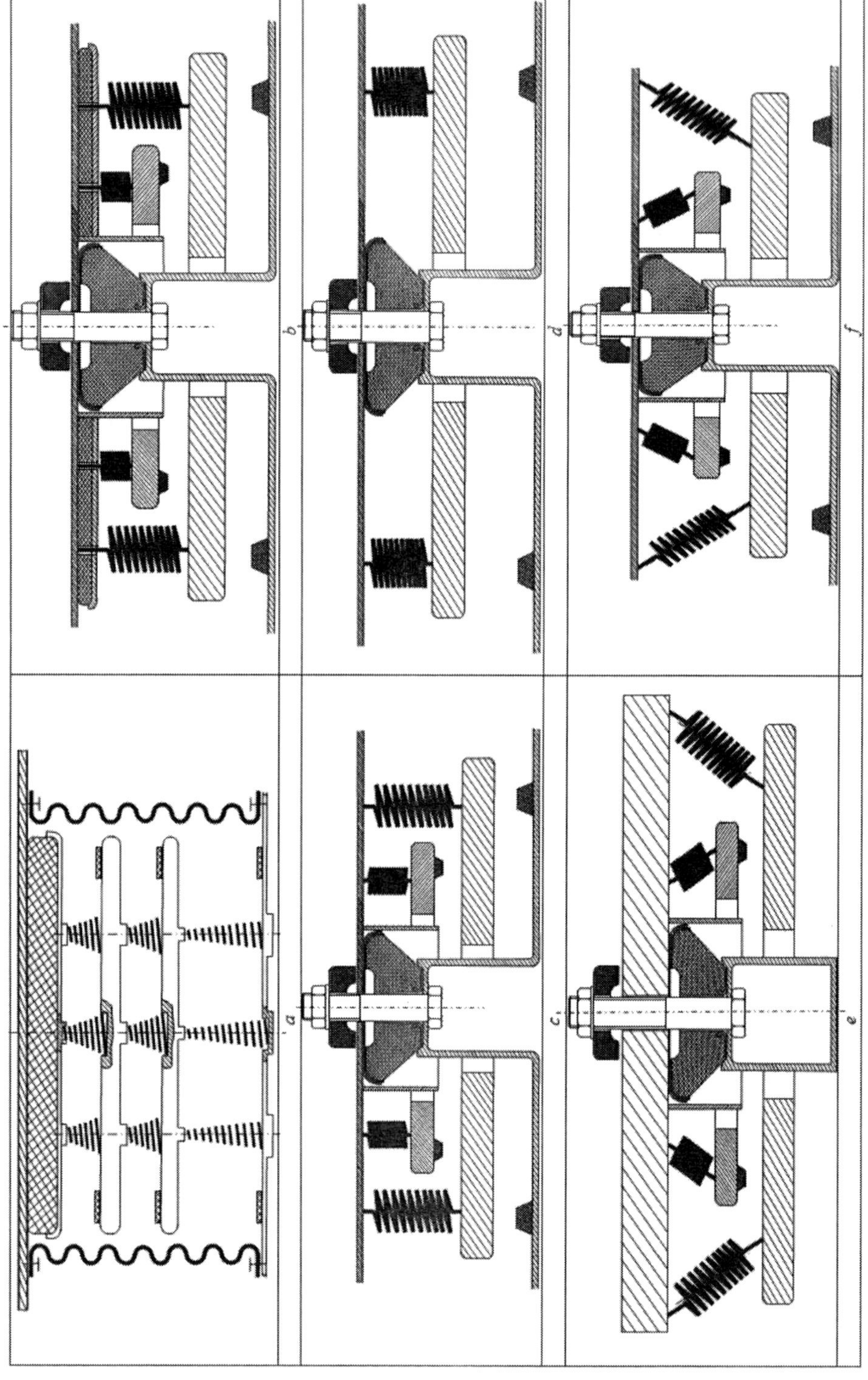

Fig. 2. Schemes of dynamic vibration isolators: a — with three consecutively arranged masses and conical springs; b — with three parallel masses, elastomer and standard vibration isolator; c — with two parallel masses and standard vibration isolator; d — with one mass and standard vibration isolator; e — with two parallel masses, cabin mass, standard vibration isolator and springs installed at an angle; f — with two parallel masses, standard vibration isolator and springs installed at an angle

and damping characteristics, but they also do not provide the most effective protection in the entire range of frequencies of operational vibration effects, but only in a certain band of frequencies, which can be changed, if the software control the pressure in the cavities of the pneumatic element, that is, its rigidity. Such control systems are developed, applied on the best samples of foreign equipment and are gradually introduced into domestic equipment. They have a complex design and quite a high cost. It is possible to unite positive qualities of vibroisolators from an elastomer and vibroisolators with pneumoelement at the expense of use of the devices providing dynamic damping of fluctuations.

Schematic solutions for dynamic vibration isolators

Dynamic vibration dampers (Fig. 2) are devices introduced into the vibrating system to dampen oscillations with certain frequencies and for partial absorption of vibrational energy [8, 9].

They include additional masses associated with the main oscillating mass of the system through elastic damping elements. The additional masses of the damper and the elastic and damping properties of its components are designed to absorb the vibrations of the main mass at certain frequencies as effectively as possible. The authors propose several variants of dynamic vibration isolators with one, two and three additional masses connected to the main mass through elastic damping elements in parallel or in series (Fig. 2). Elastomeric elements are installed in these vibration isolators to dampen high frequency vibrations, and even two such elements are installed in scheme *b*. The vibration isolator in the scheme *a* is equipped with conical screw protrusions, allowing to obtain nonlinear elastic characteristics. In vibration isolators on *e* and *f* circuits, the springs are set at an angle, which ensures that both vertical and lateral loading influences can be absorbed. In order to prevent hard impacts that may occur during operation when the moving masses are in contact with each other, the vibration isolators are equipped with dynamic stroke limiters. All vibration isolators, except those shown in diagram *d*, are equipped with two additional masses. The value of these masses, as well as the rigidity of their elastic bonds are calculated from the condition of damping of vibrations of a certain frequency band. These can be sufficiently wide bands in the spectrum of vibration effects on the frame, which are most intensely manifested operational loads. Thus, vibration isolators provide effective protection in two usually low-frequency ranges, in which monoblock elastomeric vibration isolators are ineffective. The number of moving masses can be large; in this case, vibration damping is provided in a wider frequency range, but the overall dimensions of the insulator are increased, which becomes problematic when installed between the frame and the floor of the cabin.

Dynamic model and calculation methods

In order to analyze the efficiency of using dynamic vibration isolators in the cabin suspension system with the use of the "Universal Mechanism" program package, a spatial model [19]-[28] of the tracked running system and the systems of suspension of the skeleton, engine, cabin and seat of the tracked agricultural tractor on the basis of Agromash-90TG was created (Fig. 3).

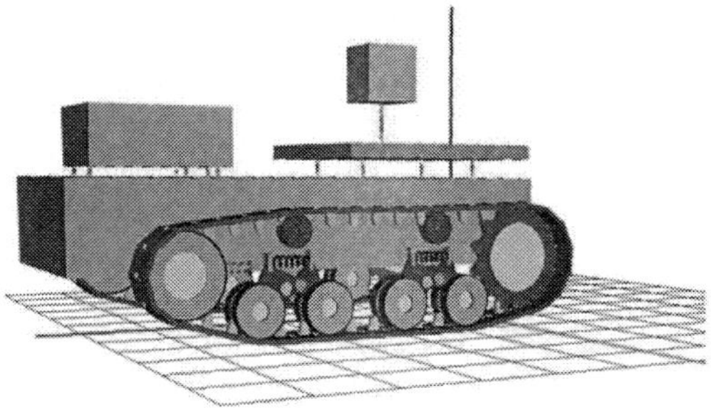

a

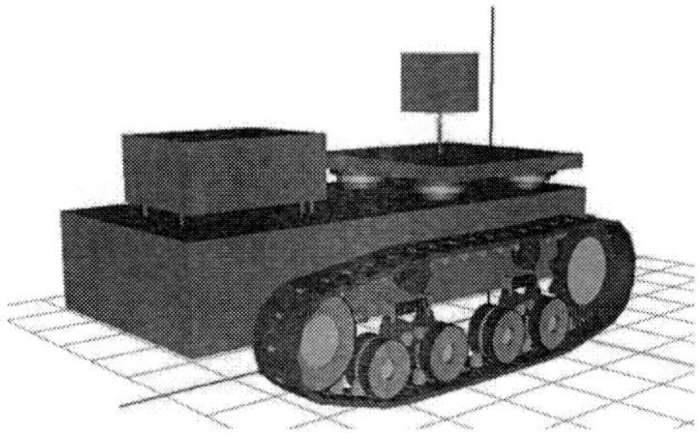

б

Fig. 3. The spatial model: a — with standard; b — with dynamic dampers

When modeling, we studied the vibrations of the sprung masses during the movement of the tractor with a hook load and without load on all the gears on a flat surface and on the polygons with single, periodic and random irregularities [20, 23].

Research results

As a result of a wide range of studies, a complete set of digital oscillograms was obtained, which show the movements, speeds and accelerations of moving masses of machine models with standard and dynamic dampers. Samples of the obtained oscillograms are shown in Fig. 4. Due to the fact that the most important for the working conditions of the operator are the accelerations of the workplace at the vertical and longitudinal angular oscillations, Fig. 4 The oscillograms of changes in cabin and seat accelerations are shown as an example. Acceleration diagrams of standard vibration isolators are marked with number 1, for dynamic — with number 2.

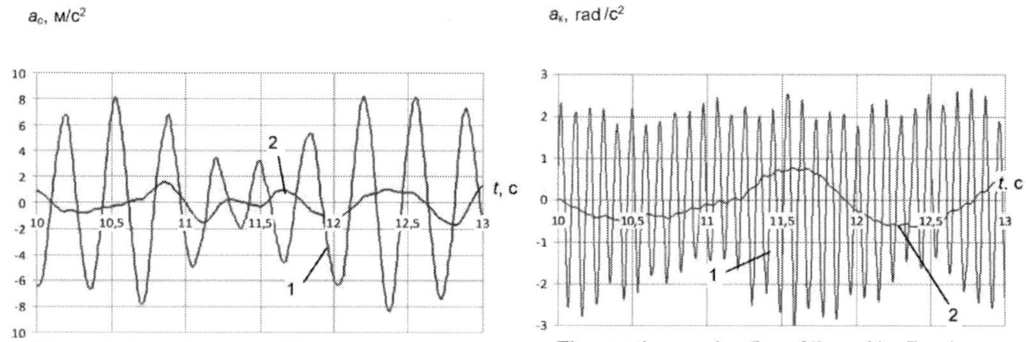

The vertical accelerations of the seat. Periodical irregularity. 7 gear with a hook load

The angular acceleration of the cabin. Random irregularities. 3 gear without a hook load

Fig. 4. Examples of compared oscillograms

Analysis of research results

The comparison of the whole set of the obtained oscillograms testifies to the fact that the vertical and longitudinal-angle accelerations of the cabin and seat decrease when installing dynamic vibration isolators of the cabin in all considered cases of motion. Table 1 and the graphs in Fig. 5 for example, the ratios of seat and cabin mass accelerations during installation of standard vibration isolators to accelerations during installation of dynamic vibrations with characteristic frequencies of vibration are given [22]:

Table 1.

The vertical vibration of the seat					
Frequency, Hz	2	3	10		
$a_{c\ stan}/a_{c\ din}$	1,5	3,5	4		
The vertical vibration of the cabin					
Frequency, Hz	4	7	11	17	18
$a_{к\ stan}/a_{к\ din}$	1,7	2,5	8,4	9,6	10,0
The angular vibration of the cabin and seat					
Frequency, Hz	3	5	11	14	17
$a_{ск\ stan}/a_{ск\ din}$	2,5	2,8	3,6	4,9	8,0

– the vertical accelerations of the seat at 2 Hz reduce by 1.5 times, at 3 Hz — by 3.5 times, at 10 Hz — by 4 times;

– the vertical acceleration of the cabin at 4 Hz decreases 1.7 times, at 7 Hz — 2.5 times, at 11 Hz — 8.4 times, at 17 Hz — 9.6 times, at 18 Hz — 10 times;

– the longitudinal-angular acceleration of the cabin and seat at the frequency of 3 Hz is reduced by 2.5 times, at the frequency of 5 Hz — by 2.8 times, at the frequency of 11 Hz — by 3.6 times, at the frequency of 14 Hz — by 4.9 times, at the frequency of 17 Hz — by 8 times.

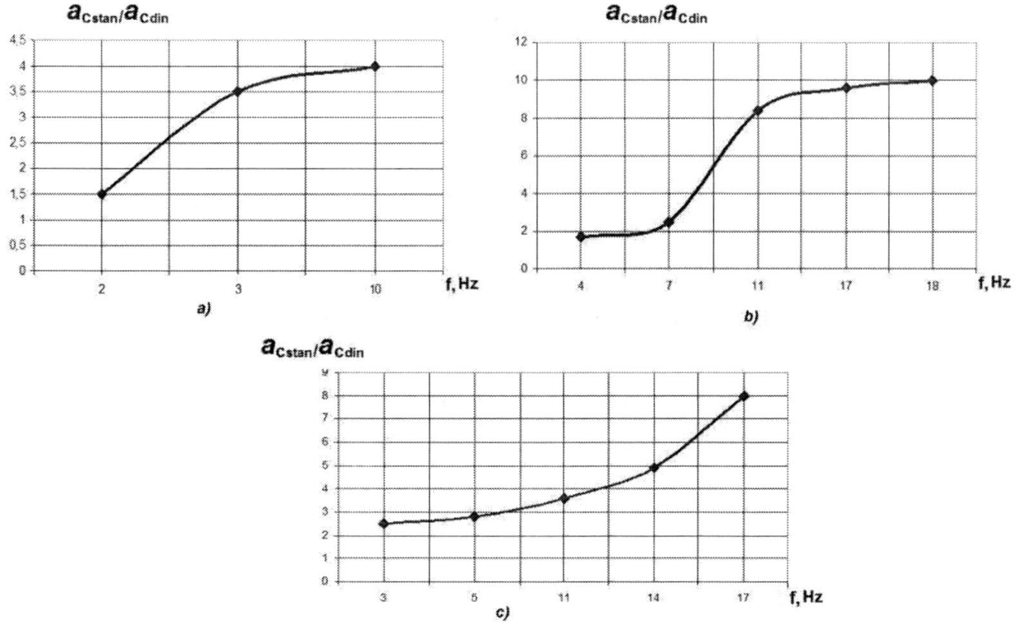

Fig. 5. The comparison of the efficiency of the cabin suspension systems with the standard and dynamic vibration isolators: *a* — reduction of the amplitudes of vertical accelerations of the seat; *b* — reduction of the amplitudes of vertical accelerations of the cabin; *c* — reduction of the amplitudes of angular accelerations of the seat and cabin

Conclusions

The results of the conducted researches testify to the undoubtedly best vibro-protective properties of cabins suspension systems with dynamic vibration isolators. When using such systems in almost all possible operating modes, the vibration level of the tractor operator's workplace is reduced by several times.

It should be noted that the this work was carried out at the Bauman Moscow State Technical University, with financial support from the government in the face of the Russian Ministry of Education under the project: №14.577.21.0272. (Identification number: RFMEFI57717X0272)

References

[1] Shekhovtsov, V.V. Technical solutions of the elastic damping devices for the tractor cab suspension / V.V. Shekhovtsov, M.V. Lyashenko, V.P. Shevchuk, N.S. Sokolov-Dobrev, K.V. Shekhovtsov // International research journal = Research Journal of International Studies. — 2013. — № 7–2. — Pages 122–125.

[2] The utility model 124340 RUS, F16/F 3/093, F16/F 1/373. Vibroizoljator / V.V. Shekhovtsov, A.V. Pobedin, V.P. Shevchuk, M.V., O.D. Kosov, M.V. Lyashenko, K.V. Shekhovtsov; VolgGTU. — 2012.

[3] Modeling, simulation, and validation of a pendulum-pounding tuned mass damper for vibration control Structural Control and Health Monitoring/Wang, W.a, Hua, X.aEmail Author, Chen, Z.a, Wang, X.b, Song, G.c Volume 26, Issue 4, April 2019.

[4] Prediction of the vibration characteristics for wheeled tractor with suspended driver seat including air spring and MR damper Zheng, E.Email Author, Fan, Y., Zhu, R., Zhu, Y., Xian, J. /Journal of Mechanical Science and Technology Volume 30, Issue 9, 1 September 2016, Pages 4143–4156.

[5] Pobedin, A.V. Testing of the vibration isolators at the stand (in Russian) / A.V. Pobedin, K.V. Shehovtsov // Izvestia VolgGTU. Ground transportation systems" series. Issue. 4: Interuniversity collection of scientific articles / VolgGTU. — Volgograd, 2011. — № 12. — Pages 41–43.

[6] Shehovtsov, K.V. Vibration Isolators`s Laboratory Testing Plant / K.V. Shehovtsov, A.V. Pobedin, O.D. Kosov // 30th Anniversary Seminar of the Students` Association for Mechanical Engineering (11-13.05.2011, Warsaw, Poland): book of Abstracts / Military University of Technology, Faculty of Mechanical Engineering. — Warsaw, 2011. — Pages 66–67.

[7] Shekhovtsov, K.V. Tests of the tractor cabin vibration isolators [Electronic resource] / K.V. Shekhovtsov // Engineering bulletin of the Don: electronic journal. — 2012. — № 1. — URL: http://ivdon.ru/magazine/latest /n1y2012/639/.

[8] Experimental determination of the characteristics of the tractor cabin vibration isolators (in Russian) / V.V. Shekhovtsov, M.V. Lyashenko, V.P. Shevchuk, N.S. Sokolov-Dobrev, K.V. Shekhovtsov // International research journal. — 2013. — № 7 (ч. 2). — Pages 118–122.

[9] The utility model 118056 RUS, G 01 M 7/02. Test bench for the vibration isolators in the mode of the forced and own vibrations (in Russian) / K.V. Shekhovtsov, A.V. Pobedin, V.P. Shevchuk, M.V. Lyashenko; VolgGTU. — 2012.

[10] The utility model 112416 RUS, G 01 M 7/02. Test bench for the vibration isolators in the forced vibration mode / K.V. Shekhovtsov, A.V. Pobedin, V.P. Shevchuk, M.V. Lyashenko, D.V. Busalayev; VolgGTU. — 2012.

[11] The utility model 112416 RUS, G 01 M 7/02. Test bench for the vibration isolators in the forced vibration mode / K.V. Shekhovtsov, A.V. Pobedin, V.P. Shevchuk, M.V. Lyashenko, D.V. Busalayev; VolgGTU. — 2012.

[12] The utility model 112415 RUS, G 01 M 7/02. Test bench for the vibration isolators under the action of loads with axial and side co-components / K.V. Shehovtsov, A.V. Pobedin, V.P. Shevchuk, M.V. Lyashenko, D.V. Busa-laev; VolgGTU. — 2012.

[13] Application and development challenges of dynamic damper in cabin booming noise elimination Cheni, R.K.Email Author, Jain, C.P.Email Author, Muthiah, R.Email Author, Gomatam, S. /SAE Technical Papers Volume 1, 2014 SAE 2014 World Congress and Exhibition; Detroit, MI; United States; 8 April 2014 до 10 April 2014.

[14] Analysis of driving seat vibrations in high forward speed tractors Servadio, P. Marsili, A., Belfiore, N.P. (2007) Biosystems Engineering.

[15] Deprez, K., Moshou, D., Anthonis, J., De Baerdemaeker, J., Ramon, H. Improvement of vibrational comfort on agricultural vehicles by passive and semi-active cabin suspensions (2005) Computers and Electronics in Agriculture, 49 (3), pp. 431-440. doi: 10.1016/j.compag.2005.08.009.

[16] Zehsaz, M., Sadeghi, M.H., Ettefagh, M.M., Shams, F. Tractor cabin's passive suspension parameters optimization via experimental and numerical methods (2011) Journal of Terramechanics, 48 (6), pp. 439–450. doi: 10.1016/j.jterra.2011.09.005.

[17] Whole body vibration exposure during rotary soil tillage operation: The relative importance of tractor velocity, draft and soil tillage depth International Singh, A., Nawayseh, N.b, Singh, L.P.a, Singh, S.a, Singh, H.c /Journal of Automotive and Mechanical Engineering Volume 15, Issue 4, 1 December 2018, Pages 5927-5940.

[18] Tractor cabin's passive suspension parameters optimization via experimental and numerical methods Journal of Terramechanics Volume 48, Issue 6, December 2011, Pages 439–450 Zehsaz, M., Sadeghi, M.H., Ettefagh, M.M., Shams, F.

[19] The utility model 136110 RUS, F16F3/087 / Vibro-isolator of a cabin of a vehicle / K.V. Shekhovtsov, N.S. Sokolov-Dobrev, A.V. Pobedin, V.P. Shevchuk, M.V. Lyashenko, V.V. Shekhovtsov; Volgograd State Technical University. — 2013.

[20] Tractor cab suspension model / A.V. Pobedin, O.D. Kosov, V.V. Shekhovtsov, K.V. Shekhovtsov // Progress of vehicles and systems - 2009: materials of the international scientific-practical conference, Volgograd, October 13–15. 2009: at 2 pm. Ч. 1 / VolgGTU [et al.]. — Volgograd, 2009. — Pages 222–223.

[21] Shekhovtsov V Sokolov-Dobrev N, Lyashenko M, Potapov P, Shekhovtsov K, Kalmykov A 2014 Influence of elements dynamic cohesiveness in power shafting on torsional vibrations spreading and dynamic equality of reducible model. Mechanika (Kaunas) 2 pp 190–196

[22] Tractor cabins suspension using the dynamic vibration dampers / K.V. Shekhovtsov, A.V. Pobedin, N.S. Sokolov-Dobrev, V.V. Shekhovtsov // Izvestia VolgGTU. Terrestrial transport systems" series. Issue. 6: Interuniversity collection of scientific articles / VolgGTU. — Volgograd, 2013. — № 10 (113). — Pages 43–46.

[23] Evaluation of the vibration-protection properties of the standard and dynamic systems of the kabin suspension in accordance with the requirements of the standards / K.V. Shehovtsov, A.V. Pobedin, N.S. Sokolov-Dobrev, M.V. Lyashenko, V.V. Shekhovtsov // Izvestia VolgGTU. Terrestrial transport systems" series. Issue. 7: Interuniversity collection of scientific articles / VolgGTU. — Volgograd, 2013. — № 21 (124). — Pages 53–55.

[24] V.V. Novikov, A.V. Pozdeev, A.S. Diakov, Research and testing complex for analysis of vehicle suspension units, Procedia Engineering, Vol. 129, pp. 465–470 (2015) DOI: 10.1016/j.proeng.2015.12.153

[25] Evseev, K.B., Kartashov, A.B., Dashtiev, I.Z., Pozdeev, A.V. Analysis viscoelastic properties of fiber-reinforced composite spring for the all-terrain vehicle, MATEC Web of Conferences 224, 02039 (2018).

[26] Diakov, A.S., Pozdeev, A.V., Novikov, V.V. The main directions of the development of snowmobiles in the Russian Federation (2018) MATEC Web of Conferences, 224, № 02080.

[27] E. Klubnichkin, A.S. Dyakov, E.E. Klubnichkin, A.Yu. Zakharov, U. Sh. Vakhidov, A.S. Suchenina and I.V. Basmanov. Experimental evaluation of stability and controllability of domestic and foreign made utility terrain vehicles. IOP Conf. Series: Journal of Physics: Conf. Series 1177 (2019) 012045. DOI:10.1088/1742-6596/1177/1/012045.

[28] V.E. Klubnichkin, A.S. Dyakov, E.E. Klubnichkin, A.Yu. Zakharov, U.Sh. Vakhidov , A.S. Suchenina and I.V. Basmanov. Experimental evaluation of speed and brake properties of domestic and foreign made utility terrain vehicles. IOP Conf. Series: Journal of Physics: Conf. Series 1177 (2019) 012048. DOI:10.1088/1742-6596/1177/1/012048.

IOP Conf. Series: Materials Science and Engineering **820** (2020) 012040 doi:10.1088/1757-899X/820/1/012040

Increase ride wheeled vehicles due to the continuous damping control in the suspension system

M Zhileykin[1,3] and I Fedotov[2]

[1]Bauman Moscow State Technical University
[2]FSUE «NAMI» State Research Center of the Russian Federation (2 Avtomotornaya St., Moscow, Russia, 125438),

[3]E-mail: jileykin_m@mail.ru

Annotation. One of the promising ways to improve the wheeled vehicles smooth running is the development of dynamic semi-active springing systems, as well as the laws of such systems control. Dynamic control systems in this paper are understood as systems operating at a real-time rate with the current (instantaneous) phase coordinates values. The aim of the work is to develop the laws of optimal control of the wheeled vehicles suspension, providing an increase in the driver and passengers comfort in a wide range of operating conditions. The of the optimal control system algorithm of the level of damping shock absorber suspension wheel machine to improve of the driver and passengers comfort. The proposed algorithm efficiency is proved by simulation methods. It is established that at rectilinear movement on different roads with different speeds, the level of the movement comfort is increased by 8 to 41% when the frequency controlling effect of 15 Hz and 13...59% at the frequency control action of 50 Hz compared to uncontrolled suspension.

Introduction

Creation of effective means of protection against external dynamic influences has always been one of the important problems of modern technology. Particularly acute issues of ensuring acceptable levels of vibration are raised in the creation of modern vehicles: aircraft, wheeled vehicles (WV), ships. With increasing speeds of these vehicles, the intensity of dynamic impacts usually increases, so the development of vehicles is accompanied by a continuous increase in the requirements for vibration protection systems.

Traction and speed properties of the car have a major impact on the formation of maximum and average speeds. Other operational properties, in particular, smooth running, act as factors limiting the further increase in operating speeds [1]. Currently, the design solutions of components and assemblies of the car, affecting the indicators of stability, handling, braking properties, turnability, patency have reached a high degree of perfection. However, despite these successes, the system of the primary cushioning inhibits further growth speeds.

One of the promising ways to improve the smoothness of the WV is the development of dynamic semi-active springing systems, as well as the laws of such systems control [2–20]. Dynamic control systems in this paper are understood as systems operating at a real-time rate with the current (instantaneous) values of phase coordinates. The aim of the work is to develop the laws of optimal control WV suspension, providing increased comfort driver and passengers in a wide range of operating conditions.

Calculation of control actions to improve the stroke smoothness

Consider a motion linearized model of the vehicle body, with respect to only three phase coordinates (fig. 1):

– relative to the vertical Z axis of the associated coordinate system;
– rotation of the body relative to the main axis of inertia X related coordinate system;
– rotation of the body relative to the main axis of inertia Y of the associated coordinate system.

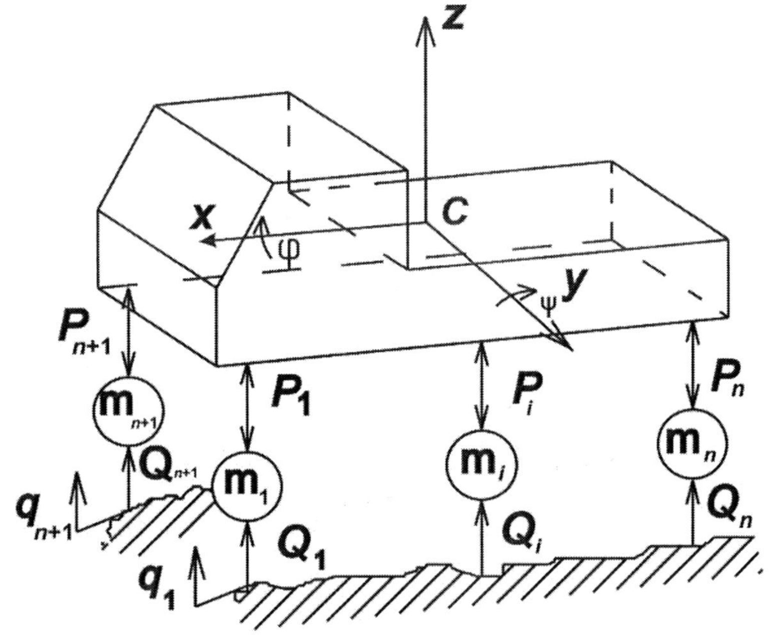

Fig. 1. A design scheme for multiple-seated oscillatory system WV:
C — vehicle body center of gravity; X, Y, Z — the coordinate system associated with the vehicle body center of gravity; φ, ψ — angles of longitudinal and transverse inclination of the body, respectively; P_i — force in i-th suspension; m_i — mass i-th wheel; n — number of axis; Q_i — force in i-thtyre; q_i — unevenness height of the support base under the i-th wheel.

Let's consider the body is absolutely solid, suspension wheels made by "McPherson" scheme.
As a result, we obtain the following equations of state for the body of WV:

$$\begin{cases} \ddot{z} = -U_1 = \dfrac{1}{M}\sum\limits_{i=1}^{2n}\Delta P_i \\ \ddot{\varphi} = -U_2 = \dfrac{1}{J_y}\sum\limits_{i=1}^{2n}\Delta P_i l_i \\ \ddot{\psi} = -U_3 = \dfrac{1}{J_x}\sum\limits_{i=1}^{2n}\Delta P_i B_i \end{cases} \quad (1)$$

where $\ddot{\varphi}$ — body angular acceleration relative to the axis X; $\ddot{\psi}$ — body angular acceleration relative to the axis Y; $\ddot{z}$ — vertical acceleration of the vehicle body center of gravity; J_x, J_y — sprung mass moments of inertia WV relative an axis X and Y respectively; M — the wheel vehicle sprung weight; l_i, B_i — longitudinal and transverse coordinates (respectively) of fastening i-th suspension to the WV body in the coordinate system associated with the machine body; ΔP_i — i-th control action (force change in i-th the damping suspension member); n — wheeled vehicle number of axles; U_1 — controlling the main vector of forces created by the suspension damping elements in the vertical direction; U_2 _ control torque generated by the suspension damping elements relative to the axis Y (applied to the vehicle body center of gravity); U_3– control torque generated by the suspension damping elements relative to the axis X (applied to the vehicle body center of gravity).

The initial system of equations of state (1) is a linear system described by a vector differential equation

$$\frac{d\boldsymbol{X}(t)}{dt} = \boldsymbol{A}(t)\boldsymbol{X}(t) + \boldsymbol{R}(t)\boldsymbol{U}(t), \tag{2}$$

where $\boldsymbol{X}(t)$— state vector; $\boldsymbol{A}(t)$, $\boldsymbol{R}(t)$—matrix functions of time according to orders $2n\times 2n$; $2n\times 3$; $U(t)$—$2n$- dimensional control vector.

Note that in the system of equations (2) all elements a_{ij} matrix $\boldsymbol{A}(t)$ equal to zero.

Vector control $\boldsymbol{U}(t) = [U_1(t), U_2(t), U_3(t)]^T$ for obvious physical reasons is bounded, i.e. belonging to a closed set

$$|U_i(t)| \le U_{imax}, i = 1,2,3$$

Control forces:

$$\Delta P_i = P_{i\partial} - P_{iT}$$

where $P_{i\partial}$— actual current force value in the suspension damping element; P_{iT}—the required force value in the damping element.

To determine values ΔP_i we formulate the criterion of optimality, on the basis of which we will determine the required variables values. The most appropriate is to obtain such values P_{iT}, which would be the least different from the actual current forces values in the damping elements $P_{i\partial}$, that is, it is necessary to strive to ensure that

$$\Delta P_i = P_{i\partial} - P_{iT} \to 0, i = 1,2,...,2n$$

We formulate the quadratic optimality criterion as follows:

$$f(\Delta P) = \sum_{i=1}^{2N} (P_{iT} - P_{i\partial})^2 = 0 \tag{3}$$

The resulting problem is an optimization problem in the presence of constraints in the form of equality. Let's introduce auxiliary functions

$$\begin{cases} g_1 = U_1 + \dfrac{1}{M}\sum_{i=1}^{2n} \Delta P_i = 0 \\ g_2 = U_2 + \dfrac{1}{J_y}\sum_{i=1}^{2n} \Delta P_i l_i = 0 \\ g_3 = U_3 + \dfrac{1}{J_x}\sum_{i=1}^{2n} \Delta P_i B_i = 0 \end{cases}$$

The necessary minimum conditions of the function (3) are written as follows [9]:

$$\begin{cases} \dfrac{df}{d\Delta P_i} + \sum_{i=1}^{3}\left[\lambda_i \cdot \dfrac{\partial g_i}{\partial \Delta P_i}\right] = 0 & j = 1,2,...,2n \\ g_i(\Delta P_i) = 0 & i = 1,2,3 \end{cases} \tag{4}$$

where: λ— indeterminate Lagrange multipliers.

In our case, equations (4) will take the following form

$$\begin{cases} 2(\Delta P_1)+\lambda_1 l_1+\dfrac{B\lambda_2}{2}+\lambda_3=0 \\ \dots\dots\dots\dots\dots\dots \\ 2(\Delta P_n)+\lambda_1 l_n+\dfrac{B\lambda_2}{2}+\lambda_3=0 \\ 2(\Delta P_{n+1})+\lambda_1 l_{n+1}+\dfrac{B\lambda_2}{2}+\lambda_3=0 \\ \dots\dots\dots\dots\dots\dots \\ 2(\Delta P_{2n})+\lambda_1 l_{2n}+\dfrac{B\lambda_2}{2}+\lambda_3=0 \\ \ddot{z}=-\dfrac{1}{M}\sum_{i=1}^{2n}\Delta P_i \\ \ddot{\varphi}=-\dfrac{1}{J_y}\sum_{i=1}^{2n}\Delta P_i l_i \\ \ddot{\psi}=-\dfrac{1}{J_x}\dfrac{B}{2}\left[\sum_{i=1}^{n}\Delta P_i-\sum_{j=n+1}^{2n}\Delta P_j\right] \end{cases} \quad (5)$$

System (5) is a linear system containing $2n+3$ algebraic equations with respect to $2n+3$ unknown ΔP_i, λ_1, λ_2, λ_3, from which are easily determined ΔP_i.

After calculating the control actions ΔP_i you must convert these values to damping factors k_i.

Limitations on control action

The regulator output must have limited amplitude for at least two reasons. First, due to the fact that in case of exceeding the maximum pressure value, the safety valve is triggered. Thus, the upper limit of permissible control actions in the equations system (5) is always determined. There is also a lower limit of control actions associated with the presence of a viscous friction minimum value in the damping device, as well as the presence of "dry" friction in the seals. The upper and lower bounds form the area of acceptable control values.We assume for certainty that the upper and lower bounds of the permissible values of forces P_a, developed shock absorber, differ from the nominal level twice.

Operation algorithm of the continuous optimal control system of springing system damping

1. For the current time t_j according to the vertical acceleration sensor installed in the vehicle body center of gravity of the body, determine the value of $\ddot{z}_j$ and differentiating the readings of the angular velocity sensors of the housing to determine the values $\ddot{\varphi}_j$ and $\ddot{\psi}_j$.

2. For the current time t_j sensor readings differentiation of suspensions deflections to determine the speed of deflection of $\dot{h}_{ij}, i=1,2,\dots,2n$.

3. For the current time t_j to determine the current power $P_{ijT}, i=1,2,\dots,2n$ in each shock absorber based on the readings of pressure sensors installed in their upper and lower cavities:

$$P_{ijT}=p_{up\,ij}S_{up\,i}-p_{low\,ij}S_{low\,i},\ i=1,2,\dots,2n,$$

where $p_{up\,ij}, p_{low\,ij}$ — pressure values in the upper and lower cavities i-th shock absorber, respectively; $S_{up\,i}, S_{low\,i}$ — shock absorber piston area in the upper and lower cavities i-th shock absorber, respectively.

4. For the current time t_j solve the system (5) and determine the required force value $P_{ijmp} = P_{ijT} + \Delta P_{ij}, i = 1,2,\ldots,2n$ in each shock absorber.

5. For the current time t_j according to pre-programmed in the memory of the on-Board computer load characteristics of the shock absorber on the basis of the values obtained in paragraph 1 $\dot{h}_{ij}$ to define a nominal current strength in the shock $P_{ij\partial}, i = 1,2,\ldots,2n$ (curve 1 on fig. 2).

6. For the current time t_j define the upper $P_{ij\text{верх}} = 2P_{ij\partial}, i = 1,2,\ldots,2n$ (curve 2 on fig. 2) and lower $P_{ij\text{ниж}} = 0,5P_{ij\partial}, i = 1,2,\ldots,2n$ (curve 3 on fig. 2) the limits of the shock absorber force variation to the obtained values $\dot{h}_{ij}$.

7. For the current time t_j determine the values of scaling factors $\Delta k_{ij}, i = 1,2,\ldots,2n,$ which will allow to proportionally change the values of the forces P_{ijmp}, to done the inequality $P_{ij\text{ниж}} \le P_{ijmp} \le P_{ij\text{верх}}, i = 1,2,\ldots,2n.$

7.1. if $P_{ijmp} > 0$ and $P_{ijmp} > P_{ij\text{верх}}$, then $\Delta k_{ij} = \left|\frac{P_{ij\text{верх}}}{P_{ijmp}}\right|, i = 1,2,\ldots,2n,$

7.2. if $P_{ijmp} < 0$ and $P_{ijmp} < P_{ij\text{ниж}}$, then $\Delta k_{ij} = \left|\frac{P_{ij\text{ниж}}}{P_{ijmp}}\right|, i = 1,2,\ldots,2n,$

7.3. if $P_{ijmp} > 0$ and $P_{ijmp} < P_{ij\text{ниж}}$, then o $\Delta k_{ij} = \left|\frac{P_{ijmp}}{P_{ij\text{ниж}}}\right|, i = 1,2,\ldots,2n,$

7.4. if $P_{ijmp} < 0$ and $P_{ijmp} > P_{ij\text{ниж}}$, then $\Delta k_{ij} = \left|\frac{P_{ijmp}}{P_{ij\text{верх}}}\right|, i = 1,2,\ldots,2n.$

8. For the current time t_j of all Δk_{ij} to choose the minimum value $\Delta k_{j\,min} = min\left[\Delta k_{ij}\right], i = 1,2,\ldots,2n.$

9. For the current time t_j adjust the values of the required force $P_{ijmp} = P_{ijT}\Delta k_{j\,min}, i = 1,2,\ldots,2n.$

10. For the current time t_j calculate damping coefficients for each shock absorber k_{ij}

10.1. if $sign\left(\dot{h}_{ij}\Delta P_{ij}\right) > 0,$ then $k_{ij} = min\left[\left|\frac{P_{ijT}}{P_{ijmp}}\right|, \left|\frac{P_{ijmp}}{P_{ijT}}\right|\right], i = 1,2,\ldots,2n,$

10.2. if $sign\left(\dot{h}_{ij}\Delta P_{ij}\right) < 0,$ then $k_{ij} = max\left[\left|\frac{P_{ijT}}{P_{ijmp}}\right|, \left|\frac{P_{ijmp}}{P_{ijT}}\right|\right], i = 1,2,\ldots,2n,$

10.3. if $sign\left(\dot{h}_{ij}\Delta P_{ij}\right) = 0,$ then $k_{ij} = 1, i = 1,2,\ldots,2n.$

11. For the current time t_j to normalize the values of the coefficients k_{ij} for the range [0,5–1,0]

$$k_{j\,max} = max\left[k_{ij}\right], i = 1,2,\ldots,2n,$$

$$k_{ij} = 0,5 + \frac{k_{j\,max}}{2k_{ij}}, i = 1,2,\ldots,2n.$$

12. Further, the steps 1 — 11 again.

Efficiency investigation and algorithm efficiency of continuous optimal control system damping springing system

The controlled suspension efficiency according to the above algorithm was tested by simulation methods in comparison with the unmanaged suspension, elastic and damping characteristics of which had a nominal damping and static deflection. The main technical characteristics of the two-axle wheeled machine are given in table 1.

Table 1. Specifications two-axle wheeled machine

Parameter	Value
Number of axis	2
Vehicle body mass, kg	2400
Wheel base, m	3,3
Wheel track, m	1,7
Wheel weight, kg	80
Free wheel radius, m	0,4
The maximum suspension deflection, relative to the wheel, m	0,24
The vehicle body inertia of the relative to the longitudinal axis passing through the center of gravity, kg×m^2	1450
The vehicle body inertia of the relative to the lateral axis passing through the center of gravity, kg×m^2	8000
Distance from body center of gravity to first axis, m	1,56
Distance from body center of gravity to second, m	1,74
Springingsystem type	independent suspension of all wheels
Elastic suspension element type	pneumatic elastic element
Damping suspension member type	controlled twin-tube shock absorber

Table 2. Study results of the suspension control system effectiveness of a two-axle wheeled vehicle

Typeofroadsurface	Movementspeed, kph	Frequencyregulation, Hz	The value of the criterion "Comfort», m/s^2	
			controlled suspension	uncontrolled suspension
Road with asphalt concrete pavement	35	15	3,78	4,55
	35	50	3,71	
	80	15	7,73	8,89
	80	50	8,19	
Dirt road of satisfactory quality	20	15	6,14	8,25
	20	50	7,55	
	35	15	24,90	60,66
	35	50	35,60	

"Comfort" criterion was used as a criterion of the wheel suspension control system efficiency [21]. The analysis of this criterion was carried out in the frequency range of 0.7–22.4 Hz.

The simulation of wheeled vehicle rectilinear motion with different speeds on three types of road surface was carried out:

1. road with asphalt concrete pavement;
2. dirt road of satisfactory quality.

To assess the control frequency influence on the control efficiency, the continuous damping control system simulation was carried out for two frequency values: 15 Hz and 50 Hz. The simulation results are presented in table. 2.

Analysis of the results given in table. 2 shows that in the case of moving two-axle wheeled vehicle with controlled suspension on various quality roads cover level of comfort movements increased by 8 to 41% when the frequency controlling effect of 15 Hz and 13...59% at the frequency control action of 50 Hz compared to uncontrolled suspension.

Conclusions

1. The continuous changes system algorithm in the level of damping shock absorber suspension wheel machine to improve the driver and passengers comfort.
2. The proposed algorithm efficiency is proved by simulation methods. It is established that at rectilinear movement on different roads with different speeds, the comfort level of the movement is increased by 8 to 41% when the frequency controlling effect of 15 Hz and 13...59% at the frequency control action of 50 Hz compared to uncontrolled suspension.

References

[1] Novikov, V.V., Pozdeev, A.V., Diakov, A.S. Research and testing complex for analysis of vehicle suspension units (2015) Procedia Engineering, 129, pp. 465–470.

[2] Du H, Sze K Y and Lam J 2005 Semi-active H∞ control of vehicle suspension with magneto-rheological dampers J. Sound Vib. 283 981–96.

[3] Poussot-Vassal C, Sename O, Dugard L, Gaspar P, Szabo ZandBokor J 2008 A new semi-active suspension control strategy through LPV technique Control Eng. Pract.161519–34.

[4] Fallah M S, Bhat R and Xie W F 2009 New model and simulation of Macpherson suspension system for ride control applications Veh. Syst.Dyn.47195–220.

[5] Laws S M 2010 An active camber concept for extreme maneuverability: mechatronic suspension design, tire modeling, and prototype development PhD Diss. StanfordUniversity.

[6] Zhileykin, M.M., Kotiev, G.O., Nagatsev, M.V. Comparative analysis of the operation efficiency of the continuous and relay control systems of a multi-axle wheeled vehicle suspension (2018) IOP Conference Series: Materials Science and Engineering, 315 (1), article № 012030..

[7] Zhileykin, M.M., Kotiev, G.O., Nagatsev, M.V. Synthesis of the adaptive continuous system for the multi-axle wheeled vehicle body oscillation damping (2018) IOP Conference Series: Materials Science and Engineering, 315 (1), article № 012031.

[8] Belousov, B.N., Merkulov, I.V., Fedotov, I.V. Controlled suspensions of automobiles(2004) Avtomobil'nayaPromyshlennost, (1), pp. 23–25.

[9] Belousov, B.N., Merkulov, I.V., Fedotov, I.V. Synthesis of dynamic control system for active mountings of multiaxis ATM (2004) Avtomobil'nayaPromyshlennost, (4), pp. 15–17.

[10] Belousov, B., Ksenevich, T.I., Naumov, S.Automated system to control steering and wheel springing parameters in vehicle locomotion module(2015) SAE Technical Papers.

[11] Belousov, B., Demik, V., Kozlova, A., Ksenevich, T., Naumov, S., Medvedev, E., Lyushnin, S., Kuzminkov, K. The schematic diagrams of actuators of an mechatronic wheel steering system (2018) FISITA World Automotive Congress 2018.

[12] Belousov, B., Ksenevich, T.I., Vantsevich, V., Naumov, S. An active long-travel, two performance loop control suspension of an open-link locomotion module for off-road applications(2014) SAE Technical Papers, 2014-January.
[13] Sarach, E., Kotiev, G., Beketov, S. Methods for road microprofile statistical data transformation (2018) MATEC Web of Conferences, 224, article № 04009.
[14] Alanoly J and Sankar S 1987 A new concept in semi-active vibration isolation J. Mech. Des.109242–7.
[15] Sohn H-C, Hong K-S and Hedrick J K 2000 'Semi-active control of the Macpherson suspension system: hardware-in-the-loop simulations. '2000. Proc. of the 2000 IEEE Int. Conf. on Control Applications, pp982–7.
[16] Karnopp D, Crosby M J and Harwood. R A 1974 Vibration control using semi-active force generators J. Manufact. Sci. Eng. 96619–26.
[17] Choi S B, Choi Y T, Chang E G, Han S Jand Kim. C S 1998 Control characteristics of a continuously variable ER damper Mechatronics 8143–61.
[18] Carlson J D, Catanzarite D M and Clair K A St 1996 Commercial magneto-rheological fluid devices Int. J. Mod. Phys. B 102857–65.
[19] Jalili N 2002 A comparative study and analysis ofsemi-activevibration-control systems J. Vib. Acoust. 124593–605.
[20] Vasiliev O. V. Optimization methods. World Federation Publishers Company, ICN, Atlanta, USA, 1996.
[21] GOST 31191.1-2004 (ISO 2631-1:1997) «Vibration and shock. Measurement of total vibration and assessment of its impact on humans. Part 1. Generalrequirements». — M.: Publishingstandards 2004. — 45 p.

IOP Conf. Series: Materials Science and Engineering **820** (2020) 012041 doi:10.1088/1757-899X/820/1/012041

Simulation of regenerative pump performing on multiphase mixture

I Zaporozhets[1,2], M Kalugina[1], A Mukhlaeva[1], N Isaev[1] and A Protopopov[1]

[1]Bauman Moscow State Technical University

[2]E-mail:moonig@ya.ru

Annotation: The features of gas-liquid mixtures hydrodynamic modeling during the separation process are described. The paper considers the model of a centrifugal pump, which is a part of mobile objects, such as wheeled and tracked vehicles etc. The simulation is carried out in the STAR-CCM + software. A comparison of the effect of gas separators various designs on the flow during the operation of the pump on a two-phase mixtureis carried out and the most suitable configuration is identified.

Introduction

Currently, regenerative pumps are widely used in rural water supply, small automatic pumping stations, in municipal services and in the chemical industry for supplying aggressive reagents [1]. This happens since regenerative pumps have simple design and the ability to work on a mixture of liquid and gas, they are easy to operate and can run in self-suction mode.

The disadvantage of the regenerative pump is its relatively low efficiency (the efficiency of existing pumps does not exceed 45% at the operating mode), this circumstance does not allow to use them at high power [2,3].

Regenerative pumps are preferred for short-term pumping of volatile liquids, working on a mixture of liquid and gas (in aircraft tankers, in automobile gas stations), on vessels for the supply of drinking water and as a nutrient for auxiliary boiler installations.

As it was mentioned earlier, one of the most important qualities of a regenerative pump is the ability to work on a mixture of liquid and gas, as well as in self-suction mode. Regenerative pumps of the closed type with an open channel, without additional devices, do not operate on gas and do not possess self- suction capacity. To ensure self- suction ability, a separating cap and a separator are installed at the outlet of the discharge pipe. From a theoretical point of view, this process of pump operation is practically not studied. Therefore, the problem of numerical simulation of a regenerative pump operation on a gas-liquid mixture is relevant. Existing design recommendations were obtained after processing only experimental data and require mathematical justification.

In this article, it has been attempted to compute the flow of a gas-liquid mixture in a flow part of a closed-type regenerative pump using computational hydrodynamics methods. Moreover, the self-suction process was described and optimal design parameters from the point of view of pump operation on a two-phase mixture were obtained.

The object under consideration is the regenerative stage with a pressure pipe and a separator.

Figure 1 represents the 3D-model of the pump flow part with the pressure cap and the separator.

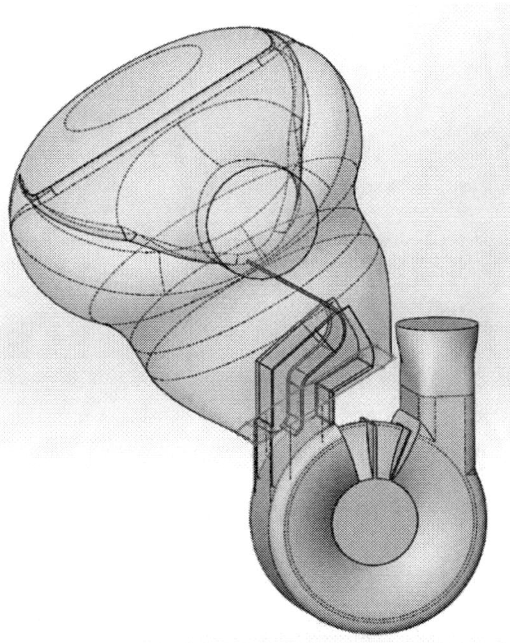

Figure1. The flow part of the pump with the pressure cap and the separator

To compare the results of numerical calculation with the experimental data, a comparative simulation of the fluid flow in the flow part with various separator designs, which were manufactured for existing regenerative pumps, was carried out. Figure 2 shows 3D-models ofseparatordesignsunderconsideration.

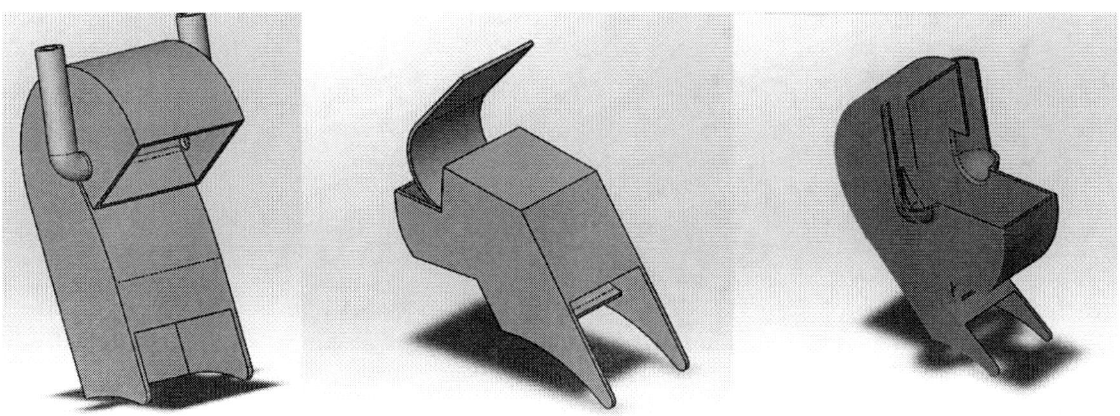

Figure2. 3D-models for several separator configurations

Methods

The method of numerical simulation is based on solving discrete analogs of the basic equations of hydrodynamics [4–8]. For incompressible fluid $(\rho = const)$ they look like:

Mass conservation equation (continuity equation)

$$\frac{\partial \overline{u_j}}{\partial x_j} = 0,$$

where $\overline{u_j}$ — the projection of fluid velocity averaged value on the j-axis (j=1,2,3);

The equation of conservation of momentum (Reynolds averaging):

$$\rho\left[\frac{\partial U_i}{\partial t}+U_i\frac{\partial U_i}{\partial x_j}\right]=-\frac{\partial P}{\partial x_i}+\frac{\partial}{\partial x_i}\left[\overline{T_{ij}^{(v)}}-\rho u_i u_j\right],\ where$$

U, P — averaged velocity and pressure;

$\overline{T_{ij}^{(v)}}=2\mu\overline{S_{ij}}$ — viscous stress tensor for incompressible fluid;

$\overline{S_{ij}}=\frac{1}{2}\left[\frac{\partial \overline{u_i}}{\partial x_j}+\frac{\partial \overline{u_j}}{\partial x_i}\right]$ — instantaneous strain rate tensor;

$\rho u_i u_j$ — Reynolds stresses.

The introduction of the Reynolds averaged Navier – Stokes equation leaves the system of equations unclosed, since additional unknown Reynolds stresses appear. To solve this system, a semi-empirical *k-ω SST* turbulence model was used. It introduces the necessary additional equations: the equations for the transfer of the turbulence kinetic energy and the relative rate of dissipation of this energy:

$$\frac{\partial k}{\partial t}+U_j\frac{\partial k}{\partial x_j}=P_k-\beta k\omega+\frac{\partial}{\partial x_j}\left[\left(v+\sigma_k v_T\right)\cdot\frac{\partial k}{\partial x_j}\right]$$

$$\frac{\partial \omega}{\partial t}+U_j\frac{\partial \omega}{\partial x_j}=\alpha\cdot S^2-\beta\cdot\omega^2+\frac{\partial}{\partial x_j}\left[\left(v+\sigma_\omega v_T\right)\cdot\frac{\partial \omega}{\partial x_k}\right]+2\cdot\left(1-F_i\right)\cdot\sigma_{\omega 2}\cdot\frac{1}{\omega}\cdot\frac{\partial k}{\partial x_j}\cdot\frac{\partial \omega}{\partial x_j}$$

To simulate the process of gas separation, a two-phase model with approach known as VOF (Volume of Fluid) [9] was used. The VOF method considers multiphase medium as a single fluid, the properties of which vary according to the volume ratio of present phases:

$$\alpha_i=\frac{V_i}{V},\ where$$

V_i — i-phase volume in the cell,

V — cell volume.

Volume ratio of each phase in the cell should satisfy the condition:

$$\sum_{i=1}^{N}\alpha_i=1,$$

where N — the number of phases.

Depending on the volume fraction, the presence of different phases or liquids in the cell can be distinguished:

$\alpha_i=0$ — there is no i-phase in the cell

$\alpha_i=1$ — the cell is filled with i-phase

$0<\alpha_i<1$ — there are several phases in the cell

The equation for the mass concentration of each phase is:

$$\frac{\partial\left(\alpha_i\rho_i\right)}{\partial t}+\nabla\cdot\left(\alpha_i\rho_i V\right)=0$$

The minimum bubble size and the minimum gas concentration in any cell are strictly limited to very small, but non-zero values. The time step was chosen based on the characteristic flow rates and cell size. The selected time step is 1e-5 s. The number of inner iterations at each time step is 15. To obtain discrete analogs of the original continuous equations STAR-CCM + software package uses the control volume method described in [10].

The computational grid built for the the regenerative pump with the separator consists of 400 000 polyhedral cells, base size is 5 mm, the grid is shown in Figure 3.

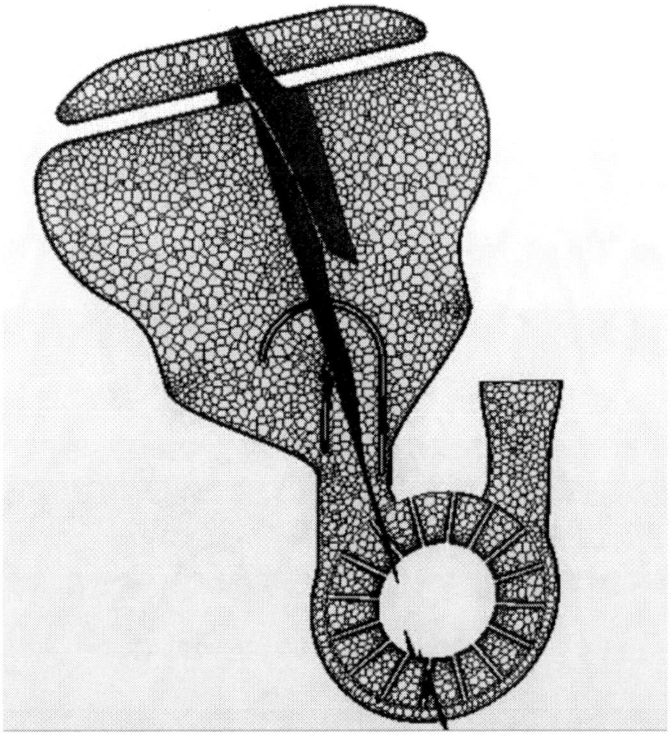

Figure3. The computational grid of the regenerative pump with the pressure cap

When the regenerative pump operates on a mixture of liquid and gas, the flow part is pre-filled with liquid. In order to recreate the actual working conditions, a function that allows filling of the pump flow part with water to the desired level was created. Figure 4 presents the distribution of the liquid and gas phase in the flow part at the initial moment of time t = 0. In the process of simulation, a mixture of gas and water enters through the inlet boundary of the pump.The flow rate of the mixture at the inlet and the pressure at the outlet were set as boundary conditions.

The criteria for assessing the quality of separator designs are the average gas flow rate at the outlet from the cap and the manufacturability of the separator.

Figure 4. Initial state of the pump

Results

Figures of water volume fraction distribution for different separator configurations were created during the two-phase simulation of the pump.

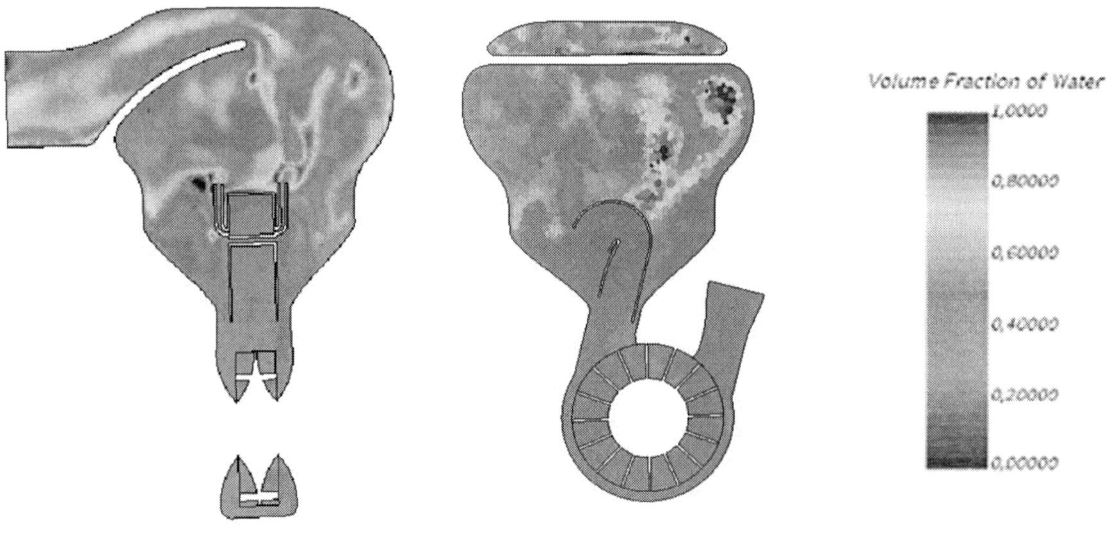

Figure 5. The distribution of water volumefractionin separator#1

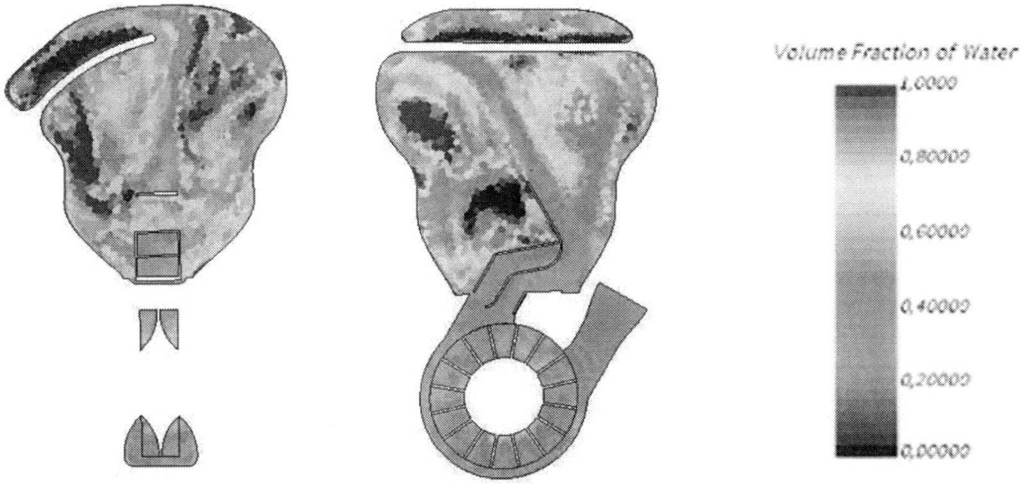

Figure 6. The distribution of water volume fraction in separator #2

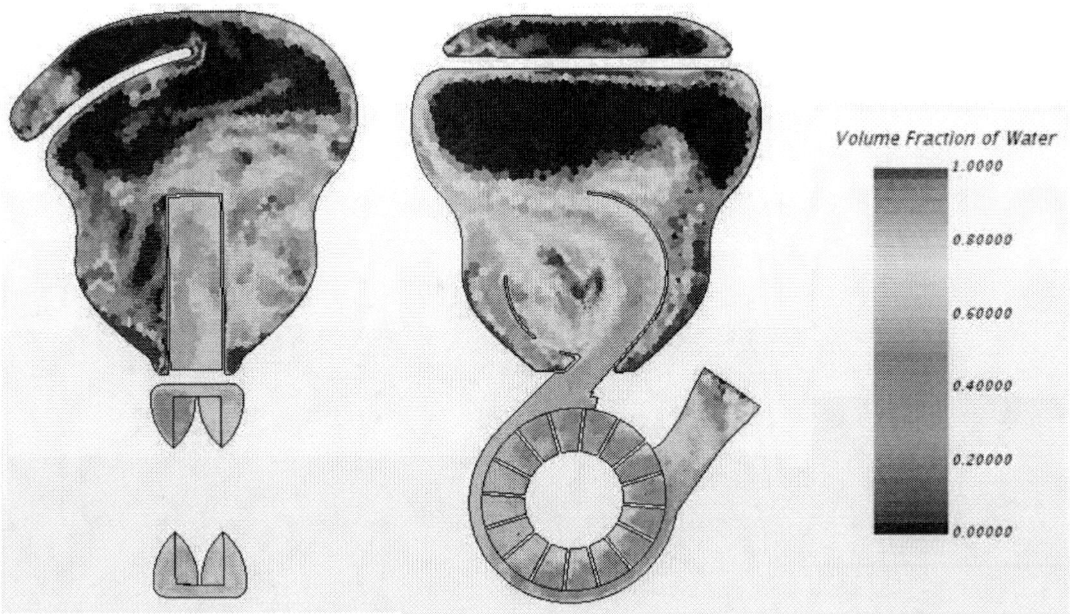

Figure 7. The distribution of water volume fraction in separator #3

According to the figures the pump works on approximately 50/50 mixture of gas and water. The results for several separator configurations are shown in table 1.

Table 1. Simulation results

Separator configuration	Head, m
1	18
2	25
3	31

IOP Conf. Series: Materials Science and Engineering **820** (2020) 012041 doi:10.1088/1757-899X/820/1/012041

Obtained distributions and results indicate the need for further research and simulation sin this field.

References

[1] V Cheremushkin and APolyakov 2019 *IOP Conf. Ser.: Mater. Sci. Eng.* **589** 012001
[2] V Tkachyk et al 2019 *IOP Conf. Ser.: Mater. Sci. Eng.* **589** 012010
[3] V Tkachyk et al 2019 *IOP Conf. Ser.: Mater. Sci. Eng.* **589** 012011
[4] E Morozova et al 2019 *IOP Conf. Ser.: Mater. Sci. Eng.* **589** 012012
[5] K Abramov 2019 *IOP Conf. Ser.: Mater. Sci. Eng.* **589** 012013
[6] A Boyarshinova et al 2019 *IOP Conf. Ser.: Mater. Sci. Eng.* **589** 012014
[7] N Sosnovsky and D Ganieva 2019 *IOP Conf. Ser.: Mater. Sci. Eng.* **589** 012016
[8] M Saprykina and V Lomakin 2019 *IOP Conf. Ser.: Mater. Sci. Eng.* **589** 012017
[9] A Boyarshinova et al 2019 *IOP Conf. Ser.: Mater. Sci. Eng.* **589** 012028
[10] A Martynyuk et al 2019 *IOP Conf. Ser.: Mater. Sci. Eng.* **589** 012029

Cryogenic pumps of mobile machinery

D Babikova[1,2], M Kalugina[1], A Petrov[1] and A Protopopov[1]

[1]Bauman Moscow State Technical University

[2]E-mail: babikova.dina@gmail.com

Abstract. Cryogenic pumps are of particular importance in mobile machinery, in particular in wheeled and tracked vehicles. This is due to the development of hydrogen energy. In the nearest future, we will see hydrogen powered mobile cars. In this regard, it would be useful to have an up-to-date overview of modern cryogenic pumps. This article discusses the main types of cryogenic pumps, identifies problems encountered in the design and operation of pumping units and formulates the requirements for these devices. A brief overview of the most common designs of cryogenic pumps of foreign and Russian production is presented.

1. Introduction

In the modern world, cryogenic products are an integral part of many technological processes in the energy sector, oil and gas industry, metallurgy, astronautics, chemical industry, medicine and many other areas of human life [1]. Cryogenic products include substances or mixtures of substances at cryogenic temperatures (below 120 °K). The most common cryogenic products are low-temperature air separation products and inert gases: nitrogen, oxygen, argon, neon, krypton, xenon, ozone, fluorine, methane, hydrogen, helium. To produce, store, transport and use cryogenic products, special devices are required that correspond to the special properties of liquefied gases and ensure the safety of personnel when working with cryogenic products. [2]

2. Overview of the types and designs of cryogenic pumps

Cryogenic pumps are designed for pumping cryogenic products and can be of piston or centrifugal type. Reciprocating cryogenic pumps are used to produce and transport liquefied gases such as oxygen, nitrogen, hydrogen, helium, argon, natural gas. Using pumps of this type, cryogenic liquids are pumped into tanks and cylinders, and piston pumps are included in gasification plants. These pumps are used for small volumes of pumped product in the medium and high pressure range. [3]

Another type of cryogenic pumps is centrifugal pump. There is huge modeling experience on centrifugal pumps [4]–[7]. Of particular interest are publications [8]–[13]. Cryogenic centrifugal pumps are used in air separation plants both for compressing a cryogenic product and for circulating a coolant in the system. In addition, with the help of centrifugal pumps, liquefied gases are pumped from tankers and storage facilities and the cryogenic products are subsequently delivered from tanks to the consumer (for example, filling gas vehicles). Centrifugal pumps can have a single-stage or multistage design. In multistage pumps, the pumped liquid passes sequentially through several impellers, between which guide vanes are located for supplying fluid from the previous stage to the next. This design allows to significantly increase the pump head in comparison with a single-stage pump. [14]

IOP Conf. Series: Materials Science and Engineering **820** (2020) 012042 doi:10.1088/1757-899X/820/1/012042

When creating cryogenic pumps, difficulties arise due to the properties of cryogenic products, and in particular their low temperature. Therefore, due to the low temperature of the pumped liquefied gas, a large temperature difference is created between the wet part of the pump and the ambient medium. Because of this, water condenses on the outer surfaces of the pump and forms a layer of frost. Heating of the cryogenic product due to external heat gain and heat generation during operation can cause cavitation, because all liquefied gases are low-boiling. To keep coolant in a liquid state, it is necessary to maintain a low temperature and high overpressure. Incorrect design of the wet part of the cryogenic pump can lead to cavitation and, as a result, to cavitation erosion and loss of pressure. Therefore, hydrodynamic modeling of flows in the wet part of the pump is a very useful design tool, as it allows one to determine the optimal pump parameters. [15] In addition, the calculation of the pump cavitation characteristics is complicated due to the presence of two phases in the flow (liquid and vapor). To solve problems in this formulation, various methods can be used, for example, those given in [16] and [17]. For cryogenic pumps, when finding a critical suction head, it is necessary to take into account corrections related to the influence of gas inclusions, the viscosity of the pumped medium, as well as its thermophysical properties. The seals in the cryogenic pump operate in conditions of insufficient lubrication, since cryogenic products seeping through the gap between the wear surfaces quickly evaporate at ambient temperature. In addition, the seal freezes, and hoarfrost falling into the friction zone leads to increased wear of the parts. Bearing units of pumps can fail much earlier than the expected resource, since the lubricant in them freezes and ceases to fulfill its functions. In addition, the low temperature of cryogenic products can lead to embrittlement of pump parts. In addition, some liquefied gases are toxic, flammable or explosive, so the sealing units must provide either complete sealing or minimal controlled leakage. [18]

Thus, a number of requirements for cryogenic pumps can be formulated.

1. Reduce heat influx to the pumped product to minimum values. To do this, it is necessary to insulate the pump casing.
2. Ensure reliable operation of seals and friction units without the use of lubricant.
3. Ensure reliable operation of bearings throughout the entire life. To do this, one can separate the flow part and the shaft bearings to reduce heat transfer between them or use bearings lubricated directly by the pumped liquid.
4. When choosing materials for pump parts, it is necessary to take into account the temperature and chemical activity of the pumped cryogenic product. The suitability of the material for operation at a given temperature is judged by the temperature viscosity margin equal to the difference between the operating temperature and the cold-shortness threshold.
5. When working on combustible, poisonous, or rare gas pumps, product leakage must be avoided.

One of the common types of pumps used for pumping cryogenic liquids is a close-coupled electric pump (electric motor and cryogenic pump combined in one unit). This type includes the pump NkpM. It is used for discharging and filling tanks in storage and transportation systems for liquid cryogenic media, as well as in air separation metallurgical plants. Pumps NkpM are of horizontal and vertical execution, one- or two-stage. Capacity range of these pumps is from 3 to 200 m^3/h and pump head range is from 25 to 90 m. Fig. 1 shows a section of a vertical close coupled centrifugal pump of NkpM type with a housing for insulating fill.

It is also worth paying attention to cryogenic centrifugal pumps from Cryostar (France). They are designed for pumping liquefied gases (nitrogen, argon, oxygen, methane, hydrogen, carbon dioxide, etc.) from one cryogenic or isothermal tank to other under pressure from 2 to 100 bar, as well as in various low-temperature technological processes. Pumps have a wide range of capacities — from 10 to 10,000 l/min. One of the most common is the GBS pump (Fig. 2) [19].

The advantages of this pump include the increased resource of the composite seal (compared with graphite), good protection against heat influx and the possibility of easy seal change due to the removable housing cover.

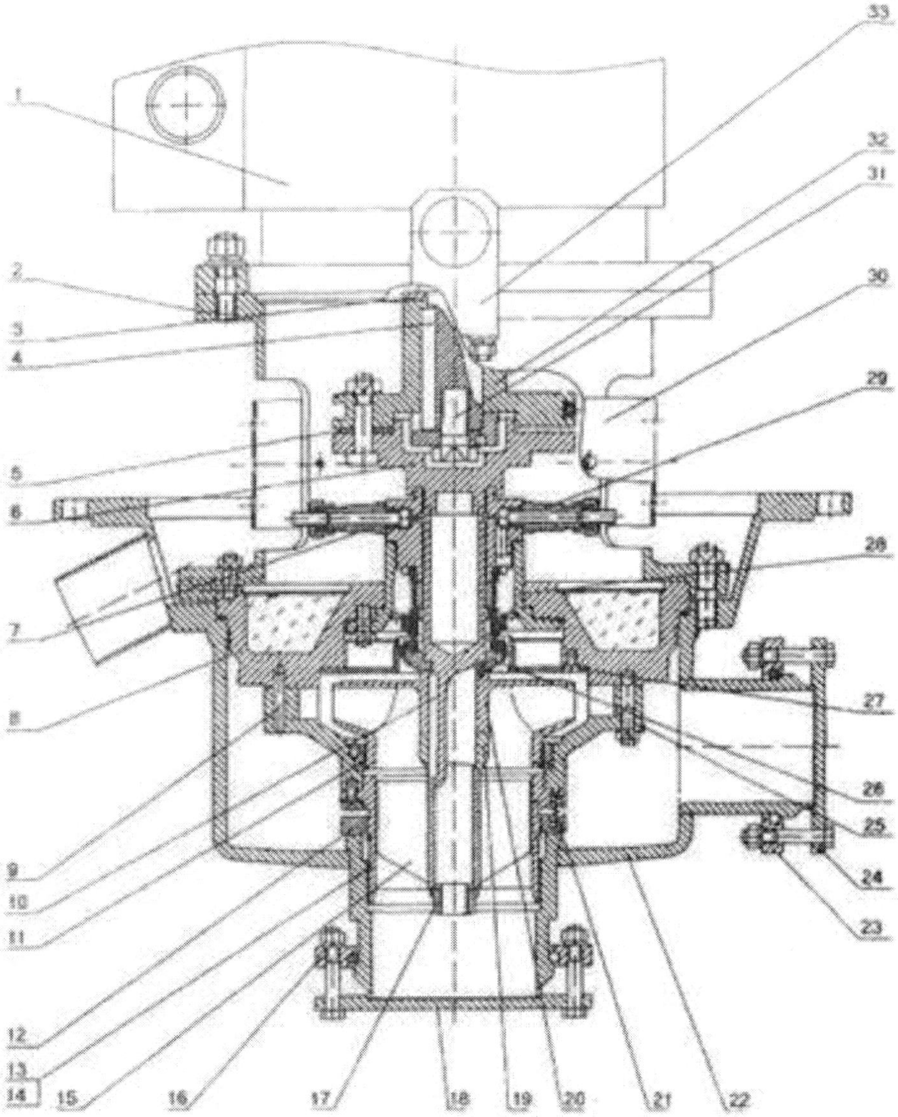

Fig. 1. Vertical close coupled centrifugal pump NkpM type with a housing for insulating fill: 1 — electric motor; 2 — motor stool; 3 — spacer ring; 4 — key; 5 — spacer washer; 6 — extension shaft; 7 — fitting; 8 — cover; 9 — pin; 10 — key; 11 — sealing ring; 12 — sealing ring; 13 — pipe; 14 — expandable ring; 15 — screw; 16 — cap flange; 17 — nut; 18 — stub; 19 — bushing; 20 — impeller; 21 — spring; 22 — housing; 23 — cap flange; 24 — stub; 25 — guide vanes; 26 — support ring.

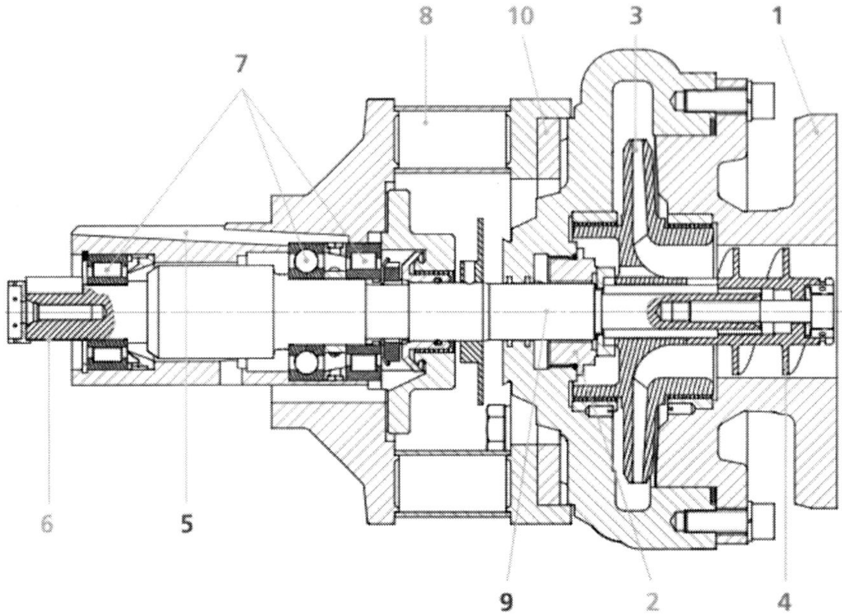

Fig. 2. Cryostar GBS pump: 1 — removable housing cover; 2 — composite mechanical seal; 3 — impeller; 4 — screw inductor; 5 — gear box; 6 — gear; 7 — bearings; 8 — hollow spokes connecting the pump to the electric motor; 9 — shaft; 10 — plate with low thermal conductivity.

Fig. 3. Flowserve WUC submersible pump.

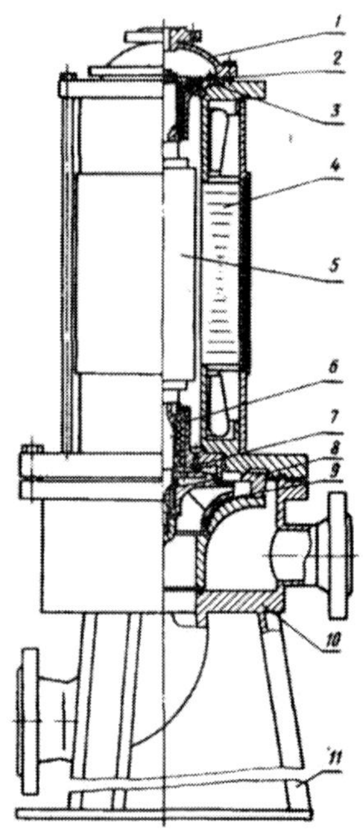

Fig. 4. Electric pump 4HGV-6-40-5: 1 — motor cover; 2 — thrust bearing; 3 — upper bearing; 4 — stator; 5 — rotor; 6 — lower bearing; 7 — hydraulic balancing device; 8 — guide vanes; 9 — impeller; 10 — pump housing; 11 — support.

Submersible centrifugal cryogenic pumps provide the most effective thermal insulation, as they are located directly in the pumped medium. This type includes the WUC pump (VS6) from Flowserve (USA) (Fig. 3) [20]. The temperature of the pumped liquid is from -200 °C to 350 °C. Pump head up to 2000 m, pressure up to 200 bar. The seal chamber is standardized and allows the installation of dual liquid, dry gas or gas injector labyrinth seals.

It is worth noting there are leak-free pumps for cryogenic liquids. In these units, external heat inflows are not eliminated as efficiently as in submersible pumps, but cryogenic product leaks are excluded. Electric pumps of the HGV series (Russia) are intended for pumping aggressive liquids containing harmful substances, including liquefied gases, at temperatures from -50 °C to 100 °C. Fig. 4 shows a 4HGV-6-40-5 pump.

3. Conclusion

Thus, we can say that the efficiency of the cryogenic pump largely depends on the design, or how the characteristic problems of cryogenic pumps were solved related to the special properties of the medium. The most rational solution to these problems determines the promising directions for the development of cryogenic pumps, which include the following:

1. The use of new materials resistant to low temperatures in the design of pumps.
2. The implementation of effective thermal insulation of pumps (for example, using powders, fibers, organic foams) [1].
3. The use of leak-free pump arrangements in order to avoid problems with seals.

References

[1] Botkina L.A. Foreign Equipment of Cryogenic Systems (Zarubezhnoe oborudovanie kriogennykh system). Moscow, *PMB TsINTIkhimneftemasha*, 1973. 46 p.

[2] Baranov A.Yu., Sokolova E.V. Storage and Transportation of Cryogenic Liquids (Khraneniye i transportirovka kriogennykh zhidkostey). Part 1: Study Guide. St. Petersburg, *ITMO University*, 2017. 95 p.

[3] Burenin V.V., Dronov V.P., Vorobev E.V. Designs of Pumps for Cryogenic Liquids (Konstruktsii nasosov dlya kriogennykh zhidkostey). Moscow, *PMB TsINTIkhimneftemasha*, 1981. — 44 p.

[4] A Petrov et al 2019 IOP Conf. Ser.: Mater. Sci. Eng. 589 012030

[5] V Tkachyk et al 2019 IOP Conf. Ser.: Mater. Sci. Eng. 589 012010

[6] V Tkachyk et al 2019 IOP Conf. Ser.: Mater. Sci. Eng. 589 012011

[7] E Morozova et al 2019 IOP Conf. Ser.: Mater. Sci. Eng. 589 012012

[8] M Ershov and V Lomakin 2019 IOP Conf. Ser.: Mater. Sci. Eng. 589 012033

[9] A Boyarshinova et al 2019 IOP Conf. Ser.: Mater. Sci. Eng. 589 012014

[10] N Sosnovsky and D Ganieva 2019 IOP Conf. Ser.: Mater. Sci. Eng. 589 012016

[11] V Grishakov et al 2019 IOP Conf. Ser.: Mater. Sci. Eng. 589 012034

[12] A Boyarshinova et al 2019 IOP Conf. Ser.: Mater. Sci. Eng. 589 012028

[13] A Martynyuk et al 2019 IOP Conf. Ser.: Mater. Sci. Eng. 589 012029

[14] Lomakin A.A. Centrifugal and Axial Pumps (Tsentrobezhnyye i osevyye nasosy). Moscow, *Mashinostroenie*, 1966. — 354 p.

[15] Petrov A.I., Lomakin V.O., Semenov S.E. Ways to increase the energy efficiency of dynamic pumps based on modern computer technology (Puti povysheniya energoeffektivnosti dinamicheskikh nasosov na osnove sovremennykh komp'yuternykh tekhnologiy). *Inzhenernyi zhurnal: nauka i innovatsii*, 2013, no. 4. URL: http://engjournal.ru/catalog/machin/hydro/689.html

[16] Aleksensky V.A., Zharkovsky A.A., Pugachev P.V. Calculation Definition of Centrifugal Pumps Cavitational Characteristics. *Bulletin of the Samara Scientific Center of the Russian Academy of Sciences*, 2011, v. 13, No. 1–2, p. 411–414.

IOP Conf. Series: Materials Science and Engineering **820** (2020) 012042 doi:10.1088/1757-899X/820/1/012042

[17] Lomakin V.O., Petrov A.I., Kuleshova M.S. Investigation of Two-Phase Flow in Axial Centrifugal Impeller by Hydrodynamic Modeling Methods. Science and Education: Scientific Publication. 2014;(9):45–64. DOI:10.7463/0914.0725724

[18] Low temperature machinery. Cryogenic machines and tools: textbook (Mashiny nizkotemperaturnoy tekhniki. Kriogennyye mashiny i instrumenty) / [A.M. Arkharov, I.A. Arkharov, A.N. Antonov, etc.]; ed. A.M. Arkharov and I.K. Butkevich. Moscow, Publishing House of Bauman MSTU, 2011. 582 p.

[19] Center for VM Technologies | Official Representative of CRYOSTAR S.A.S. in Russian Federation and Republic of Belarus / Electronic resource / http://www.cryostar-russia.ru/c-gbs.htm

[20] WUC API 610 (VS6) Vertical Multistage Process Pump For Cryogenic Applications (or Services) // Flowserve Corporation / Electronic Resource / https://www.flowserve.com/sites/default/files/2016-07/ pss-40-9-1-e.pdf

Analysis of the centrifugal pump operation inunsteady modeas apart of a mobile machine

A Protopopov[1,2], A Anisimova[1], A Mukhlaeva[1], M Kalugina[1] and D Babikova[1]

[1]Bauman Moscow State Technical University

[2]E-mail: proforg6@yandex.ru

Abstract. Mobile vehicles, both tracked and wheeled, are often used to refuel/unload stationary tanks. In this case, the unsteady mode of operation of such centrifugal pumps is often observed. A constant change of operation conditions, in particular the static component of the pump head, leads to a change in its operating mode, as a result of which most of the time the pump operates in off-design mode. A methodology has been developed for determining the actual pump efficiency for a tank filling-draining cycle with two variable parameters: the distance from the level of the pumped fluid in the well to the bottom of the tank (to the on float) and the tank height (or the distance between on and off floats).

Introduction

Tracked or wheeled mobile vehicles are often used to refuel/unload stationary tanks. In this case, the operating mode of the auxiliary discharge pump as part of such machines will be unsteady due to the unstable stationary pressure of the pump. Especially, the problem of determining the efficiency of such pumps should be noted. In general, various methods for modeling centrifugal pumps are presented in works [1–6], but they all consider the pump in a stationary or quasi-stationary pressurestate. Of particular interest are the works [7–15], however, there are no answers to the questions we have, in particular, they do not provide a methodology for assessing the actual system efficiencyper work cycle.

To create a methodology for determining the actual efficiency of the pump per cycle, we consider the basic arrangement of the system of interest to us (Fig. 1).

In this system, the variable parameters are$z$ — the distance from the bottom of the tank to its top (or the distance between the on and off floats) and H_0 —the distance from the liquid level in the well to the bottom of the tank (to the on float).

Methods and results

When the pump is operating to fill such a reservoir, the static head will continuously change within the limits of $H_0 \ldots H_0+z$. To determine the actual pump efficiency per cycle, we consider its predicted pressure characteristic and the dependence of the pump efficiency on its supply (Fig. 2).

To solve the problem, we will approximate these two graphs with second-order parabolas (Fig. 3).

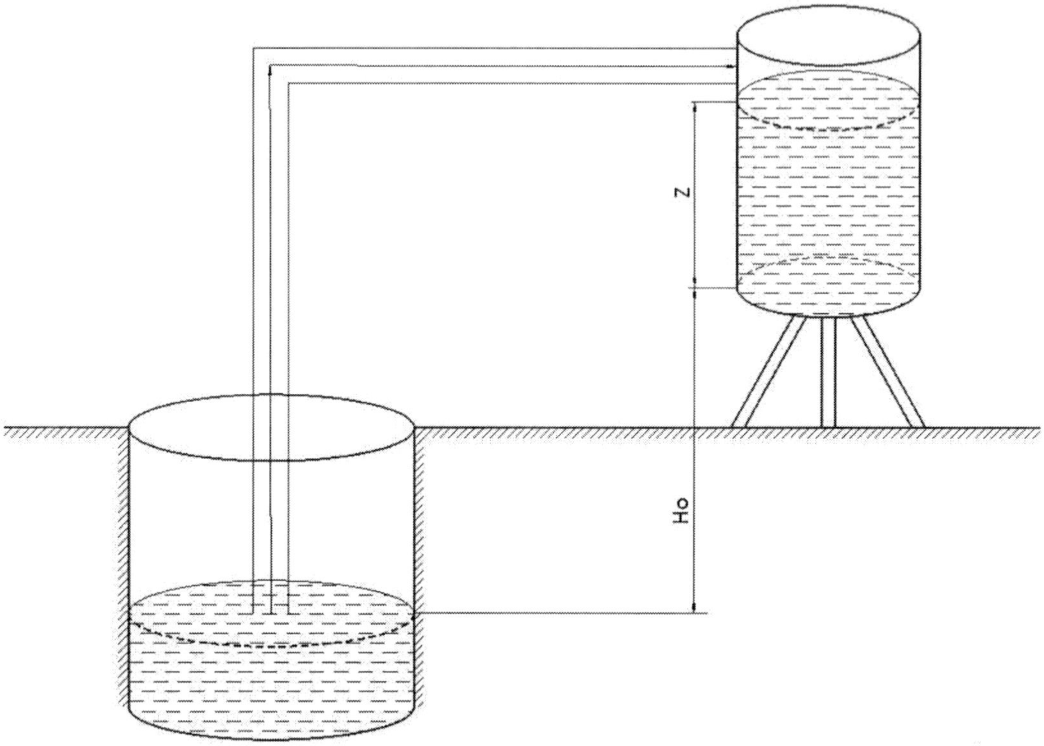

Fig. 1. Basic arrangement of the system

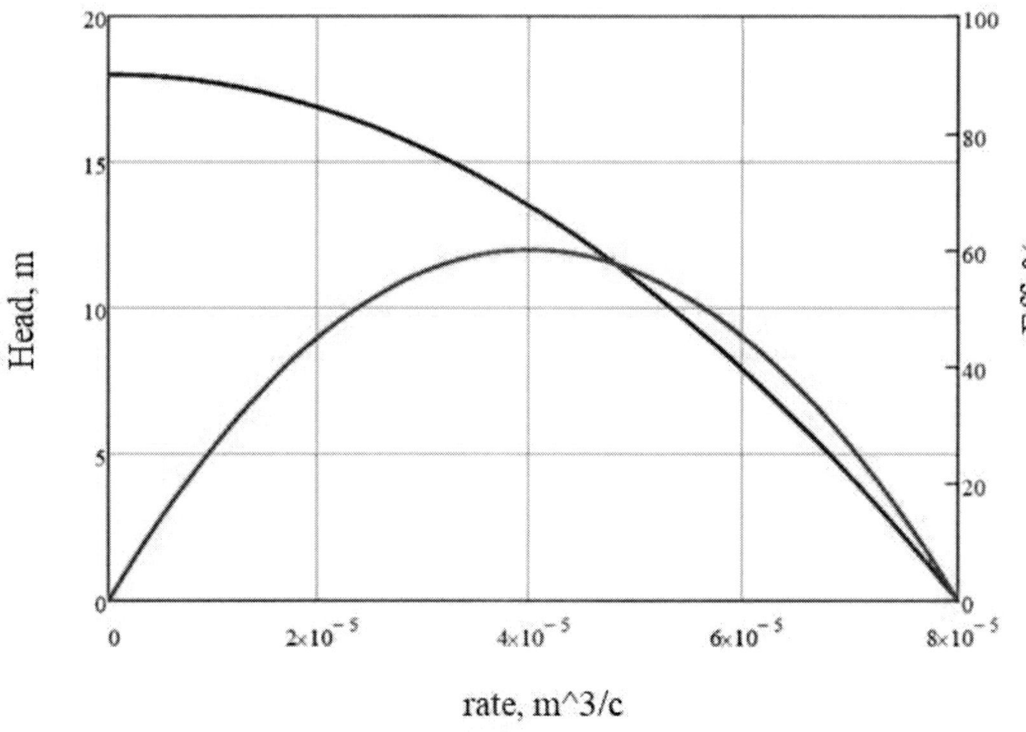

Fig. 2. Predicted pump pressure characteristic and the dependence of pump efficiency on flow rate

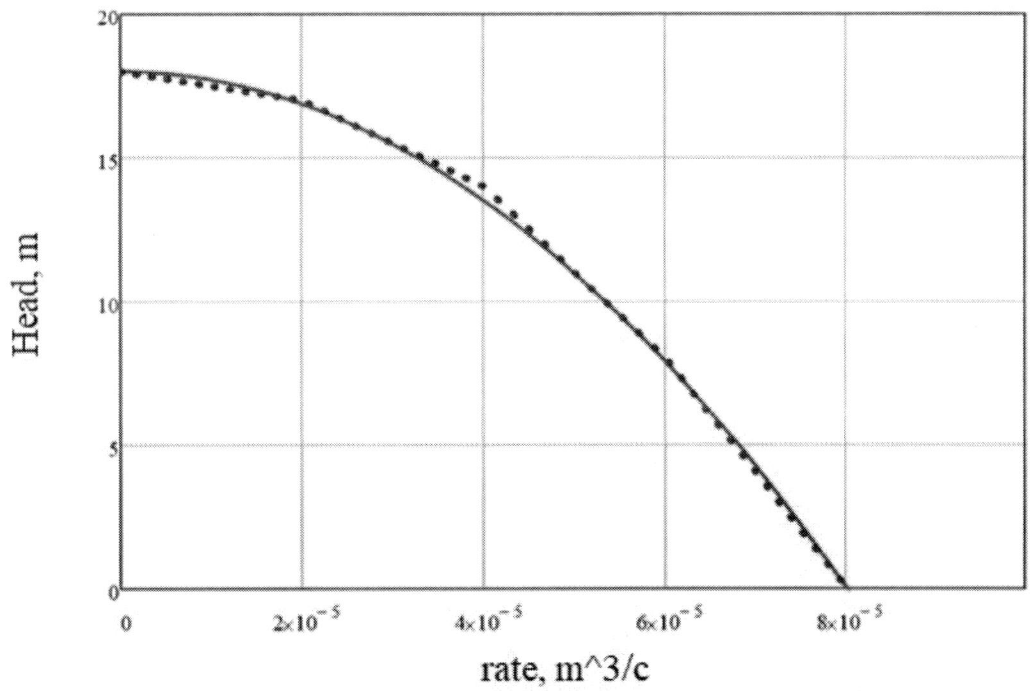

Fig. 3. Approximation of the predicted pressure characteristic

Then the pressure characteristic can be represented as

$$H(Q) = a_1 + a_2Q + a_3Q^2,$$

where $H(Q)$ —predicted pump head; a_1, a_2, a_3 — approximation coefficients; Q — pump flow rate.
Similarly, for efficiency we obtain the expression

$$\eta(Q) = b_1 + b_2Q + b_3Q^2,$$

where$\eta(Q)$ — current efficiency (corresponding to a given point in time); b_1,b_2,b_3 — approximation coefficients.
Then the actual efficiency per cycle is calculated as

$$\eta_a = \frac{\int_{t_1}^{t_2} Q\left(a_1 + a_2Q + a_3Q^2\right)dt}{\int_{t_1}^{t_2}\left(b_1 + b_2Q + b_3Q^2\right)dt}, \tag{1}$$

where$t_1$andt_2— start and end time of the tank filling process.
To determine t_1 and t_2 we use the following equations:

$$-Fdz = Qdt\ ; \tag{2}$$

$$a_1 + a_2Q + a_3Q^2 = H_0 + z + K_pQ^2, \tag{3}$$

where F — tank cross-sectional area (which in this paper is taken as a known constant); K_p — pipe drag coefficient, as a percentage of squared flow rate.

Conclusion

We combined equations (1)–(3) in the system

$$\begin{cases} \eta_a = \dfrac{\int_{t_1}^{t_2} Q\left(a_1 + a_2 Q + a_3 Q^2\right) dt}{\int_{t_1}^{t_2} \left(b_1 + b_2 Q + b_3 Q^2\right) dt}; \\ -F dz = Q dt; \\ a_1 + a_2 Q + a_3 Q^2 = H_0 + z + K_p Q^2. \end{cases} \quad (4)$$

The system of equations (4) allows one to find the actual pump efficiency per cycle.

Thus, a technique has been created for determining the actual pump efficiency per cycle for a cyclically operating system, provided that the tank cross section is constant.

References

[1] A Protopopov and D Bondareva 2019 *IOP Conf. Ser.: Mater. Sci. Eng.* **492** 012002
[2] A Protopopov and V Vigovskij 2019 *IOP Conf. Ser.: Mater. Sci. Eng.* **492** 012003
[3] P Chaburko and Z Kossova 2019 *IOP Conf. Ser.: Mater. Sci. Eng.* **492** 012011
[4] V Lomakin et al 2019 *IOP Conf. Ser.: Mater. Sci. Eng.* **492** 012012
[5] A Gouskov et al 2019 *IOP Conf. Ser.: Mater. Sci. Eng.* **492** 012013
[6] N Egorkina and A Petrov 2019 *IOP Conf. Ser.: Mater. Sci. Eng.* **492** 012015
[7] K Dobrokhodov and A Petrov 2019 *IOP Conf. Ser.: Mater. Sci. Eng.* **492** 012016
[8] N Isaev 2019 *IOP Conf. Ser.: Mater. Sci. Eng.* **492** 012026
[9] S Korsakova and A Protopopov 2019 *IOP Conf. Ser.: Mater. Sci. Eng.* **492** 012032
[10] A Protopopov and C Jakovich 2019 *IOP Conf. Ser.: Mater. Sci. Eng.* **492** 012034
[11] A Shablovskiy and E Kutovoy 2019 *IOP Conf. Ser.: Mater. Sci. Eng.* **492** 012035
[12] A Petrov et al 2019 *IOP Conf. Ser.: Mater. Sci. Eng.* **492** 012036
[13] V Lomakin and O Bibik 2019 *IOP Conf. Ser.: Mater. Sci. Eng.* **492** 012037
[14] T Valiev and A Petrov 2019 *IOP Conf. Ser.: Mater. Sci. Eng.* **492** 012038
[15] V Cheremushkin and V Lomakin 2019 *IOP Conf. Ser.: Mater. Sci. Eng.* **492** 012039

IOP Conf. Series: Materials Science and Engineering **820** (2020) 012044 doi:10.1088/1757-899X/820/1/012044

Modeling of viscoelastic properties of composite springs in wheeled vehicles suspension systems

K Evseev[1,2] and A Kartashov[1]

[1]Bauman Moscow State Technical University

[2]E-mail: kb_evseev@bmstu.ru

Abstract. The article provides a review of models allowing to take into account the viscoelastic properties of springs made using polymer composite materials. The features of solving equations with fractional derivatives that describe the considered models are considered. The equation of the stress-strain state of the composite spring is obtained, which takes into account the rheological properties of the material. A model has been developed in MATLAB / Simulink, which allows calculating composite springs. A composite spring was made and tested, which made it possible to compare the elastic characteristics of the spring obtained as a result of modeling and experiment. The authors of the article determined the scope of such springs in the automotive industry and the direction of improvement of composite springs used in wheeled vehicles.

Introduction

The analysis of application of polymeric composite materials (PCM) in the automotive industry establishes the trends for increasing use of PCM parts in vehicle parts and components every year [1–2]. One of the perspective directions of use of the parts made with the use of PCM is their application in wheeled vehicle suspension systems. PCM is most commonly used in wheeled vehicle suspension systems to create elastic elements and guiding system elements. Nowadays PCM is actively used in leaf spring design, besides, recently there have been tendencies to use helical springs made with PCM as elastic elements of suspension systems.

A topical issue is the reduction of the weight of the vehicle as a whole, as well as unsprung masses. Fiberglass springs allow to achieve not only reduction of masses, but also a lot of other advantages in comparison with steel ones: provision of large suspension strokes, high durability, absence of corrosion, cost-effective production, good fatigue resistance, low creep, good vibration and noise insulation properties, possibility to provide interchangeability with steel springs, possibility to obtain complex shapes and progressive loading characteristics, availability of damping properties [3–7]. Therefore, the use of springs made with the use of PCM as elastic elements of the vehicle is a logical and actual direction of research.The wide application of composite springs in the automotive industry requires the study of new design methods, as such elements should take into account the internal hysteresis losses that result from the viscoelastic properties of the material, so the actual issue is the creation of a method for calculating such springs.

To take into account the properties of viscoelasticity, different models were used, which were developed by Maxwell, Voigt and Kelvin [8]. These models can be used not only to take into account the properties of solid substances (dense solutions, elastic bodies and slurries), but also to describing and taking into account the properties of PCM.

One of the simplest models to describe the properties of viscoelasticity is Maxwell model, which consists of a combination of elastic and viscous elements (Fig. 1). Hooke's Law for the elastic medium and Newton's Law for the viscous medium are combined into one viscoelastic medium equation that has the form:

$$\sigma(t)+\tau\ D_t\,\sigma(t)=E\tau\ D_t\,\varepsilon(t),$$

where $\sigma(t)$ — stresses; $\varepsilon(t)$ — deformations; E — Young 's modulus.

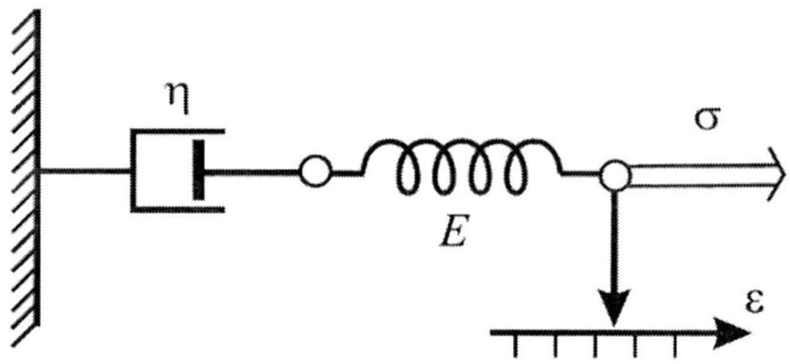

Figure 1. Maxwell model

Another classic model for describing a viscoelastic medium is the Kelvin-Feugt model, which consists of a spring and damper connected in parallel (Fig. 2). The structural equation for this model of viscoelastic medium is:

$$\sigma(t)=E\left[\tau\ D_t\,\varepsilon(t)+\tau\ D_t\,\varepsilon(t)\right].$$

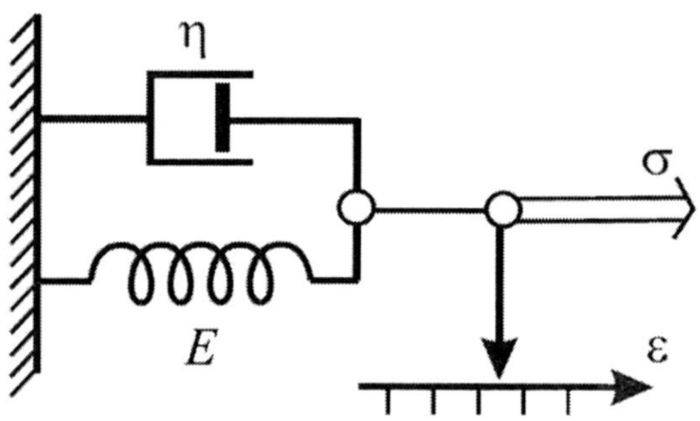

Figure 2. Kelvin-Feugt model

Both above classic models do not give the most accurate description of viscoelastic properties, so for a more complete description of the properties of the model of Zener (Fig. 3), fractional-differential form of this model (three-parameter model) is described by the equation:

$$\sigma(t)+\tau D_t\,\sigma(t)=E\left[\varepsilon(t)+\theta D_t\,\varepsilon(t)\right].$$

For the most accurate description of the material properties, a generalized model is used. The equations of state of the material have the following form:

$$\alpha_0\sigma(t)+\alpha_1\frac{d\sigma(t)}{dt}+\alpha_2\frac{d^2\sigma(t)}{dt^2}+\alpha_3\frac{d^3\sigma(t)}{dt^3}+...=$$
$$=\beta_0\varepsilon(t)+\beta_1\frac{d\varepsilon(t)}{dt}+\beta_2\frac{d^2\varepsilon(t)}{dt^2}+\beta_3\frac{d^3\varepsilon(t)}{dt^3}+...$$

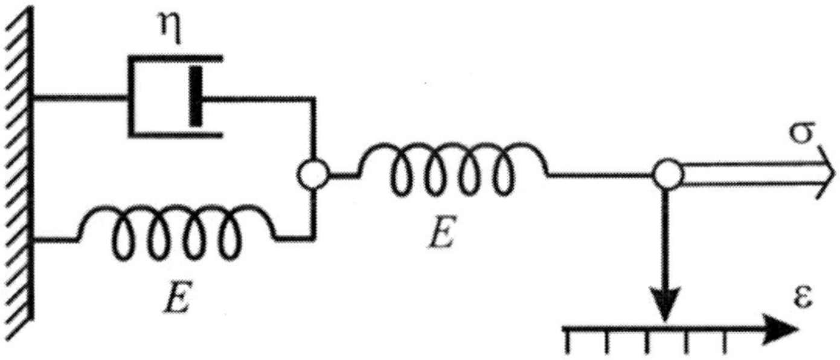

Figure 3. Model of Zener

The above equations with integer derivatives describing different models have insufficient adequacy in terms of the quality of the model or have a large number of components, so the apparatus of fractional derivatives is used to describe the models under consideration [8–9]. With this approach, the fractional-differential generalization of a standard three-parameter model with integer derivatives is the four-parameter Zener model, which is described by the equation:

$$\sigma(t)+\tau^{\alpha}\,{}_0D_t^{\alpha}\sigma(t)=E\left[\varepsilon(t)+\theta^{\alpha}\,{}_0D_t^{\alpha}\varepsilon(t)\right],$$

where ${}_aD_x^{\nu}f(x)$ — fractional derivative $x(t)$.

There are several different methods to generalize the definition of a fractional derivative, but all of them are the same as an usual derivative in the case of a natural order [10]. One of the ways to find a fractional derivative is a function called a Riemann-Liouville derivative, for which the expression of a fractional derivative of the order $0\le\nu<1$ is written as follows:

$${}_aD_x^{\nu}f(x)=\frac{1}{\Gamma(1-\nu)}D_x\int_a^x(x-\xi)^{-\nu}f(\xi)d\xi,$$

where $\Gamma(z)=\int_0^{\infty}t^{z-1}e^{-t}dt$ — gamma function.

In addition to the Riemann-Liouville fractional derivative, another definition, called the Caputo derivative, is often used. The fractional derivatives of Riemann-Liouville and Caputo generally do not match, as they have mathematical principles and are used as a tool for the most accurate description of the physical processes in the considered objects, as well as for the matching of calculation and experiment results. In addition, there is a form of fractional derivative of Marchaux, which is used to differentiate smooth functions without jumps and differentiable.

As a result of the numerical methods used in the research, another definition of a fractional derivative is used, which is called a fractional derivative of Grunwald-Letnikov with a finite limit. The fractional derivatives of Riemann-Liouville, Caputto and Marchaux reviewed above are applicable for known functions $f(x)$, so their application to random functions and in the case of real-time calculations and simulations is limited or impossible. The definition of Grunwald-Letnikov's fractional derivative is obtained by considering finite differences, i.e. the definition is obtained from considering the increment of the function and the argument, and the type of the function may be random. In the MATLAB Simulink environment, calculations are performed discretely with a constant or variable step. In addition, the external power or kinematic disturbance in real objects is often random, so the definition of a fractional derivative of Grunwald-Letnikov, best suited to solve this type of problems. Grunwald-Letnikov's derivative can be written down as follows:

$$_aD_t^{\alpha}f(t)=\lim_{h\to 0}\frac{1}{h^{\alpha}}\sum_{j=0}^{\frac{t-a}{h}}(-1)^k\frac{\Gamma(\alpha+1)}{j!\Gamma(\alpha-j+1)}f(t-jh).$$

To solve a fractional derivative in MATLAB Simulink environment, an S-function was written. It consists of a subsystem inside which there is an S-function with two inputs: an input for the function to be differentiated and an input of the order of the fractional derivative. At the output of the S-function, a fractional derivative of the input function of the corresponding order is determined.

On the example of the sin(x) function, we debugged the model under consideration. The results of the numerical calculations using the model described above fully coincide with the results obtained in the analytical calculation. Figure 4 shows the plots of the initial function sin(x) and its fractional derivative of the order of 0.5.

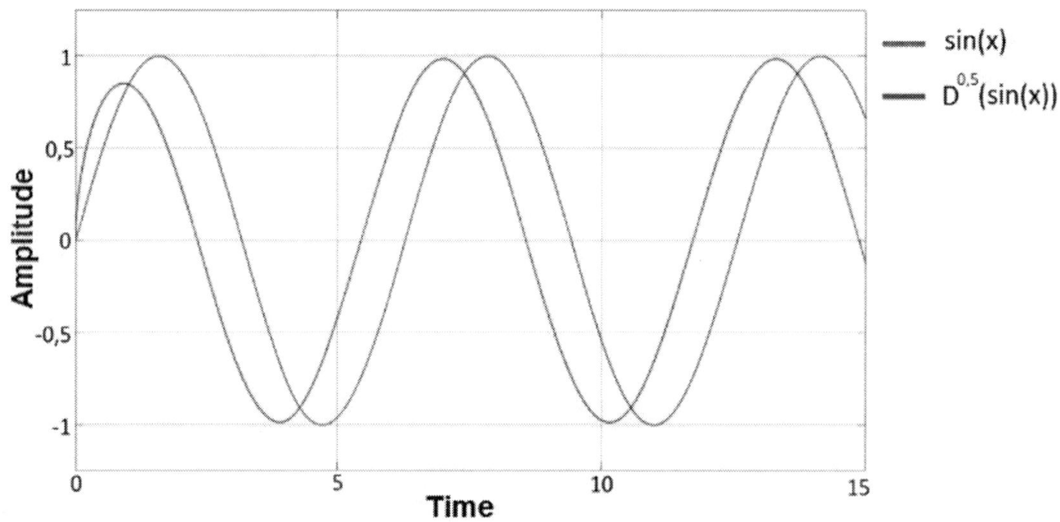

Figure 4. Plots of the sin(x) function and its fractional derivative of the order 0.5

If the order of the fractional derivative tends to zero, the function we are looking for will fully repeat the original function, and if the order of the fractional derivative tends to be one, the function we are looking for will tend to the integer derivative of the original function — cos(x).

According to the Voltaire principle, the problem solution for linear system elasticity can be used to solve problems with regard to viscoelastic properties by replacing the elastic parameters of the system with the parameters of the system with regard to viscoelasticity. For this purpose, the integral Laplace transform is used.

To take into account the properties of viscoelasticity, an equation of the stress-strain state in the linear statement was made, according to Hooke's law, then a transition to the operator's representation was made, the elastic parameters of the system were replaced by their operator's representation, and the equation in the fractional-differential statement, which describes the viscoelastic properties of the system under consideration, was returned to the originals.

We will consider in detail the approach described above on the example of composite springs. Figure 5 shows the spring, which is affected by an external disturbance in the form of force F(t).

IOP Conf. Series: Materials Science and Engineering **820** (2020) 012044 doi:10.1088/1757-899X/820/1/012044

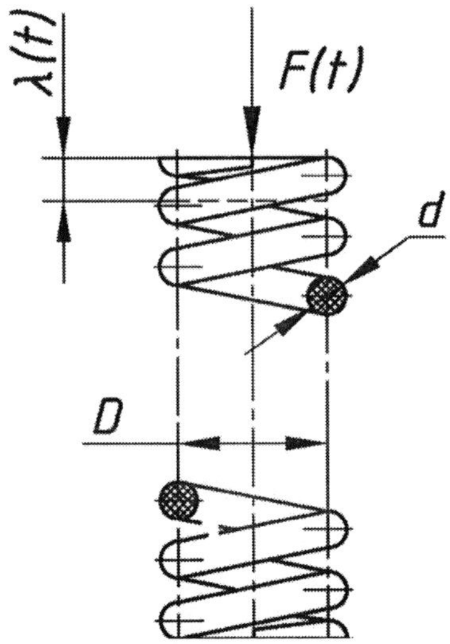

Figure 5. General parameters of a helical spring and its loading scheme

The deflection of such a spring can be determined using Mora integral:

$$\lambda = \int_l \frac{M_k M_{k1}}{GJ_k} dz = \frac{8FD^3 n}{Gd^4},$$

where D — mean diameter spring;
n— number of active coils;
d — spring wire diameter;
G — shear modulus of material.

The four-parameter model [10] of a standard linear body for a torsion case is described by the equation:

$$\tau(t) + \varsigma^\alpha D^\alpha \left[\tau(t)\right] = G\left[\gamma(t) + \theta^\alpha D^\alpha \left[\gamma(t)\right]\right].$$

Operator polynomials that correspond to the four-parameter viscoelastic body model under consideration:

$$Q(s) = \varsigma^\alpha s^\alpha + 1,$$

$$R(s) = \theta^\alpha s^\alpha + 1.$$

The elastic operator, which corresponds to the shear modulus when taking into account the viscoelastic properties, will look like:

$$G(s) = \frac{GR(s)}{Q(s)}.$$

Using the Laplace transformations and replacing the elastic parameters with the corresponding parameters when taking into account the viscoelastic properties, we obtain a differential equation with fractional derivatives that connects the spring deflection $\lambda(t)$ with a force acting on the spring $F(t)$:

$$\lambda(t) + \theta^\alpha D^\alpha \left[\lambda(t)\right] = \frac{8D^3 n}{Gd^4}\left[F(t) + \varsigma^\alpha D^\alpha \left[F(t)\right]\right].$$

Using this equation, a model in the MATLAB Simulink environment, shown in Figure 6, was created to describe the elastic properties of the spring taking into account rheological features. The

IOP Conf. Series: Materials Science and Engineering **820** (2020) 012044 doi:10.1088/1757-899X/820/1/012044

model is designed in the form of a subsystem whose input parameter is the spring deflection, i.e. the kinematic effect, and the output parameter is force.

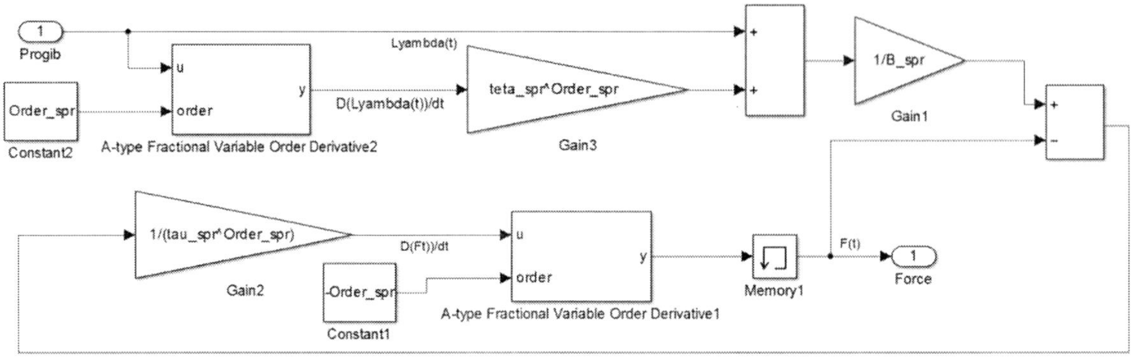

Figure 6. Composite spring model describing its viscoelastic properties

To check the adequacy of the model under consideration, a composite spring made of fiberglass was made as shown in Figure 7 (a). The spring was mounted on the test bench and the loading and unloading characteristics of the spring were obtained, which are shown in Figure 7 (b). At a force of 2500 N, the spring deflection is 100 mm. Figure 7 (b) shows the hysteresis loop, which is further studied.

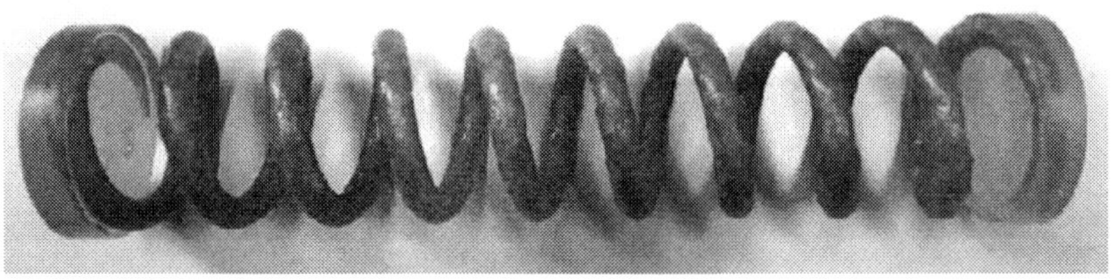

(a)

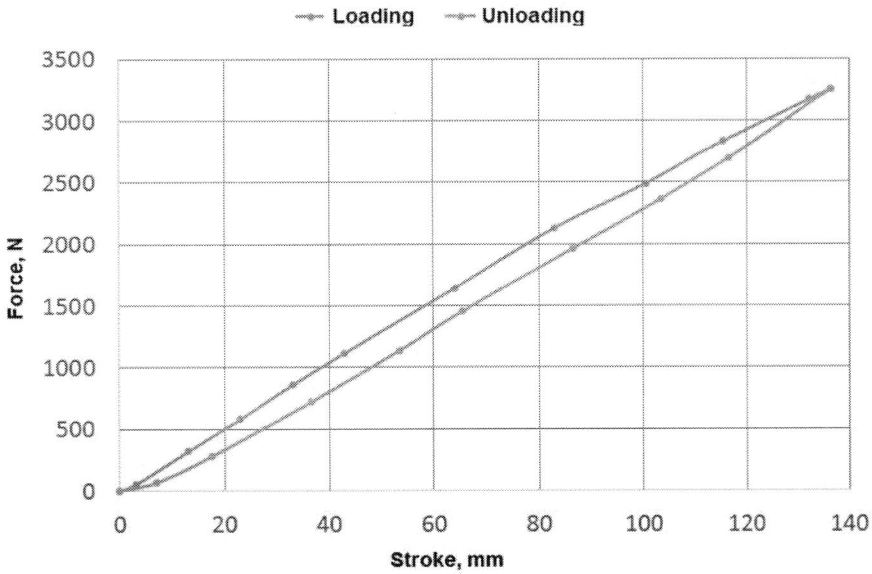

(b)

Figure 7. Fiberglass spring: (a) — general view of the spring; (b) — load and unload characteristics of this spring

As the compression spring wire operates on torsion [11], the optimum winding angles of the composite spring wire are +45° and −45°. The main geometrical parameters of such a spring are given in Table 1.

Table 1. Main geometrical parameters of the fiberglass spring

Spring parameter	Value
Spring stroke	*160 mm*
Spring freelength	*340 mm*
Spring innerdiameter	*56 mm*
Spring wire diameter	*16 mm*
Number of active coils	*8 pcs.*
Spring stiffness	*25 N/mm*

The kinematic perturbation in the form of a sinusoid with a frequency of 3Hz and an amplitude of 100 mm in the form of an input parameter acting on the composite spring. The loading characteristic obtained by modeling such a spring is shown in Figure 8. The figure shows that the maximum deflection (abscissa axis) of such a spring force (ordinate axis), arising in the spring is the same as the experimentally removed load characteristic. Of particular interest is the hysteresis loop obtained by modeling with the use of fractional differentiation and observed in the experiment with the use of springs made with the use of PCM. This loop requires an additional series of experimental research, on the base of which it is possible to specify the parameters included in the four-parameter model and the fractional-differential equation describing

IOP Conf. Series: Materials Science and Engineering **820** (2020) 012044 doi:10.1088/1757-899X/820/1/012044

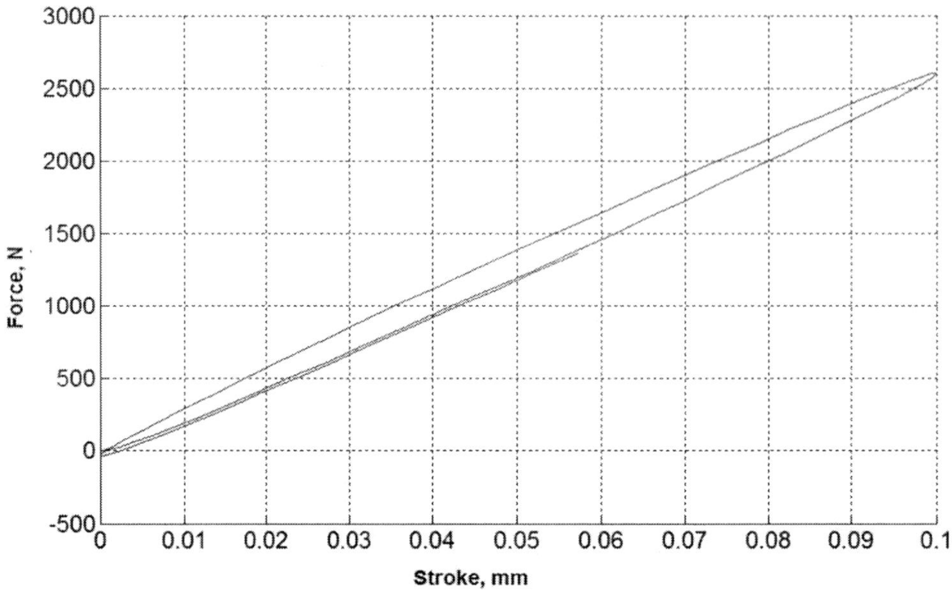

Figure 8. Composite spring load characteristic from simulation

The composite spring model can be used for simulation purposes. For a wheeled vehicle, the suspension force depends on two components: the relative deflection and the suspension deflection rate. The expression for determining the force in the independent suspension can be written in the following form [12,13]:

$$F_{ij} = Py\_spr(h_{ij}) + P_{y\_dempij}\left(\frac{dh_{ij}}{dt}\right),$$

where $Py\_spr(h_{ij})$ — is the force in the ii-th elastic element of the j-th side;

$P_{y\_dempij}\left(\frac{dh_{ij}}{dt}\right)$ — force in the i-th damping element of the j-th side.

Figure 9 shows such a simple model of the suspension system. In such a suspension system, the damping characteristic is determined by the dependence of the deflection velocity on the force in the shock absorber, and the elastic characteristic of the composite spring is calculated with the use of equations with derivatives of fractional order.

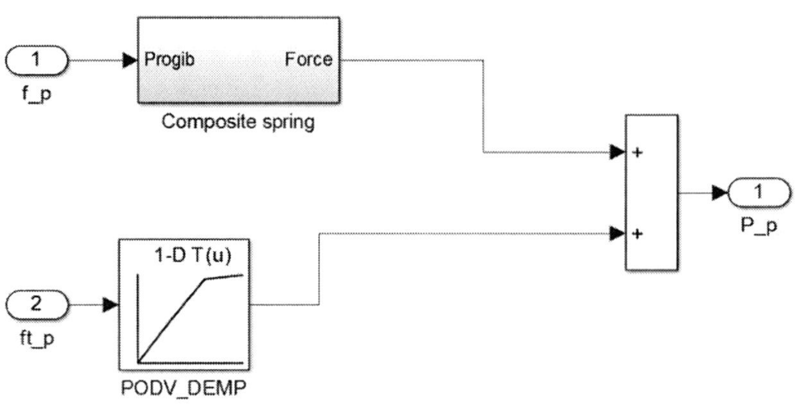

Figure 9. Model of the suspension system with composite spring

The composite spring model developed allows to predict the rheological properties of PCM. To predict such properties it is necessary to realize a number of experiments which will allow to specify and systematize the parameters which are a part of the four-parameter model of a composite spring. A perspective direction is the use of springs with progressive load characteristic. The technology of composite springs production and the use of PCM in their manufacture allow to obtain progressive loading characteristics of composite springs, which are impossible or difficult to obtain when using steel springs. Such loading characteristics can be obtained by creating springs with different reinforcement angles and variable wire diameter. The method considered in the article can be applied in determining the elastic properties of composite springs with progressive characteristics. The obtained model of composite springs can be used in the simulation of the vehicle [14–16] as a whole, as well as to assess the impact of such a spring on the operational parameters of the vehicle.

It should be noted that this work was carried out at the Bauman Moscow State Technical University, with financial support from the government in the face of the Russian Ministry of Education under the project: № 14.577.21.0287. (Identification number: RFMEFI57718X0287)

References

[1] B.A. Afanas'ev, I.Z. Dashtiev, *Proektirovanieehlementovavtomobilyaizpolimernyhkompozicionnyhmaterialov* (BMSTU, Moscow, 2006)
[2] K.B. Evseev, A.B. Kartashov, *Aktual'nost' primeneniyavintovyhpruzhin, vypolnennyhsispol'zovaniempolimernyhkompozicionnyhmaterialov, vsistemahpodressorivaniyasovremennyhavtomobilej,* ZHurnalAvtomobil'nyhinzhenerov, #4 (99), pp. 6–11, (2016)
[3] Kotiev, G.O., Padalkin, B.V., Kartashov, A.B., Diakov, A.S,*Designs and development of Russian scientific schools in the field of cross-country ground vehicles building,* ARPN Journal of Engineering and Applied Science, Vol. 12, Issue 4, 2017, pp. 1064–1071
[4] V.V. Novikov, A.V. Pozdeev, A.S. Diakov, *Research and testing complex for analysis of vehicle suspension units*, Procedia Engineering, Vol. 129, pp. 465–470 (2015) DOI: 10.1016/j.proeng.2015.12.153
[5] A.V. Pozdeev, V.V. Novikov, A.S. Diakov, A.V. Pohlebin, I.M. Ryabov, K.V. Chernishov, Reguliruemye*pnevmaticheskieipnevmogidravlicheskieressorypodvesokavtotransportnyhsredstv* (VGSTU, Volgograd, 2013)
[6] A.S. Diakov, V.V. Novikov *Ehksperimental'noeissledovanieharakteristikpnevmogidravlicheskojressorydlyakolyosnogoshassispecial'nogonaznacheniya*, Gruzovik, #7, pp. 7–11 (2015)
[7] Novikov, V.V., Pozdeev, A.V., Diakov, A.S.,*Research and testing complex for analysis of vehicle suspension units,* Procedia Engineering, 129, pp. 465–470
[8] Ahid D. Nashif, David I. G. Jones, John P. Henderson, *Vibration Damping*, (John Wiley & Sons, Inc., New York, 1985)
[9] N.N. Yacenko*Pogloshchayushchayaisglazhivayushchayasposobnost shin. Mashinostroenie*, (Moscow, 1978)
[10] Uchaikin V.V. *Metoddrobnyxproizvodnyx*— Ulyanovsk: Artichoke Publishing, 2008. — 512 p.
[11] Birger I.A., Mavlyutov R.R. *Soprotivleniematerialov*— Moscow.: Science, 1986. —560 s.
[12] Nashif A., Jones D., Henderson J. *Dempfirovaniekoleanij*— M.: Mir, 1988. — 448 p.
[13] Zhileikin M.M. *Teoreticheskieosnovypovysheniyapokazatelejustojchivostiiupravlyaemostikolesnyxmashinnabazemetodovnechetkojlogiki* / M.M. Zhileikin. — Moscow: Publishing House MSTU. N.E. Bauman, 2016. — 238 p.
[14] Kotiev G.O., Miroshnichenko A.V., Stadukhin A.A., Kositsyn B.B. *Estimating operation modes for the individual wheel electric drive of the all-wheel drive vehicle with the use of the*

driving simulator. Volume 534, Issue 1, 12 June 2019, International Automobile Scientific Forum: Intelligent Transport System Technologies and Components, IASF 2018

[15] Kotiev G.O., Miroshnichenko A.V., Stadukhin A.A., Kositsyn B.B. *Determination of mechanical characteristics of high-speed tracked vehicles traction motor with individual drive wheels*. Journal of Physics: Conference Series. Volume 1177, Issue 1, 10 April 2019, №012058, 3rd Science and Technology Seminar on Mobility of Transport and Technological Machines, MTTM 2018; Nizhny Novgorod; Russian Federation; 28 August 2018

[16] Kotiev G.O., Butarovich D.O., Kositsyn B.B. *Energy efficient motion control of the electric bus on route*. Volume 315, Issue 1, 22 February 2018, 2017 International Automobile Scientific Forum, IASF 2017

AUTHOR INDEX

Institute of Physics
Dirac House, Temple Back
Bristol BS1 6BE UK

ISSN: 1757-8981
ISBN 978-1-7138-3161-7

CURRAN ASSOCIATES INC.
proceedings.com